Hypersonic Meteoroid Entry Physics

IOP Series in Plasma Physics

Series Editors

About the series

The IOP Plasma Physics ebook series aims at comprehensive coverage of the physics and applications of natural and laboratory plasmas, across all temperature regimes. Books in the series range from graduate and upper-level undergraduate textbooks, research monographs and reviews.

The conceptual areas of plasma physics addressed in the series include:

- Equilibrium, stability and control
- Waves: fundamental properties, emission, and absorption
- Nonlinear phenomena and turbulence
- Transport theory and phenomenology
- Laser-plasma interactions
- Non-thermal and suprathermal particle populations
- Beams and non-neutral plasmas
- High energy density physics
- Plasma-solid interactions, dusty, complex and non-ideal plasmas
- Diagnostic measurements and techniques for data analysis

The fields of application include:

- Nuclear fusion through magnetic and inertial confinement
- Solar-terrestrial and astrophysical plasma environments and phenomena
- Advanced radiation sources
- Materials processing and functionalisation
- Propulsion, combustion and bulk materials management
- Interaction of plasma with living matter and liquids
- Biological, medical and environmental systems
- Low temperature plasmas, glow discharges and vacuum arcs
- Plasma chemistry and reaction mechanisms
- Plasma production by novel means

Hypersonic Meteoroid Entry Physics

Gianpiero Colonna, Mario Capitelli and Annarita Laricchiuta

IOP Publishing, Bristol, UK

ISBN 978-0-7503-1668-2 (ebook)
ISBN 978-0-7503-1666-8 (print)
ISBN 978-0-7503-1667-5 (mobi)

DOI 10.1088/2053-2563/aae894

Version: 20190301

IOP Expanding Physics
ISSN 2053-2563 (online)
ISSN 2054-7315 (print)

British Library Cataloguing-in-Publication Data: A catalogue record for this book is available from the British Library.

Published by IOP Publishing, wholly owned by The Institute of Physics, London

IOP Publishing, Temple Circus, Temple Way, Bristol, BS1 6HG, UK

US Office: IOP Publishing, Inc., 190 North Independence Mall West, Suite 601, Philadelphia, PA 19106, USA

This book and the 61st course Hypersonic Meteoroid Entry Physics *of the Ettore Majorana Foundation are dedicated to the memory of Professor Sergio Martellucci, Director of the International School of Quantum Electronics, excellent scientist and wonderful man.*

Contents

Preface

The study of meteoroids and meteorites is a field that inspires great interest in different research communities. The prediction of the trajectory of large meteoroids and of the fragment swarm, created after impact with the atmosphere, is dedicated to estimating the damage produced in a collision with the Earth's surface. Moreover, meteorites can provide information on their extraterrestrial origins, disclosing to astrophysicists information on the formation of the solar system. These two aspects have been developed separately: while hypersonic physics privileges the fluid-dynamics of atmospheric entry, with more or less sophisticated chemical models, the second aspect focuses on the experimental characterization of fragments as well as on the optical tracking of meteors made visible by the ablation process.

Advanced chemical physics models are being developed to understand the phenomenology of meteoroid entry into Earth's atmosphere at a velocity in the range of 10–70 km s^{-1} and an altitude of about 70–100 km. Entry velocities vary from Earth parabolic entry (11 km s^{-1}) to the maximum speed a body may have to be a member of the solar system (72 km s^{-1}). Meteorites originate from the asteroid belt or are expelled from planets through collisions with large asteroids, in contrast to comets which probably originate from the Oort cloud and Kuiper belt.

In the impact of a meteoroid with the Earth's atmosphere, its kinetic energy is transferred to molecules heating translational–rotational degrees of freedom at a temperature, dependent on the entry velocity, that can easily reach 100 000 K just behind the shock front. Approaching the body surface this energy is exchanged with other internal degrees of freedom, such as vibrational and electronically excited levels, activating also dissociation and ionization, decreasing the temperature of the gas. A given amount of this energy is emitted as radiation, that can be either dispersed, reducing the amount of energy stored in the surrounding plasma, or contribute to surface heating. Another mechanism contributing to cooling is the ablation of the surface, which, together with the energy dissipated by evaporation, increases the energy lost by radiation.

In this context the thermal relaxation of the gas around the shock wave evolves through non-equilibrium states, where the internal energy of atoms and molecules is in synergy with the gas composition and chemical processes. This problem has been widely studied theoretically for space vehicles entering planetary atmospheres by using multi-temperature or state-to-state vibrational kinetics. For a meteor, the problem is more complex due to fragmentation and evaporation, which completely change the physical and chemical properties of the gas surrounding the entering body, and as a consequence the trajectory.

There are different hypersonic ground test facilities that have been involved in measuring the behaviors of meteorites in high-enthalpy flows. Another important aspect is the characterization of the meteorites to determine their composition, from which it is possible to deduce their origin, and even the presence of organic material to support the idea that life on Earth has an extraterrestrial origin.

All these aspects are investigated by researchers who operate in different fields. For this reason, in October 2017, we organized the 61st course of *Hypersonic Meteoroid Entry Physics* (HyMEP) in the International School of Quantum Electronics of the Ettore Majorana Foundation and Centre for Scientific Culture, in Erice, Sicily, bringing together researchers from different fields to give a unified view of the problems related to meteor science.

This book collects the contributions of the lecturers of the course. The first chapter gives an overview of the different aspects related to meteoroid entry physics, discussing open problems in the field. The other chapters are grouped into three parts: I. Meteoroid and meteorite science, including four chapters describing the Chelyabinsk impactor, techniques for tracking meteoroid trajectories and the delivery of chondritic material to Earth; II. Hypersonic entry physics, discussing the radiative gas-dynamic simulation of a small meteor, numerical reconstruction of artificial meteoroids and the Chelyabinsk meteorite, high-enthalpy neutral and ionized flows, experimental apparatus for hypersonic entry, and ablation and fragmentation mechanisms; III. Elementary processes in hypersonic flows, where dynamical data for kinetic modeling of high-enthalpy flows are reported, discussing also the computational methods to obtain state resolved data.

Gianpiero Colonna
Mario Capitelli
Annarita Laricchiuta

Acknowledgments

The editors want to thank all the contributing authors for their efforts and invaluable work, including private communications used in the preparation of chapter 1.

About the editors

Gianpiero Colonna

Gianpiero Colonna graduated in physics in 1991 with a thesis on the stationary solution of the Boltzmann equation, obtained a PhD in chemistry in 1995 with a thesis on the state-to-state kinetics in the boundary layer of hypersonic flows, under the supervision of Mario Capitelli, and held a post-doctoral position at the University of British Columbia in Vancouver, under the supervision of Bernie Shizgal. After a period as researcher at the Pharmacy Faculty of the University of Bari, he became a researcher at the Italian National Research Council (CNR). Currently, he is a senior researcher at the same institution. His research activities are focused on plasma modeling, state-to-state self-consistent kinetics in gas discharges and hypersonic flows, the thermodynamic and transport properties of plasmas, and modeling the plasma plume produced by nanosecond laser pulses. He is the author of about two hundred research papers and conference proceedings. He has also co-authored two books in the series *Fundamental Aspects of Plasma Chemical Physics* (published by Springer). He was the editor of the book *Plasma Modeling: Methods and Applications*, published in the IOP *Plasma Physics* series. In addition, together with the co-editors of this book, he was a director of the HyMEP course in Erice.

Mario Capitelli

Mario Capitelli is a full professor of inorganic chemistry at the University of Bari and has been director of the Research Center of Plasma Chemistry CNR in Bari. From 2001 to 2011 he was responsible for the Bari section of the Institute IMIP CNR. He was the delegate of the rector of the University of Bari for research from 2006–2011. His research activity has spanned from the fundamental aspects of plasma physics and plasma chemistry —including the dynamics of elementary processes, the construction of advanced chemical–physical models for the kinetic simulation of non-equilibrium plasma, and the thermodynamic and transport characterization of equilibrium plasmas—to the investigation of plasma systems for application in the fields of fusion, aerothermodynamics, laser–plasma interaction, CO_2 activation and plasma-assisted combustion. A member of international advisory committees of plasma conferences and of editorial boards of plasma journals, he has chaired international conferences and was the director of two NATO-ASI (Advanced Study Institute) courses and of four courses of the International School of Quantum Electronics of the Ettore Majorana Foundation in Erice. He received the *Laurea Honoris Causa* of the Russian Academy of Sciences, Moscow, in 2013, and the title of *docteur honoris causa* of the Université Paris 13, in 2015. He is the author of about 500 papers published in international journals of chemistry, physics and engineering, and of numerous communications at national and international conferences, and has co-authored four books in the *Springer Series on Atomic, Optical, and Plasma Physics*. He was also the editor of seven multi-author books published by Springer, AIP, Kluwer and NOVA.

Annarita Laricchiuta

Annarita Laricchiuta graduated in chemistry in 1997, obtained her PhD in environmental chemistry in 2000 at the University of Bari and has been a CNR Researcher since 2001. She has focused her research activity on the theoretical calculation of dynamical data for elementary processes in plasmas, investigating the role of excitation of vibrational and electronic excitation in target molecules in affecting the probability of electron-impact-induced excitation, dissociation and ionization processes, and deriving elastic and charge-exchange cross sections for the estimation of collision integrals for transport properties of plasmas. She is the co-author of 90 papers in JCR journals, many communications at international conferences and two books of the *Springer Series on Atomic, Optical, and Plasma Physics*.

Contributors

Alina A Alexeenko
alexeenk@purdue.edu
Purdue University, School of Aeronautics and Astronautics
West Lafayette, IN, USA

Daniil A Andrienko
daniila@tamu.edu
Texas A & M University Aerospace Engineering
College Station, TX, USA

Vincenzo Aquilanti
vincenzoaquilanti@yahoo.it
Università di Perugia—Dipartimento di Chimica, Biologia e Biotecnologie
Perugia, Italy

Patricia R P Barreto
prpbarreto@gmail.com
Laboratorio Associado de Plasma (LAP)
Istituto Nacional de Pesquisas Espaciais (INPE)/MCT
Sao Jose dos Campos, Sao Paulo, Brazil

James Beck
james.beck@belstead.com
Belstead Research Ltd
Ashford, Kent, TN25 4PF, UK

Francesco Bonelli
francesco.bonclli@poliba.it
Polytechnic University of Bari—Department of Mechanics, Mathematics and Management
Bari, Italy

Jiří Borovička
jiri.borovicka@asu.cas.cz
Astronomical Institute of the Czech Academy of Sciences
Department of Interplanetary Matter
Ondřejov, Czech Republic

Salvatore Borrelli
s.borrelli@cira.it
CIRA Italian Center for Aerospace Research
Capua (CE), Italy

Iain D Boyd
iainboyd@umich.edu
University of Michigan—Aerospace Engineering Department
Ann Arbor, MI, USA

Francesco Capitelli
francesco.capitelli@ic.cnr.it
Istituto di Cristallografia, CNR
Roma, Italy

Mario Capitelli
mario.capitelli@nanotec.cnr.it
PLASMI Lab @ CNR-NANOTEC
Bari, Italy

Giuseppe Ceglia
g.ceglia@cira.it
CIRA Italian Center for Aerospace Research
Aerothermodynamic and Combustion Measurements Laboratory
Capua (CE), Italy

Roberto Celiberto
roberto.celiberto@poliba.it roberto.celiberto@nanotec.cnr.it
Polytechnic University of Bari—DICATECh
and PLASMI Lab @ CNR-NANOTEC
Bari, Italy

Gianpiero Colonna
gianpiero.colonna@cnr.it
PLASMI Lab @ CNR-NANOTEC
Bari, Italy

Luigi Cutrone
l.cutrone@cira.it
CIRA Italian Center for Aerospace Research—Aerothermodynamic
Capua (CE), Italy

Antonio D'Angola
antonio.dangola@unibas.it
Scuola di Ingegneria SI—Università della Basilicata
Potenza, Italy
and PLASMI Lab @ CNR-NANOTEC
Bari, Italy

Massimo D'Orazio
massimo.dorazio@unipi.it
Università di Pisa—Dipartimento di Scienze della Terra
Pisa, Italy

Mario De Cesare
m.decesare@cira.it
CIRA Italian Center for Aerospace Research
Aerothermodynamic and Combustion Measurements Laboratory
Capua (CE), Italy

Antonio Del Vecchio
A.DelVecchio@cira.it
CIRA Italian Center for Aerospace Research
Aerothermodynamic and Combustion Measurements Laboratory
Capua (CE), Italy

George Duffa
gduffa@club-internet.fr
CEA
France

Fabrizio Esposito
fabrizio.esposito@cnr.it
PLASMI Lab @ CNR-NANOTEC
Bari, Italy

Kazuhisa Fujita
fujita.kazuhisa@jaxa.jp
JAXA, Research and Development Directorate
Research Unit II, 7-44-1 Jindaiji-Higachi-Machi, Chofu-shi
Tokyo 182-8522, Japan

Carolyn Jacobs
carolyn.jacobs@centralesupelec.fr
CentraleSupélec Laboratoire EM2C (CNRS UPR288)
Gif-sur-Yvette, France

Ratko K Janev
ratkojanev@gmail.com
Macedonian Academy of Sciences and Arts
Skopje, Macedonia

Elena Kustova
elena_kustova@mail.ru
Saint Petersburg State University
Saint Petersburg, Russia

Hervé Lamy
herve.lamy@aeronomie.be
Royal Belgian Institute for Space Aeronomy
Brussels, Belgium

Vincenzo Laporta
vincenzo.laporta@univ-lehavre.fr
Laboratoire Ondes et Milieux Complexes CNRS
Université du Havre-Université Normandie
Le Havre, France
and
Department of Physics and Astronomy
University College London
London, United Kingdom

Annarita Laricchiuta
annarita.laricchiuta@cnr.it
PLASMI Lab @ CNR-NANOTEC
Bari, Italy

Christophe O Laux
christophe.laux@centralesupelec.fr
CentraleSupélec Laboratoire EM2C (CNRS UPR288)
Gif-sur-Yvette, France

Adrien Lemal
adrien.lemal@star-ale.com
Astro Live Experiences (ALE)
Research and Development department
Simulations, Science and Data section
Kawamoto Building 2nd floor
2-21-1 Akasaka, Minato-ku
Tokyo 107-0052, Japan

Andrea Lombardi
ebiu2005@gmail.com andrea.lombardi@unipg.it
Università di Perugia
Dipartimento di Chimica, Biologia e Biotecnologie
Perugia, Italy

Han Luo
luo160@purdue.edu
Purdue University, School of Aeronautics and Astronautics
West Lafayette, IN, USA

Megan E MacDonald
megan.e.macdonald@nasa.gov
CentraleSupélec Laboratoire EM2C (CNRS UPR288)
Gif-sur-Yvette, France

Robyn Macdonald
macdona3@illinois.edu
University of Illinois at Urbana-Champaign
Department of Aerospace Engineering
Urbana, IL, USA

Sergey Macheret
macheret@purdue.edu
Purdue University, School of Aeronautics and Astronautics
West Lafayette, IN, USA

Pierre E Mariotto
pierre.mariotto@centralesupelec.fr
CentraleSupélec Laboratoire EM2C (CNRS UPR288)
Gif-sur-Yvette, France

Sean D McGuire
sean.mc-guire@centralesupelec.fr
CentraleSupélec Laboratoire EM2C (CNRS UPR288)
Gif-sur-Yvette, France

Zsolt J Mezei
mezeijzs@gmail.com
Laboratoire Ondes et Milieux Complexes CNRS
Université du Havre-Université Normandie
Le Havre, France
and
Laboratoire des Sciences des Procédés et des Matériaux
CNRS—Université Paris 13-USPC, Villetaneuse, France
and Institute of Nuclear Research, Hungarian Academy of Sciences
Debrecen, Hungary

Richard G Morgan
r.morgan@uq.edu.au
Center for Hypersonics, The University of Queensland
St Lucia, Queensland, Australia

Félix Mouzet
felixmouzet@wanadoo.fr
ISA
France

Alessandro Munafò
munafo@illinois.edu
University of Illinois at Urbana-Champaign
Department of Aerospace Engineering
Urbana, IL, USA

Ekaterina Nagnibeda
e_nagnibeda@mail.ru
Saint Petersburg State University
Saint Petersburg, Russia

Kevin Neitzel
kevin.neitzel@gmail.com
University of Michigan—Aerospace Engineering Department
Ann Arbor, MI, USA

Satoshi Nomura
nomura.satoshi2@jaxa.jp
JAXA, Aeronautical Technology Directorate
Aerodynamics Research Unit, 7-44-1 Jindaiji-Higachi-Machi, Chofu-shi
Tokyo 182-8522, Japan

Fabrizio Paganucci
fabrizio.paganucci@unipi.it
Università di Pisa—Dipartimento di Ingegneria Civile ed Industriale
Pisa, Italy

Federico Palazzetti
fede_75it@yahoo.it
Università di Perugia—Dipartimento di Chimica, Biologia e Biotecnologie
Perugia, Italy

Marco Panesi
mpanesi@illinois.edu
University of Illinois at Urbana-Champaign
Department of Aerospace Engineering
Urbana, IL, USA

Giuseppe Pascazio
giuseppe.pascazio@poliba.it
Polytechnic University of Bari—Department of Mechanics, Mathematics and Management
Bari, Italy

Lucia Daniela Pietanza
luciadaniela.pietanza@cnr.it
PLASMI Lab @ CNR-NANOTEC
Bari, Italy

Fernando Pirani
pirani.fernando@gmail.com
pirani@dyn.unipg.it
Università di Perugia—Dipartimento di Chimica, Biologia e Biotecnologie
Perugia, Italy

Philippe Reynier
Philippe.Reynier@isa-space.eu
ISA
France

Luigi Savino
l.savino@cira.it
CIRA Italian Center for Aerospace Research
Aerothermodynamic and Combustion Measurements Laboratory
Capua (CE), Italy

Antonio Schettino
a.schettino@cira.it
CIRA Italian Center for Aerospace Research—Aerothermodynamic
Capua (CE), Italy

Ioan F Schneider
ioan.schneider@univ-lehavre.fr
Laboratoire Ondes et Milieux Complexes CNRS
Université du Havre—Université Normandie
Le Havre, France
and
Laboratoire Aimé-Cotton, CNRS—Université Paris-Sud ENS Cachan—Université Paris-Saclay
Orsay, France

Giulio Seller
giulio.seller@gmail.com
Bari, Italy

Giorgio S Senesi
giorgio.senesi@cnr.it
PLASMI Lab @ CNR-NANOTEC
Bari, Italy

Kelly A Stephani
ksteph@illinois.edu
University of Illinois at Urbana-Champaign—Department of Mechanical Science and Engineering
Urbana, IL, USA

Sergey T Surzhikov
surg@ipmnet.ru
Ishlinsky Institute for Problems in Mechanics (IPMech), Russian Academy of Sciences
Moscow, Russia

Jonathan Tennyson
j.tennyson@ucl.ac.uk
University College of London—Department of Physics and Astronomy
London, UK

Augustin Tibère-Inglesse
augustin.tibere-inglesse@centralesupelec.fr
CentraleSupelec, Laboratoire EM2C
Chatenay-Malabry, France

Josep M Trigo-Rodríguez
trigo@ice.cat
Institute of Space Sciences (CSIC-IEEC)
Cerdanyola del Vallés (Barcelona), Catalonia, Spain

Michele Tuttafesta
micheletuttafesta@gmail.com
Bari, Italy

Jogindra M Wadehra
wadehra@wayne.edu
Department of Physics and Astronomy
Wayne State University
Detroit, MI, 48202, USA

Fabian Zander
Fabian.Zander@usq.edu.au
Center for Hypersonics, The University of Queensland
St Lucia, Queensland, Australia

IOP Publishing

Hypersonic Meteoroid Entry Physics

Gianpiero Colonna, Mario Capitelli and Annarita Laricchiuta

Chapter 1

Considerations on meteoroid entry physics

Mario Capitelli, Salvatore Borrelli, Gianpiero Colonna and Annarita Laricchiuta

Hypersonic meteoroid entry physics (HyMEP) is a subject of continuous interest, either for understanding the relevant fluid-dynamics phenomenology (see chapters 6, 7, 9 and 10) or to study the composition of the fragments coming from the meteoroid–Earth impact containing primordial elements characterizing the life of the Universe (chapters 2–5 and 11).

Historically these two aspects have been developed separately, the hypersonic physics emphasizing the fluid-dynamics of the process with more or less sophisticated chemical models, while the second approach has focused on the experimental characterization of the fragments as well as of the emitting meteors following the ablation process.

Advanced chemical physics models are being developed to understand the phenomenology of meteoroid entry into Earth's atmosphere at a velocity in the range of 10–70 km s^{-1} and an altitude of about 70–100 km. Entry velocities vary from Earth parabolic speed (11 km s^{-1}) to about the maximum Earth approach speed a body may have and still be a member of the solar system (72 km s^{-1}). Meteoroids originate from asteroids and comets. Asteroids originate from the asteroid belt while comets are believed to originate from the Oort Cloud and Kuiper Belt. Meteorites on, the other hand, originate from the fragmentation of meteoroids in the atmosphere.

In the impact the kinetic energy of a meteoroid is transformed into translational–rotational energy of the surrounding air which can reach, depending on the entry velocity, translational/rotational temperatures higher than 100 000 K. As an example, a meteoroid of non-spherical form with a 2 cm cross section radius entering at a velocity of 30 km s^{-1} and an altitude of 70 km has a translational temperature along the stagnation line that can reach a maximum temperature of 160 000 K at a distance of 0.2 cm from the stagnation point, strongly decreasing from 0.2 cm onward to 2000–3000 K. At the same time the vibrational temperatures of the different air components present flat values of about 10 000 K [1–4].

doi:10.1088/2053-2563/aae894ch1

During the impact meteoroids are decelerated and ablated, producing a luminous phenomenon called meteor. Significant studies of the ablation of meteoroids in Earth's atmosphere are currently being undertaken, including the interpretation of meteor and fireball emission spectra and the quantification of meteoroid elemental abundances averaging those measured along the meteor column [5, 6]. The meteor spectra allow the study of the ablation of a hypersonic flying meteoroid with the Earth's atmosphere, adjusting an equilibrium model to infer chemical elemental abundances in the ionized column, and also allow one to infer the existence of non-equilibrium processes occurring in the shock wave and within the expanding meteor column [7].

The high values of the vibrational temperatures during the impact promote dissociation and ionization processes, as well as excitation processes of electronic excited states. The electron temperature is strongly coupled with the vibrational temperature so that to a first approximation one can consider that the vibrational and electron temperatures are equal. A better approach is to use the multi-temperature models developed by Park [8]. These models, however, imply that the internal distributions (rotational, vibrational and electronic) of the relevant neutral and ionized molecular and atomic species are in equilibrium at the respective temperatures, an assumption that has been widely debated in recent years.

An alternative method is to describe the non-equilibrium chemical kinetics occurring in the high-temperature medium on the basis of a state-to-state (StS) approach [9–11]. The StS approach decomposes any degree of freedom of the molecules and atoms into internal levels (rotational, vibrational and electronic). Each level is considered as an independent species described by an appropriate continuity equation and its own cross sections and rate coefficients. Under many circumstances the internal distributions are far from the Boltzmann distribution, so the concept of temperature loses its meaning. The difficulty in this case is the increase of computational times from the introduction of StS kinetics in robust fluid-dynamic codes. Very recently, however, the coupling between the non-equilibrium StS vibrational kinetics and dissociation–recombination processes in nitrogen has been inserted into a 3D Euler code [12]. The next step in this approach will be to insert an StS description of air in the dissociation–recombination regime in the 3D Navier–Stokes code.

Numerous contributions of StS kinetics are reported in the current book (chapters 8, 12, 13, and 18), including StS cross sections entering StS kinetics (chapters 15–18).

Concerning the StS cross sections, particular importance should be given to heavy-particle–heavy-particle collisions describing vibration–vibration (VV) and vibration–translation (VT) energy transfer processes. StS dissociation processes as well as other reactive processes must be considered, emphasizing the role of the vibrational excitation of reactants in enhancing the relevant rates [9–11] (see also chapters 16–18).

The same importance can be attributed to the dependence of electron molecule cross sections (resonant and direct processes) including excitation, dissociation and ionization processes on the vibrational quantum number [10–12] (see also chapter 15).

The relevant cross sections must be transformed into rates by using a suitable electron energy distribution function (EEDF) obtained by coupling the Boltzmann equation for the EEDF with the kinetics of vibrationally and electronically excited states. Thus far this problem has been considered by inserting the StS kinetics in an Euler 1D solver [13, 14] (see chapter 8). The ionization considered in this study was limited to the electron molecule ionization processes. However, the precursor ionization processes are of central importance in ionizing the medium, as reported in chapter 9.

Under many situations this coupling generates structures in the EEDF due to collisions of the second kind (superelastic) between excited states and cold electrons. A new paradigm has to be considered to introduce these effects in 2D–3D Navier–Stokes codes.

Particular attention in meteoroid entry physics is paid to the fact that gas radiation energy becomes a significant part of the total energy of the flow, so it is no longer possible to analyze the flow fields with the assumption that the energy of elemental volumes remains constant, since for such cases a transfer of energy by radiation from an elemental volume to another one significantly alters the flow. This implies the necessity of studying the non-equilibrium radiative gas-dynamics of small meteoroids self-consistently coupled with computational fluid-dynamics (CFD) solvers. Radiation transfer under large Mach number conditions is responsible for the ablation of meteoroids and therefore for the radiation signature of some ablated components of the meteoroid. This last consideration introduces a new phenomenology linked to the fact that the population of electronic excited states of atoms and ions coming from ablation as well as from the excited surrounding air is the result of electron-impact excitation processes. The rate coefficients of these processes depend on the relevant cross sections and the actual EEDF. The latter, as anticipated, contains structures due to second kind collisions, which in turn depend on the concentration of excited states. Again these concentrations depend on the re-absorption of optical lines by the relevant atoms which generate two types of plasmas, i.e. thin and thick plasmas. These extreme conditions are characterized by no re-absorption and complete re-absorption of radiation with large consequences on the concentrations of excited states and therefore on the EEDF. A more satisfactory approach is to couple the Euler equations, the level kinetics and the Boltzmann solver for free electrons with a radiation transport equation, capable of quantifying the absorption coefficients for each spectral frequency. The corresponding model has recently been developed and used in the hypersonic re-entry of vehicles in the Jupiter atmosphere (He/H_2). The results show, either in the shock wave or in the expansion zone, a strong coupling of the excited states from atomic hydrogen with the EEDF [14–17] (see chapters 6 and 8).

These approaches, while very difficult to implement in robust CFD codes, are nonetheless important for testing some points of the complex phenomenology occurring during the meteoroid entry fully described by the different existing CFD platforms. In particular, NERAT(2D) + ASTEROID, discussed in chapter 6, has been adapted to study the aero-physics of meteoroids and space vehicles entering planetary atmospheres and for diagnostics of trajectory parameters through spectral signatures of the objects.

The high temperature induces ablation and fragmentation/demise processes which become of paramount importance in the relevant phenomenology (chapter 10).

In many approaches the ablation process is taken into account by considering the composition of ablated materials as an initial condition for the full fluid-dynamics approach. The corresponding initial condition can be estimated from the thermodynamic equilibrium at a given surface temperature. This approximation can be improved by trying to develop more fundamental ablation mechanisms based on numerical simulation and plasma chemical physics models. These are useful for rationalizing results obtained in laser–plasma interaction laboratories, and also for the improvement of entry models and the analysis of real meteor spectra, thereby also contributing to data classification. Existing analytical models [18] as well as the recent chemical ablation model, CAMOD [19] (which includes the following processes: sputtering through inelastic collisions with air molecules before the meteoroid melts; evaporation of atoms and oxides from molten particles; diffusion-controlled migration of the volatile constituents (Na and K) through the molten particles; and impact ionization of the ablated fragments by hyper-thermal collisions with air molecules), should be mentioned.

The electron component in local thermodynamic equilibrium (LTE) and non-LTE conditions should also be considered [20]. In particular, a Boltzmann solver for the EEDF and a suitable collisional–radiative model for ablated metals should be considered [17] (see chapter 8).

Note that the ablated materials entering in the flow can change the thermodynamic and transport properties of the medium (chapter 14), so this phenomenon needs to be taken into account in CFD platforms

Fragmentation and demise processes are linked to the response of the meteorite to aerodynamic forces and radiative and convective heating. Moreover, depending on the forces acting on the meteoroid, fragmentation and demise can occur, introducing a scenario that needs to be controlled to decrease Earth risks. In this context, a significant result of the CFD simulation of meteoroids is the determination of the surface stress tensor components, which are directly connected with the problem of mechanical de-fragmentation. The final fragmentation process is then evidenced by the formation of meteorites, which are then recovered at the Earth's surface.

Recovered meteorites often show a stable aerodynamic shape similar to an entry capsule. Recovering and characterizing the rock-forming minerals and the ablation properties of meteorites from falls can lead to their orbital reconstruction, as shown by the Villalbeto de la Peña and Puerto Lápice meteorites in Spain [21, 22], and also the Cali and Berduc chondrite falls in Colombia and Argentina, respectively [23, 24].

These problems have been investigated within the extant Belstead Research Limited (BRL) platform, which consists of a set of coupled in-house Java modules which were originally designed for modeling the destructive re-entry of spacecraft (see chapter 10).

The ground-state experiments include shock wave interaction and plasma and laser interaction with meteorites, their terrestrial analogues as well as with materials with the same composition as the meteorite. In this case the selected meteorite is submitted in the appropriate facility to the action of a high-temperature environment (chapter 11), and x-ray based analysis and LIBS techniques (chapter 5) can be used to detect the changes in the meteorite structure before and after the interaction. The complex geochemical nature of chondritic meteorites can be investigated by means of multiple techniques: SEM + EDX, UHRTEM, micro-Raman, ICP-MS and x-ray microprobes. In addition, modern single-crystal and powder x-ray diffractometers (SCXD, PXD) can be employed for the structural characterizations of different mineral phases (silicates, carbonates, phosphates and oxides) within meteorites [25, 26].

Laser-induced breakdown spectroscopy (LIBS) is a powerful analytical technique to detect the composition of atomic species in complex matrices [27, 28]. It consists of a pulsed laser impinging the surface of the meteorite or its terrestrial analogue, creating a plasma. The subsequent spectral emission is then analyzed to obtain qualitative and quantitative information on the meteorite composition. LIBS is currently considered an important tool for space applications, as proved by its insertion in the Mars Curiosity Rover for the analysis of Martian rocks [29–35].

Space experiments can be classified as follows:

- Sample return capsules (SRCs), to be considered as artificial meteoroids [36, 37].
- Natural meteoroid entry.

SRCs are not natural bolides, but they have similar flow conditions and they provide simpler and well-known experimental starting conditions for studying various aspects of the ablation process as well as radiation problems. Their size (0.8–1.5 m) is in the range of asteroid fragments at the peak of mass influx (≈4 m). The SRC entry speed (11.0–12.9 km s^{-1}) is high enough to have a significant contribution from radiative heat flux and samples the lower end of the range for natural asteroid entries (11 to ~30 km s^{-1}, peaking around 15 km s^{-1}).

Interesting radiation results have been obtained by the re-entry capsules Genesis, Stardust and Hayabusa (chapter 7), that to a given extent can be considered as low-velocity entry (about 11 km s^{-1}) meteoroids, thus generating a bridge between capsule and meteoroid entry phenomenology [36, 37].

In these kinds of experiments we can also include the STONE 6 artificial meteorite experiment, studying the effect of heat shock during atmospheric entry on organic matter embedded in carbonaceous meteorites [38, 39].

Large natural meteoroids, having initial velocities ranging from ≈11–73 km s^{-1} to be bound to the solar system, can produce strong shock waves upon entry into Earth's atmosphere while ablating, fragmenting, decelerating, and producing copious amounts of luminosity, with heat-transfer flow regimes ranging from the extremes of free-molecular to continuum flow. In this process, we must deal with their generally unknown characteristics of shape, radius, speed, composition, degree of porosity, rotation, tumbling, and so forth.

A lot of diagnostics have been developed in the last century to obtain information about the re-entry of these objects, including radiation measurements (chapters 3 and 4).

For decades, meteor observations were typically made with photographic and TV cameras and small meteor radars, and much was learned about the continuous impact of meteoroids ranging in size from sand grains to large boulder-sized meteoroids.

Tracking records of bright meteors is a technology that has been used for many years by the astrophysical community to follow meteoroid entry in the atmosphere. The old system consisted of one or more pairs of cameras located in the order of 20 km apart along a known baseline. Each camera is provided with a rotating shutter in constant operation during those times when good meteor tracking is possible. When a meteor flight occurs within the range of these cameras photograph records are obtained. By careful triangulation it is possible to accurately determine the trajectory and from the interrupted shutter record the velocity and deceleration of the meteor, including the luminous intensity history. All of this information can be used to obtain information on heat transfer [4, 40].

Over the past decade, large radars such as the European Incoherent Scatter (EISCAT) Radar and the Arecibo Observatory in Puerto Rico have been pointed towards the sky to measure meteoroid impacts. These radars have observed two types of radar meteor reflections that are now widely studied. Moreover, a radar can monitor the speed of the plasma formed in the interaction, an observable which will be extensively studied in this book.

Examples of natural meteoroid entries include the Příbram meteor, originating from a 1000 kg meteoroid entering with a speed of 20.88 km s^{-1}. Its luminosity was recorded in the altitude range 98–22 km. At lower altitude the meteor was observed to shatter in a progressive manner. The chondrite stone broke into 17 pieces as a result of aerothermodynamic stresses.

A second example is the Meanok meteor, the integrated luminosity of which was followed in the altitude range 67.39–38.21 km, speed range 17.42–9.16 km s^{-1} and observation times of 0–2.2 s.

More recent results were obtained for the California meteor, 3 m in diameter and with an entry speed of 28.6 km s^{-1}, and the recent super bolide Chelyabinsk subjected to ablation and fragmentation processes (entry speed 19.1 km s^{-1}) [41] (chapters 2 and 7).

Another possibility is to better understand the Tunguska meteor (entry velocity 17.9–20 km s^{-1}). The explanation of soil effects was provided either by invoking a nuclear reaction during the impact or chemical reactions forming NO and therefore fertilizers [42, 43].

As a final consideration in this introduction, we hope to have emphasized the main problems still existing in explaining the complex phenomenology of meteoroid entry. This book aims to be a stimulus for mixing the activities coming from the different disciplines necessary for a complete understanding of the relevant problems. We believe that only interdisciplinary experimental and theoretical approaches will be able to provide a complete description of meteoroid entry physics.

References

[1] Surzhikov S 2014 Non-equilibrium radiative gas dynamics of small meteor, *44th AIAA Fluid Dynamics Conf.* AIAA paper 2014–2636

[2] Surzhikov S 2011 Numerical simulation of spectral signature of bolide entering into Earth atmosphere, *3rd AIAA Atmospheric Space Environments Conf.* AIAA paper 2011–3152

[3] Surzhikov S 2012 Spectral signature of ablating bolide entering into Earth atmosphere, *4th AIAA Atmospheric and Space Environments Conf.* AIAA paper 2012–2943

[4] Surzhikov S 2015 Numerical simulation of radiating re-entry flows around orbital space vehicle: comparison with observed data, *53rd AIAA Aerospace Sciences Meeting* AIAA paper 2015–1701

[5] Trigo-Rodríguez J M, Llorca J, Borovička J and Fabregat J 2003 Chemical abundances determined from meteor spectra: I. Ratios of the main chemical elements *Meteorit. Planet. Sci.* **38** 1283–94

[6] Trigo-Rodríguez J M, Llorca J and Fabregat J 2004 Chemical abundances determined from meteor spectra: II. Evidence for enlarged sodium abundances in meteoroids *Mon. Not. R. Astron. Soc.* **348** 802–10

[7] Trigo-Rodrıguez J 2013 Meteor emission spectroscopy: clues on the delivery of primitive materials from cometary meteoroids *METEOROIDS 2013, Proc. of the Int. Conf. held at the Adam Mickiewicz University in Poznań, Poland* **2013** 105

[8] Park C 1990 *Non-Equilibrium Hypersonic Aerothermodynamics* (New York: Wiley)

[9] Capitelli M (ed) 1986 Nonequilibrium vibrational kinetics *Topics in Current Physics* vol 39 (Berlin: Springer)

[10] Capitelli M, Ferreira C M, Gordiets B F and Osipov A I 2001 Plasma kinetics in atmospheric gases *Plasma Phys. Control. Fusion* **43** 371

[11] Capitelli M, Celiberto R, Colonna G, Esposito F, Gorse C, Hassouni K, Laricchiuta A and Longo S 2015 *Fundamental Aspects of Plasma Chemical Physics: Kinetics* (Springer Series on Atomic, Optical, and Plasma Physics vol 85) (Berlin: Springer)

[12] Cutrone L, Tuttafesta M, Capitelli M, Schettino A, Pascazio G and Colonna G 2014 3D nozzle flow simulations including state-to-state kinetics calculation *AIP Conf. Proc.* **1628** 1154–61

[13] Colonna G and Capitelli M 2001 Self-consistent model of chemical, vibrational, electron kinetics in nozzle expansion *J. Thermophys. Heat Transfer* **15** 308–16

[14] Colonna G and Capitelli M 1996 Electron and vibrational kinetics in the boundary layer of hypersonic flow *J. Thermophys. Heat Transfer* **10** 406–12

[15] Colonna G, D'Ammando G, Pietanza L and Capitelli M 2014 Excited-state kinetics and radiation transport in low-temperature plasmas *Plasma Phys. Control. Fusion* **57** 014009

[16] Capitelli M, Colonna G, Pietanza L D and D'Ammando G 2013 Coupling of radiation, excited states and electron energy distribution function in non-equilibrium hydrogen plasmas *Spectrochim. Acta Part* B **83-84** 1–13

[17] D'Ammando G, Capitelli M, Esposito F, Laricchiuta A, Pietanza L D and Colonna G 2014 The role of radiative reabsorption on the electron energy distribution functions in H_2/He plasma expansion through a tapered nozzle *Phys. Plasmas* **21** 093508

[18] ReVelle D O 1979 A quasi-simple ablation model for large meteorite entry: theory vs observations *J. Atmos. Terr. Phys.* **41** 453–73

[19] Vondrak T, Plane J, Broadley S and Janches D 2008 A chemical model of meteoric ablation *Atmos. Chem. Phys.* **8** 7015–31

[20] Casavola A R, Colonna G and Capitelli M 2009 Kinetic model of titanium laser induced plasma expansion in nitrogen environment *Plasma Sources Sci. Technol.* **18** 025027
[21] Trigo-Rodríguez J M, Borovička J, Spurný P, Ortiz J L, Docobo J A, Castro-Tirado A J and Llorca J 2006 The Villalbeto de la Peña meteorite fall: II. Determination of atmospheric trajectory and orbit *Meteorit. Planet. Sci.* **41** 505–17
[22] Trigo-Rodríguez J M, Borovička J, Llorca J, Madiedo J M, Zamorano J and Izquierdo J 2009 Puerto Lápice eucrite fall: strewn field, physical description, probable fireball trajectory, and orbit *Meteorit. Planet. Sci.* **44** 175–86
[23] Trigo-Rodríguez J M, Llorca J, Rubin A E, Grossman J N, Sears D W, Naranjo M, Bretzius S, Tapia M and Sepúlveda M H G 2009 The Cali meteorite fall: a new H/L ordinary chondrite *Meteorit. Planet. Sci.* **44** 211–20
[24] Trigo-Rodríguez J M, Llorca J, Madiedo J M, Tancredi G, Edwards W N, Rubin A E and Weber P 2010 The Berduc L6 chondrite fall: meteorite characterization, trajectory, and orbital elements *Meteorit. Planet. Sci.* **45** 383–93
[25] Capitelli F, Chita G, Ghiara M and Rossi M 2012 Crystal-chemical investigation of $Fe_3(PO_4)_2 \cdot 8H_2O$ vivianite minerals *Z. Kristallogr.* **227** 92–101
[26] Rossi M, Ghiara M R, Chita G and Capitelli F 2011 Crystal-chemical and structural characterization of fluorapatites in ejecta from Somma-Vesuvius volcanic complex *Am. Mineral.* **96** 1828–37
[27] Dell'Aglio M, De Giacomo A, Gaudiuso R, De Pascale O, Senesi G S and Longo S 2010 Laser induced breakdown spectroscopy applications to meteorites: chemical analysis and composition profiles *Geochim. Cosmochim. Acta* **74** 7329–39
[28] De Giacomo A, Dell'Aglio M, De Pascale O, Longo S and Capitelli M 2007 Laser induced breakdown spectroscopy on meteorites *Spectrochim. Acta* B **62** 1606–11
[29] Cousin A, Forni O, Maurice S, Gasnault O, Fabre C, Sautter V, Wiens R and Mazoyer J 2011 Laser induced breakdown spectroscopy library for the Martian environment *Spectrochim. Acta* B **66** 805–14
[30] Fabre C, Maurice S, Cousin A, Wiens R, Forni O, Sautter V and Guillaume D 2011 Onboard calibration igneous targets for the Mars science laboratory Curiosity rover and the chemistry camera laser induced breakdown spectroscopy instrument *Spectrochim. Acta* B **66** 280–9
[31] Harmon R, Russo R and Hark R 2013 Applications of laser-induced breakdown spectroscopy for geochemical and environmental analysis: a comprehensive review *Spectrochim. Acta* B **87** 11–26
[32] Fortes F, Moros J, Lucena P, Cabalín L and Laserna J 2013 Laser-induced breakdown spectroscopy *Anal. Chem.* **85** 640–69
[33] Sallé B, Lacour J-L, Mauchien P, Fichet P, Maurice S and Manhès G 2006 Comparative study of different methodologies for quantitative rock analysis by laser-induced breakdown spectroscopy in a simulated Martian atmosphere *Spectrochim. Acta* B **61** 301–13
[34] Tucker J, Dyar M, Schaefer M, Clegg S and Wiens R 2010 Optimization of laser-induced breakdown spectroscopy for rapid geochemical analysis *Chem. Geol.* **277** 137–48
[35] Senesi G 2014 Laser-induced breakdown spectroscopy (LIBS) applied to terrestrial and extraterrestrial analogue geomaterials with emphasis to minerals and rocks *Earth-Sci. Rev.* **139** 231–67
[36] ReVelle D O and Edwards W N 2007 Stardust—an artificial, low-velocity 'meteor' fall and recovery: 15 January 2006 *Meteorit. Planet. Sci.* **42** 271–99

[37] Jenniskens P *et al* 2005 Preparing for Hyperseed MAC: an observing campaign to monitor the entry of the Genesis Sample Return Capsule *Modern Meteor Science: An Interdisciplinary View* (Berlin: Springer) pp 339–60

[38] Foucher F, Westall F, Brandstätter F, Demets R, Parnell J, Cockell C S, Edwards H G, Bény J-M and Brack A 2010 Testing the survival of microfossils in artificial Martian sedimentary meteorites during entry into Earth's atmosphere: the STONE 6 experiment *Icarus* **207** 616–30

[39] Parnell J, Bowden S A, Muirhead D, Blamey N, Westall F, Demets R, Verchovsky S, Brandstätter F and Brack A 2011 Preservation of organic matter in the STONE 6 artificial meteorite experiment *Icarus* **212** 390–402

[40] Millman P M and Cook A F 1959 Photometric analysis of a spectrogram of a very slow meteor *Astrophys. J.* **130** 648

[41] Chodas P and Chesley S 2013 An overview of the Chelyabinsk impact event, *AGU Fall Meeting Abstracts* pp NH21D–05

[42] D'Alessio S and Harms A 1989 The nuclear and aerial dynamics of the Tunguska event *Planet. Space Sci.* **37** 329–40

[43] Park C 1978 Nitric oxide production by Tunguska meteor *Acta Astronaut.* **5** 523–42

Part I

Meteoroid and meteorite science

IOP Publishing

Hypersonic Meteoroid Entry Physics

Gianpiero Colonna, Mario Capitelli and Annarita Laricchiuta

Chapter 2

The trajectory, structure and origin of the Chelyabinsk impactor

Jiří Borovička

On 15 February 2013, 9:20 local time (nearly at sunrise; 3:20 UT), the citizens of Chelyabinsk, a Russian city with more than one million inhabitants, and its wider surroundings were surprised by a bright bolide in the clear morning sky (figure 2.1). The bolide steadily brightened and reached a maximum intensity 30× higher than the Sun for witnesses located south of Chelyabinsk. After the bolide disappeared, a bright and long sun-illuminated dust trail remained visible in the sky. A couple of minutes later, people looking at the trail were surprised by the arrival of a damaging blast wave, which caused many windows to be broken and other structural damage. In total, 7230 buildings were affected [1]. In the city of Chelyabinsk itself, about 40% of all buildings were damaged and ~10% of all windows were broken [2]. More than 1600 people were injured, mostly from broken glass.

In the next few days, an 8 m wide hole caused by meteorite impact was observed in the ice of lake Chebarkul, 70 km west of Chelyabinsk. The impact was reportedly

Figure 2.1. Left: Views of the Chelyabinsk bolide recorded by a dashcam from Kamensk-Uralsky, north of Chelyabinsk (by Aleksandr Ivanov, see https://www.youtube.com/watch?v=kFlpCT3v12E). Right: The dust trail of the Chelyabinsk bolide (photo by Konstantin Kudinov licensed under the Creative Commons Attribution-Share Alike 3.0 Unported).

doi:10.1088/2053-2563/aae894ch2

seen by local fishermen. Small meteorite fragments were found on the ice around the hole. Several weeks later, a record from a low-resolution security camera showing the impact from a distance of 2.5 km was revealed. The record shows just a plume of snow and ice. In addition, numerous, mostly small, meteorites were being extracted from a 75 cm thick snow layer in an extended region to the south of Chelyabinsk. Further meteorites, including a 4 kg piece, were recovered in the spring after the snow melted. The meteorites were classified as ordinary chondrites of type LL5 [3, 4]. The largest ~650 kg meteorite fragment was recovered from the bottom of the lake on 16 October 2013 [1].

Meteorites are not unusual but the bolide brightness and the strength of the blast wave indicated that Chelyabinsk witnessed the largest impact of a cosmic body since the Tunguska event in 1908. Fortunately for science, this time the event was well documented. Hundreds of casual video records from dashboard cameras in moving or standing cars, traffic cameras and security cameras appeared on the Internet. About 400 videos show the bolide in flight, the other videos show the illuminated landscape, the dust trail and the arrival of the blast wave, or they document the damage [5]. The audio records of many videos taken by mobile phones also contain the arrival of secondary sonic booms after the main blast wave. Sonic effects were also documented in seismic records in the region [6] and infrasonic records from around the world [7]. There are also satellite data. US Government sensors captured the bolide [2] and many meteorological satellites imaged the dust trail in the following minutes and hours [8, 9]. The dust from the trail dispersed around the northern hemisphere within several days [10]. Additional information comes from the distribution and properties of the recovered meteorites and from the effects of the blast wave.

2.1 Trajectory

The determination of bolide trajectory and velocity is important for many reasons. These data are needed as inputs for modeling the entry. Knowledge of the heights of fragmentation is needed to evaluate the strength and structure of the body, while knowledge of the heights of energy deposition is important for damage assessment. Knowledge of fragment trajectories can help in finding further meteorites. Finally, the trajectory and entry velocity are needed to compute the pre-impact orbit, which enables one to judge the origin of the body and to search for possible pre-impact images in the archives of large telescopes.

The Chelyabinsk bolide trajectory was computed by several authors using casual video records. Since no celestial objects are visible in the records, a celestial coordinate system could not be obtained directly. Moreover, in most cases the geographical coordinates of the recording place were not known. It was therefore necessary to locate the places where useful bolide recordings were taken. This was done with the help of maps and Google Earth imagery using the information posted by the authors. Only provisional bolide azimuths and elevations can be obtained through comparison with the streets, buildings and other objects visible on the maps. For precise work, *in situ* image calibration is needed. Nocturnal calibration images showing both stars and terrestrial objects must be taken from the same sites as the original records. Trajectories based on such calibrations were published by Borovička *et al* [11] and Popova *et al* [1]. Popova *et al* used a graphical method to transfer the coordinate system to the original video, while Borovička *et al* used mathematical transformations [12]. Both groups agree reasonably well on the trajectory (table 2.1). Borovička *et al* also considered the bending of the trajectory due to gravity and computed the trajectories of individual fragments visible in the final parts of the videos.

The bolide trajectory, locations from where it was videotaped and the locations of recovered meteorites are shown in figures 2.2 and 2.3. The observed luminous trajectory was about 270 km long and started at the height of 95 km near the border of Russia and Kazakhstan. The bolide traveled generally to the west (deviating only by 13° to the north from a purely westward direction) on a shallow trajectory with a slope of 18.5° to the horizontal. The velocity relative to Earth's surface was slightly above 19 km s^{-1}. The bolide passed only 35 km to the South (horizontal distance) from the center of Chelyabinsk city. At that point the bolide height was 28 km, very close to the point of maximum brightness, which was reached at the height of 30 km. Here the meteoroid was already disrupted into many fragments (see the analysis below). The largest fragment, which was later recovered from the lake, deviated by 1.3° to the north from the original trajectory. All other fragments were much smaller, three had a predicted mass in the range 15–30 kg and all others below 5 kg [11]. Using the predicted impact positions of Borovička *et al* [11], one 24.3 kg meteorite was actually found in December 2013.

2.2 Structure

From the impactor energy of ~500 kt TNT (2×10^{15} J) measured by various methods [1, 2] and a velocity of 19 km s^{-1}, a mass of 12 000 metric tons and a

Table 2.1. Bolide trajectory and velocity according to two sources.

		Borovička *et al* [11]	Popova *et al* [1]
Beginning	Longitude	64.477	64.565
	Latitude	54.545	54.445
	Height (km)	95.0	97.1
	Velocity (km s^{-1})	19.03	19.16
	Time (UT)	3:20:21.07	3:20:20.8
End	Longitude	60.5883	60.625
	Latitude	54.9361	54.931
	Height (km)	12.57	13.6
	Velocity (km s^{-1})	3.2	4.9
	Time (UT)	3:20:37.80	3:20:36.8
Entry radiant	Azimuth	103.5	103.2
	Elevation	18.55	18.3

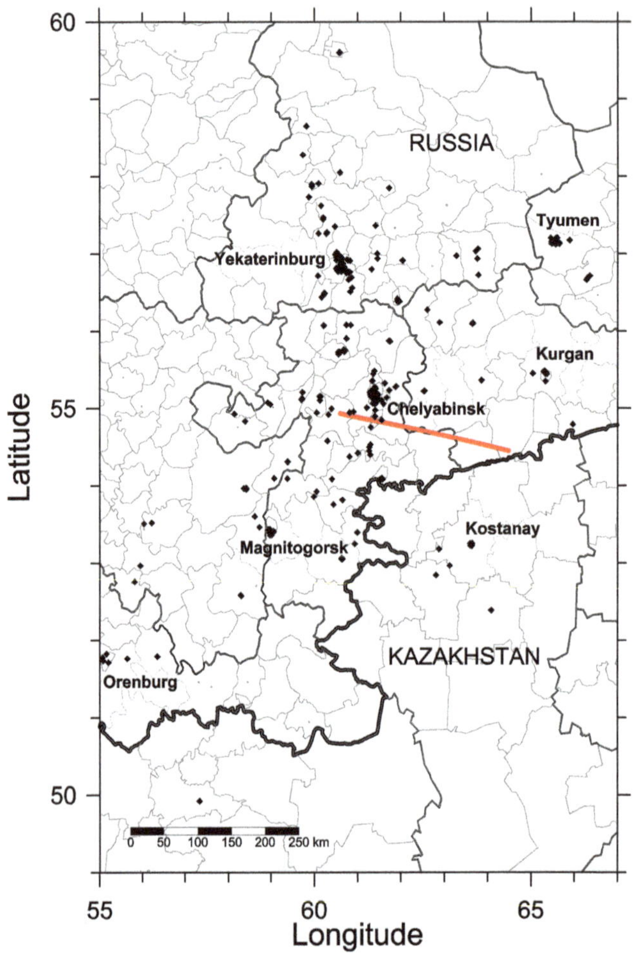

Figure 2.2. A schematic map showing the ground projection of the bolide trajectory (red line) and the locations from where the bolide was videotaped (diamonds). Background lines are administrative boundaries (source: http://www.diva-gis.org).

diameter of 19 m were determined. In order to infer information about its internal structure we need to study the behavior of the asteroid during its atmospheric passage, in particular its fragmentation. The known data are the trajectory and velocity, the observed deceleration toward the end, the trajectories and decelerations of individually observed fragments, the bolide light curve and the arrival times of sonic booms. Individual fragments can be seen on the videos only in the final stages of the trajectory, below a height of 26 km [11]. At that time the fragments separated enough to be resolved, but the more important aspect was that the enormous bolide glare already decreased at that time, enabling details to be seen on the videos. The main fragmentation, however, occurred higher in the atmosphere. The bolide light curve was crucial for the study of the main fragmentation.

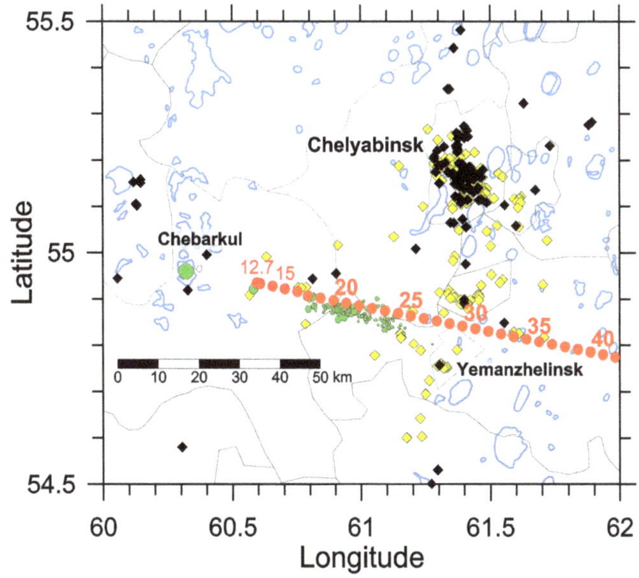

Figure 2.3. Schematic map of the terminal part of the bolide trajectory. Ground projections of bolide positions [11] at 1 km height intervals are plotted as red circles with labels showing the height in kilometers. Below 17 km the positions of the largest fragments, which deviated from the main trajectory, are given, including the bolide terminal point at a height of 12.7 km. Black diamonds are the locations of video records showing the bolide (at least part of it) and yellow diamonds are the locations of videos containing the arrival of the blast wave. The positions of recovered meteorites [1] are plotted as green circles with sizes proportional to meteorite mass. Blue lines are lake shorelines; gray lines are administrative boundaries (source: www.diva-gis.org).

The bolide light curve was measured most reliably by Brown *et al* [2] from selected video records. It was calibrated using the radiated energy measured by US Government sensors. The light curve, displayed in figure 2.4, has a broad asymmetric main peak lasting two seconds.

The decrease of brightness was steeper than the increase. The maximum occurred at a height of 30 km and the absolute (as seen from a 100 km distance) stellar magnitude reached −28, i.e. the bolide was three times brighter than the Sun (magnitude −26.7). For people just below the bolide, i.e. at a distance of 30 km, the object was 30× brighter than the daytime Sun (the actual Sun was close to the horizon and thus dimmer). There were two other peaks on the light curve, one before the main peak at a height of 42 km and one after it, at a height of 24 km. The former reached a magnitude of −23, i.e. it was 100× fainter than the main peak. The third peak was the most difficult to measure. It probably reached a magnitude of −25 after the magnitude dropped to −24 at a height of 26 km.

Several fragmentation models were used to fit the light curve. Borovička *et al* [11] applied their semi-empirical model [13]. The model uses the formalism of normal and eroding fragments. Eroding fragments are continuously releasing dust and small fragments until they disintegrate completely. The released particles ablate and evaporate independently. Adjusting the mass of the eroding fragment, the erosion rate and the size distribution of released particles, a hump on the light curve can be

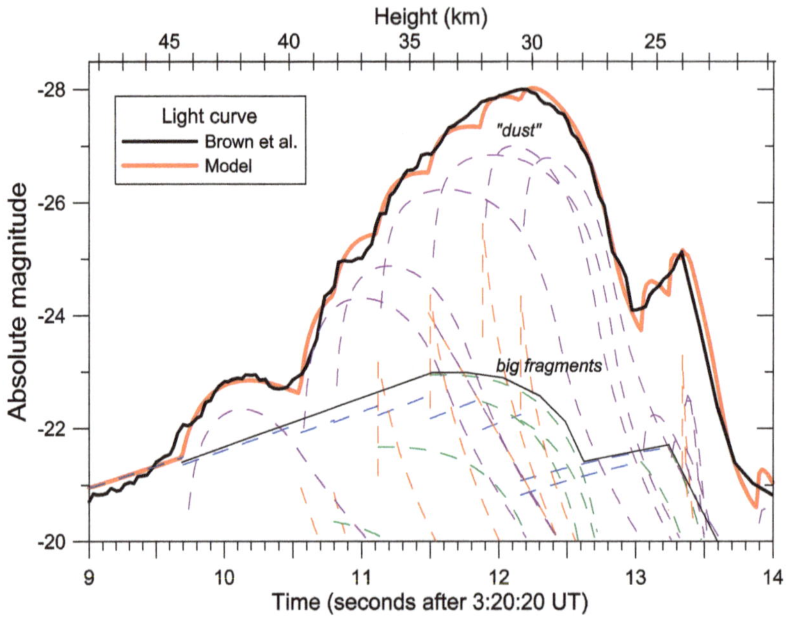

Figure 2.4. The brightest part of the Chelyabinsk bolide light curve (thick black line) as measured by Brown *et al* [2] and its fit by the semi-empirical model (thick red curve). Individual components contributing to the modeled light curve are plotted by the dashed lines: the dust, including small fragments, is shown in purple and orange, and big fragments (~meter-sized and larger) are shown in blue and green. The thin black line shows the summary light produced by the big fragments. The height on the upper axis refers to the leading fragment.

fitted. The first peak at 42 km could be explained by a gradual release of 0.5% of the original mass in the form of gram to kilogram fragments. To explain the main peak, 95% of the mass needed to be converted into sub-kilogram fragments between heights of 38–30 km. In figure 2.4 the contribution of this 'dust' to the radiation is plotted by purple dashed lines. Big fragments contributed much less. Six eroding fragments were used in the model but the exact fragmentation procedure cannot be revealed. The steep drop of brightness below 30 km, nevertheless, shows that no more than few percent of the fragmented mass survived as sizeable (>100 g) fragments.

Below the height of 30 km the bolide showed measurable deceleration, which indicates that the mass of the leading fragment was $\sim 2 \times 10^4$ kg (size ~3 m, 0.2% of the original mass). Nearly 10–20 fragments of this size were needed to produce the third flare during the subsequent fragmentations, which left only one fragment larger than 100 kg to emerge.

Other models were able to fit the light curve as well. Popova *et al* [1] used a hybrid model, in which part of the mass went into independent fragments and part into a spreading debris cloud. Debris clouds were modeled as single bodies with increasing cross sections. To explain the light curve, more than 96% of mass was put into the debris clouds. A similar but more general concept was used by Wheeler *et al* [14]. Three disruptions, each into four almost equal fragments, which then continued to fragment progressively according to a prescribed law, and a debris cloud were able to explain the light curve, except for the first peak. At each disruption, 86% of mass went into the debris cloud (in total therefore more than 99%).

Robertson and Matthias [15] used a hydrocode simulation in which the shear strength determines the burst height. Fragmentation occurs when the ram dynamic pressure (=ρv^2, where ρ is atmosphere density and v is velocity) exceeds the shear strength. Comparing the model result with an actual light curve they found that the Chelyabinsk asteroid shear strength was lower than 5 MPa. Although models with low strengths (<1 MPa) caused disruption at higher altitudes, pressure at lower altitudes was needed to disperse the rubble. The maximum of the light curve was thus also reached near a 30 km height. Shuvalov *et al* [16] even modeled Chelyabinsk as a strengthless liquid-like body and were able to reproduce the light curve reasonably, except for the third peak (and of course, no meteorites on the ground were produced).

Nevertheless, sonic booms suggest that fragmentations indeed occurred at the heights indicated by the light curve. The initial (and strongest) sonic boom was produced by supersonic motion of the body and the formation of a cylindrical blast wave [1, 2]. The secondary sonic booms that arrived later are interpreted as arrivals of spherical waves originating at various fragmentation points along the trajectory. From the time delay of the arrival, the height of origin can be determined. This analysis indicates a fragmentation at a height around 21 km, numerous events at 25–26 km as well as at 30–37 km, and some others at about 43 km. These heights correspond to the ascending parts of the light curve peaks (figure 2.4).

The dynamic pressure acting at fragmentation points provides information about the strength of the asteroid. While the (tensile) strength of stony meteorites is typically about 50 MPa, the strengths of meter-sized meteoroids were found to be much lower by their atmospheric fragmentation analysis [17]. The first significant fragmentation of Chelyabinsk occurred at a height of about 45 km, where the dynamic pressure was ~0.5 MPa. Less than 1% of the mass was lost here. Large-scale disruption occurred at heights of 39–30 km under pressures of 1–5 MPa. All models agree that at least 95% of mass was lost here in the form of dust and small fragments, probably of sub-kilogram mass (<10 cm). The rest emerged as 10–20 boulders of sizes 1–3 m. These boulders represented the strongest parts of the asteroid. They finally disrupted at heights of 26–21 km under pressures of 10–18 MPa. Only one large fragment (~60 cm) reached the ground after being decelerated and falling vertically on the ice. We know that it partly fragmented during the impact since one large piece and several smaller ones were later recovered from the bottom of the lake. Even the largest piece was not particularly strong; it broke apart during manipulation on shore.

The analyses of the meteorites revealed that they contain three different lithologies: light-colored, dark-colored and impact melt [4, 18]. All have the same mineralogical composition as LL5 chondrites. The impact-melt lithology is close to whole-rock melt and the dark-colored lithology is shock-darkened due to partial melting of iron and sulfides. Both the light- and dark-colored parts contain abundant micro-fractures. It is evident that the material was subject to shock and partial melting during a large collision in interplanetary space. This collision (or more collisions) evidently also caused numerous large-scale fractures in the asteroid, which led to its disintegration into mostly small fragments and dust particles under only moderate dynamic pressures of 1–5 MPa.

The dust trail left behind the bolide was quite long and started at a height of around 70 km. Dust release therefore already started at these heights. There were no flares or other irregularities in the light curve at heights above 50 km [19], so there is no indication of sudden fragmentation there. The dust was probably released gradually from the surface of the body. There may have been some regolith on the surface or the surface had been corroded by meteoroid impacts in space.

2.3 Origin

From the known trajectory and entry velocity (see section 2.1) the pre-encounter heliocentric orbit of the asteroid was computed. Orbital elements derived by two teams [1, 11] are given in table 2.2.

They are in good agreement. The orbit, plotted in figure 2.5, is a typical near-Earth asteroid orbit with the perihelion in the vicinity of the orbit of Venus, the aphelion approximately in the middle between the orbits of Mars and Jupiter, and a low inclination of 5°. Statistical analysis puts the most likely origin of the Chelyabinsk asteroid in the inner asteroid belt. The most probable transport route to near-Earth space was via the action of ν_6 secular resonance [1]. The most likely source of Chelyabinsk is the Flora asteroid family or the related Baptistina family [18]. These families seem to be common sources of LL chondrites [20, 21]. The measured cosmic ray exposure (CRE) age for the Chelyabinsk meteorites, 1.2 Ma [1, 22], is, however, much shorter than the typical CRE age of LL chondrites, which is 15 Ma [23]. The latest break-up, which released the asteroid from a larger body, therefore occurred more recently than is usual.

Borovička *et al* [11] noted the orbital similarity of Chelyabinsk and asteroid 86039 (1999 NC43) (see figure 2.5). 86039 is a large near-Earth asteroid with size of 2.2 km and spectral type Q corresponding to ordinary chondrites. There is only a 1:10^4 chance that such an orbital similarity with an asteroid of such size will occur by chance [11, 24]. However, detailed comparison of reflectance spectra did not confirm

Table 2.2. Bolide orbit according to two sources. Angular elements are given in equinox J2000.0.

	Borovička *et al* [11]	Popova *et al* [1]
Semi-major axis (AU)	1.72 ± 0.02	1.76 ± 0.08
Eccentricity	0.571 ± 0.006	0.581 ± 0.009
Perihelion distance (AU)	0.738 ± 0.002	0.739 ± 0.010
Inclination (°)	4.98 ± 0.12	4.93 ± 0.24
Longitude of ascending node (°)	326.459 ± 0.001	326.4422 ± 0.0014
Argument of perihelion (°)	107.67 ± 0.17	108.3 ± 1.9
Perihelion date	2012-12-31.39 ± 0.17	2012-12-31.9 ± 1.0

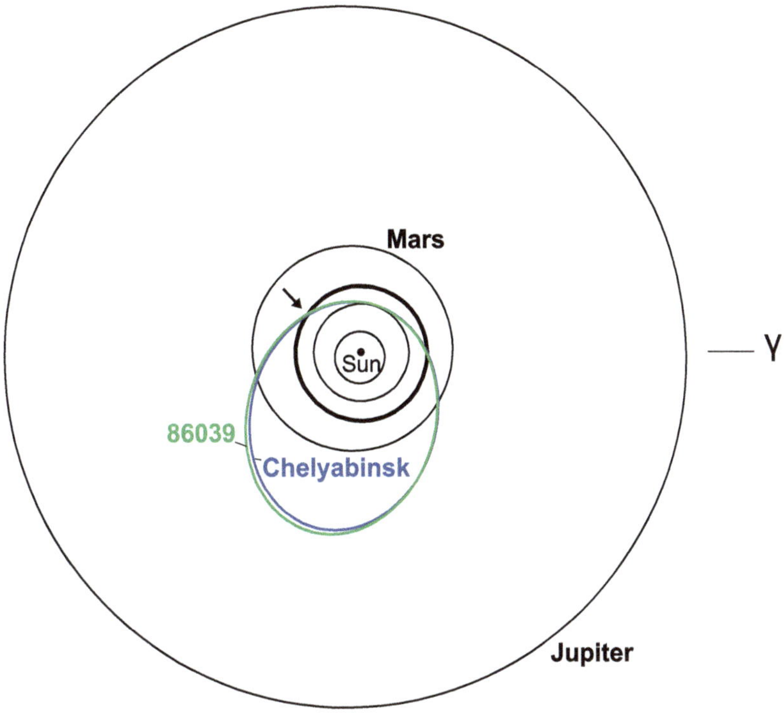

Figure 2.5. Projection of the orbits of the Chelyabinsk impactor and asteroid 86039 (1999 NC43) into the plane of ecliptic. The location of the Earth's impact is marked by an arrow. All objects orbit the Sun counterclockwise. The direction to vernal equinox is to the right.

compositional similarity. Asteroid 86039 seems to be composed of L-type chondritic material [24].

The Chelyabinsk impact occurred as a surprise. The asteroid was not known before its collision with the Earth. It approached our planet from the direction of the Sun, being projected on the day sky (within 45° of the Sun) and thus remaining unobservable from the ground for six weeks before the collision. Earlier than that it was too far and thus too faint to be detected [11].

2.4 Implications

With a mass of 12 000 metric tons and a size of 19 m Chelyabinsk was the largest Earth impactor since the 1908 Tunguska event. Despite the fact that the body was relatively fragile and disrupted almost completely 30 km above the surface, it produced a blast wave that caused significant damage. Fortunately, there were no causalities. The event came as a surprise and many people in Chelyabinsk were disoriented but, fortunately, the authorities evaluated the situation correctly and did not misidentify the bolide with something else, e.g. a military attack. It is estimated that impacts of this energy occur on average two times per century on Earth [25]. New generations of telescopic surveys will enhance our chance to detect such bodies before impact, but space borne instruments would be needed to also cover the daytime sky.

The outcome of impacts of asteroids of these sizes will always depend on their internal structure. The material (ordinary chondrite) and pre-impact orbit of Chelyabinsk were rather typical, but the degree of internal breakage can vary from case to case. Since it is difficult to reveal the internal structures of asteroids from astronomical observations, analyses of bright bolides caused by meter-sized objects provide an opportunity to study the properties of asteroidal materials in a statistical way.

Acknowledgements

This work was supported by grant no. 16-00761S from GA ČR.

References

[1] Popova O P *et al* 2013 Chelyabinsk airburst, damage assessment, meteorite recovery, and characterization *Science* **342** 1069–73

[2] Brown P G *et al* 2013 A 500-kiloton airburst over Chelyabinsk and an enhanced hazard from small impactors *Nature* **503** 238–41

[3] Galimov E M, Kolotov V P and Nazarov M A 2013 Analytical results for the material of the Chelyabinsk meteorite *Geochem. Int.* **51** 522–39

[4] Kohout T *et al* 2014 Mineralogy, reflectance spectra, and physical properties of the Chelyabinsk LL5 chondrite—Insight into shock-induced changes in asteroid regoliths *Icarus* **228** 78–85

[5] Borovička J *et al* 2016 A catalog of video records of the 2013 Chelyabinsk superbolide *Astron. Astrophys.* **585** A90

[6] Tauzin B, Debayle E, Quantin C and Coltice N 2013 Seismoacoustic coupling induced by the breakup of the 15 February 2013 Chelyabinsk meteor *Geophys. Res. Lett.* **40** 3522–6

[7] Le Pichon A *et al* 2013 The 2013 Russian fireball largest ever detected by CTBTO infrasound sensors *Geophys. Res. Lett.* **40** 3732–7

[8] Proud S R 2013 Reconstructing the orbit of the Chelyabinsk meteor using satellite observations *Geophys. Res. Lett.* **40** 3351–5

[9] Miller S D *et al* 2013 Earth-viewing satellite perspectives on the Chelyabinsk meteor event *Proc. Natl Acad. Sci. USA* **110** 18092–7

[10] Gorkavyi N, Rault D F, Newman P A, Silva A M and Dudorov A E 2013 New stratospheric dust belt due to the Chelyabinsk bolide *Geophys. Res. Lett.* **40** 4728–33

[11] Borovička J *et al* 2013 The trajectory, structure and origin of the Chelyabinsk asteroidal impactor *Nature* **503** 235–7

[12] Borovička J 2014 The analysis of casual video records of fireballs, *Proceedings of the International Meteor Conference, Poznan, Poland* 22–25 August 2013 101–5

[13] Borovička J *et al* 2013 The Košice meteorite fall: atmospheric trajectory, fragmentation, and orbit *Meteorit. Planet. Sci.* **48** 1757–79

[14] Wheeler L F, Register P J and Mathias D L 2017 A fragment-cloud model for asteroid breakup and atmospheric energy deposition *Icarus* **295** 149–69

[15] Robertson D K and Mathias D L 2017 Effect of yield curves and porous crush on hydrocode simulations of asteroid airburst *J. Geophys. Res. Planets* **122** 599–613

[16] Shuvalov V, Svetsov V, Popova O and Glazachev D 2017 Numerical model of the Chelyabinsk meteoroid as a strengthless object *Planet. Space Sci.* **147** 38–47
[17] Popova O *et al* 2011 Very low strengths of interplanetary meteoroids and small asteroids *Meteorit. Planet. Sci.* **46** 1525–50
[18] Reddy V *et al* 2014 Chelyabinsk meteorite explains unusual spectral properties of Baptistina Asteroid Family *Icarus* **237** 116–30
[19] Borovička J 2016 Are some meteoroids rubble piles? *IAU Symp.* **318** 80–5
[20] Vernazza P *et al* 2008 Compositional differences between meteorites and near-Earth asteroids *Nature* **454** 858–60
[21] Vokrouhlický D, Bottke W F and Nesvorný D 2017 Forming the Flora family: implications for the near-Earth asteroid population and large terrestrial planet impactors *Astron. J.* **153** id172
[22] Povinec P P *et al* 2015 Cosmogenic radionuclides and mineralogical properties of the Chelyabinsk (LL5) meteorite: what do we learn about the meteoroid? *Meteorit. Planet. Sci.* **50** 273–86
[23] Eugster O, Herzog G F, Marti K and Caffee M W 2006 Irradiation records, cosmic-ray exposure ages, and transfer times in meteorites *Meteorites and the Early Solar System II* ed D S Lauretta and H Y McSween Jr. (Tucson, AZ: University of Arizona Press) 829–52
[24] Reddy V *et al* 2015 Link between the potentially hazardous asteroid (86039) 1999 NC43 and the Chelyabinsk meteoroid tenuous *Icarus* **252** 129–43
[25] Harris A W and D'Abramo G 2015 The population of near-Earth asteroids *Icarus* **257** 302–12

Chapter 3

Properties of meteoroids from forward scatter radio observations

Hervé Lamy

3.1 Radio meteor theory

When a meteoroid enters the Earth's atmosphere with velocities in excess of ~10 km s^{-1}, it impacts the atoms and molecules of the upper atmosphere with a kinetic energy large enough to ionize them. It also heats up to high temperatures and starts to ablate. As a result, a trail of ions and electrons forms along the trajectory path behind the meteoroid.

If a radio wave of a given frequency is sent towards space from a ground-based transmitter, it can be temporarily reflected by the meteor trail towards the ground. The incident radio wave is scattered only by electrons as ions are too heavy. Under some geometrical conditions (see below), it can be recorded by a receiver tuned to the same frequency. The signal recorded at the receiving station is called a meteor trail echo. Its duration is directly related to the lifetime of the meteor trail as the electrons tend to quickly scatter into the neutral ambient atmosphere. A meteor radar corresponds to the case where the transmitter and the receiver are located at the same place. When the transmitter and receiver(s) are not on the same site, we talk about forward scatter observations (see figure 3.1) which is the topic of this chapter.

Meteor echoes are usually classified into two categories called underdense and overdense, according to the value of the line electron density α (expressed in number of electrons per meter). This is based on McKinley's [1] classical radio meteor theory, where it is first assumed that the trail is a stationary straight line of electrons, a hypothesis made because the longitudinal extent of the trail (typically a few kilometers to a few tens of kilometers) is much larger than its radial extent (from a few tens of centimeters to a few meters). Additional factors are then included, such as the initial radius effect, to take into account the finite radial extent of the trail (assumed to be cylindrical), and the ambipolar diffusion to model the scattering of trail electrons into the ambient neutral atmosphere. The typical duration of

doi:10.1088/2053-2563/aae894ch3

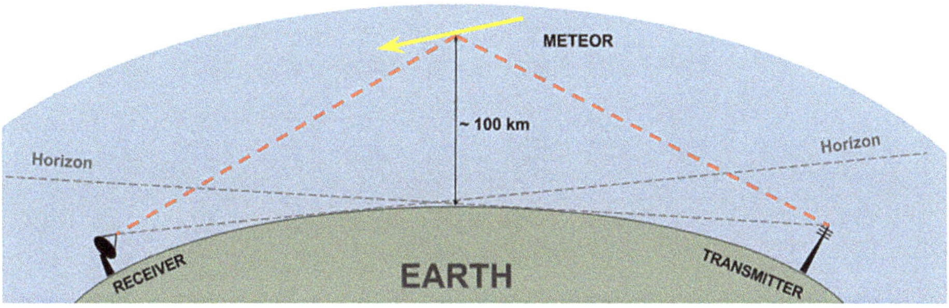

Figure 3.1. Sketch of the general principle of radio forward scatter observations.

underdense meteor echoes is a fraction of a second while overdense meteor echoes can typically last from ~one second to a few tens of seconds.

Underdense meteor echoes have typical line electron densities $\alpha < 10^{13}$ e − m^{-1} while overdense meteor echoes have $\alpha > 10^{15}$ e − m^{-1}. There are of course a lot of meteor echoes with intermediate electron line densities for which the physics is more complex. Some recent developments (e.g. [2]) are trying to model these meteor echoes, but this is beyond the goal of this chapter. From a physics point of view, the radio wave can penetrate underdense meteor trails and is scattered by individual electrons while for overdense meteor trails, the electron density is large enough that the dielectric constant of the medium becomes negative and the trail behaves as a plasma. The classical theory of McKinley [1] models the overdense meteor echoes as if the radio wave was reflected on the surface of an expanding metallic cylinder. As we will see below, this theory for overdense meteor echoes has limited applications as it is too simplistic and does not include additional phenomena.

One fundamental property of underdense meteor echoes is that the reflection of the radio wave is specular, which means that the majority of the received power reflected off the trail occurs when the so-called specular reflection point is created on the meteoroid path. The position of this reflection point is easy to determine as it is the point along the meteoroid path that is tangential to an ellipsoid whose foci are the transmitter and the receiver (see figure 3.2).

This property simply results from adding elemental small contributions from each electron individually along the path. The vast majority of the power actually comes from a small region centered on the specular reflection point and called the first Fresnel zone, whose size depends mostly on the meteor speed and on the wavelength of the radio wave. Contributions from other parts of the meteor trail can add constructively or destructively to this main signal, leading to so-called Fresnel oscillations in the signal.

An important consequence of the specular reflection is that a given system made of a transmitter and one receiver cannot detect all meteors. This is illustrated in figure 3.3. If the reflection point is located too high in the atmosphere (bottom case), the atmosphere is too thin at these altitudes such that the created ionization is not large enough to reflect a detectable amount of power. On the other hand, if the reflection point is located too low in the atmosphere (top case), the meteoroid might be completely ablated before this point is created and therefore no signal at all will be detected at the receiver. In figure 3.3, what is called the 'meteor zone' refers to a

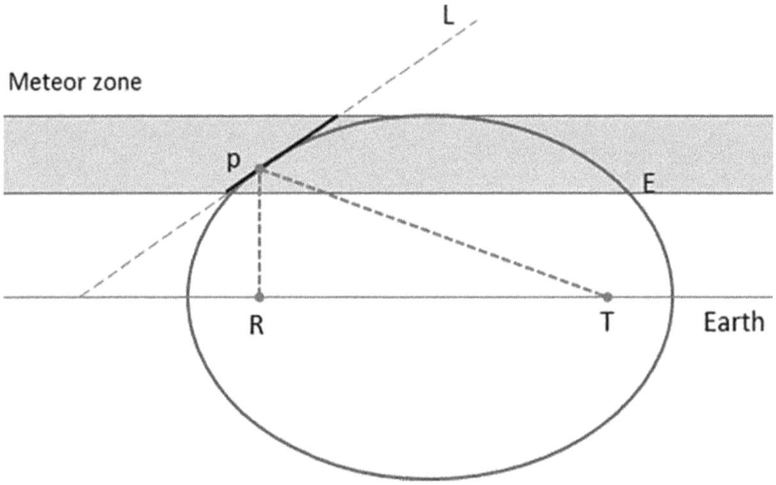

Figure 3.2. Illustration of the specularity condition. The specular reflection point p is the point along the meteoroid path L that is tangential to the ellipsoid E whose foci are the transmitter T and receiver R.

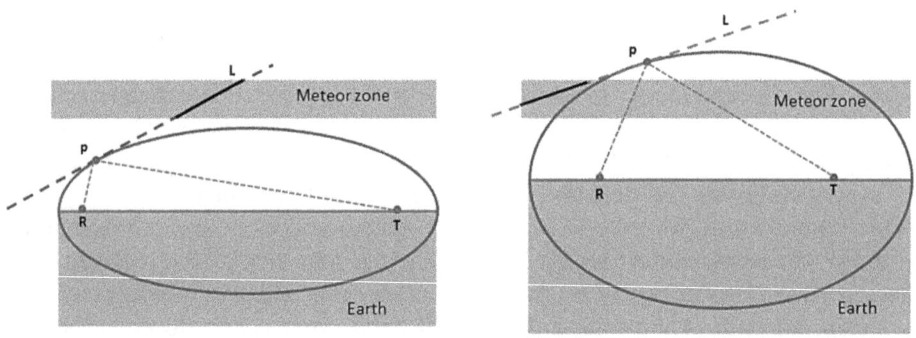

Figure 3.3. Consequences of the specularity condition. For given L, T and R, if the specular point p is too low (left) or too high (right), no signal will be recorded at T.

region in altitude where most meteor echoes should occur and can roughly be considered as 85–110 km.

The theory of forward scatter of radio waves has been developed by McKinley [1] as an extension of the classical theory for back scatter/radar systems. It provides formulas for the power profile, namely the amount of power in the meteor echo as a function of time.

For underdense meteor echoes, the power at the receiver is given by

$$P_R = \frac{P_T G_T G_R \lambda^3 r_e^2 \alpha^2 \sin^2(\gamma)}{16\pi^2 R_T R_R (R_T + R_R)(1 - \sin^2(\phi)\cos^2(\beta))} \exp\left[-\frac{8\pi^2 r_0^2}{\lambda^2 \sec\phi}\right] \exp\left[-\frac{32\pi^2 D_a t}{\lambda^2 \sec\phi}\right] \{C^2 + S^2\}, \tag{3.1}$$

where P_T is the power sent by the transmitter, G_T and G_R are the antenna gains for the transmitter and receiver in the directions to the reflection point, λ is the wavelength, r_e is the classical radius of the electron, α is the electron line density, γ is the polarization of the radio wave, R_T and R_R are the distances between, respectively, the transmitter/receiver and the specular reflection point, ϕ is half of the scattering angle of the radio wave and β is the inclination of the meteor trail with respect to the propagation plane of the radio wave. The geometrical parameters are illustrated in figure 3.4.

The first term in equation (3.1) is the peak value obtained when the specular reflection point is created and assuming all electrons lie on a straight line. The actual peak value is corrected by the initial radius r_0 that takes into account that trail electrons are not located on a single line but inside a cylinder, which reduces the strength of the signal and leads to the first exponential in equation (3.1). The second exponential depends on time t and is due to the diffusion of trail electrons in the neutral atmosphere. D_a is the ambipolar diffusion coefficient. The final terms in equation (3.1), C and S, are Fresnel integrals:

$$C = \int \cos\frac{\pi x^2}{2}dx \quad S = \sin\frac{\pi x^2}{2}dx, \tag{3.2}$$

where $x = 2s/\sqrt{R_F\lambda}$ is the Fresnel length, s is the distance along the trail measured from the specular reflection point, counted positively in the direction where the meteoroid is moving, and R_F is the size of the first Fresnel zone. This term is at the origin of the Fresnel oscillations which superimposes on the main signal. Indeed, every time a new Fresnel zone is created along the meteoroid path, it creates signals

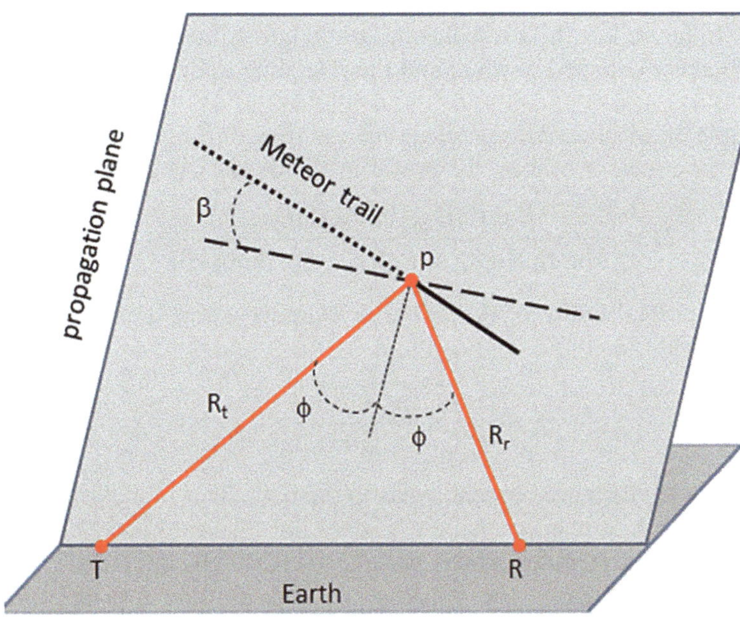

Figure 3.4. Geometrical parameters for forward scattering of radio waves.

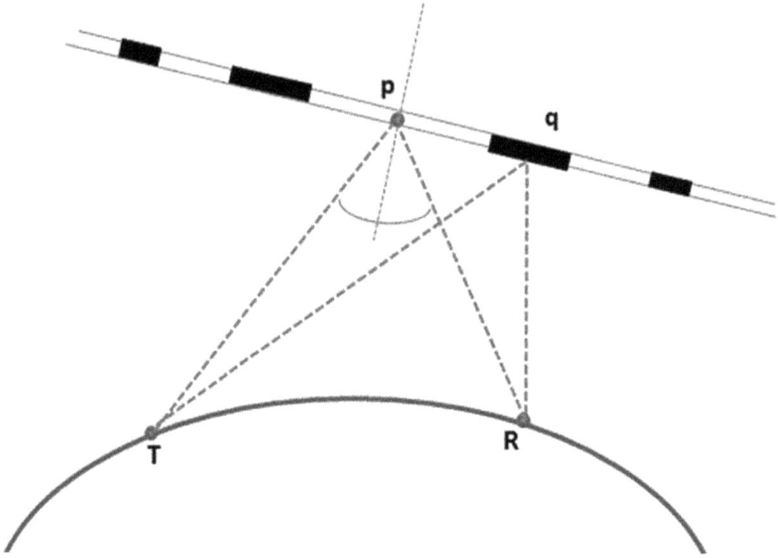

Figure 3.5. Illustration of Fresnel zones. The main Fresnel zone is centered on the specular reflection point p. Fresnel zones with constructive interference with the main signal are in white while those leading to destructive interference are in black.

which are alternatively in phase and out of phase with the main signal and leads to an increase/decrease of the total power. An example is given in figure 3.5 where the contribution from point q adds up to the main signal coming from point p in a destructive way, leading to a slight decrease of the power. Along a given meteoroid path and knowing the speed of the object, the positions of the Fresnel zones can be determined. In figure 3.5, the Fresnel zones leading to constructive interference with the main signal are depicted in white and those leading to destructive interference in black.

An example of an underdense meteor echo is shown in figure 3.6.

For overdense meteor echoes, the power at the receiver is given by

$$P_R = \frac{P_T G_T G_R \lambda^2 \cos\phi \sin^2\gamma}{32\pi^2 R_T R_R (R_T + R_R)(1 - \sin^2\phi\cos^2\beta)} r_c, \tag{3.3}$$

where r_c is the critical radius of the 'metallic' cylinder where the dielectric constant becomes negative,

$$r_c = \sqrt{r^2 \ln \frac{\alpha r_e \lambda^2 \sec^2\phi}{\pi^2 r^2}}, \tag{3.4}$$

where r is the radial distance perpendicular to the path. This model is rather poor as it neglects a number of phenomena, e.g. it does not take into account the effect of external parts of the cylinder which become increasingly underdense with time. More important, it neglects the influence of high-altitude mesospheric/thermospheric shear winds which can reach speeds from tens to a few hundreds of meters

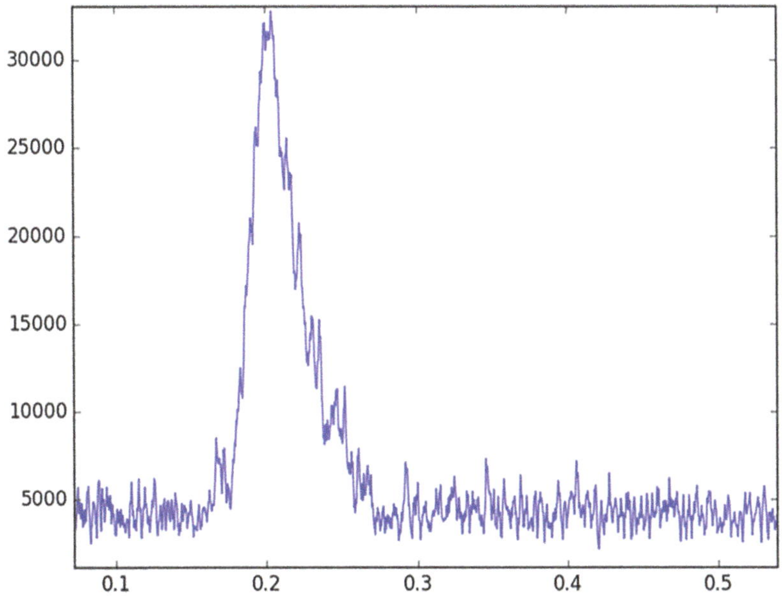

Figure 3.6. Typical underdense meteor echo. The vertical axis is power in arbitrary units while the horizontal axis is time in seconds.

per second. They can break the cylindrical trail into multiple parts and hence create multiple specular reflection points. The multiple reflections can then interfere constructively or destructively and lead to a much more complex power profile than the one modeled by equations (3.3) and (3.4).

An example of an overdense meteor echo is shown in figure 3.7.

Meteor trail echoes form the vast majority of meteor echoes detected by meteor radars such as, e.g., the Canadian Meteor Orbit Radar (CMOR, e.g. [3]) or by the Belgian Radio Meteor Stations (BRAMS) network that will be described in detail in the next section. Another type of meteor echo, mostly observed with larger objects, is the head echo which occurs when the incoming radio wave is reflected upon the ionized region forming in front of the meteoroid. This region moves at the same high speed ($>\sim 10$ km s^{-1}) as the meteoroid itself while the speed of the meteor trail depends mostly on the speed of high-altitude mesospheric/thermospheric winds. Consequently, the Doppler effects associated with these two types of meteor echoes are very different and allow one to easily discriminate them. Head echoes will not be discussed in detail in this chapter. Only a few examples of observations will be given in the section about optical versus radio observations of meteors.

3.2 The BRAMS project

3.2.1 The BRAMS network

BRAMS is a Belgian radio network using forward scatter techniques to detect and study meteoroids. It uses a dedicated transmitter located at the Geophysical Center in Dourbes in the south of Belgium and 26 identical receiving stations spread all over

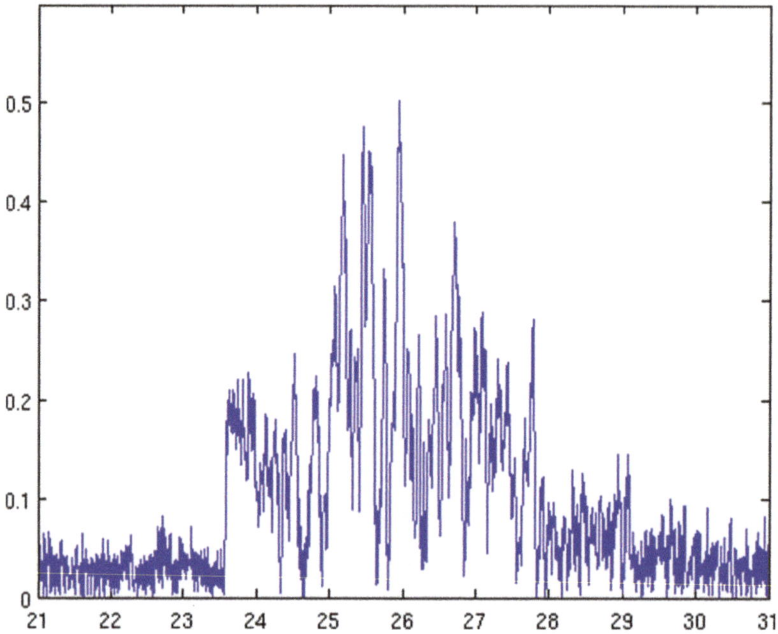

Figure 3.7. A typical overdense meteor echo. The vertical axis is power in arbitrary units while the horizontal axis is time in seconds.

Belgian territory. Figure 3.8 shows the position of the transmitter and of the 26 receiving stations in August 2018.

The transmitter is a crossed-dipole antenna with an 8 m × 8 m metallic grid acting as the reflector. It emits a pure sinusoidal wave with no modulation at a frequency of 49.97 MHz and with a power of approximately 150 W. The choice of the frequency was based on several physical and practical reasons. First, the frequency is high enough to avoid any reflection on the ionospheric layers (the E region peak is approximately at the same height as the meteor zone). Second, as can be seen from equation (3.1), the power of an underdense meteor echo scales as λ^3 and its duration (measured by the time constant of the exponential decay) scales as λ^2. Consequently, for a given meteor and trajectory, if the frequency was 150 MHz instead of ~50 MHz, the received signal would be 27 times less powerful and its duration 9 times shorter. Finally, the frequency must be available and protected.

The decision to use an 8 m × 8 m metallic grid for the reflector was prompted by the desire to emit the maximum amount of power toward the zenith, to have a relatively broad lobe to cover a large portion of the sky and to not emit too much power horizontally. The simulated vertical and horizontal patterns of the transmitter are shown in figure 3.9. Details of the electromagnetic simulations can be found in [4].

Each receiving station uses the same technology, which is depicted in figure 3.10.

The antenna is a three-element Yagi antenna set up vertically and tilted in azimuth to the direction of the transmitter. It has a broad lobe as well in order to cover a large portion of the sky and capture as many meteor echoes as possible.

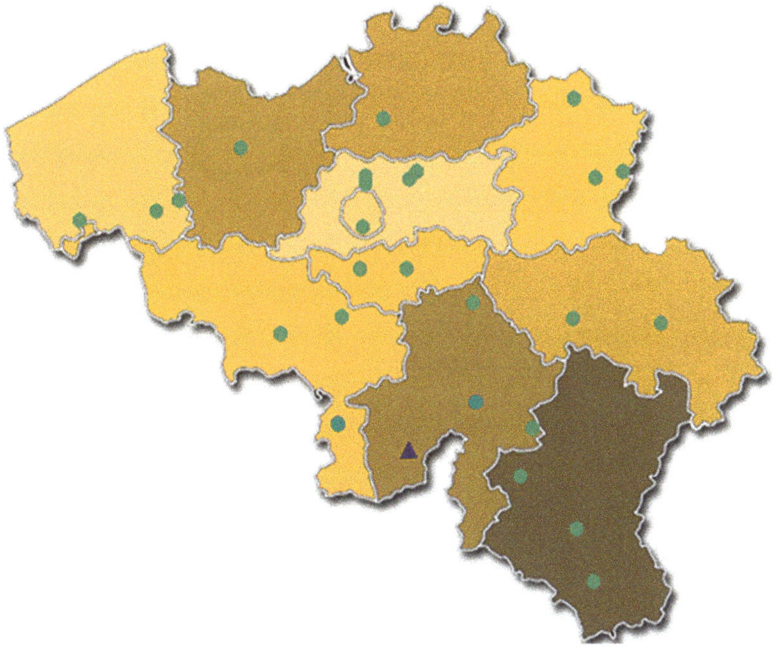

Figure 3.8. Map of the BRAMS network on 30 August 2018. The transmitter is the blue triangle while the green dots are the 26 identical receiving stations.

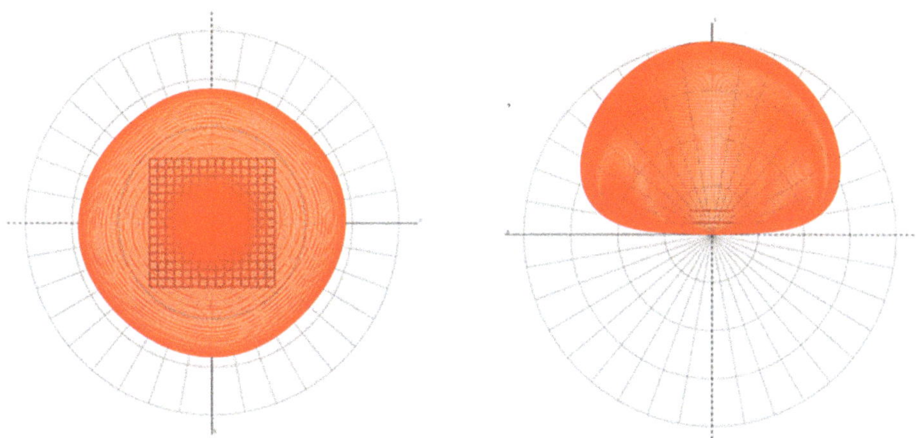

Figure 3.9. Simulations of the horizontal (left) and vertical (right) patterns of the BRAMS transmitter [4].

It has nulls along the direction of the elements of the antenna and the reflector protects from unwanted reflections on the ground. The simulated horizontal and vertical patterns of the antenna are shown in figure 3.11.

The antenna is connected to a commercial ICOM-R75 receiver whose local oscillator frequency is set to 49.969 MHz in order to shift the frequencies from

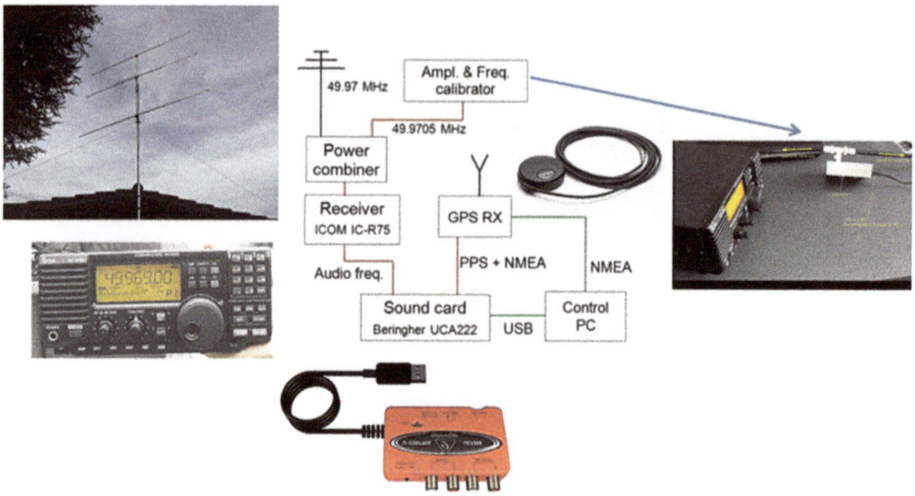

Figure 3.10. Technology used by each BRAMS receiving station.

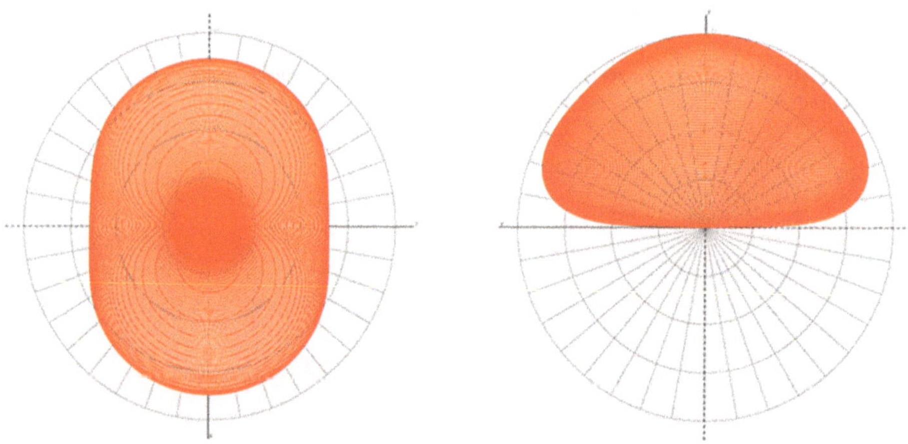

Figure 3.11. Simulations of the horizontal (left) and vertical (right) patterns of the BRAMS receiving antennas [4].

around 49.97 MHz to around 1 KHz using a frequency mixer. The advantage of shifting down the signal to lower frequencies is that it can then be sampled using a cheap external Behringer UCA222 soundcard. The sampling frequency is 5512 Hz. Each BRAMS station is also equipped with a GPS clock which provides a very accurate pulse per second (PPS) signal, allowing all the BRAMS stations to be synchronized. The signals coming from the receiver and from the GPS clock are sampled simultaneously using the two stereo input entries of the soundcard. The soundcard is controlled by free software called Spectrum Lab (SL) running on Windows on a local PC. An additional device called the BRAMS calibrator is added to the signal coming from the antenna via a Tee. This device is designed in order to produce a signal at a unique frequency which is very stable both in frequency and amplitude. By construction this signal is emitted approximately 500 Hz above the signal of our transmitter, hence at 49.9705 MHz. It provides an accurate reference for frequency measurements (useful, for example, to measure a Doppler effect) and a continuous measurement of the gain of the receiving chain (receiver + soundcard).

3.2.2 The BRAMS data

BRAMS data are saved locally on a PC by SL under an audio WAV format. Data are saved every 5 min which means that 288 WAV files are saved every day at each station. An example of BRAMS data is shown in figure 3.12. Amplitude is plotted in arbitrary units as function of the sample number. Since each file lasts approximately 300 s, the total number of samples in a WAV file is of the order of $300 \times 5512 \sim 1.6$ million. Note that this is the amplitude that is recorded (as a voltage measurement in the receiver) which therefore can take positive and negative values. We are interested in the power profile of the meteor echoes, which is simply the square of these data.

Figure 3.12 clearly indicates that the recorded signal is very noisy and it is hard to distinguish meteor echoes from other spurious signals. A much better representation is to generate a spectrogram from the raw BRAMS data. For that, a fast Fourier transform (FFT) is carried out on 16 384 (2^{14}) samples and by stacking consecutive FFT, the spectrogram can be obtained and displays how the power of the signal is distributed among frequencies as a function of time. The power is color-coded. The

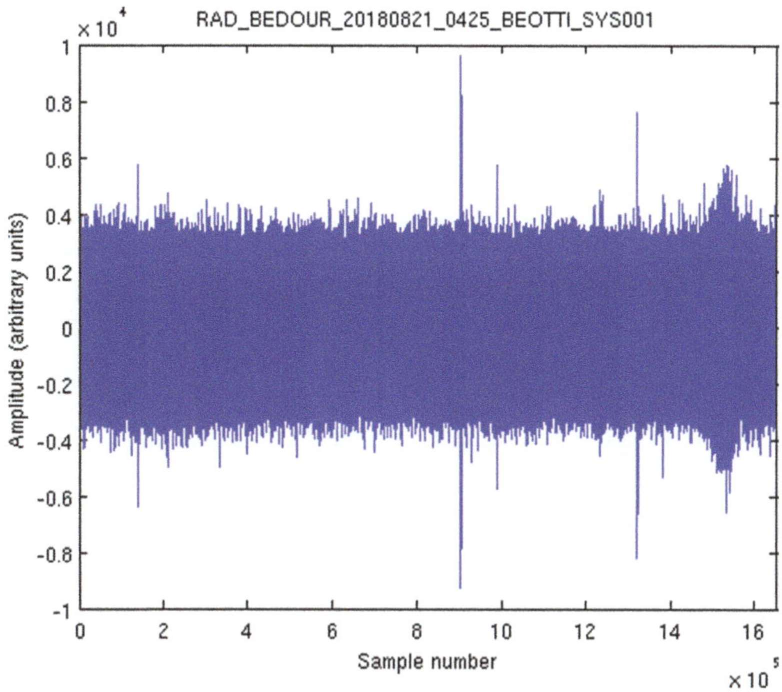

Figure 3.12. Example of raw BRAMS data. Station: BEOTTI. Date: 21/08/2018 at 04:25 UT.

time resolution Δt of the spectrogram is much worse than in the raw data and is equal to 16 384/5512 ~ 2.97 s. Conversely, the frequency resolution Δf is quite good and equal to 5512/16 384 ~ 0.34 Hz. In BRAMS data, however, an overlap of 90% is added in order to smooth the signals in the spectrogram. This means that each FFT includes only 10% of new points and 90% of common points with the previous FFT. By doing this, the apparent time resolution $\Delta f'$ becomes equal to ~0.3 s. In practice the power is simply spread over consecutive columns of the spectrogram so a meteor echo will appear on several consecutive columns of the spectrogram. In figure 3.13, the spectrogram corresponding to the raw data from figure 3.12 is displayed. The spectrogram is constructed to show a 200 Hz range centered on the horizontal signal, which is the direct (tropospheric) signal coming from the transmitter. In this example it appears at around 1262 Hz and not 1 KHz as previously explained. The reason is that the local oscillator (LO) of the ICOM-R75 receiver is not perfectly stable and is strongly influenced by the temperature of the receiver. As a consequence, the LO drifts and appears at a larger frequency. These LO drifts can be easily monitored by measuring the frequency of the calibrator signal. The long lasting signals in figure 3.13, although sometimes discontinuous, are reflections on airplanes flying near the transmitter. Since the distances between the transmitter and all BRAMS stations are lower than 250 km, airplanes flying at an altitude of around 10 km will never appear below the horizon. These reflections are spurious signals which strongly complicate the analysis of BRAMS data. The nearly vertical signals are all underdense meteor

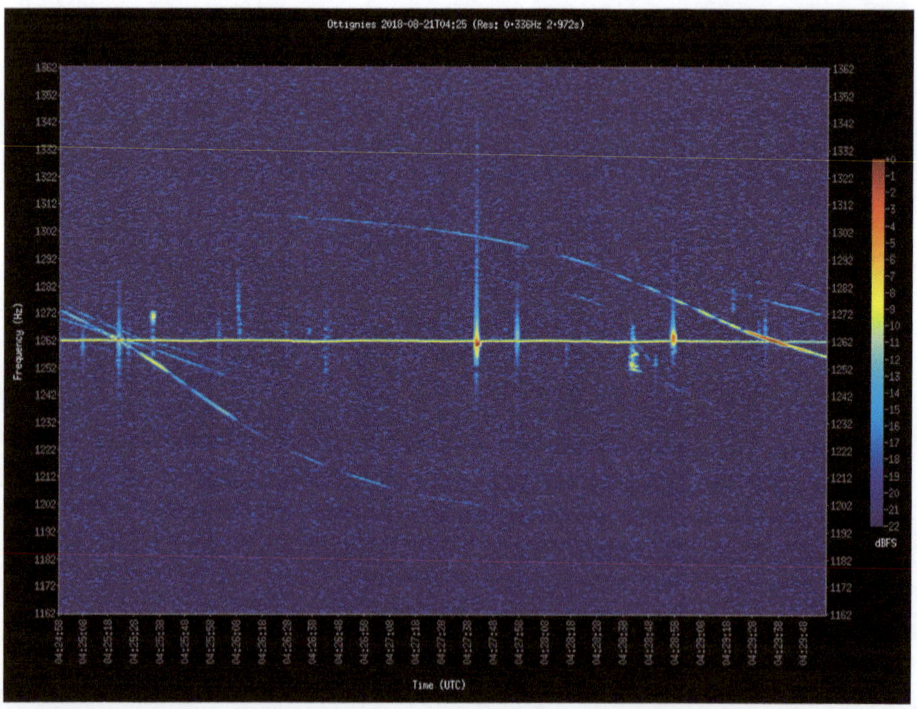

Figure 3.13. Example of BRAMS spectrogram. Station: BEOTTI. Date: 21/08/2018 at 04:25 UT.

echoes while the one with a more complex shape on the right-hand side of the spectrogram is an overdense meteor echo.

Figure 3.13 displays only the 200 Hz range because that range is sufficient to encompass all of the trail meteor echoes. Indeed, the vertical position of a meteor echo in a spectrogram depends on the Doppler effect due to the speed of the meteor trail, hence on the speed v_w of mesospheric/thermospheric winds. If we take a typical value of 100 m s^{-1} for v_w, the maximum Doppler associated effect is $\Delta f = 2 \times f \times v_{wr}/c$, where $f = 49.97$ MHz, v_{wr} is the radial component of the wind speed and c is the speed of light. The factor 2 is a rough estimate and comes from the fact that there is a double Doppler effect, first between the transmitter and the moving trail, then between the moving trail and the receiver. This provides a value of ~33 Hz. So, trail meteor echoes will never appear outside a range of ±100 Hz from the direct signal. The situation is different for head echoes for which the associated speeds and Doppler effects are much larger. Nevertheless these are rarely detected with BRAMS and will not be discussed here.

With a sampling frequency $f_s = 5512$ Hz, Nyquist's theorem tells us that we have in theory access to all frequencies between 0 and $f_s/2 = 2756$ Hz. Figure 3.14 displays the same spectrogram as figure 3.13 but showing a larger frequency span going from 900 to 1800 Hz. The signal just below 1750 Hz is the signal from the BRAMS calibrator.

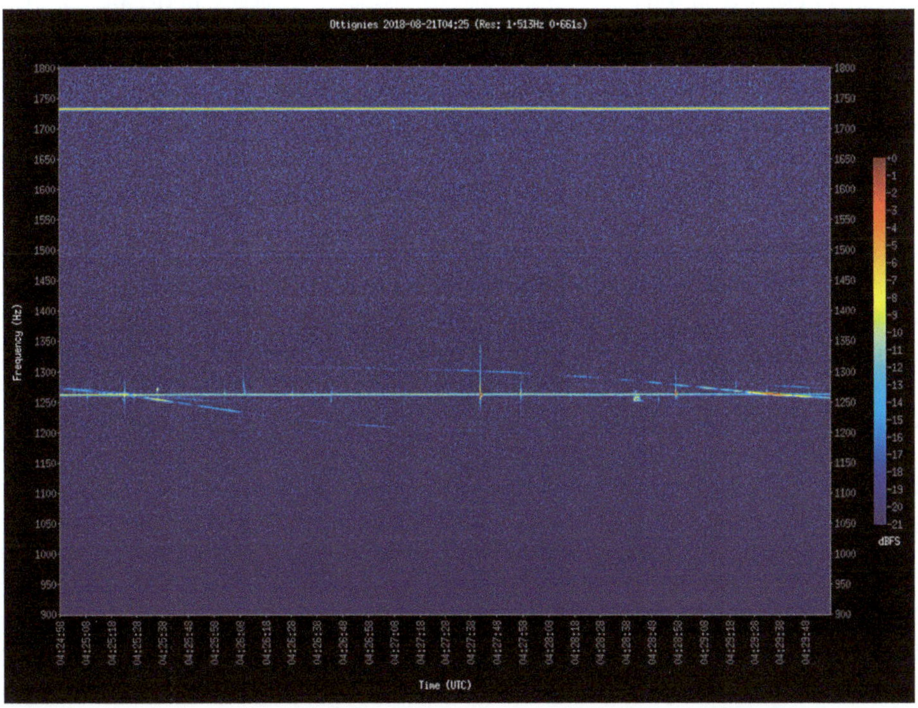

Figure 3.14. The same spectrogram as in figure 3.13 but with a larger frequency range. The signal near 1740 Hz is the BRAMS calibrator.

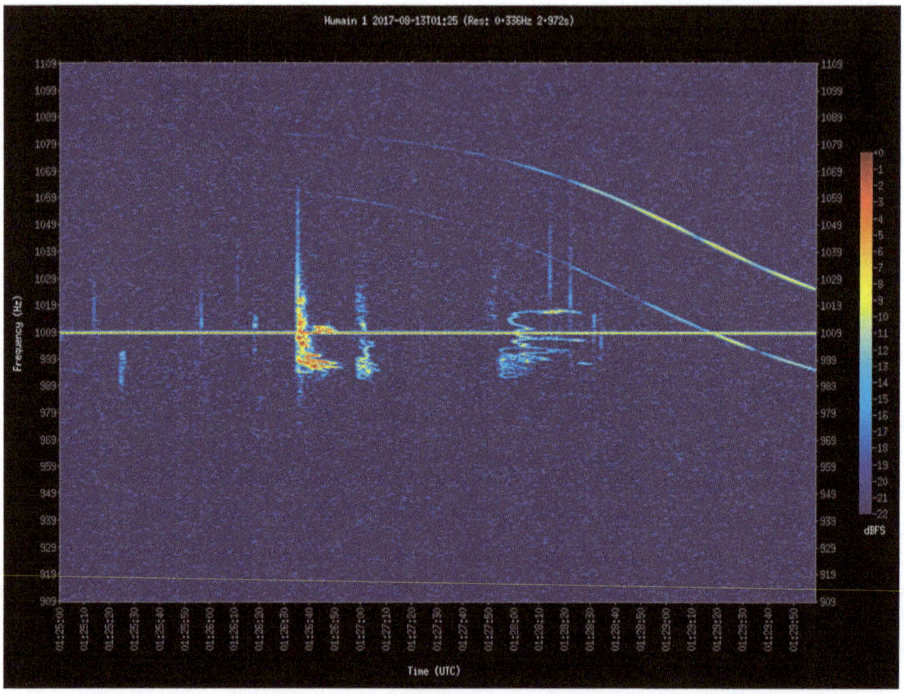

Figure 3.15. Example of BRAMS spectrogram during the Perseids 2017. Station: BEHUMA. Date: 13/08/2017 at 01:25 UT.

During meteor showers, the number of overdense meteor echoes strongly increases, as can be seen for example in figure 3.15, obtained during the Perseids 2017.

3.2.3 Determination of meteoroid trajectories using BRAMS data

One of the main objectives of the BRAMS project is to reconstruct meteoroid trajectories. Due to the specular reflection, a single receiving station is only sensitive to one point of the trajectory. However, since the position of each specular reflection point depends on the geometry and will therefore vary along the meteoroid path for each station, it is in principle possible to reconstruct the trajectory from observations of the same meteor at various stations. To reconstruct the trajectory at least six stations need to detect the same meteor in order to obtain the coordinate of one point (three unknowns), one direction (two angles) and the speed (assuming the deceleration is negligible, otherwise additional parameters should be included). This is a reason why the BRAMS network has to be relatively dense in order to maximize the chances of obtaining at least six multiple detections.

In [5], an attempt is made to use a suggestion proposed by Nedeljkovic [6]. The idea is the following: when the meteoroid trajectory L is known, finding the position of the specular point for a given transmitter (T) and receiver (R) pair requires only to find the point on the trajectory that is tangent to the ellipsoid with T and R as foci (see figure 3.1). It is called the direct problem and has a simple analytical solution.

The inverse problem of retrieving a trajectory that is tangential to a set of ellipsoids with foci T and R_1, R_2, ... is much more complex (R_i being the position of receiver i). Instead, a set of possible trajectories is generated and only the ones tangential to a number of ellipsoids corresponding to the number of receiving stations with detections are selected. The selection criterion is that the altitude of the specular reflection point must be within the meteor zone, otherwise the corresponding trajectories are rejected. The higher the number of stations considered, the lower the number of remaining trajectories. The advantage of this method is that it uses only the direct problem. The disadvantage is that a very large number of trajectories have to be generated first. Among all possible remaining trajectories, an assumption on the speed of the meteoroid can be made to generate time delays between two receiving stations. Indeed, for two stations R_1 and R_2, the specular reflection points are located at different positions on the meteoroid trajectory. Therefore, the meteor echo will appear first at station R_1 then at station R_2 when the meteoroid has traveled the distance between the two reflection points. This additional condition of the time delays allow one to reduce the number of remaining possible trajectories. This work is still on-going.

One station, located in the radio-astronomical site of Humain, in the south-east of Belgium, is a radio interferometer which, unlike all other BRAMS stations, is able to retrieve the direction of arrival of a meteor echo to an accuracy of the order of 1°. It is made of five Yagi antennas, three of them aligned along two orthogonal axes, roughly aligned N–S and E–W, with the central antenna common to the two axes. The principle of the interferometer is based on measuring phase differences between two pairs of antennas from the three co-aligned antennas. Then, using the method proposed by Jones *et al* [7], it is possible to determine accurately and unambiguously the projections of the angle of arrival of the meteor echo in the N–S and E–W planes (see [8] for more details about the technique and the interferometer itself). The principle is illustrated in figure 3.16 with three antennas named '0', '1' and '2'. The projection of the angle of arrival in the plane of the three antennas is called ξ.

From the projections ξ_1 and ξ_2 in the two orthogonal planes, the azimuth α and elevation β of the angle of arrival of the meteor echo can be determined using (e.g. [9])

$$\alpha = \cos^{-1}\left(\frac{\cos\xi_1}{\cos\beta}\right) = \cos^{-1}\left(\frac{\cos\xi_2}{\cos\beta}\right) \tag{3.5}$$

$$\beta = \tan^{-1}\left(\frac{\cos\xi_2}{\cos\xi_1}\right). \tag{3.6}$$

An example of results is shown below for an underdense meteor echo obtained on 5 December 2016, corresponding to the second white rectangle counted from the left in the spectrogram shown in figure 3.17. It corresponds to a bright meteor echo that does not overlap in frequency with any other signal, such as a reflection on an airplane or the direct signal coming from the transmitter.

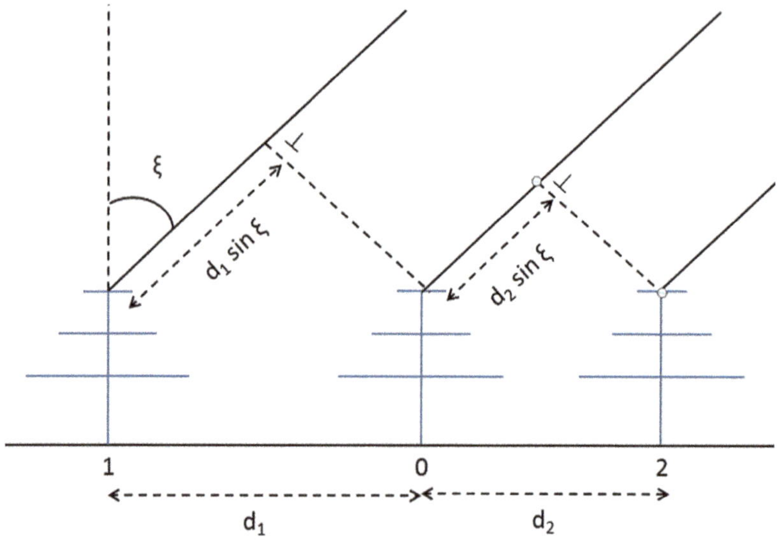

Figure 3.16. Sketch of a linear array of three antennas with the central antenna 0 being the phase reference. This principle is used along two orthogonal axes for the radio interferometer in Humain (with antenna '0' being common to the two axes).

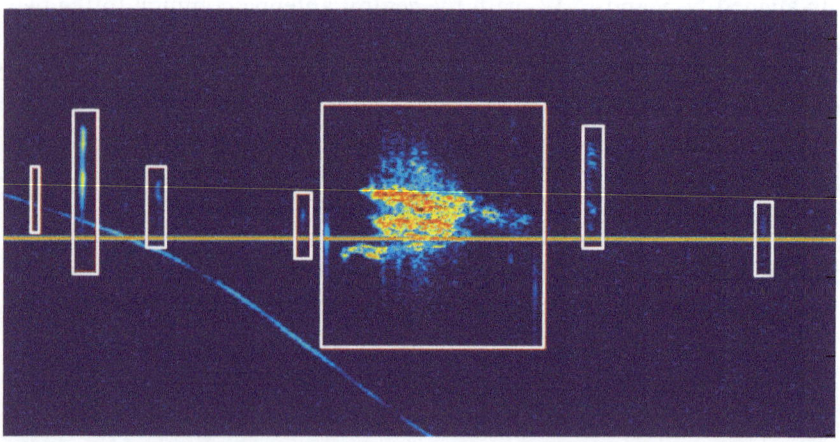

Figure 3.17. Spectrogram obtained at Humain on December 5, 2016, 00h35 UT. Rectangles are aggregated results from individual contributions from a Citizen Science project called the Radio Meteor Zoo (http://www.radiometeorzoo.eu).

Spectrograms are useful here in order to determine the best frequency to use to calculate the phase. Since the phase of the meteor echo should not depend on frequency, the best procedure would consist in selecting the frequency bin in which the signal-to-noise ratio (S/N) is the highest. The results for the phase differences between antenna pairs are shown in figure 3.18.

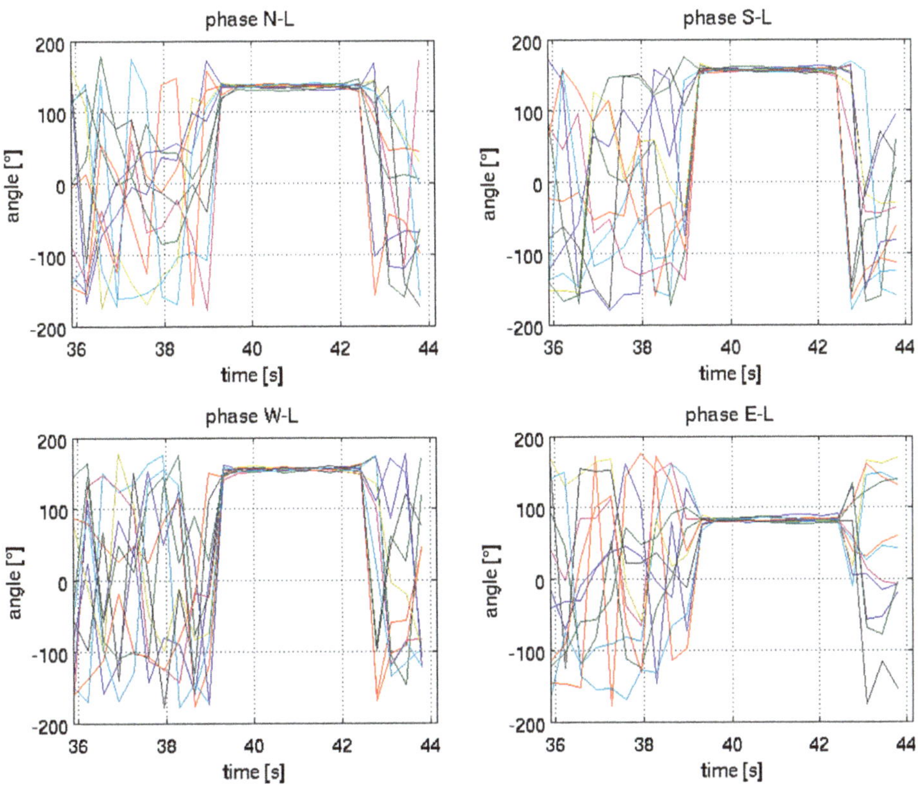

Figure 3.18. Phase differences between the central antenna L and (top left) the north (N) antenna; (top right) the south (S) antenna; (bottom left) the east antenna; and (bottom right) the west antenna, for data from figure 3.16. The ten different curves correspond to adjacent frequency bins centered on the upper bright spot of the meteor echo shown in the second rectangle of figure 3.16.

For the selected frequencies, just before and after the meteor echo, only noise is recorded and, therefore, the phase differences vary completely randomly as expected. During the meteor echo, however, the phase differences become strongly coherent. The Jones method is applied to these results to obtain the two angles of arrival ξ_{N-S} and ξ_{E-W} in the two perpendicular planes, which are then combined to provide the elevation angle and the azimuth angle of the meteor echo using equations (3.5) and (3.6). The results are shown in figure 3.19 for the frequency with the highest S/N (called FreqOfi). The results at other adjacent frequencies are very similar. Again, the results are very stable during the meteor echo which gives confidence in the method.

The example presented here consists of an 'isolated' meteor echo which does not overlap with any other spurious signal. In practice, an overlap occurs quite often with, e.g., an airplane reflection or with the direct signal from the beacon. For these meteor echoes, an automated procedure is currently being developed in order to select the frequency bin with the highest S/N and that belongs to the meteor echo.

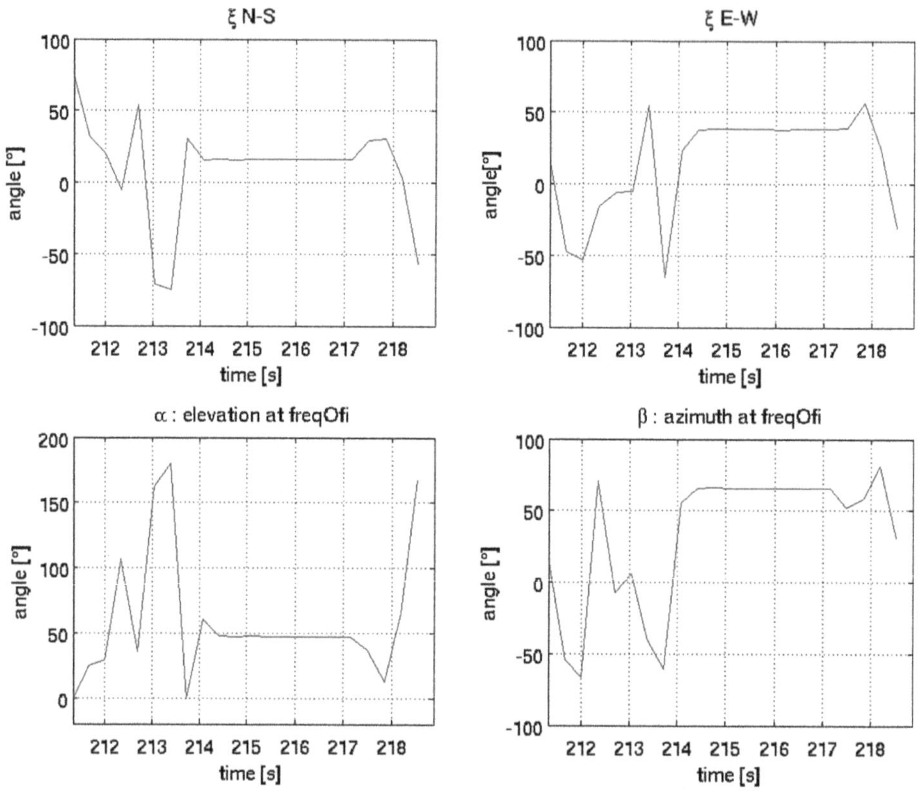

Figure 3.19. Top: Angles of arrival, ξ_{N-S} and ξ_{E-W}, as a function of time for data from figure 3.16. Bottom: Corresponding elevation (α) and azimuth (β) angles.

The directions of arrival computed for the meteor echoes with the BRAMS interferometer are thus far not calibrated. There are a number of systematic errors that need to be taken into account and corrected for. This includes a small difference in length between the cables going from antennas to receivers, an imperfect alignment of the antennas (in *X*, *Y* and *Z*), and an imperfect alignment of the axes with the N–S and E–W directions. A precise determination of these systematic errors was carried out in 2018, but has not yet been implemented in the results presented here. The algorithms are currently modified to correct for these systematic errors. Once they are taken on board, the calibration itself can be performed with one of the following methods: (1) using the BRAMS calibrator [10] as a transmitter and a calibrated antenna both attached to a drone flying in the far-field of the interferometer; (2) using the signal reflected from a plane whose position can be accurately determined (e.g. using websites such as Flight Tracker); or (3) using data from optical cameras located next to the interferometer. The first two methods are currently under investigation. The third one was used, e.g., by Madkour *et al* [9].

The results from the radio interferometer in Humain will be extremely important for the retrieval of individual meteoroid trajectories as only data from three additional traditional BRAMS receiving stations are then needed [11].

3.2.4 Comparisons of optical observations with BRAMS data

When the algorithms to retrieve meteoroid trajectories are operational, they need to be carefully checked. One possibility is to compare trajectories retrieved with BRAMS observations to those obtained with optical networks. For that purpose, BRAMS has partnered with the CAMS-BeNeLux network. CAMS (Cameras for Allsky Meteor Surveillance) is a network of optical video cameras that is able to measure the meteoroid trajectory very accurately, speed and deceleration in the Earth's atmosphere. With these measurements, accurate orbits can be computed with the end goals of validating unconfirmed meteor showers, detecting new ones and identifying their parent bodies [12, 13]. CAMS was initially developed in the US and funded by NASA, but an equivalent network has been developed in the BeNeLux since March 2012, mostly funded by motivated amateurs [14]. CAMS-BeNeLux detects a lot of meteoroid trajectories passing above or near Belgium and therefore potentially detectable by the BRAMS network. An example of such common detections is shown in figure 3.20.

In addition, another study can be done combining CAMS-BeNeLux and BRAMS observations. Indeed, with the trajectories provided by CAMS-BeNeLux, the calculations of the theoretical specular reflection points for all

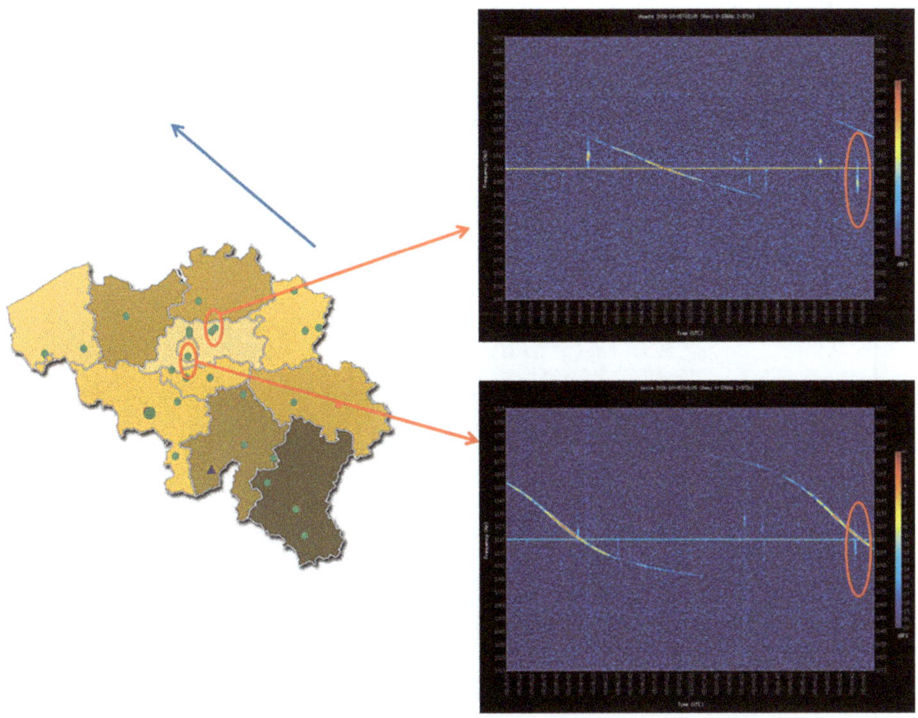

Figure 3.20. Comparison between BRAMS and CAMS observations. The blue line is the projection on the ground of a visual CAMS trajectory obtained on 5 October 2016. On the right two spectrograms from BRAMS stations BEHAAC and BEUCCL are shown. The meteor echoes are clearly visible. Note the different Doppler effect due to the geometry.

BRAMS stations are straightforward using the forward model described previously. With the speed and deceleration measurements, the time delays expected between the appearances of the meteor echoes at all BRAMS stations can also be computed. From the power profiles of the corresponding meteor echoes, the peak values can be measured. For underdense meteor echoes, equation (3.1) can be used, since in this case the whole geometry is known (so R_T, R_R, ϕ and β can be computed). Using simulated antenna gains for G_T and G_R in the direction to the specular reflection point, and making a reasonable assumption on the polarization of the radio wave, an estimate of the line electron density α can be obtained. This can be done for every BRAMS receiving station which detects the meteor, hence several measurements of α at different locations (specular reflection points) on the meteoroid path can be obtained. This set of values can then be compared to an ablation model using the same parameters for the trajectory and entry speed of the meteoroid and assuming a typical composition. By adjusting the mass as the last free parameter, a fit of the model results to the set of electron line density values can be carried out to obtain an estimate of the initial mass of the object.

Another optical network of interest is FRIPON (Fireball Recovery and InterPlanetary Observation Network, https://www.fripon.org/, see e.g. [15]). FRIPON is a dense French network with one hundred all-sky optical cameras set all over France with an average distance of ~100 km. The main objective is to

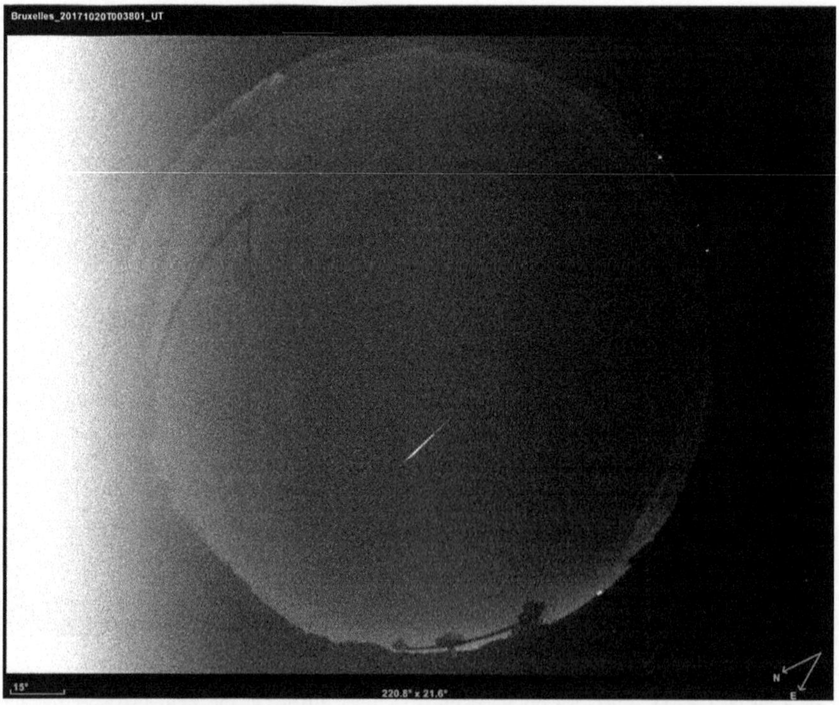

Figure 3.21. Example of a fireball detection with the FRIPON camera in Brussels on 20 October 2017 at 00H38 UT (Credit: François Colas).

determine trajectories of fireballs and try to recover on the ground potential associated meteoroids. Recently, FRIPON has been extended to various European countries including Italy, Spain, Austria, the Netherlands, Germany and Belgium. One of the FRIPON cameras was set up in Brussels in 2016. Again a comparison between FRIPON optical observations and BRAMS radio observations is planned. Note that in the case of fireballs, the associated radio meteor echoes are always overdense and a head echo is very often observed before the trail echo. An example of a fireball on 20 October 2017 by the FRIPON camera in Brussels is shown in figure 3.21.

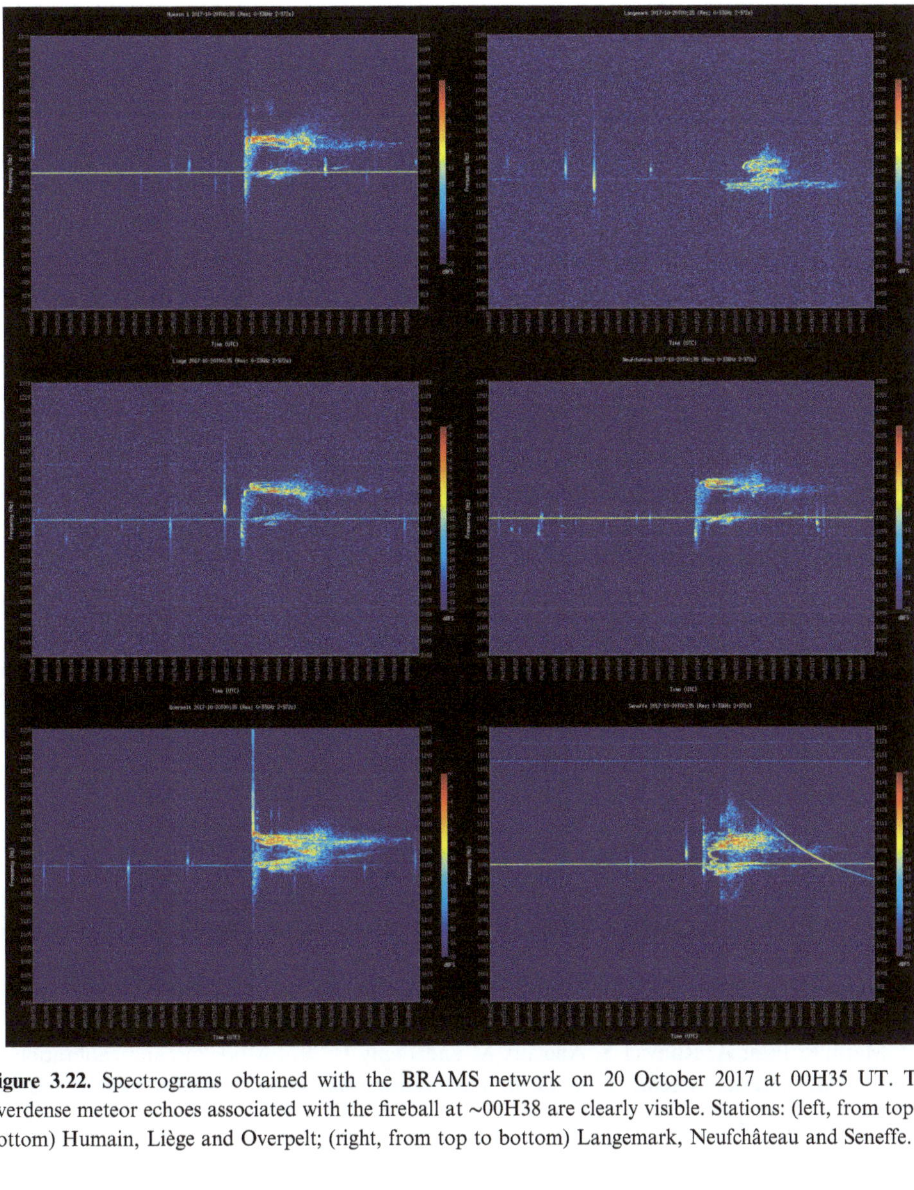

Figure 3.22. Spectrograms obtained with the BRAMS network on 20 October 2017 at 00H35 UT. The overdense meteor echoes associated with the fireball at ~00H38 are clearly visible. Stations: (left, from top to bottom) Humain, Liège and Overpelt; (right, from top to bottom) Langemark, Neufchâteau and Seneffe.

Six examples of corresponding radio observations with the BRAMS network are shown in figure 3.22. Note the variety of the complex shapes of the meteor echoes at different stations due to the different geometries. A head echo is also clearly visible in the data from the Overpelt station (bottom left spectrogram).

3.3 Conclusions

Radio observations of meteor echoes have the double advantage over optical observations that they can be carried out continuously and that they are sensitive to smaller objects that produce enough ionization but not enough light. BRAMS is a fairly recent network using forward scatter radio observations and provides a lot of useful data, mostly about meteor trail echoes. In the future, comparison of radio data with optical data provided by either the CAMS-BeNeLux or the FRIPON network, or with results from complex modeling of the meteoroid ablation, will undoubtedly produce very important new results.

Acknowledgements

BRAMS is a project of the Royal Belgian Institute for Space Aeronomy (BIRA-IASB) and has been funded mostly by the Solar-Terrestrial Center of Excellence (STCE).

The author would like to thank the organizers of the 61st Course of the International School of Quantum Electronics on *Hypersonic Meteoroid Entry Physics* for their kind invitation and giving him the opportunity to visit such a wonderful place as the Ettore Majorana Centre in beautiful Erice.

The author is indebted to all the people at BIRA-IASB who have actively contributed to the development of the BRAMS project and therefore to this publication: Sylvain Ranvier, Michel Anciaux, Emmanuel Gamby, Stijn Calders, Cédric Tétard, Antoine Calegaro and Johan De Keyser. The author would also like to thank Antonio Martinez Picar and Cis Verbeeck from the Royal Observatory of Belgium for their constant support and suggestions in developing the tools to analyze the BRAMS data. Finally, the author would like to thank the CAMS-BeNeLux team and the FRIPON team at IMCCE (Institut de Mécanique Céleste et de Calcul des Éphémérides) for their active collaboration.

References

[1] McKinley D W R 1961 *Meteor Science and Engineering* (New York: McGraw-Hill)

[2] Pecina P 2015 An analytical theory of radio-wave scattering from meteoric ionization–I. Basic equation *Mon. Not. R. Astron. Soc.* **455** 2200–6

[3] Webster A R, Brown P G, Jones J, Ellis K J and Campbell-Brown M 2004 Canadian meteor orbit radar (CMOR) *Atmos. Chem. Phys.* **4** 679–84

[4] Martínez Picar A, Ranvier S, Anciaux M and Lamy H 2014 Modeling and calibration of BRAMS antenna systems, *Proc. of the Int. Meteor Conf., (Giron, France)* Rault J-L and Roggemans P pp 201–6

[5] Lamy H and Tétard C 2016 Retrieving meteoroids trajectories using BRAMS data: preliminary simulations, *Int. Meteor Conf., (Egmond, the Netherlands)* Roggemans A and Roggemans P pp 149–52

[6] Nedeljkovic S 2006 Meteor forward scattering at multiple frequencies, *Proc. of the Int. Meteor Conf., 24th IMC, (Oostmalle, Belgium)* Verbeeck C and Wislez J-M pp 108–16

[7] Jones J, Webster A R and Hocking W K 1998 An improved interferometer design for use with meteor radars *Radio Sci.* **33** 55–65

[8] Lamy H, Tétard C, Anciaux M, Ranvier S, Martinez Picar A, Calders S and Verbeeck C 2018 First observations with the BRAMS radio interferometer, *Proc. of the Int. Meteor Conf., (Petnica, Serbia)*

[9] Madkour W, Yamamoto M-y, Kakinami Y and Mizumoto S 2016 A low cost meteor observation system using radio forward scattering and the interferometry technique *Exp. Astron.* **41** 243–57

[10] Lamy H, Anciaux M, Ranvier S, Calders S, Gamby E, Martinez Picar A and Verbeeck C 2015 Recent advances in the BRAMS network, *Int. Meteor Conf., (Mistelbach, Austria)* Rault J-L and Roggemans P pp 171–5

[11] Wislez J-M 2006 Meteor astronomy using a forward scatter set-up, *Proc. of the Radio Meteor School, (Oostmalle)* Verbeeck C and Wislez J-M pp 84–106

[12] Jenniskens P, Gural P S, Dynneson L, Grigsby B J, Newman K E, Borden M, Koop M and Holman D 2011 CAMS: Cameras for Allsky Meteor Surveillance to establish minor meteor showers *Icarus* **216** 40–61

[13] Jenniskens P, Nénon Q, Albers J, Gural P S, Haberman B, Holman D, Morales R, Grigsby B J, Samuels D and Johannink C 2016 The established meteor showers as observed by CAMS *Icarus* **266** 331–54

[14] Roggemans P, Johannink C and Breukers M 2016 Status of the CAMS-BeNeLux network, *Int. Meteor Conf., (Egmond, the Netherlands)* Rault J-L and Roggemans P pp 254–60

[15] Colas F *et al* 2015 French fireball network FRIPON, *Int. Meteor Conf., (Mistelbach, Austria)* Rault J-L and Roggemans P pp 37–40

Chapter 4

The flux of meteoroids over time: meteor emission spectroscopy and the delivery of volatiles and chondritic materials to Earth

Josep M Trigo-Rodríguez

Every night the apparently immutable night sky can be observed, but suddenly a shooting star can cross our field of view. Perhaps, in that moment, Mother Nature reminds us that our existence is linked to the continuous influx of extraterrestrial materials. The truth is that even when it is not noticeable, except for the appearance of meteors, the current flux of interplanetary matter to Earth is about 100 000 Tm yr^{-1} [1]. Obviously the flux at the top of the atmosphere must be several orders of magnitude higher, providing a continuous rain of elements to the upper atmosphere and, in the process, generating interesting chemistry between highly reactive phases (see for example the review of Plane *et al* [2] and references therein). It causes the formation of layers of metal atoms and ions in the mesosphere and lower thermosphere leading to the formation noctilucent clouds and other chemical interactions with stratospheric aerosols [3]. As well as the ablated materials, meteoroid fragments and ablation condensates reach the ground as micrometeorites and remain in the substrate allowing its flux to be quantified [1]. The mass influx was quantified and compiled in a histogram that reveals the bimodal contribution in the flux distribution (see figure 4.1): one peak corresponds to particles of about 100 μm and larger, producing visual meteors, while the other is associated with asteroids and comets with sizes of tens to one hundred meters that produce meteorite falls or even excavate craters [4].

Most of this interplanetary material coming to the Earth originates from undifferentiated bodies, small asteroids and comets that produce porous aggregates containing primordial minerals. These materials are formed by minerals condensing from the gas surrounding the proto-Sun about 4.6 Ga ago and accreted into fragile aggregates that were thermally processed to form planetesimals [5]. Small amounts were incorporated into the protoplanetary disk from nearby stars as we know by

doi:10.1088/2053-2563/aae894ch4

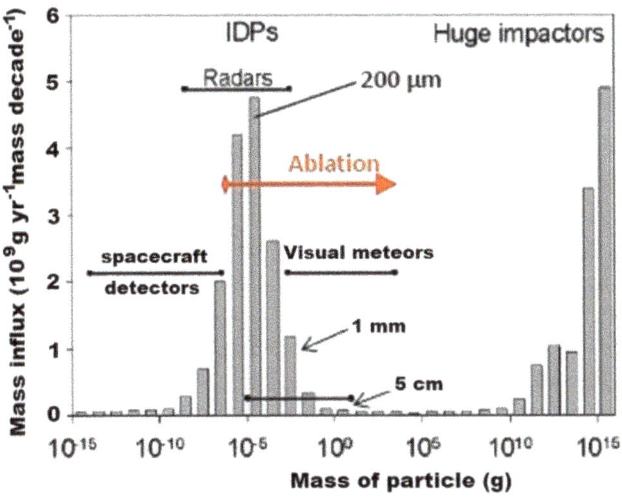

Figure 4.1. Mass influx of interplanetary materials [4].

peculiar isotopic signatures [6–8]. The unusual chemical signatures of tiny presolar grains provide clues on the peculiar formation environment of our star, and it is clear evidence that our Sun formed in a stellar association [9]. A significant fraction of these asteroidal bodies survived the heavy accretion of planetary bodies and remained undifferentiated thus producing meteorites called chondrites, and more fragile tiny aggregates that have been collected by dedicated planes in the stratosphere, called interplanetary dust particles (IDPs) [1]. The components of chondrites and IDPs represent the primordial starting materials from which the undifferentiated bodies formed. To complete the flux of extraterrestrial materials to Earth there is a much smaller contribution of meteoroids that are associated with differentiated bodies. When they survive atmospheric passage, they produce different types of differentiated meteorites that are called achondrites. Some common sources of achondritic meteorites are the Moon, Mars or Vesta (see e.g. [10]), but some small asteroids might be just fragments of large planetary bodies [11] and also be a possible source.

Undifferentiated bodies are composed of unequilibrated components that formed part of a primordial planetary disk from which the small bodies were accreted. Consequently, any view of solar system formation, as well as considering the Sun and planets, should include the many small bodies that populate different regions and have a key relevance in the chemical evolution of planetary bodies [12]. The minor bodies are smaller than the planets and are subjected to continuous erosion and decay by cosmic irradiation and impacts with other objects (a process known as gardening). As a consequence of these processes the space between the planets becomes populated by billions of particles that follow heliocentric orbits, and are usually associated with asteroids and comets. This system is known as the Zodiacal cloud and is in constant replenishment because millimeter-sized meteoroids tend to fall into the Sun in time-scales of tens of millions of years (Ma) as a consequence of the loss of kinetic energy caused by mutual collisions and non-gravitational effects [13, 14]. Obviously, the mere existence of the Zodiacal cloud requires a continuous replenishment of the interplanetary space by small fragments of other solar system bodies [15, 16]. These particles orbiting the Sun are called meteoroids and were defined by the International Astronomical Union (IAU) as a particle larger than a micron and smaller than one meter in diameter that follows a heliocentric orbit in our solar system.

Some meteoroids originated through the natural collisions of asteroids, while outgassing dominates the release of cometary meteoroids onto heliocentric orbits. In repeated approaches to the Sun a comet sublimates abruptly and releases tons of meteoroids due to the weak gravitational field and the gas outgassing. In fact, the gas pressure from the cometary surface makes the process of injection of micrometric to millimetric particles into heliocentric orbits very efficient [15]. So, it is not surprising that comets are important in producing meteoroids. These will be fragile undifferentiated objects formed in the outer solar system and composed of a weak mixture of ices, organic materials and micrometric mineral grains with solar composition [17]. These volatile-rich objects suffer significant ice sublimation when approaching the Sun. Then, volatile-rich regions produce jets of gas that drive out tons of meteoroids with diameters from decimeters to tens of microns [17]. These released particles are gravitationally distributed around the Sun, forming meteoroid streams that produce meteor showers when the Earth crosses them. Studying meteor showers using different instrumentation gives insight into the physico-chemical properties of their parent bodies. These particles are often aggregates whose constitutive mineral grains exhibit typical diameters of a few microns that are considered to be dust, so when they dynamically and collisionally evolve they become part of the Zodiacal dust [18]. An additional process releasing chondritic meteoroids is the catastrophic disruption of rubble-piles by tidal forces in close approaches to planets [19].

Although interplanetary space is populated by meteoroids originating in the decay of asteroids and comets, planetary bodies also contribute. Achondrites are meteorites coming from differentiated bodies, usually larger than about 1000 km in diameter. Lunar or Martian achondrites are also reaching the Earth, but they are not so abundant because they can only escape the gravitational field through a grazing impact [20].

To summarize, most meteoroids coming from asteroids were released by impacts, while outgassing is the main force driving cometary meteoroids to heliocentric orbits [15]. The volatile nature and small gravitational field of comets makes them dominant contributors to the so-called Zodiacal cloud. These bodies are fragile objects composed of a mixture of ices, organic materials and mineral grains with an average tensile strength close to 10 Pa [21]. The nature of weakly bounded aggregates, being a mixture of crystalline silicates, organics and dirty ices, was also revealed by the Stardust (NASA) and Rosetta (ESA) space missions [17, 22]. These objects suffer significant ice sublimation when approaching the Sun, so volatile-rich regions produce jets of gas that drive out tons of meteoroids with diameters in the typical range of centimeters to tens of microns [16]. These meteoroids form meteoroid streams that remain stable in their orbits for thousands of years [15]. Despite their large numbers, most of these particles do not undergo atmospheric entry and indirect systems are needed to understand their origin and composition.

Meteoroids can have very diverse origins, as meteor studies reveal. By obtaining their heliocentric orbits from multiple-station meteor monitoring plus meteor spectra chemical information, it is possible to better understand the delivery mechanisms and nature of exogenous material to Earth. This paper will summarize the role of emission spectroscopy of meteors and fireballs in gaining insight into the bulk elemental chemistry of meteoroids. The progress made during the last decades in reference to the role of chondritic flux in chemical evolution of the biosphere and origin of life will be reviewed. Meteor spectroscopy can be regarded as an added-value technique in order to understand the astrobiological significance and relevance of the delivery of volatiles to terrestrial planets from the continuous meteoroid flux.

4.1 The meteor phenomenon and the origin of Earth's volatiles

It follows from the formation processes described above that meteoroids are particles that moved around the Sun with typical velocities of a few tens of km s^{-1}. When they collide with a planetary atmosphere they are decelerated and ablate producing a luminous phenomenon called a meteor. The relative geocentric velocity to the Earth (hereafter V_g), the velocity with which the meteoroid enters the top of the atmosphere (before suffering significant deceleration), lies in the range $11 < V_g < 72$ km s^{-1}. When these particles penetrate into the atmosphere at these supersonic velocities the meteoroids suffer increasing collisions with atmospheric atoms or molecules and their surfaces are quickly heated to become incandescent. Then, a physical process called ablation takes place which produces the vaporization, fragmentation and sputtering of the meteoroid, forming minerals. From the ground the observed meteor phenomenon consists of three differentiated parts: the *head* or the region around the meteoroid in which the interaction is taking place, the *wake* left just behind and the *train* or meteor column. The meteor head is the part where the more energetic collisions take place and mostly contributes to light production. It is also the main source of emission lines as the meteoroid-forming minerals are progressively vaporized and the elements are released, suffering excitation and/or ionization as a consequence of the exposure of the released particles to the collisions, producing a cloud of ions and free electrons which form a significant part of the ablated material. The electrons are transmitting energy, and the ions emit light through well-defined emission lines, while the surviving dust contributes to a continuum emission. In general, emission lines can be reproduced assuming chemical equilibrium, while the emission lines coming from the *meteor train* are out of equilibrium. This is explained in further detail in the next section.

Meteoroids penetrating the Earth's atmosphere are experiencing collisions with air components, and are heated progressively. As the meteoroid minerals are heterogeneous and exhibit different vaporization temperatures, a step-by-step ablation process takes place. First, the hydrous (if any) and organic phases are ablated, second, the silicates and metals, and finally the refractory minerals. As a consequence, the moderately volatile elements, such as Na and K, are released at greater heights than the refractory ones. As a consequence the so-called differential ablation process takes place (figure 4.2).

A comparison between terrestrial rocks and chondrites suggests that enstatite and ordinary chondrites were the dominant building blocks of the early Earth. Most of these chondrites were formed in reduced conditions, so the origin of terrestrial volatiles is debated. The existence of a N-rich terrestrial atmosphere from direct outgassing seems likely, and it has been hypothesized that some of this N arrived as nitrides. Despite this, during the late accretion period large impacts eroded significant amounts of the Earth's volatile inventory. In this sense, a depletion in N and Xe has been noted with respect to other volatiles that are in chondritic proportions. Such depletion could be explained as a consequence of large impacts producing a thermal escape and affecting the atmospheric components. In fact, a giant impact was probably responsible for the origin of the Moon and eroded the early atmosphere of the Earth [23–24]. In consequence, the early atmosphere of our planet could have evolved as consequence of high-energetic impacts [25]. The last of those impacts probably had a special role in the final evolution of the entire Earth–Moon system. In fact, it has been recently demonstrated that the similarity in composition between the Earth and Moon could be a natural consequence of a projectile exhibiting similar composition to Earth in a late giant impact [26]. Recent

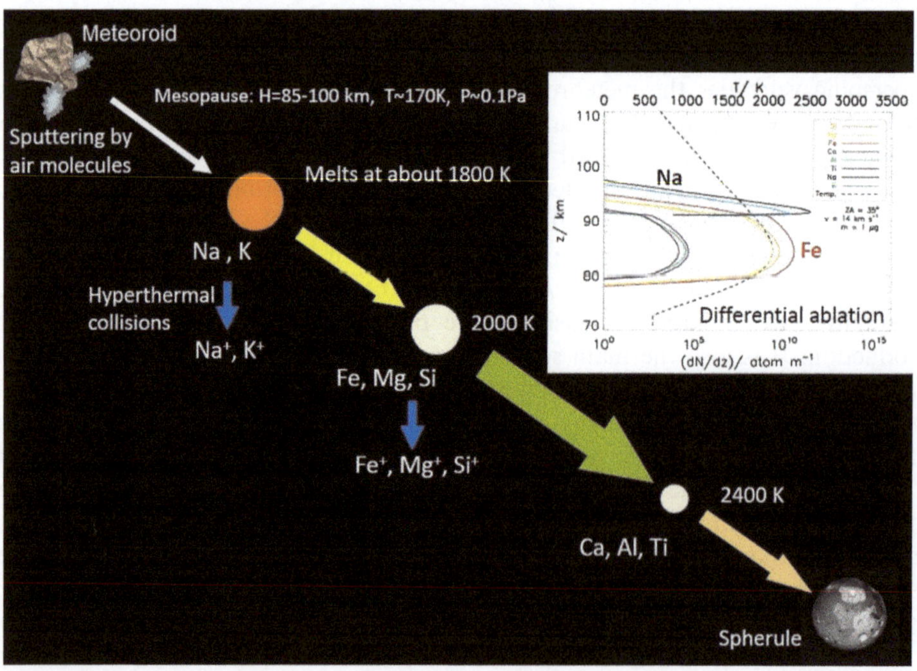

Figure 4.2. This schematic represents meteoroid ablation in a planetary atmosphere. Differential ablation is a direct consequence of the selective ablation at different heights of meteoroid minerals having different vaporization temperatures [3].

evidence is also constraining the composition of the projectile, called Theia, that could be preserved in the Moon [27].

4.2 Meteor spectroscopy: an added value to Meteoritica

Meteor spectroscopy is a technique to delve into the physical processes taking place during meteoroid ablation, but it is also a pathway to study the delivery processes of exogenous materials to Earth. From the very beginning of this field early in the 20th century, photographic plates were used to capture emission spectra that allowed the identification of emission lines from rock-forming elements (see e.g. [28, 29]). Emission spectroscopy was developed to understand the components of cometary meteoroids that did not survive atmospheric interaction. These fragile materials represent a significant fraction of the delivery of primordial materials coming from undifferentiated bodies [1]. Meteoroids of cometary origin are typically centimeter- or millimeter-sized aggregates that are fragile in nature, and are weakly bonded aggregates formed by fine micron-sized dust, organics and volatiles [30]. In consequence, they fragment and do not survive atmospheric interaction (see e.g. [31, 32]).

Obviously, there is a significant degree of difficulty in capturing and reducing the recorded meteor and fireball spectra. The light emitted during the unpredictable entry of a meteoroid must be decomposed with the right observational geometry by a prism or diffraction grating set up in front of imaging systems. In such a way the luminous column is separated in emission lines for each elemental transition. Photographic systems were the only way to obtain reliable spectral information during most of the 20th century, but new CCD and video imaging techniques now provide additional clues on the meteor phenomenon. The emission lines of the main rock-forming elements were identified by pioneers [33], but they did not obtain a clear model of the light generation process (see e.g. [29]). A tentative model was proposed by Ceplecha [34] who developed a complex cylindrical model for the radiating column, assuming local thermal equilibrium. In that approach, the theoretical curve of growth was built up, also describing the self-absorption of the lines and obtaining some physical parameters (see e.g. [35]). Unfortunately, the resulting computed number of Fe atoms in the radiating volume and the involved mass determined from the meteor luminous efficiency were not always accurate. A simpler model was proposed by Borovicka [36] and tested on the excellent Cechtice fireball photographic spectrum. With such an extraordinary emission spectrum, also obtained during routine sky monitoring from the Ondrejov Observatory, Borovicka [36] obtained for the first time a computed meteor synthetic spectrum. It adjusted exceptionally well to the observed spectrum, even though the physical approach was very simple: thermal equilibrium and constant temperature and density in the whole volume. Meteor spectra consist of two different components: the main spectrum characterized by a temperature of about 4500 K and a second spectrum that usually reaches 10 000 K [37, 38]. The second component originates in the front wave where high-energy collisions can produce the excitation of atoms increasing the ionization of the meteoroid components. It is important to remark that the high-temperature component produced in the shock wave is not detected for meteoroids penetrating into the atmosphere at velocities below $V < 35$ km s^{-1} [35].

Table 4.1. The main atoms and ions found in meteor spectra. The wavelength, multiplet number and binding energy is given for each line. Adapted from [40]. Spectral components: *p*—principal, *s*—secondary, *a*—atmospheric line.

Atom or ion	λ (Å)	Multiplet	E_2 (eV)	Relative intensity		Reference
H I	6563	1	12.09	3	*s*	[33, 41, 42]
Li I	6708	1	1.85	2	*p*	[43, 44]
N I	8680	1	11.76	4	*s, a*	[33, 41]
	6465	21	13.66	1	*s, a*	[33, 42, 45]
	4110	10	13.70	1	*s, a*	[33, 42, 45]
N II	5680	3	20.66	1	*a*	[33, 44]
O I	7772	1	10.74	5	*s, a*	[33, 41]
	6158	10	12.75	3	*s, a*	[33, 42, 45]
Na I	5890	1	2.11	5	*p*	[33, 37, 42]
Mg I	5184	2	5.11	5	*p*	[33, 37, 42]
	3838	3	5.94	5	*p*	[33, 37, 42]
Mg II	4481	4	11.63	4	*s*	[33, 42, 46]
Al I	3962	1	3.14	3	*p*	[37, 42, 44]
Si I	3906	3	5.08	3	*p*	[33, 37]
Si II	6347	2	10.07	3	*s*	[33, 37, 42]
	4131	3	12.83	2	*s*	[33]
Ca I	4227	2	2.93	4	*p*	[33, 37, 42]
	6162	3	3.91	3	*p*	[33, 37, 42]
Ca II	3934	1	3.15	5	*s, p*	[33, 37, 42]
	8542	2	3.15	4	*s, p*	[33, 41]
Ti I	4982	38	3.33	2	*p*	[37, 42, 44]
Ti II	3349	1	3.74	3	*s, p*	[46, 47]
Cr I	4254	1	2.91	4	*p*	[33, 37, 42]
	5208	7	3.32	3	*p*	[33, 37, 42]
	3593	4	3.44	3	*p*	[33, 47]
Cr II	3125	5	6.42	2	*s*	[47]
Mn I	4031	2	3.08	3	*p*	[33, 37, 42]
Fe I	3860	4	3.21	5	*p*	[33, 37, 42]
	4384	41	4.31	4	*p*	[33, 37, 42]
	5270	15	3.21	4	*p*	[33, 37, 42]
Fe II	5018	42	5.36	2	*s*	[33, 42]
	3228	6	5.51	2	*s*	[47]
Co I	4813	158	5.79	1	*p*	[42]
	4121	28	3.93	1	*p*	[37]
	3506	21	4.05	1	*p*	[48]
Ni I	5477	59	4.09	2	*p*	[37, 42]
	3462	17	3.60	2	*p*	[47]
Si II	4078	1	3.04	2	*s, p*	[33]

Consequently, the most important contribution of meteor spectroscopy is its ability to extract very relevant chemical information from the relative intensity of the emission lines contributing to each meteor spectrum (see table 4.1) [39]. Depending on the disperser system the spectral resolution will be variable, but a synthetic spectrum can always be obtained and compared with the recorded spectrum. This is made by adding the contribution of the involved chemical element lines. The main emission lines contributing to meteor spectra are compiled in table 4.1 [39, 40].

The spectral profile can be fitted to a physical model that allows the temperature and elemental abundances of their rock-forming to be computed, having taken into account the intensity of the lines [36]. The model assumes thermal equilibrium in the meteor head, and considers as a simplification that the radiating volume is a prism where the physical parameters and the chemical abundances can be determined. Then, the chemical abundances of incoming meteoroids can be inferred from the sequential recording of the variable line intensity along the luminous trajectory of a meteor or fireball [35]. That paper for the first time applied a systematic study to obtain averaged chemical abundances along the meteor columns that provided chemical clues on the chondritic composition of asteroidal and cometary meteoroids [18]. While most elements have chondritic ratios that were later proved to be consistent with particular mineral mixtures that could form chondritic aggregates [49], it was also discovered that meteoroids of cometary origin were exhibiting a peculiar feature: being overabundant in Na [21]. Consequently, emission meteor spectroscopy first pointed towards the overabundance of Na recently confirmed from Rosetta studies of comet 67P/Churyumov–Gerasimenko dust [50].

Many photographic spectra became available from the Ondrejov Observatory at the end of the 20th century, but most of them remained unstudied. Trigo-Rodríguez [39] selected some and these are shown in table 4.2. He also presented an innovative idea to infer the average elementary abundances of the incoming particles from the sequential study of the meteor spectrum. The basic idea is simple, most relative abundances of the elements present in the meteor column are many orders of magnitude over that expected in the atmosphere at the meteor ablation height (see e.g. [18]). Consequently, it is possible to estimate the chemical abundances relative to one reference element (such as Fe, which has many omnipresent emission lines) at different heights, as shown in figure 4.3 [18, 39]. Then, it is statistically possible to infer an averaged composition for the particles as inferred from the independent study of sequential ablation patterns made by Rietmeijer [49].

The intensity of emission lines recorded on the Ondrejov photographic plates was accurately measured by using a microdensitometer, an obsolete instrument to quantify the density of signal in photographic plates that is no longer needed because current detectors are digital. Trigo-Rodríguez [39] computed a synthetic spectrum for each fireball and compared it directly to the observed one until he found the best match. Then, Trigo-Rodríguez [39] obtained the relative chemical abundances of incoming meteoroids from the sequential spectroscopy of the luminous trajectories of photographic fireballs during their entry into the terrestrial atmosphere. In that work, the averaged chemical abundances for the main rock-forming elements of

Table 4.2. The photographic emission spectra analyzed by Trigo-Rodríguez [39] indicating the meteoroid source, geocentric velocity, semi-major axis, inclination and type of spectrum.

Assigned code	Recording date	Stream	V_g (km s^{-1})	a (UA)	i (°)	Spectrum type
GEM	14/12/1961	Geminid	37.8	1.7	39	Grating
PER1	2-3/8/1962	Perseid	59.9	23.0	112	Prism
PER2	12-13/8/1967	Perseid	60	11	113	Prism
PER3	11-12/8/1969	Perseid	60.9	250	114	Prism
PER4	11-12/8/1969	Perseid	60.7	19.0	114	Prism
SPO2	6-7/6/1970	Sporadic	26.4	3.0	39	Prism
PER5	12-13/8/1970	Perseid	60.6	115	113	Prism
KCIG1-1r	18-19/8/1971	κ Cignid	25.6	4	36	Prism and grating
SPO1	17-18/9/1974	Sporadic	68	∞	148	Prism
AND	8-9/10/1977	Andromedid	24.3	2.9	4	Prism
LEO	17-18/11/1980	Leonid	72.4	13	162	Prism
SPO3-3r	30-01/11-12/1989	Sporadic	25.6	2.4	5	Prism and grating
SPO4	19-20/5/1974	Sporadic	57.1	12.9	103.9	Prism

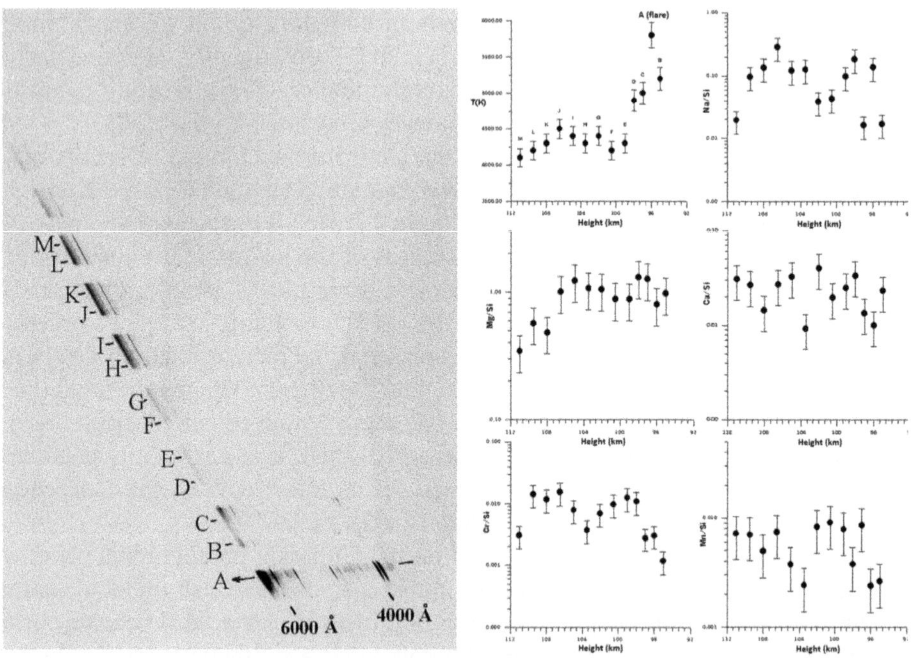

Figure 4.3. The temperature (in kelvins) and the elemental abundances relative to Fe and as a function of height for the sporadic fireball spectrum shown on the left; SPO1 in table 4.2 [39].

meteoroids were obtained, and were later discussed in [18, 21, 35]. In the next section the observational methodology and data reduction are explained.

4.3 Relative elemental abundances and cosmochemical ratios from photographic, video and CCD spectroscopy

Several research groups have been exploring the use of video and CCD spectroscopy during the last decade [9, 51]. The new techniques represent a significant improvement because the sensitivity and temporal resolution can be much higher than in photographic techniques. CCD video spectrometers are achieving a temporal resolution capable of observing physical processes during meteoroid ablation (figure 4.4) [19, 31, 52] .

As described in previous sections, there are many difficulties in recording and reducing meteor and fireball spectra. To obtain chemical information, the light emitted during the unpredictable entry of a meteoroid must be decomposed by a prism or diffraction grating set up in front of imaging systems. The system records the light emitted by the meteor as separated emission lines, but significantly reduces the limiting magnitude of the imaging system. The usual systems are only able to record fireball spectra that are bright enough to be efficiently recorded, at least during a significant part of the meteor column. Fireball spectra are currently recorded using CCD or digital video cameras, but in the past the pioneers did so using photographic plates [28, 33].

The Spanish Meteor and Fireball Network (http://www.spmn.uji.es), that has been operating with an increasing degree of completeness since 1999, has been exploring CCD and video techniques. Video emission spectroscopy has allowed the detection of moderately bright fireballs, but also extremely bright events, such as the superbolide produced by a damocloid observed in Spain on 13 July 2012 [51]. The rock-forming materials of solar system bodies reflect significant differences as a consequence of different formation times and diverse building blocks. This fact

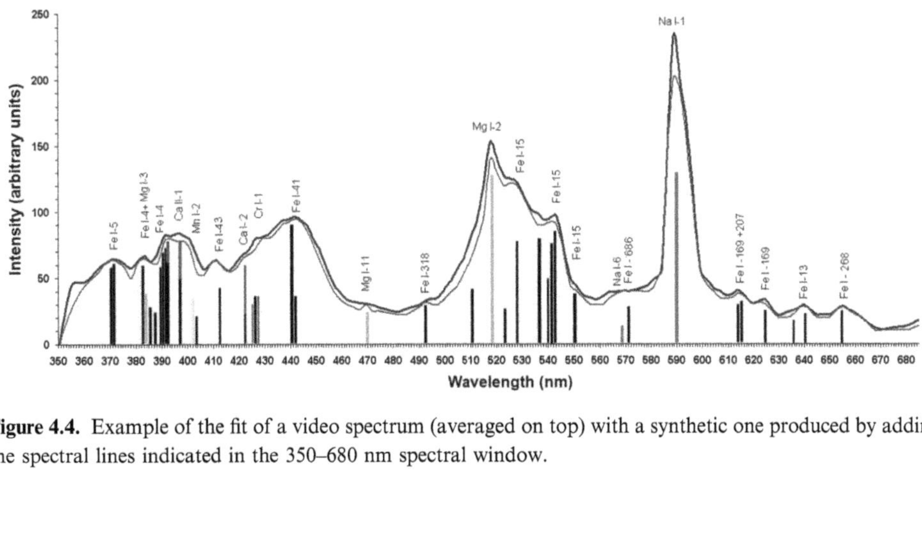

Figure 4.4. Example of the fit of a video spectrum (averaged on top) with a synthetic one produced by adding the spectral lines indicated in the 350–680 nm spectral window.

produces chemical signatures that are distinctive of achondritic materials, such as Martian rocks or the HED clan of achondrites associated with asteroid 4 Vesta (see e.g. [53]). Consequently, a comparative analysis of the inferred meteoroid elemental ratios relative to Si (such as Mg/Si, Na/Si, Ca/Si, Mn/Si) with different meteorites can exemplify that not all meteoroids reaching the top of the atmosphere are chondritic. That was the case for the bright fireball associated with PHA 2012XJ112 that was inferred to exhibit an achondritic nature [54]. The Mg/Fe or Mg/Si ratios are particularly informative to identify achondritic meteoroids. For example, two sporadic meteors studied in [35] exhibited a non-chondritic chemical composition with rock-forming phases particularly poor in Mg.

4.3.1 Obtaining meteoroid chemical abundances

As previously explained, photographic or digital spectroscopy allows the comparison of the recorded lines with synthetic spectra. As these procedures can be repeated for bright meteors along several points all along the atmospheric trajectory, it is possible to infer the abundance ratio relative to Fe for several elements [18, 35, 39].

In emission spectra produced by meteoroids with a high geocentric velocity, usually with $V_g > 40$ km s^{-1}, the second (high-temperature) component appears and the Si abundance can be inferred by fitting the intensity of the Si II lines. Unfortunately, for slow meteors it is not possible to determine the Si abundance from the spectra because that second component is very faint or non-existent [32]. This can be solved to a certain extent if high-resolution spectra, capable of separating the contribution of the Si I line (multiplet 3 at $\lambda = 3905.5$ Å) are taken. It is certainly not easy because the spectra can be noisy and this Si I line is placed in a spectral region where Fe I and Ca II lines are very prominent. As a compromise solution, when the Si abundance cannot be measured, a typical chondritic ratio of Si/Fe = 1.16 can be assumed [35].

4.3.2 Some examples of elemental abundance ratios

As previously stated, interesting conclusions about the elemental abundances can be reached by studying fireball spectra. Here we discuss the averaged abundance along the trajectories of the fireballs described in Trigo-Rodríguez *et al* [35]. The composition of most fireballs was measured in more than ten segments but the exact number of selected segments depended on the characteristics of each fireball. The exact number of segments and the values averaged in each fireball are given in Trigo-Rodríguez' PhD thesis [39].

In figure 4.5 the relative chemical abundances for selected rock-forming elements are plotted against the geocentric velocity.

Na/Si: In general the sodium ratio (Na/Si) was found to be larger than that of typical chondritic meteorites and IDPs [55], but also than the sodium content in the dust of 1P/Halley [56]. These results support that this volatile element is probably trapped in the meteoroid components, in the matrix. Trigo-Rodríguez *et al* [18] proved that the observed Na is not due to the terrestrial atmosphere and neither has it been overestimated due to our assumptions in the model to determine abundances.

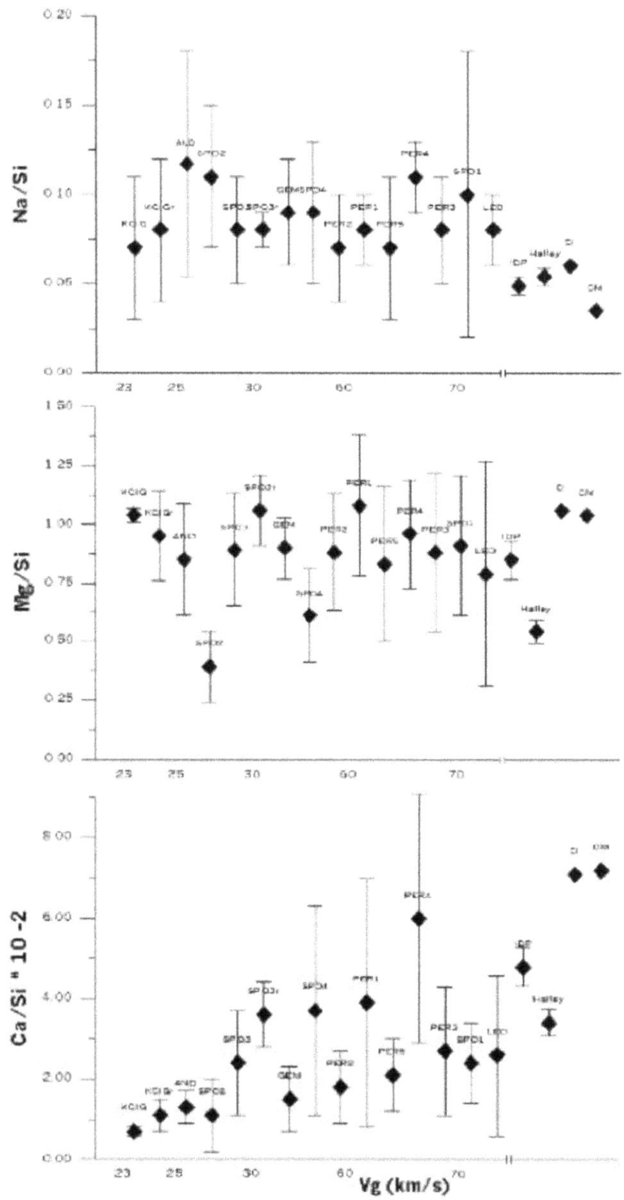

Figure 4.5. Elemental abundances relative to Si for Na, Mg and Ca for the fireball spectra studied in [39]. The average values for IDPs, Halley dust, and CI and CM are shown out of scale for comparison. Note the Na overabundance for most cometary meteors and the lack of Ca producing light in the vapor phase, an example of incomplete evaporation for its presence in refractory minerals [18, 35].

Mg/Si: This is a key cosmochemical ratio because primordial silicates were mafic in origin, but incorporated Fe as consequence of thermal processing in the protoplanetary disk or the parent bodies. The values found by Trigo-Rodríguez *et al* [35] and Madiedo *et al* [31] were within the expected range for IDPs and chondrites, but were far off the 1P/ Halley fine dust composition studied using the Giotto spectrometer [56].

Ca/Si: It is well known that calcium forms refractory mineral phases exhibiting special resistance to volatilization. In most meteor spectra the Ca lines are present without any ambiguity. In the main component the line of Ca I belonging to multiplet 2 is always evident at 4227 Å. Moreover, the lines of Ca II doublet originating by multiplet 1 appear at 3934 and 3968 Å, which are also clearly visible, particularly in high-velocity meteors. In figure 4.5 we can see the clear dependence between the observed Ca abundance and the geocentric velocity of the meteoroids. The explanation of this fact is the effect of incomplete evaporation first proposed by Boročicka [36, 37]. When the geocentric velocity increases, the main temperature reached in the meteoric column also increases. Due to the location of Ca in refractory phases, when the temperature is higher the volatilization is more efficient, contributing more to the luminous spectra. In general we note that for meteoroids with geocentric velocity between 20 and 40 km s^{-1} the effect of incomplete evaporation is that only ~30% or 50% of the calcium contributes to the luminous spectra. For fast meteors the relative abundance observed for Ca reaches larger values, but rarely the expected values for IDPs or chondritic meteorites. This is a clear trend of incomplete evaporation that produces ablation products: cosmic spherules or refractory fragments [57].

Ti/Si: Titanium is a minor element that in meteor spectra has several lines along the main spectrum and also in the high-temperature component. Usually the brightest line of Ti I is at 4982 Å and at 4550 Å for Ti II, both quite close to Fe lines and quite faint in comparison. This makes it difficult to extract conclusions in noisy segments, but it can be separated quite well for bright spectra. The derived Ti abundances are between the expected values for IDPs and chondritic meteorites.

Cr/Si: This transition element is easy to identify in meteor spectra since Cr exhibits three intense lines at 4254, 4275 and 4290 Å. In addition the multiplet 7 generates another bright emission line at 5206 Å. As many lines can be identified the fit of the synthetic spectra is easy, and permits good accuracy in the determination of Cr elemental abundances. Trigo-Rodríguez found that the Cr abundance for the photographic spectra analyzed was between the values of IDPs, chondritic meteorites and 1P/Halley dust. Some differences in the amount of Cr in cometary meteoroids were also found. Perseid meteoroids from comet 109P/Swift-Tuttle meteoroids are richer in Cr than sporadic meteoroids.

Mn/Si: Manganese can be easily identified in meteor spectra. Multiplet 1 has an important line at 4033 Å and other secondary lines at 4750 and 4850 Å. The estimated abundances of Mn are clearly below the expected values for IDPs. The incoming meteoroids had Mn/Si ratios very similar to chondritic meteorites, although three meteoroids (KCIG, PER1 and PER2) had abundances similar to those estimated by the Giotto spacecraft for 1P/Halley dust.

Fe/Si: The abundance of iron can be extracted from the second component by comparing the lines of Si II and Fe II. Using the second component, the accuracy is slightly less than for the other elements and the process is not applicable to low-velocity meteors where the Si II lines are missing. In any case, the results show that the Fe/Si ratio is clearly between the expected values for IDPs and chondritic meteorites, all very far from 1P/Halley dust values.

Co/Si: The abundance of this transition metal is low and only appears clearly in the bright segments of some detailed spectra (AND, LEO and SPO3), where the estimated Co/Si ratio is around 3×10^{-3}.

Ni/Si: Nickel lines are difficult to observe in meteor spectra. A line of Ni I lies at 5477 Å. Its intensity could be measured in only seven spectra and yielded values of Ni/Si around 2.5×10^{-2}. This is around half of the solar abundance, 5×10^{-2}. It is interesting to note that some meteoroids such as AND and PER4 showed low Ni/Si, very close to that observed for 1P/Halley dust by the Giotto spacecraft.

4.3.3 Other ratios of cosmochemical interest

In addition to the chemical abundances with reference to silicon cited above, other ratios are of particular interest in cosmochemistry: Mg/Na, Fe/Na and Mg/Fe, in particular.

Mg/Na: The observed values for this ratio are of great interest. Most of them are around the typical values for 1P/Halley dust although some cometary meteoroids (KCIG, PER1, PER2, PER3 and SPO3) have Mg/Na ratios closer to IDPs or CI typical values. Considering the possibility that we overestimated the abundance of Na, all ratios could be closer to IDP values.

Fe/Na: This ratio shows again the similarity of the incoming meteoroids to IDPs and CI and CM chondrites. We also note that the largest meteoroids (AND, KCIG and SPO3) have nearly identical averaged values. The remaining values are very close to those expected for IDPs or carbonaceous chondrites.

Mg/Fe: Again this ratio confirms the general differences with the 1P/Halley dust chemical composition. All meteoroids are within the expected value for IDPs or CI and CM chondrites, showing a typical Mg/Fe value of 1.2. Again two sporadic meteors show an anomalous composition due to their low content of magnesium.

Among the elements that were present in the primeval inner solar nebula, Si, Mg and Fe were more abundant. Their individual responses to high temperatures were probably responsible for their different abundances in the bodies condensed there [58]. These three chemical elements are the major components of inner solar system bodies because is accepted that they all formed from the condensation of a vapor rich in Mg–Fe–SiO–H_2–O_2 and small particles of silicates that were forming the protoplanetary nebula 4.5 Mya [55]. Consequently, the relative proportions of these three elements is usually a good tool to find possible relationships between different objects. Figure 4.6 shows the ternary diagram Mg–Fe–Si where the chemical similitude between the meteoroids producing the spectra analyzed in this work and the IDPs and chondritic meteorites is again implicit. All spectra are clustered between the typical IDPs and chondritic values, except for two sporadic meteoroids (SPO2 and SPO4), characterized by being poor in Mg. It is important to note that both meteoroids are unique, in that they are close to the Mg abundance of 1P/Halley dust although their content of Fe and Si is different. But are the particles of this comet really anomalous or could this be an artefact of the measurements? To answer

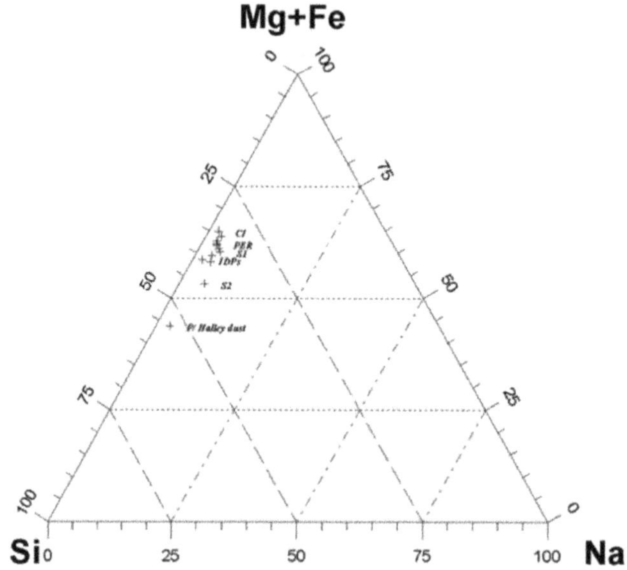

Figure 4.6. This ternary diagram compares the Mg + Fe, Si and Na abundances in the selected sample of cometary particles compared to the CI chondritic, aggregate IDPs and 1P/Halley dust abundances. The Na abundance of most cometary meteoroids is larger than in chondritic meteors. Additional labels are: S2 (SPO2) and PER (Perseids).

this we must note that Fomenkova *et al* [59], combining the mass and the composition of each particle detected by the Giotto mass spectrometers, concluded that the dust of this comet is formed by three kinds of particles: (i) the so-called CHON particles; (ii) particles rich in carbonaceous compounds and silicates; and (iii) mineral particles with Fe, Mg and Si as the main components. It is important to note that the CHON particles have not yet been recovered in the terrestrial atmosphere, probably due to their fragility to solar radiation during their sojourn in the interplanetary medium or, as was suggested by Rietmeijer [60], due to being melted during their fast entry into the terrestrial atmosphere. The Halley mass spectrometers detected only very small particles that have a mass equivalent to the so-called principal components (PCs) that are inside the matrix of IDPs. Neither PCs nor any other constituent (e.g. mineral grains) have chondritic element proportions. In consequence, the Giotto measurements were biased towards small cometary particles and from our results we conclude that they not are representative of other cometary particles arriving at Earth.

From meteor spectroscopy we usually infer that most meteoroids ablate, producing a luminous phase with characteristic chondritic bulk elemental chemistry [18, 35]. This is a direct consequence of the fact that the chondritic bodies are dominant in the current flux on Earth [1]. Laboratory experiments are also needed, and the ablation of chondrites can provide clues about the pathways for the delivery of volatile and moderately volatile elements to Earth (see e.g. [3]).

4.4 The Na overabundance: clues on the delivery of volatiles from fragile meteoroids and IDPs

From the analysis of ablation columns of cometary meteoroids using meteor spectroscopy, it appears that Na is mainly contributing to the meteor light during the first stages of ablation [35]. The inferred Na overabundance in reference to the chondritic ratios was evident in Trigo-Rodríguez *et al* [35]. The study of the specific abundance of Na in meteor columns and its comparison to Na abundance in the upper atmosphere pointed to the fact that cometary meteoroids must have significant enrichment of this moderately volatile element [18, 21]. This was explained by Na not only forming part of the rocky minerals, but also being trapped in organics and/or ices in the outer disk regions where comets formed. It was envisaged that during the early stages of the protoplanetary disk the materials fell continuously onto the protostar, causing vaporization of rocky components. At that stage the intense solar wind from the young Sun depleted the content in some lithofile elements such as Na, K or Mn in chondritic meteorites [61, 62]. Today the process of sodium depletion of the inner solar system continues due to the solar wind removing this element from young comets and tenuous atmospheres where it is accumulating by vaporization of meteoroids. For these reasons, Na abundances are probably higher in comets formed in the outer region of the disk [62]. It is likely that the Na became incorporated into the volatile-rich interstitial fine-grained matrix, consistent with recent Rosetta observations [50]. Such a model is consistent with our meteoroid bulk chemical data for the main rock-forming elements in which Na was shown to be overabundant ([18] figure 4.6).

Sodium has also been widely observed in cometary comae, but there is an intrinsic difficulty in establishing the origin of Na and its relative abundance from those data [63]. The measured Na/Si atomic ratio of 3×10^{-5} in the coma of comet Hale–Bopp was explained as being produced from sputtering of the particles' surfaces only. The sodium present in the tail of some comets forms a well-defined Na-tail [64]. These authors suggested that this sodium could be being produced in a near-nucleus region, probably by cometary degassing or through an extended source such as the break-up of Na-bearing molecules, ions or dust particles. Detailed studies are required in order to infer the exact mechanism of neutral sodium release from dust particles. If sodium was embedded into organic material or ices associated with low-temperature melting phases, as suggested by our meteor spectrum results, the amount of sodium released from some comets could be easily explained. This overabundance has also been explained as a consequence of aqueous alteration processes in comets [65].

4.5 Astrobiological implications of the continuous arrival of chondritic components to Earth's surface

As previously mentioned, the carbonaceous chondrites (CCs) contain highly reactive minerals and have been continuously reaching the terrestrial surface, although the flux has significantly changed over time. The study of CCs reveals that they formed part of the undifferentiated parent bodies that formed in the outer protoplanetary disk, and their composition reinforces the idea of a continuum between asteroids and comets (figure 4.2). Some CCs have experienced little thermal homogenization so they are considered highly unequilibrated meteorites that contain the primordial components of the protoplanetary disk, just weakly compacted [8]. The parent bodies of these accretion aggregates were highly porous [66] and retained significant amounts of water, organics and volatile compounds that were widely available in the outer disk formation regions. Small asteroids and comets are formed from primordial materials, and at the very beginning were subjected to planetary perturbations and fragmentations during close approaches to planets, so were probably easily disrupted [8]. Such processes later extended during the late heavy bombardment [67]. Consequently, it seems plausible that at early times and subsequent later periods of time the Earth was subjected to a meteoritic flux at least 5–6 orders of magnitude greater than the current one [68]. Consequently, a large amount of chondritic materials reached the Earth's surface at an annual rate of thousands of billions of metric tons. Hence the amount of volatiles delivered under such meteoroid high-flux circumstances was also very significant, probably playing a key role in fertilizing the Earth's surface [69]. To support the previously outlined hypothesis, significant progress has recently been made in understanding the role of chondrites in prebiotic evolution [69]. The catalytic effect of six CCs (table 4.1) was analyzed in the presence of water and formamide, namely Allan Hills 84 028 (group and petrologic type: CV3), Elephant Moraine 92 042 (CR2), Miller Range 05 024 (CO3), Larkman Nunatak 04 318 (CK4), Grosvenor Mountains 95 551 (C-ung) and Grosvenor Mountains 95 566 (C2-ung). The carbonaceous chondrites (CCs) were requested from the Johnson Space Center facility in the framework of two Spanish research projects (AYA2011-26522 and AYA2015-67175-P) to identify pristine meteorites in the NASA Antarctic collection and study their properties [69]. We found that, once the intrinsic organics hosted by the matrix of the carbonaceous chondrites are removed, the minerals forming these meteorites are able to catalyze complex organics under warm (100 °C) interaction with water and formamide. Thus, in view of the unique reactive properties of carbonaceous chondrites, it can be deduced that the reactive minerals forming CCs reached the surface of Earth and other planetary bodies, and when exposed to a warm and water-rich environment could have started catalytic reactions to promote organic complexity (figure 4.7). These reactions, independent of the ability of complex organics to survive atmospheric deceleration, probably initiated the first steps in increasing the organic complexity necessary to promote the origin of life [69, 70].

4.6 Conclusions and future work

The study of undifferentiated bodies is able to provide clues on the role of these primordial materials in the formation of planets, the delivery of volatiles and the origin of life. The main conclusions of this work are:

- Not all meteoroids have enough strength to survive long periods of time in the interplanetary medium, or the loading pressure experienced during their fast

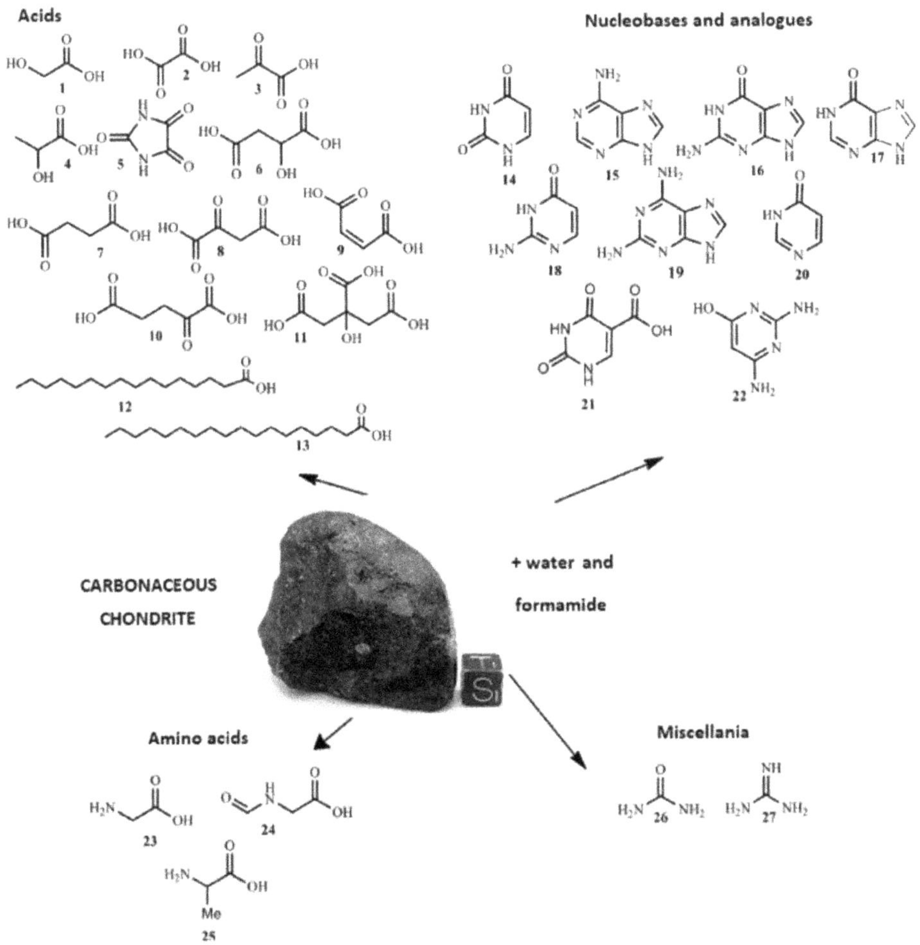

Figure 4.7. Catalysis of organic compounds from chemical reactions promoted by carbonaceous chondrite minerals in the presence of water and formamide [69].

deceleration in the atmosphere. These materials are probably associated with transitional C-rich asteroids or comets that cannot produce meteorites, but some of them that reach our planet with velocities lower than 12 km s^{-1} are able to survive as micrometeorites or IDPs. This produces a continuous flux of micrometric materials reaching the Earth's surface that, at the present time, is estimated to be about 40 000 Tm, but which was probably much higher in the past, particularly at the time of the late heavy bombardment.

- For all these materials that reach the top of the atmosphere, but leave no surviving meteor, spectroscopy can be a promising way to understand the processes taking place during ablation. The study of meteor ablation columns provides a pathway to a better understanding of the decay of these materials in the atmosphere and the pathways of volatile delivery to our planet.

- Future fireball spectra with new instrumentation will provide additional insight into some key spectral windows (UV and NIR, for example) where we can find spectral lines of interest to learn about the formation of the vapor cloud, and the decay of organic compounds.

From the very beginning of this field early in the 20th century, emission spectra allowed the identification of emission lines from rock-forming components. A systematic and statistical approach allows meteor emission spectroscopy to provide direct evidence on the chondritic materials and volatiles participating in the continuous delivery of primitive materials from undifferentiated bodies. In fact, most meteoroids are fragile and are severely fragmented in the atmosphere. These materials that are unable to survive atmospheric interaction are of key importance. We know now that this is a direct consequence of the fragile nature of these bodies that are weakly bonded aggregates formed of fine micron-sized dust, organics and volatiles. IDPs represent a fascinating and enriched continuous bathing of our planet in the Cosmic Ocean.

Acknowledgements

Support from the Spanish Ministry of Science and Innovation under research project AYA2015-67175-P is acknowledged. The author thanks the careful English revision made by Professor Iwan Williams.

References

[1] Brownlee D E 2001 The origin and properties of dust impacting the Earth *Accretion of Extraterrestrial Matter Throughout Earth's History* ed B Peucker-Ehrenbrink and B Schmitz (New York: Kluwer Academic/Plenum) pp 1–12

[2] Plane J M, Feng W and Dawkins E C 2015 The mesosphere and metals: chemistry and changes *Chem. Rev.* **115** 4497–541

[3] Gómez-Martín J C, Bones D L, Carrillo-Sánchez J D, James A D, Trigo-Rodríguez J M, Fegley B Jr and Plane J M C 2017 Novel experimental simulations of the atmospheric injection of meteoric metals *Astrophys. J.* **836** 212

[4] Hughes D W 1993 Meteoroids—an overview *Meteoroids and their Parent Bodies, Proc. of the Int. Astronomical Symp.* ed J Stolh and I P Williams (Bratislava: Astronomical Institute Slovak Academy of Sciences) p 15

[5] Testi L *et al* 2014 Dust evolution in protoplanetary disks *Protostars and Planets VI* vol 914 ed H Beuther, R S Klessen, C P Dullemond and T Henning (Tucson, AZ: University of Arizona Press) pp 339–61

[6] Zinner E 2003 Presolar grains Treatise on Geochemistry *Meteorites, Comets and Planets* 1st edn. ed A M Davis (Amsterdam: Elsevier)

[7] Boss A P 2004 From molecular clouds to circumstellar disks *Comets II* ed M C Festou, H U Keller and H A Weaver (Tucson, AZ: The University of Arizona Press) pp 67–80

[8] Trigo-Rodríguez J M 2015 Aqueous alteration in chondritic asteroids and comets from the study of carbonaceous chondrites *Planetary MineralogyEMU Notes in Mineralogy* vol 15 ed M R Lee and H Leroux (Twickenham: The Mineralogical Society of Great Britain and Ireland) pp 67–87

[9] Trigo-Rodríguez J M, García-Hernández D A, Lugaro M, Karakas A I, Van Raai M, Lario P G and Manchado A 2009 The role of massive AGB stars in the early solar system composition *Meteorit. Planet. Sci.* **44** 627–39

[10] McSween H Y Jr. and Huss G R 2010 *Cosmochemistry* (Cambridge: Cambridge University Press)

[11] Bland P A *et al* 2009 An anomalous basaltic meteorite from the innermost main belt *Science* **325** 1525–7

[12] Alexander C M O'D, Newsome S D, Fogel M L, Nittler L R, Busemann H and Cody G D 2010 Deuterium enrichments in chondritic macromolecular material—implications for the origin and evolution of organics, water and asteroids *Geochim. Cosmochim. Acta* **74** 4417–37

[13] Trigo-Rodríguez J, Castro-Tirado A and Llorca J 2005 Evidence of hydrated 109P/Swift-Tuttle meteoroids from meteor spectroscopy, *36th Annual Lunar and Planetary Science Conf.* **vol 36** abst. #1485

[14] Nesvorný D, Bottke W F Jr, Dones L and Levison H F 2002 The recent breakup of an asteroid in the main-belt region *Nature* **417** 720

[15] Jenniskens P 1998 On the dynamics of meteoroid streams *Earth Planets Space* **50** 555–67

[16] Williams I P 2002 The evolution of meteoroid streams *Meteors in the Earth's Atmosphere* ed E Murad and I P Williams (Cambridge: Cambridge University Press) 13

[17] Brownlee D *et al* 2006 Comet 81P/Wild 2 under a microscope *Science* **314** 1711–16

[18] Trigo-Rodríguez J M, Llorca J and Fabregat J 2004 Chemical abundances determined from meteor spectra—II. Evidence for enlarged sodium abundances in meteoroids *Mon. Not. R. Astron. Soc.* **348** 802–10

[19] Trigo-Rodríguez J M *et al* 2007 Asteroid 2002NY40 as a source of meteorite-dropping bolides *Mon. Not. R. Astron. Soc.* **382** 1933–9

[20] Grady M M, Hutchison R and Graham A 2000 *Catalogue of Meteorites* vol 1 (Cambridge: Cambridge University Press)

[21] Trigo-Rodríguez J M and Llorca J 2007 On the sodium overabundance in cometary meteoroids *Adv. Space Res.* **39** 517–25

[22] Capaccioni F *et al* 2015 The organic-rich surface of comet 67P/Churyumov-Gerasimenko as seen by VIRTIS/Rosetta *Science* **347** aaa0628

[23] Cameron A G W and Benz W 1991 The origin of the Moon and the single impact hypothesis IV *Icarus* **92** 204–16

[24] Canup R M and Asphaug E 2001 Origin of the Moon in a giant impact near the end of the Earth's formation *Nature* **412** 708–12

[25] Trigo-Rodríguez J M and Martín-Torres F J 2013 Implication of impacts in the young Earth Sun paradox and the evolution of Earth's atmosphere *The Early Evolution of the Atmospheres of Terrestrial Planets* ed J M Trigo-Rodriguez, F Raulin, C Muller and C Nixon (Astrophysics and Space Science Proceedings, vol 35) (New York: Springer)

[26] Mastrobuono-Battisti A, Perets H B and Raymond S N 2015 A primordial origin for the compositional similarity between the Earth and the Moon *Nature* **520** 212–5

[27] Herwartz D, Pack A, Friedrichs B and Bischoff A 2018 Identification of the giant impactor Theia in lunar rocks *Science* **344** 1146–50

[28] Ceplecha Z 1961 Determination of wave-lengths in meteor spectra by using a diffraction grating *Bull. Astron. Inst. Czech.* **12** 246

[29] Millman P M 1980 One hundred and fifteen years of meteor spectroscopy *Solid Particles in the Solar System, Symp. of the Int. Astronomical Union* vol 90 ed I Halliday and B A McIntosh (Cambridge: Cambridge University Press) pp 121–8
[30] Trigo-Rodríguez J M and Blum J 2009 Tensile strength as an indicator of the degree of primitiveness of undifferentiated bodies *Planet. Space Sci.* **57** 243–9
[31] Madiedo J M, Trigo-Rodríguez J M, Konovalova N, Williams I P, Castro-Tirado A J, Ortiz J L and Cabrera-Caño J 2013 The 2011 October Draconids outburst—II. Meteoroid chemical abundances from fireball spectroscopy *Mon. Not. R. Astron. Soc.* **433** 571–80
[32] Trigo-Rodríguez J M *et al* 2011 The October Draconids outburst—I. Orbital elements, meteoroid fluxes and 21P/Giacobini–Zinner delivered mass to Earth *Mon. Not. R. Astron. Soc.* **433** 560–70
[33] Halliday I 1961 A study of spectral line identifications in Perseid meteor spectra *Publ. Dominion Obs. Ottawa* **25** 3–16
[34] Ceplecha Z 1964 Study of a bright meteor flare by means of emission curve of growth *Bull. Astron. Inst. Czech.* **15** 102
[35] Trigo-Rodríguez J M, Llorca J, Borovička J and Fabregat J 2003 Chemical abundances determined from meteor spectra: I. Ratios of the main chemical elements *Meteorit. Planet. Sci.* **38** 1283–94
[36] Borovička J 1993 A fireball spectrum analysis *Astron. Astrophys.* **279** 627–45
[37] Borovička J 1994 Line identifications in a fireball spectrum *Astron. Astrophys. Suppl. Ser.* **103** 83–96
[38] Borovička J 1994 Two components in meteor spectra *Planet. Space Sci.* **42** 145–50
[39] Trigo-Rodríguez J M 2002 Spectroscopic analysis of cometary and asteroidal fragments during their entry into the terrestrial atmosphere *PhD Thesis* University of Valencia
[40] Ceplecha Z, Borovička J, Elford W G, ReVelle D O, Hawkes R L, Porubčan V and Šimek M 1998 Meteor phenomena and bodies *Space Sci. Rev.* **84** 327–471
[41] Millman P M and Halliday I 1961 The near-infra-red spectrum of meteors *Planet. Space Sci.* **5** 137–40
[42] Ceplecha Z 1971 Spectral data on terminal flare and wake of double-station meteor No. 38421 (Ondrejov, April 21, 1963) *Bull. Astron. Inst. Czech.* **22** 219
[43] Borovička J and Zamorano J 1995 The spectrum of fireball light taken with a 2-m telescope *Earth Moon Planets* **68** 217–22
[44] Borovička J and Spurný P 1996 Radiation study of two very bright terrestrial bolides and an application to the comet S-L 9 collision with Jupiter *Icarus* **121** 484–510
[45] Harvey G A 1997 Air radiation in photographic meteor spectra *J. Geophys. Res.* **82** 15–22
[46] Millman P M, Cook A F and Hemenway C L 1971 Spectroscopy of Perseid meteors with an image orthicon *Can. J. Phys.* **49** 1365–73
[47] Halliday I 1969 A study of ultraviolet meteor spectra *Publ. Dominion Observatory Ottawa* **25** 313–22
[48] Harvey G A 1973 Spectral analysis of four meteors *Evolutionary and Physical Properties of Meteoroids NASA SP* vol 319 ed C L Hemenway, P M Millman and A F Cook (Washington, DC: NASA) pp 103–129
[49] Rietmeijer F J M 2004 Interplanetary dust and carbonaceous meteorites: constraints on porosity, mineralogy and chemistry of meteors from rubble-pile planetesimals *Earth Moon Planets* **95** 321–38

[50] Schulz R *et al* 2015 Comet 67P/Churyumov-Gerasimenko sheds dust coat accumulated over the past four years *Nature* **518** 216
[51] Madiedo J M *et al* 2014 Analysis of a superbolide from a damocloid observed over Spain on 2012 July 13 *Mon. Not. R. Astron. Soc.* **436** 3656–62
[52] Trigo-Rodriguez J M, Madiedo J M, Williams I P and Castro-Tirado A J 2009 The outburst of the κ Cygnids in 2007: clues about the catastrophic break up of a comet to produce an Earth-crossing meteoroid stream *Mon. Not. R. Astron. Soc.* **392** 367–75
[53] McSween H Y, Mittlefehldt D W, Beck A W, Mayne R G and McCoy T J 2011 HED meteorites and their relationship to the geology of Vesta and the Dawn mission *The Dawn Mission to Minor Planets 4 Vesta and 1 Ceres* (Berlin: Springer) pp 141–74
[54] Madiedo J M, Trigo-Rodríguez J M, Williams I P, Konovalova N, Ortiz J L, Castro-Tirado A J, Pastor S, De Los Reyes J A and Cabrera-Caño J 2014 Near-Earth object 2012XJ112 as a source of bright bolides of achondritic nature *Mon. Not. R. Astron. Soc.* **439** 3704–11
[55] Rietmeijer F J M and Nuth J A 2000 Collected extraterrestrial materials: constraints on meteor and fireball compositions *Earth Moon Planets* **82** 325–50
[56] Jessberger E K, Christoforidis A and Kissel J 1988 Aspects of the major element composition of Halley's dust *Nature* **332** 691
[57] Genge M J 2008 Micrometeorites and their implications for meteors *Earth Moon Planets* **102** 525–35
[58] Ozawa K and Nagahara H 2000 Kinetics of diffusion-controlled evaporation of Fe–Mg olivine: experimental study and implication for stability of Fe-rich olivine in the solar nebula *Geochim. Cosmochim. Acta* **64** 939–55
[59] Fomenkova M N, Kerridge J F, Marti K and McFadden L A 1992 Compositional trends in rock-forming elements of comet Halley dust *Science* **258** 266–9
[60] Rietmeijer F J 2002 The earliest chemical dust evolution in the solar nebula *Chem. Erde-Geochem.* **62** 1–45
[61] Wasson J T and Kallemeyn G W 1988 Compositions of chondrites *Philos. Trans. R. Soc. Lond.* A **325** 535–44
[62] Despois D 1992 Solar system-interstellar medium: a chemical memory of the origins *Astrochemistry of Cosmic Phenomena: Proc. of the 150th Symp. of the Int. Astronomical Union (Campos do Jordao, Sao Paulo, Brazil, August 5–9, 1991)* vol 150 ed P D Singh (Dordrecht: Kluwer) pp 451–8
[63] Cremonese G, Huebner W F, Rauer H and Boice D C 2002 Neutral sodium tails in comets *Adv. Space Res.* **29** 1187–97
[64] Cremonese G, Boehnhardt H, Crovisier J, Rauer H, Fitzsimmons A, Fulle M, Licandro J, Pollacco D, Tozzi G and West R M 1997 Neutral sodium from comet Hale-Bopp: a third type of tail *Astrophys. J. Lett.* **490** L199
[65] Ellinger Y, Pauzat F, Mousis O, Guilbert-Lepoutre A, Leblanc F, Ali-Dib M, Doronin M, Zicler E and Doressoundiram A 2015 Neutral Na in cometary tails as a remnant of early aqueous alteration *Astrophys. J. Lett.* **801** L30
[66] Blum J, Schräpler R, Davidsson B J R and Trigo-Rodríguez J M 2006 The physics of protoplanetesimal dust agglomerates. I. Mechanical properties and relations to primitive bodies in the solar system *Astrophys. J.* **652** 1768
[67] Gomes R, Levison H F, Tsiganis K and Morbidelli A 2005 Origin of the cataclysmic late heavy bombardment period of the terrestrial planets *Nature* **435** 466

[68] Trigo-Rodríguez J M, Llorca J and Oró J 2004 Chemical abundances of cometary meteoroids from meteor spectroscopy *Life in the Universe: From the Miller Experiment to the Search for Life on Other Worlds* ed J Seckbach *et al* (Berlin: Springer) pp 201–4
[69] Rotelli L, Trigo-Rodríguez J M, Moyano-Cambero C E, Carota E, Botta L, Di Mauro E and Saladino R 2016 The key role of meteorites in the formation of relevant prebiotic molecules in a formamide/water environment *Sci. Rep.* **6** 38888
[70] Trigo-Rodríguez J M, Saladino R, Di Mauro E, Rotelli L, Moyano-Cambero C E, Carota E and Botta L 2017 The catalytic role of chondritic meteorites in the prebiotic enrichment of earth and other planetary-rich surfaces, under high meteoritic flux, *Lunar and Planetary Science Conf.* **vol 48**

IOP Publishing

Hypersonic Meteoroid Entry Physics

Gianpiero Colonna, Mario Capitelli and Annarita Laricchiuta

Chapter 5

Compositional, mineralogical and structural investigation of meteorites by XRD and LIBS

Giorgio S Senesi and Francesco Capitelli

Meteorites are solid objects originating from fragments of asteroids, comets and even planets, which descend onto the Earth, commonly via airbusts, through the atmosphere. Although the composition and mineralogy of meteorites reflects the original material of their source, they may vary due to the temperature and pressure the meteorites are subjected to during their journey to Earth. Thus, the study of these objects is uniquely suited for inferring their composition, the presence of water and organic compounds related to possible life on the extraterrestrial body from which they originate, and the temperature and pressure conditions encountered during their journey to and arrival on the Earth. In addition, they can be used as historic extraterrestrial tracers for gaining information on the origin of the Earth, and solar system evolution and processes. Furthermore, meteorites discovered and analyzed by the NASA Mars Science Laboratory Curiosity in the Gale crater on Mars have provided unique information, unveiling the past atmospheric and environmental conditions and the weathering processes they were subjected to on the surface of the planet.

Currently, of the total of about 1060 retrieved meteorite samples in the Meteoritical Bulletin Database [1], 174 samples are classified as having been ejected from Mars by impacts and subsequently fallen to Earth. Of these, 142 consist of basaltic to lherzolitic igneous rocks classified as shergottites, 18 are classified as nakhlites and consist of olivine-clinopyroxene cumulates [2], 3 are dunitic cumulates classified as chassignites and 11 are not classified. Among these samples of the Martian surface, known as the Shergotty, Nakhla and Chassigny (SNC) meteorites [3], nakhlites are the most studied. These materials belong to the same lithologic unit with a common basaltic crystallization age of 1.3 Ga [3] and contain hydrated secondary minerals, indicating fluvial alteration on Mars [4–6]. Thus, the detailed mineralogical analysis of these meteorites is important as the presence of certain minerals may indicate a past habitable environment on Mars.

doi:10.1088/2053-2563/aae894ch5

Meteorites have been widely analyzed to infer their compositional, structural, geological and mineralogical features by employing various destructive or semi-destructive techniques, which include inductively coupled plasma-mass spectrometry (ICP-MS), ICP-atomic emission spectroscopy (ICP-AES) and x-ray diffraction (XRD) [7]. Although these techniques can provide very useful and important information, they consume or destroy the sample during analysis, at least partially. Thus, the so-called non- or micro-destructive and non- or micro-invasive techniques are assuming great importance in meteorite analysis as they do not damage the analyzed sample (which is often of high economic and scientific value) and preserve its integrity as much as possible. Among these techniques, the most commonly used are: particle-induced x-ray emission (PIXE) and micro-PIXE, secondary ion mass spectrometry (SIMS) and nano-SIMS, scanning electron microscopy equipped with energy dispersive x-ray spectroscopy (SEM–EDS), electron microprobe analysis (EMPA), x-ray fluorescence spectroscopy (XRF), alpha-particle x-ray spectrometry (APXS), laser-ablation inductively coupled plasma-mass spectrometry (LA-ICP-MS), Mössbauer spectroscopy, Fourier transform infrared spectroscopy (FTIR), Raman spectroscopy and laser-induced breakdown spectroscopy (LIBS) [7].

In this chapter a review is provided of the results obtained using two techniques, i.e. the commonly used semi-destructive XRD and the relatively new micro-destructive LIBS, both of which allow us to gain unique and complementary information on the elemental, mineralogical and structural composition of meteorites found on the surfaces of Earth and Mars.

In particular, the XRD technique can provide a number of types of molecular and structural information on the major mineral phases present in a meteorite, and is particularly appropriate for probing the phase transitions that may occur in the object during the ejection impact, the successive possible impacts through its journey in space and the arrival impact on Earth. The value of the careful investigation of meteoritic mineral phases in understanding the genesis and provenance of meteorites has been recognized for a long time. Minerals occurring in meteorites include phosphides and phosphates, sulfides and sulfates, carbides and carbonates, and silicides and silicates. Although most meteoritic phases also occur in terrestrial rocks, many meteoritic minerals, e.g. those occurring in reduced samples, are extremely rare or absent on Earth. In particular, a number of refractory meteoritic minerals (e.g. hibonite, perovskite) found in carbonaceous chondrite inclusions formed at high temperatures in the solar nebula are relatively rare on Earth, where they may be present in metamorphosed limestones, nepheline syenites, carbonatites and ultramafic rocks. Furthermore, some meteoritic minerals not found on Earth, e.g. ringwoodite, can be produced by shock metamorphism while other phases, e.g. cassidyite and schollhornite, may be formed by terrestrial weathering of primary meteoritic minerals. Finally, other typical terrestrial minerals formed under conditions of high static pressure and temperature unlikely to have occurred on asteroids may be found in meteorites [8]. The XRD technique, however, has some disadvantages, e.g. the need of prior milling and consumption of the sample. Furthermore, if amorphous or poorly diffracting materials (PDM) are present in the sample, the diffraction signal results can be very poor and indistinguishable from the noise.

Among the micro-destructive techniques available for meteorite elemental analysis, LIBS is gaining a very important role in the field of meteorite research. One of the characteristics that differentiates this technique from those mentioned above is that LIBS is very sensitive to light elements such as H, Li, B, C, N and O [9], and the analysis can be performed under reduced pressure conditions, thus avoiding any atmospheric interference, or by using a blank and subtracting the atmospheric input. Furthermore, LIBS offers several other benefits, such as ease-of-use, no need for sample pretreatment, robustness and versatility. However, LIBS is considered a micro-destructive technique as each laser pulse removes part of the sample, which amounts to a few nanograms per pulse, creating a crater on the surface around 200 mm in diameter.

5.1 The XRD technique

XRD is a well-established and versatile technique able to analyze the crystalline phases of materials, either in crystalline powders, consisting of a multitude of isotropically oriented micro-crystals, or in single crystals in which the isotropic orientations of the crystals are assured by complex goniometric cradles which make the sample rotate in the space. XRD works in both laboratory set-ups and portable instruments.

Powder XRD is typically used to study minerals, inorganic, organic and hybrid compounds, alloys, etc, whose powder pattern is considered the 'fingerprint' of the sample and provides qualitative information on the phases and crystallinity of the material and quantitative information in the presence of multi-phase matrices such as rocks, soils, alloys, meteorites, etc. Single-crystal XRD is used to study minerals and other inorganic and organic compounds in order to obtain structural information such as atomic positions within the unit cell, bond distances, angular values, coordination polyhedrals, etc. Due to the difficulty of retrieving a single crystal from complex rock matrices, in meteorite investigations powder XRD is used much more commonly than single-crystal XRD. Thus, in the following sections powder XRD and its applications to meteorite analysis will primarily be discussed, whereas the use of single-crystal XRD will be specified where appropriate.

5.1.1 Principles and basic instrumentation

Diffraction occurs when light is scattered by a periodic array of atoms with long-range order in a crystal, thus producing constructive interferences at specific angles. As the wavelength of x-rays is similar to the distance between atoms within solids, the XRD technique uses this principle to investigate the crystalline nature of materials. In particular, the scattering of x-rays from atoms produces a diffraction pattern that contains information about the atomic arrangement in the crystal. XRD analysis is based on Bragg's law, which describes the general relationship between the wavelength of the incident x-rays, the incident angle of the beam and the spacing between the crystal lattice planes of atoms. When a beam of x-rays of wavelength λ enters a crystal, the maximum intensity of the reflected ray occurs when $2d(\sin\theta) = n\lambda$, where θ is the complement of the angle of incidence, n is an integer number and d is the distance between the layers of atoms within the solid. In contrast, amorphous materials, such as glasses, that do not have a periodic array with a long-range order, do not produce any significant peak in the diffraction pattern.

The powder x-ray diffractometer is a compact instrument that is able to analyze the diffraction pattern characteristic of the powder sample under investigation. Usually, a powder XRD set-up consists of an x-ray generator that delivers a high-tension current to the x-ray source, i.e. a vacuum-sealed x-ray tube, a sample holder where the sample is placed, a detector capable of measuring x-ray photons scattered by the sample (which may contain multiple channels in a 1D or 2D arrangement) and a goniometer that allows the precise relative angular positioning of the x-ray source, sample and detector in the 'equatorial' plane with the axis of rotation perpendicular to the equatorial plane 'axial' direction (figure 5.1).

The diffraction pattern, i.e. the diffractogram, obtained from a powdered material consisting of a multitude of micro-crystallites (usual size 10–50 μm) oriented isotropically in the space, plots the intensity versus the 2θ angle of the detector. In a diffractogram the peak position depends upon the wavelength of the x-ray tube, whereas its absolute intensity (the number of x-rays in the given peak) may vary as a function of various instrumental and experimental parameters.

A diffractometer can work either in a transmission or, more commonly, in a reflection configuration. When Bragg's law is satisfied the interaction between the incident x-ray beam and the sample produces intense reflected x-rays by constructive interference that occurs when the difference in the travel path of the incident x-rays is equal to an integer multiple of the wavelength.

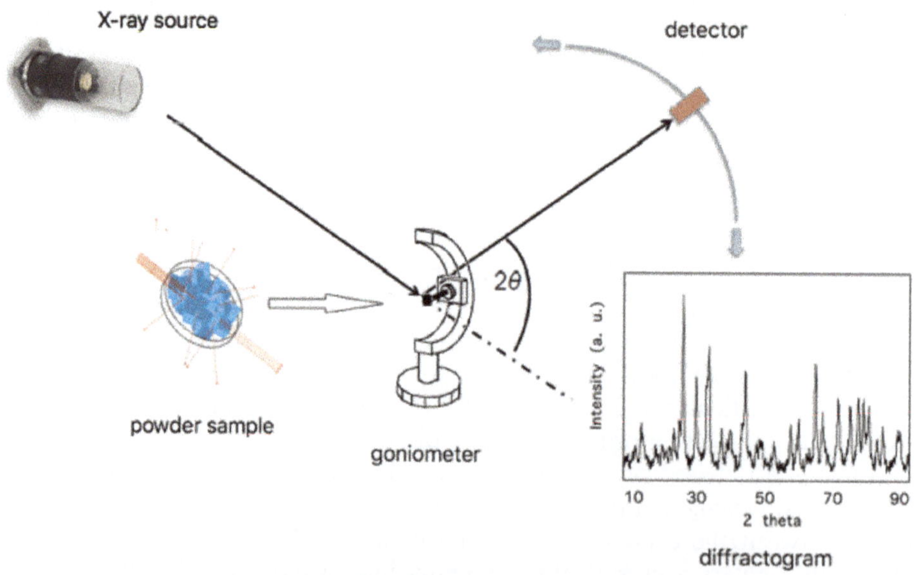

Figure 5.1. Typical set-up for powder XRD.

The single-crystal diffractometers are similar to those working with powders, the main difference being the position of the sample holder at the center of a complex of three goniometric circles in order to provide the required orientation to the crystal (figure 5.2). These circles refer to the four angles (2θ, χ, ϕ and Ω) that define the relationship between the crystal lattice, the incident ray and the detector. Samples (usual size 0.02–0.2 mm) are mounted on thin glass fibers attached to brass pins which in turn are mounted within suitable goniometer heads that allow the setting of the X, Y and Z orthogonal directions in order to center the crystal within the x-ray beam. The x-rays are either transmitted through the crystal, reflected off the surface, or diffracted by the crystal lattice. A beam stop is located directly opposite the collimator to block transmitted rays and prevent burning-out of the detector. The reflected rays are not picked up by the detector due to the angles involved, whereas the diffracted rays are collected at the correct orientation. Modern single-crystal diffractometers use charge-coupled device (CCD) technology to transform the x-ray photons into an electrical signal that is transmitted to a computer for processing.

The qualitative identification and quantitative determination of the crystalline phases within a polycrystalline matrix are the main objectives of the powder XRD technique, which has gained increasing interest in many scientific fields, e.g. in organic and inorganic chemistry, pharmaceutics, archaeometry, mineralogy and meteorite phase investigations.

5.1.2 Field instrumentation

In the last two decades, *in situ* non-destructive qualitative and quantitative analyses have been required in a large number of sectors, such as artistic/museal [10, 11], environmental [12] and planetary sciences [13]. Thus many efforts have been dedicated to the development of portable XRD instruments, of which there are of two types. One type uses area detectors such as CCD or imaging plates and does not employ detector/tube movements, such as the well-known museal devices developed by NASA [14]. The other type of instrument adopts a conventional goniometer-based diffractometer, such as the PT-APXRD III, through which data are acquired by scanning the detector and/or x-ray source [11]. However, the success of portable

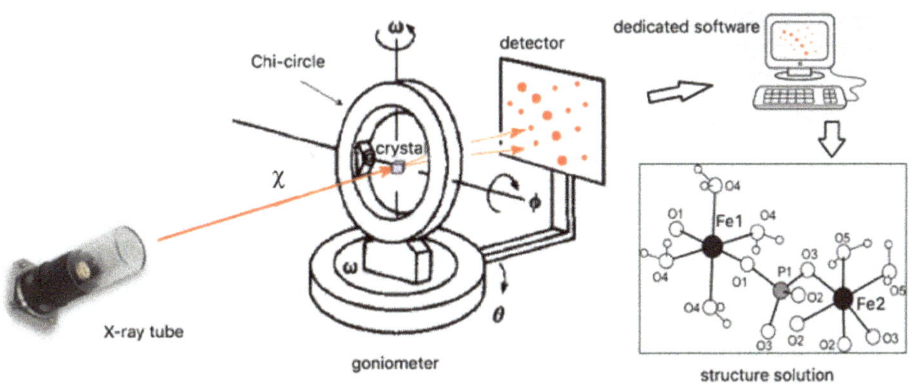

Figure 5.2. Typical set -up for single-crystal XRD.

x-ray powder diffractometers is still significantly lower than that of portable XRF spectrometers in determining the elemental composition of the samples [15, 16].

A portable XRD device was successfully sent to the surface of Mars in August 2012, when the Curiosity rover (within The Mars Science Laboratory) began field studies on its drive toward Mount Sharp, a central peak made of ancient sediments. CheMin, one of the rover instruments, consists of a miniaturized XRD (coupled to XRF) instrument that uses transmission geometry with an energy-discriminating CCD and onboard standards for calibration. Four Martian samples, i.e. soil, two mudstones and one sandstone, have been analyzed by CheMin [17].

5.1.3 Qualitative analysis

The main objective of XRD qualitative analysis is the determination of the major and minor phases in a powdered sample. The quality of analysis depends strictly on the appropriate alignment (line position, intensity), the preparation of the powder sample and the use of pertinent reference patterns. Many phases may be present within a sample, some of which at low abundance, which may cause line overlapping and/or interference problems, determine the decrease of intensity with concentration and/or influence intensities by mass absorption [18]. Thus usually it is important, and often critical, to possess preliminary supplementary information, both chemical and physical, about the sample. Furthermore, a very important condition is the match, i.e. adequately accounting for the distribution of intensity in the diffraction pattern, whose calculated values produce the figures of merit (FOM) of search–match programs. In particular, two or more compounds of the same space group and with similar cell parameters will produce very similar diffraction patterns, so that any search/match approach usually reaches a stage where a number of possibilities exist and a choice must be made. The search–match methods include manual approaches that consider significant lines, computer-based approaches that use FOM and more recent graphically based full-pattern methods for phase identification [19].

5.1.4 Quantitative analysis

Powder XRD is an excellent tool for obtaining quantitative information from a multi-phase polycrystalline sample. Quantitative analysis methodologies can be divided into two groups: (i) traditional standard methods [20, 21], which have been the most widely applied for a long time and require pure phases as internal standards and time consuming procedures to define calibration curves and/or reference intensity ratio values for each analyzed phase; and (ii) standardless methods, which are currently the most popular and usually involve the modeling of the diffraction pattern so that the calculated pattern duplicates the experimental one [22, 23]. The latter methods are based on the Rietveld approach that exploits the whole information pattern and offers great advantages, including no need for calibration curves or pattern decomposition in integrated intensity values, and no reflection effects overlapping in the results. However, this approach requires, for each phase, prior knowledge of the structural model, which is usually kept fixed during refinement, i.e. approximate models can be refined if the experimental information is sufficient [24].

Quantitative XRD analysis requires the accurate determination of the diffraction pattern of the sample in terms of both peak positions and intensities. While some kinds of analyses (i.e. particle shape and clay structure) rely on the existence of a preferred orientation, most analyses require a uniformly sized, randomly oriented, fine (ideally 1–5 μm) powder specimen to produce intensities that accurately reflect the structure and composition of the phase analyzed. Furthermore, the successful performance of quantitative methods requires accurate sample preparation, and a good quality of data and data collection procedures.

5.1.5 Advantages and disadvantages of powder XRD analysis

Powder XRD features a number of intrinsic advantages, such as easy sample preparation, its non-destructive nature, high sensitivity and reliability, a user-friendly system, fast speed, efficient resolution, proper automation, and the capacity for both qualitative and quantitative analysis. On the other hand, the technique has a few disadvantages, mainly related to the use of harmful radiation and the requirement for standard references, in particular for quantitative investigations.

In association with other techniques, such as differential scanning calorimetry (DSC), thermogravimetric analysis (TGA), differential thermal analysis (DTA), FTIR and Raman spectroscopy, XRD can solve numerous problems encountered in the process of material development in industry. To date, no practical application of XRD coupled with LIBS exists, apart from the Curiosity rover that landed on Mars in 2012, where both the XRD (CheMin) and LIBS (ChemCam) devices operate independently [25].

5.1.6 Meteorite characterization using XRD

The XRD technique is a powerful tool in meteorite investigation as every single natural phase of a meteorite rock is defined by a characteristic crystal structure that provides a unique x-ray diffraction pattern. The technique allows not only the rapid identification of the mineral phases present within the sample, but also the determination of their proportions in the meteorite.

For a long time meteorites were subject to several classifications based mainly on their chemical, petrographical and mineralogical features [26–29]. Currently, however, meteorites are divided into two major categories based on their bulk composition and texture, i.e. chondrites and non-chondritic meteorites, which are subdivided into achondrites, stony irons and iron meteorites. Based on the degree of melting experienced by asteroidal achondrites, they are often subdivided into primitive and igneously differentiated achondrites. Meteorites are further classified into groups based on their chemistry, oxygen isotopes, mineralogy and petrography. These classification schemes provide descriptive labels for classes of meteorites with similar origins or formation history which can derive from the same asteroidal or planetary body, and can reveal possible genetic links between various classes [30].

Focusing on the mineralogy of meteorites, the most abundant phases retrieved in meteorites are silicates, including: pyroxene (Ca,Mg,Fe)(SiO_3), olivine $(Mg,Fe)_2SiO_4$, plagioclase $(Na,Ca)(Si,Al)_3O_8$ and feldspar $(K,Na,Ca)(Si,Al)_4O_8$; kamacite α-(Fe,Ni) and taenite γ-(Fe,Ni); and small amounts of sulfide troilite FeS, phosphide schreibersite $(Fe,Ni)_3P$ and carbide cohenite (Fe,Ni)C. Silicate minerals (pyroxenes, olivines and feldspars) are abundant in stony meteorites, while α-(Fe,Ni) kamacite and γ-(Fe,Ni) taenite with small amounts of schreibersite and cohenite are abundant in iron meteorites. Other secondary but no less important phases are oxides, hydroxides and phosphates, such as apatite $Ca_5(PO_4)_3(OH,Cl,F)$, whitlockite $Ca_9(Mg,Fe)(PO_4)_6[PO_3(OH)]$ and merrillite $Ca_{18}Na_2Mg_2(PO_4)_{14}$. In particular, phosphates and phosphides are very important phases as possible sources of P essential for the origin of terrestrial life [31, 32].

The number of recognized phases was about 275 in 1997 [8], but has increased up to about 435 in the last two decades, despite their complexity and thanks to the improvement of analytical techniques [33]. The current review chapter will focus on those meteoritic phases that have been specifically characterized by XRD. The main crystallographical features of the phases, i.e. the crystal system, space group (i.e. the set of symmetry operations involving the crystal, defined by an initial capital letter representing the lattice, and the three symbols corresponding to the symmetry operators, i.e. axes, glide planes, etc) [34] and unit cell constants (a, b, c, α, β, γ and volume V), will be reported as they appear in the original papers.

The first pioneering meteorite investigations by means of XRD were performed about six decades ago when Dawson *et al* (1960) [35] studied the Abee meteorite that had fallen in Canada in 1952, a polymict breccia with local inclusions of black chondrite. This meteorite was found to be composed of kamaeite–taenite mixed with silicates, mostly enstatite $MgSiO_3$, and a minor presence of SiO_2 both as quartz and α-cristobalite polymorphs. Around this time, Fuchs (1962) was the first author to use XRD to detect whitlockite phosphate in several chondrite samples, in association with the Cl-apatite suggested previously to be present in these meteorites [36]. Lipschutz provided the first XRD evidence of diamonds in meteorites by analyzing samples retrieved from the Dyalpur achondrite, which fell in 1872 in India [37]. Frondel and Klein Jr (1965) detected, using single-crystal XRD, the presence of the ureyite pyroxene ($NaCrSi_2O_6$) in the iron meteorite Coahuila, retrieved in Mexico in 1837, by proving isostructurality with clinopyroxene [38]. Lipschutz and Jaeger (1966) used XRD to analyze the minerals from various iron meteorites, finding a pronounced shock-induced alteration in the minerals' crystallographic characters. The extent of alteration appeared to be dependent on the degree of shock, therefore serving as a possible shock intensity indicator. The changes were interpreted as a consequence of the direct recrystallization of the minerals during the passage of the shock wave [39]. Bass (1971) found serpentine $Mg_3(Si_2O_5)(OH)_4$ and the clay mineral montmorillonite as the bulk constituent phases of the Orgueil carbonaceous chondrite, which fell in France in 1864 (figure 5.3) [40].

Later, Putnis and Price (1979) [41] showed the presence of high-pressure $(Mg, Fe)_2SiO_4$ olivine phases in the Tenham chondritic meteorite (which fell in Australia in 1879), and that at high pressure olivines transform to a solid solution series of (Mg,Fe) spinels (cubic γ-phase, called ringwoodite), whereas 'modified' spinels with an orthorhombic structure (β-phase) were observed in olivines with Mg/(Mg + Fe) ratios greater than 0.85. In chondritic meteorites high-pressure (Mg, $Fe)_2SiO_4$ polymorphs appeared to be confined to shock-produced phases. Furthermore, these authors provided definitive EMPA, TEM and XRD evidence on the nature of the ringwoodite purple grains in the Tenham chondrite.

One of the first applications of the Rietveld refinement on powder XRD data of meteoritic minerals was provided by Post and Buchwald (1991), who refined the crystal structure of akaganeite [$Fe_{7.6}Ni_{0.4}O_{6.35}(OH)_{9.65}Cl_{1.25}$] (figure 5.4) [42], an oxide phase found as a corrosion crust on the Campo del Cielo meteorite found in Argentina in 1576. The study, performed also with the contribution of Mössbauer spectroscopy, questioned the tetragonal symmetry reported previously, and indicated a monoclinic *I2/m* structure with unit cell constants a = 10.600(2) Å, b = 3.0339(5) Å, c = 10.513(2) Å, β = 90.24(2)° and V = 338.09 Å^3 [42].

The improvement in the performance of XRD devices in the last two decades, in particular the development of advanced area detector revelators, generated an

Figure 5.3. Representation of the Orgueil meteorite fall published in 1865 in the *Annuaire Mathieu de la Drome*.

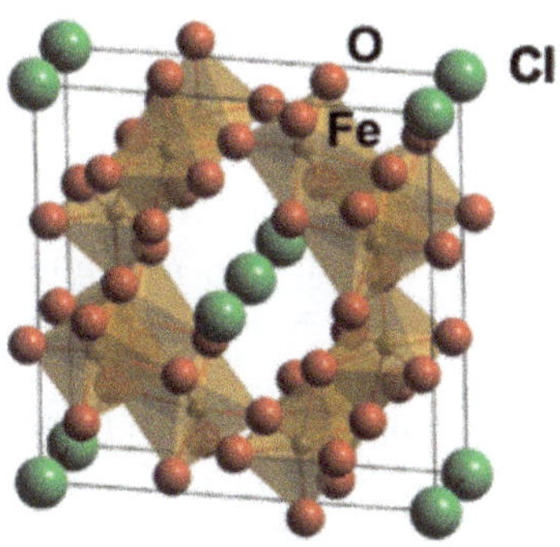

Figure 5.4. Simplified structure of akaganeite $Fe_{7.6}Ni_{0.4}O_{6.35}(OH)_{9.65}Cl_{1.25}$ FeO_6 octahedra and Cl anions (green).

increasing number of XRD studies on meteorite phases. In the Onello iron meteorite retrieved in Russia in 1997, Britvin *et al* (2002) [43] found the presence of allabogdanite, whose ideal crystal formula is $(Fe,Ni)_2P$, a new and very rare phosphide occurring as thin lamellar crystals disseminated in kamacite–taenite intergrowths (*plessite*). The phase was recognized as orthorhombic *Pnma*, while the unit-cell parameters refined from XRD data were found to be $a = 5.748(2)$ Å, $b = 3.548(1)$ Å, $c = 6.661(2)$ Å and $V = 135.8(1)$ Å^3 [43]. Another very rare phosphide, andreyivanovite (whose ideal crystal formula is FeCrP), was found by Zolensky *et al* (2008) [44] in the Kaidun meteorite which fell in South Yemen in 1980, a complex breccia containing both chondritic and achondritic lithologies. The examination of micrograins of andreyivanovite using patterns collected by an *in situ* synchrotron XRD did not allow single crystals to be found of sufficient quality to perform a complete structural analysis. However, the calculated powder XRD pattern confirmed the crystal arrangement as orthorhombic *Pnma*, $a = 5.833(1)$ Å, $b = 3.569(1)$ Å, $c = 6.658(1)$ Å and $V = 138.61$ Å^3, which was previously observed using a synthetic isotypic analogue [44] with an allabogdanite phase.

Chukanov *et al* (2009) [45] found, in a fragment of the Dronino iron meteorite discovered in Russia in 2000, a new halide phase, droninoite, ideally $Ni_3Fe^{3+}Cl(OH) \cdot 2H_2O$, which is trigonal, with space group $R\bar{3}m$, $a = 6.206(2)$ Å, $c = 46.184(18)$ Å and $V = 1540.4(8)$ Å^3. Furthermore, reflections of the powder-XRD spectrum were indexed by analogy with hydrotalcite, $Mg_6Al_2(OH)_{16}(CO_3){\cdot}4H_2O$. The isostructural character of the two phases was proven from the very similar intensity of most reflections of the experimental XRD powder pattern of droninoite and the theoretical XRD pattern of hydrotalcite calculated from structural data (ICDD database, card 89-0460 [46]).

Ma and Rossman (2009) [47] studied a number of phases retrieved from the Allende meteorite that fell in Mexico in 1969. In a chondrule from the meteorite sample they showed first the presence of tistarite, ideally Ti_2O_3, a new member of the corundum–hematite group, which was found as a subhedral crystal in a cluster of micro-sized refractory grains together with khamrabaevite (TiC), rutile and corundum crystals. The XRD analysis revealed a rhombohedral structure, $R3c$: $a = b = 5.158$Å, $c = 13.611$ Å and $V = 313.61$ Å^3 [47]. Furthermore, these authors showed the presence of two new pyroxenes of the Ca clinopyroxene group: (i) davisite, ideally $CaScAlSiO_6$, occurring as micro-sized crystals together with perovskite and spinel in an ultra-refractory inclusion, which is monoclinic in $C2/c$, with $a = 9.884$Å, $b = 8.988$ Å, $c = 5.446$ Å, $\beta = 105.86°$ and $V = 465.39$ Å^3 [48]; and (ii) grossmanite, $Ca(Ti^{3+}, Mg, Ti^{4+})AlSiO_6$, occurring as micrometer-sized crystals together with spinel and perovskite in a melilite host in Ca-,Al-rich refractory inclusions, which is monoclinic in $C2/c$, with $a = 9.80$Å, $b = 8.85$ Å, $c = 5.36$ Å, $\beta = 105.62°$ and $V = 447.70$ Å^3 [49].

Kubo *et al* [50] proposed using plagioclase $(Na,Ca)(Si,Al)_3O_8$, both in the amorphous form and in the high-pressure phase, as an indicator of meteorite shock conditions. The proposal was based on the hypothesis that when shocked meteorites are formed, their parent body undergoes shock metamorphism, usually forming plagioclases. These authors performed *in situ* XRD analyses of two plagioclase samples in conditions of increasing pressure and temperature, and found that the amorphization pressure of plagioclase decreases with increasing temperature. Thus, they proposed that the study of plagioclase breakdown can constrain the pressure–temperature–time history of shock events, thus helping to reconstruct the collisional history of asteroids in the early solar system.

Gismelseed *et al* [51] used a multi-methodological approach consisting of Mössbauer spectroscopy, XRD and EMPA to study the iron phases found in the Almahata Sitta meteorite, which fell in 2009 in Sudan. In particular, the modal abundance of mineral phases was determined by applying the Rietveld refinement to the powder XRD data, based on Mössbauer results that showed the presence of small amounts of troilite (FeS) and cohenite ($[Fe,Ni,Co]_3C$). The relative abundance of each Fe-bearing phase was also determined by and compared with the results obtained through XRD [51].

Tschauner *et al* [52] discovered, in the shocked Tenham chondrite, the presence of bridgmanite, formerly known as $MgSiO_3$-perovskite, the most abundant mineral on Earth. Meteorites exposed to high pressures and temperatures during impact-induced shock, such as the Tenham meteorite, often contain natural phases usually found in the deeper portions of Earth's mantle. The associated phase assemblage indicated peak shock conditions of ~24 GPa and 2300 K. The Rietveld refinement of XRD data based on the experimental crystal formula $(Mg_{0.75}Fe_{0.20}Na_{0.03}Ca_{0.02}Mn_{0.01})SiO_3$ (figure 5.5) assigned to bridgmanite an orthorhombic *Pnma* perovskite-structure with unit cell parameters $a = 5.02(3)$ Å, $b = 6.90(3)$ Å, $c = 4.81(2)$ Å and $V = 167(2)$ Å^3 [52].

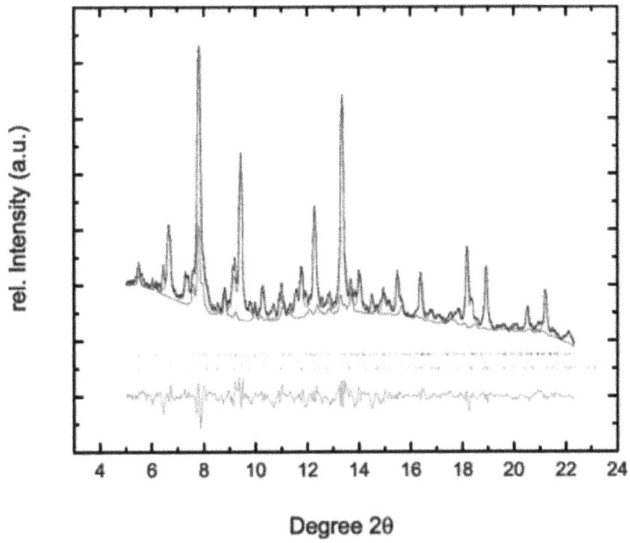

Figure 5.5. Powder diffraction pattern and Rietveld refinement of bridgmanite [52]. The figure shows the observed diffraction pattern (black line and symbols) of bridgmanite-bearing shock-melt vein material in thin section USNM 7703, whole pattern refinement (red), a refined pattern of bridgmanite (blue), residual offit (green), and positions of observed reflections of bridgmanite, akimotoite and ringwoodite (blue, red and green tick marks, respectively). Reproduced with permission from [52]. Copyright 2014 The American Association for the Advancement of Science.

Gabrig Turbay Rangel *et al* (2017) [53] highlighted the chondrite nature of the Varre-Sai meteorite that fell in Brazil in 2010, on the basis of petrographycal and geochemical data based on major elements and rare earth elements (REEs). Powder XRD analysis indicated that forsteritic olivine and enstatitic orthopyroxene are the most abundant phases in this meteorite (figure 5.6). The enstatite was identified using PDF card 22-714, while forsterite was identified using PDF card 31-795 [46]. To evaluate the unit cell parameters, a Rietveld refinement was performed on both phases starting from the crystallographic parameters of pertinent PDF cards. The results indicated an orthorhombic symmetry for both phases with unit cell parameters in good agreement with the two PDF cards: (i) enstatite *Pbca*, a = 18.270(20) Å, b = 8.910(10) Å and c = 5.210(20) Å; and (ii) forsterite *Pbnm*, a = 4.7929(8) Å, b = 10.348(2) Å and c = 6.0270(7) Å.

Calcium phosphate minerals, i.e. apatite $Ca_5(PO_4)_3(F,Cl,OH)$, whitlockite $Ca_9(Mg,Fe)(PO_4)_6[PO_3(OH)]$ and merrillite $Ca_{18}Na_2Mg_2(PO_4)_{14}$, represent very important phases in meteorites as they are the primary phosphate minerals retrieved in both terrestrial [54, 55] and extraterrestrial rocks [56]. Further, these minerals are also present in natural biological systems, e.g. hydroxyapatatite $Ca_5(PO_4)_3OH$ and whitlockite in human bones and teeth [57], which act as the natural counterparts of synthetic phosphates used as materials for multiple purposes, e.g. respectively, the class of apatite-based biomaterials [58] and the class of β-$Ca_3(PO_4)_2$ tricalcium phosphates (β-TCP) [59–60].

Hughes *et al* [61] showed that lunar merrillite present in the original samples returned from the Apollo missions and terrestrial whitlockite have similar atomic arrangements, but differ in the presence or absence of H. In whitlockite (rhombohedral *R*3*c*, $a = b$ = 10.3612(6) Å, c = 37.096(4) Å and V = 3448.88 Å^3 [59]), H is incorporated into the atomic arrangement by disordering of one of the phosphate

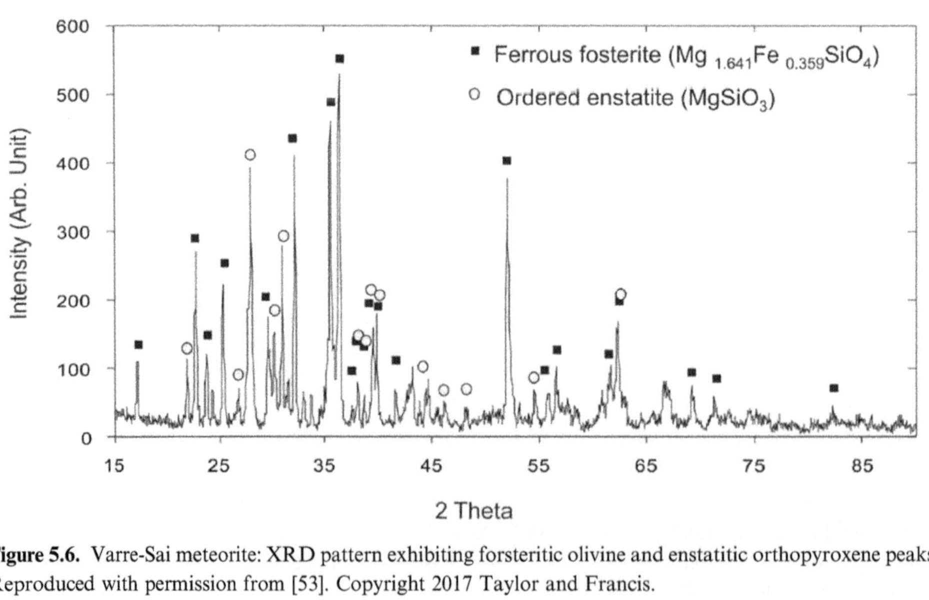

Figure 5.6. Varre-Sai meteorite: XRD pattern exhibiting forsteritic olivine and enstatitic orthopyroxene peaks. Reproduced with permission from [53]. Copyright 2017 Taylor and Francis.

tetrahedra and formation of a $PO_3(OH)$ group to allow charge balance. In contrast, the single-crystal XRD structure showed that lunar merrillite ($R3c$, $a = 10.291(1)$ Å and $c = 36.875(7)$ Å) lacks H, and thus no disorder exists within the tetrahedral groups. The charge balance for substituents Y and REE (for Ca) is maintained by Si ↔ P tetrahedral substitution and a Na-vacancy at the Na site [61]. Xie *et al* (in 2003 and 2013) hypothesized the formation of tuite γ-$Ca_3(PO_4)_2$, chemically $(Ca_{2.51}Mg_{0.29})_{2.80}Na_{0.28}(P_{1.01}O_4)_2$, trigonal, $R\bar{3}m$, $a = 5.258$Å and $c = 18.727$ Å, which was previously discovered as a high-pressure polymorph of whitlockite in shock veins of the Suizhou meteorite that fell in China in 1986 [62], from high-pressure decomposition of chlorapatite $Ca_5(PO_4)_3Cl$. The tuite genesis from chlorapatite decomposition under pressure was also confirmed by the results of synthesis experiments at high-pressure using chlorapatite as the starting material. The XRD pattern (figure 5.7) and the calculated unit cell parameters of synthetic tuite, $a = b = 5.251(1)$ Å, $c = 18.675(6)$ Å and $V = 445.9(2)$ Å^3, were very similar to those of the natural counterpart. These authors also suggested that the Na_2O, MgO and Cl contents in natural tuite could be employed as indicators for distinguishing the precursor phosphate mineral, chlorapatite or whitlockite, from which the tuite was formed [63].

Recently, Adcock *et al* (2017) [64], using synchrotron micro-XRD, studied the shock-transformation of whitlockite to merrillite in Martian meteorites by performing shock pressure experiments on synthetic Mg-whitlockite crystals embedded in a powdered copper matrix, which was used because its bulk impedance is similar to that of Martian basaltic meteorites. The diffraction images and Rietveld refinements revealed that at the contact between the Cu matrix and the crystal up to 36% of whitlockite was transformed into merrillite, whereas towards the interior of the grain the amount of merrillite decreased and polycrystalline whitlockite increased [64].

Finally, mention should be made of the natural quasicrystals that represent an important breakthrough in mineralogy and in condensed matter physics. Quasicrystals are solids which display unconventional atomic arrangements that violate the mathematical constraints of conventional crystallography, i.e. they exhibit rotational symmetry forbidden to crystals, such as five-fold, seven-fold and higher-order symmetry axes. The structure of these compounds is, by definition, not

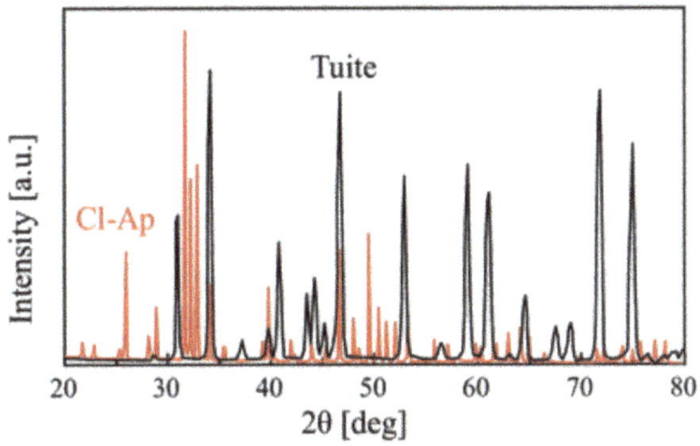

Figure 5.7. XRD patterns of synthesized tuite and the starting material chlorapatite [63].

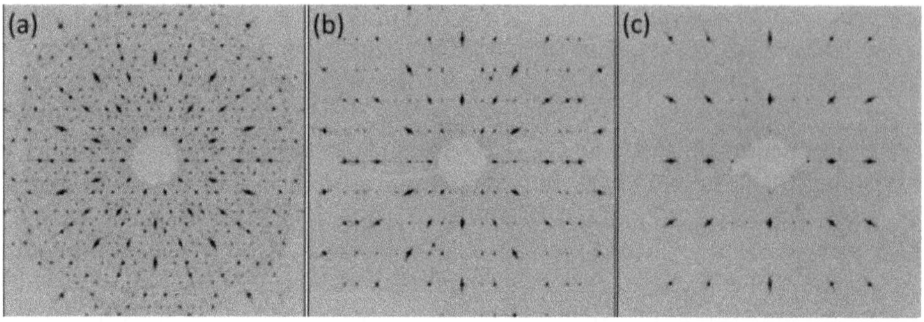

Figure 5.8. Reconstructed precession images along the ten-fold symmetry axis (a) and perpendicular to the ten-fold direction (b),(c), obtained using the single-crystal x-ray dataset (MoKa radiation) collected from decagonite. Reproduced with permission from [67]. Copyright 2017 Mineralogical Society of America.

reducible to a single three-dimensional unit cell, so unit cell parameters cannot be obtained [65]. The first evidence of a quasicrystalline natural phase was found by Bindi *et al* in 2011 [66] in the Khatyrka meteorite, a carbonaceous chondrite found in Russia in 2011. The phase consisted of icosahedrite $Al_{63}Cu_{24}Fe_{13}$, which displayed a five-fold symmetry in two dimensions and an icosahedral symmetry in three dimensions. In the same meteorite these authors discovered another extremely rare natural quasicrystal, decagonite, ideally $Al_{71}Ni_{24}Fe_5$. A combined TEM and single-crystal XRD study revealed the signature for this quasicrystal of a crystallographically forbidden decagonal symmetry, i.e. a pattern of sharp peaks arranged in straight lines with a ten-fold symmetry together with periodic patterns perpendicular to the ten-fold direction (figure 5.8) [67].

5.2 The LIBS technique

Laser-induced breakdown spectroscopy (LIBS) is an advanced and versatile atomic emission spectroscopy analytical technique in which a high-energy laser pulse is focused on a sample surface, thus causing its ablation, evaporation, atomization and ionization, and excitation, i.e. the generation of a high-temperature plasma that emits radiation whose spectral wavelengths are typical of its components, while the intensity is proportional to the concentration of the corresponding component in the target material. In the last few decades LIBS has been employed widely and successfully for the qualitative and quantitative analysis of various elements in materials of diverse origin and nature in the solid, liquid or gaseous phases [68–70].

5.2.1 Instrumentation, principles, mechanisms and processes of plasma formation and dynamics

A LIBS instrument basically consists of a laser source, an optical set to focus the laser, a unit for the light detection of plasma, and a computer for control and data acquisition processing (figure 5.9). In most LIBS applications the preferred laser source is the reliable and compact neodymium doped yttrium/aluminum garnet (Nd: YAG) pulsed laser, which can operate at output wavelengths of 1064, 532, 355, 266 or 213 nm. The high-power pulsed laser is focused through a lens on the sample

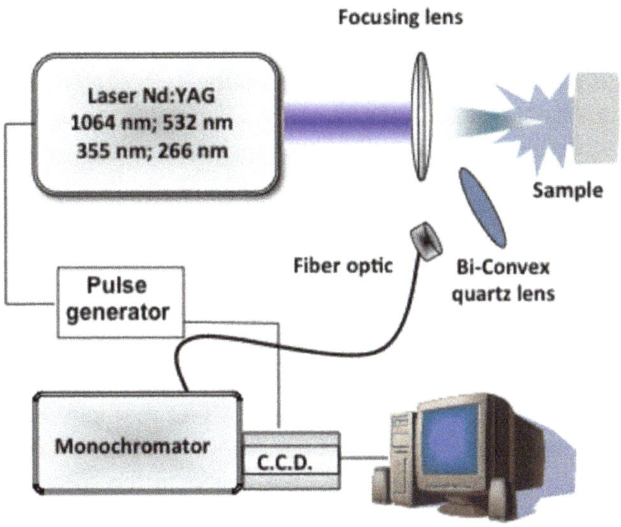

Figure 5.9. Basic experimental set-up of LIBS apparatus. Reproduced with permission from [9]. Copyright 2014 Elsevier.

placed a few centimeters to micrometers from the laser source. The typical energy of LIBS laser pulses ranges from 1 to 150 mJ, depending on the specific requirements. The laser blasts the sample with one or more nanosecond-wide laser shot pulses of high intensity, each lasting between 5 and 20 ns. The free electrons collide with the lattice phonons on the sample surface, thus heating the material and leading to avalanche ionization. A small portion of the target, rich in free electrons, ions and neutrons originating from the dissociation of molecules, is ablated and vaporized. Then, the ablated region expands, generating a shock wave. During its expansion, the vapor continues to absorb optical energy from the laser and ignites a luminous plasma plume with temperatures that can reach 100 000 K, and a range of 5000–20 000 K during data collection. An irradiance threshold of 10^{10}–10^{12} W cm^{-2} is needed to obtain a plasma that is representative of the target composition [68–70].

In the very early stages of its expansion the plasma is characterized by a high electron-number density and the main spectral feature is the continuum radiation due to processes involving free electrons, i.e. bremsstralung processes, radiative recombination of electrons with ions, and photoionization. In these stages element emission lines are absent due to the limitation in the number of accessible levels when the Debye effects are not negligible. The continuum radiation provides little or no spectroscopic information from the merely analytical aspect, although it can be useful for nanomaterial production.

When the time of the processes described above is shorter than that of the plasma expansion, a balance in the elementary processes is locally established, so that the fulfillment of local thermodynamic equilibrium (LTE) conditions can be assumed, i.e. the Maxwell, Boltzmann and Saha distributions are expected to hold for all species [68–70]. The Boltzmann plot linearity is one of the criteria to check the validity of the LTE assumption, because it guarantees that the Boltzmann distribution is held within the energy range selected for the analysis, which should be as large as possible. Thus, the assumptions to be fulfilled in order to obtain a plasma chemical composition corresponding to that of the sample are the existence of an optically thin and homogeneous plasma in LTE conditions and of a Boltzmann distribution of excited levels. Generally, these assumptions are valid for a delay time from hundreds of nanoseconds to a few microseconds after the laser pulse ends. In such a temporal window, the plasma plume collides with the surrounding species and begins to cool and emit high-intensity radiation through emission and recombination processes. The entire process lasts up to tens of microseconds.

During relaxation, excited atoms, ions and molecular fragments in the plasma emit radiation at wavelengths distinctive to the elements present in the sample under study. The emitted light is collected, spectrally resolved and then detected by a charge-coupled detector (CCD) or intensified (I)CCD that can delay the detection with respect to plasma formation and be gated, so to optimize the signal-to-noise ratio of the acquired spectrum. The individual peaks in the spectrum are typical of the elements present in the target material and feature three main parameters, i.e. wavelength, intensity and shape, which depend on the structure, surrounding environment and amount of emitting elements.

In conclusion, the objective of LIBS is to obtain an optically thin plasma in the LTE state whose elemental composition is equal to that of the sample, i.e. a stoichiometric ablation can be assumed to occur. In this condition the spectral line intensities are related to the actual element concentrations in the sample [68–70].

5.2.2 Field LIBS instrumentations

Field LIBS instrumentations adapted to *in situ* analysis are an attractive option, particularly in industrial, military and security applications where access to the sample is difficult, and for geological exploration and environmental monitoring in hostile environments when the object/material cannot be transported to the laboratory. In the past decade the development of compact and robust transportable LIBS instruments has featured continuous progress in allowing and improving *in situ* analyses [71, 72]. In particular, the apparatus size and weight have been reduced, the performance of lasers has been increased with the use of fiber lasers, spectrographs and detectors powered by batteries, and the use of compact spectrometers and fiber optics for guiding plasma emission have offered greater flexibility, while reducing the risk of instrument failure.

This instrumentation is available in 'portable', 'remote' and 'stand-off' configurations [71]. Currently, 'portable' handheld self-contained commercial LIBS instruments are able to measure most major, minor and trace elements in a broad variety of minerals, which cannot be achieved by other common field-portable instruments. For example, LIBS allows one to measure light elements such as Li, C, B and Na that cannot be measured by XRF, and shows improved detection limits for Mg, Al and Si [73–75].

In 'remote' LIBS analysis the laser beam and the signal are transmitted through a fiber-optic cable, so that the operator can be located far from the target, and targets not directly accessible or located in extreme environments can be measured [71, 76]. The fiber can be tens of meters long so that the device itself can stand, for example, in a car and samples can be analyzed up to the distance of the fiber length. Most 'remote' LIBS applications have specifically focused on geomaterial analysis and environmental monitoring [77].

Where large areas must be analyzed and/or the sample is not at a touchable distance, a 'stand-off' LIBS configuration in which both the laser radiation and the returning signal are transmitted along an open path can be used. However, the attenuation of light by the atmosphere is one of the major problems associated with stand-off LIBS [78], which requires higher quality optics in the device and a laser system with a good beam quality for laser-induced plasma formation at long distances [79]. Recently, a stand-off LIBS system was deployed on the Mars Curiosity rover as a part of the ChemCam analytical system currently being used to study the geochemistry of the Martian surface [80].

5.2.3 Qualitative analysis

The plasma spectrum contains all qualitative and quantitative analytical information, i.e. the emission line wavelengths and intensities, needed to identify and determine the elemental composition of the sample [68–70]. If the sample composition is known approximately, the set-up can be adapted to highlight the optimal spectral range where the emission lines of the elements of interest are located, and ignore the emission lines of elements unlikely to be present in the sample. In the case of expected interferences the lines in question should be assigned to the main elements and not to the minor ones that are unlikely to be found, unless the sample is contaminated [68–70]. The ionization status of each element is also important, e.g. in the case that two elements equally likely to be found in a sample yield overimposed lines that can be assigned either to a neutral species or to a double or triple ionized species, it is most correct to assign the line to the neutral species, because higher ionization states are unlikely to be detected in air [68–70].

The LIBS experimental conditions and parameters may influence the species detected. For example, in air atmosphere the lines from Fe(I) and Fe(II) are measured, whereas in a vacuum the Fe(III) line may also appear because the ionization potential of the second ionization stage of Fe is lower in a vacuum than in air [68–70]. Furthermore, many elements feature several strong emission lines, thus if one line is observed the other strong lines of the element are expected to be present. Although repetitive shots on the same location can be used to ablate below the surface to reach underlying layers and obtain an in-depth profiling analysis, LIBS is essentially a surface analysis technique, thus the condition of the surface is another important factor that affects the correct identification of elements in a sample. Many parameters, including the geometry of the measurement system, affect the precision and accuracy of an LIBS measurement. Some of these, such as the stability of the laser pulse energy, the repetition rate, the detector parameters and the lens-to-sample distance (LTSD), can be controlled. In contrast, other parameters that are dependent on the sample (e.g. sample matrix) and sampling procedure are difficult to control [68–70]. Furthermore, depending on the application, some instrumental parameters can be held constant during data collection so to maximize the analytical performance of LIBS.

5.2.4 Quantitative analysis

Quantitative LIBS analysis requires the development and use of a calibration curve that relates the observed analyte signal to its amount (concentration or mass) in the sample. The curve is constructed by reporting the data of known concentrations of a set of standard samples and the unknown analyte concentration as a function of the corresponding values of the sensor output, i.e. the intensity (or the area) of the LIBS emission peak of the element under analysis. Ideally, the standard samples should have the same matrix composition as the sample, which in most cases is unknown, and even if it is known, it is unlikely to match that of the calibration standards.

The sample matrix is a major factor affecting the accuracy of LIBS quantitative results. The 'matrix effects' are generally related to differences in ablation and/or excitation processes in the matrices, and can be distinguished in physical and chemical effects [81–83]. The 'physical' matrix effects are due to the physical properties (e.g. grain size, texture, reflectivity, roughness and hardness) of the sample surface and may result in an alteration of the amount of ablated mass which, in turn, may cause a variation of the line emission intensity of an element even if its concentration is the same in the various matrices. The 'chemical' matrix effects occur when the emission behavior of an element is altered by the presence of another one, so that an element present in equal concentration in two different matrices exhibits different LIBS emission intensities [84, 85].

These effects make it very challenging to find matrix-matched standards to be used to perform quantitative LIBS analysis of natural samples. However, the influence of matrix non-homogeneities can be reduced by one or more of the following approaches: (a) fine homogenization of the sample; (b) utilization of an algorithm that is able to discard anomalous spectral lines that are non-representative of the bulk sample; (c) using hundreds, or even thousands, of laser pulses distributed in a grid pattern to analyze the sample [86, 87]; (d) using the so-called 'internal' standardization procedure, which is based on keeping the concentration of a particular element in the sample constant, e.g. by its addition to the sample, and using its value to correct the response of the other elements; and (e) application of the so-called 'external' standardization procedure that uses an indirect physical magnitude to correct (i.e. normalize) the spectra, thus also reducing the impact of other effects, such as unwanted changes in shot-to-shot laser pulse parameters, on the accuracy of the quantitative analysis.

A conceptually different approach used to overcome sample matrix issues arising in LIBS experiments is the calibration-free (CF)-LIBS method, which attempts to provide the concentrations of all elemental constituents of the sample without using any calibration standards. This procedure is based on the assumptions of the occurrence of stoichiometric ablation and formation of an optically thin and homogeneous plasma in the LTE condition, which allows the calculation of

Boltzmann plots of plasma temperature and electron density for essentially all relevant species in the plasma, from which the elemental composition of the sample can be determined [88–90]. Furthermore, assuming that the radiation source is thin and the plasma composition corresponds to the sample composition, the CF algorithm allows the calculation of the line integral intensity between two energy levels of an atomic species. Thus, using different lines at different energy levels, the concentration of each element in the sample can be calculated without calibration of the integral intensities of each line. Theoretically, the CF algorithm can calculate the concentration of all the elements in the sample up to the detection limit of the method [89, 90]. The accuracy of CF-LIBS is generally quite good for major elements, i.e. those above 10% concentration, whereas the technique can provide only an approximate estimation of minor and trace components. This is because accuracy is strongly affected by (a) the instability of the signal when the laser intensity fluctuates within 1%–5%, (b) the variation of geometrical parameters during the measurements and (c) the laser scattering light.

The presence or absence of a specific spectral line in the spectrum, or whether the ratio of specific spectral line intensities exceeds a given threshold, can be used as predictors for classifying the samples analyzed [91]. However, the success of these univariate approaches is based on a correct previous line assignment supported by adequate spectral resolution and limits of detection [92]. More advanced, multivariate calibration models and chemometric methods, particularly when using compact portable LIBS systems, are more robust and more efficient in quantitative LIBS and in sample discrimination [93] and classification [94], as they simultaneously consider several features of the spectrum. The most used of these approaches include: principal component analysis (PCA), partial least square regression (PLSR) combined with discriminant analysis (DA), projection to latent structures (PLS) [95, 96], random forest (RF), support vector machines (SVM) and artificial neural networks (ANN) [97]. More advanced methods and models available to analyze LIBS data are the least absolute shrinkage and selection operator (LASSO) [98] and the sparse multivariate selection regression with covariance estimation (MRCE) [99]. The detailed treatment of these methods is, however, outside the scope of this chapter.

Finally, systematic recent investigations performed by the ChemCam team at Los Alamos [100, 101] suggest that multivariate quantitative chemometric strategies can automatically correct for matrix effects, if an appropriately large and diverse training sample set is used for setting up the model. By testing five different clustering and training selection algorithms on LIBS spectra acquired for 195 rock samples and 31 pressed powder standards, a root mean squared error (RMSE) of approximately 3 wt% was achieved in the quantitative analysis without any prior knowledge of the unknown samples.

The classification model generated should then be correctly validated, and its performance documented by presenting in full the confusion matrix of the results and describing the success of classification and identification by providing the overall accuracy, robustness, and number of true and false-positive cases.

5.2.5 Advantages and disadvantages of LIBS analysis

The LIBS technique features a number of important advantages with respect to other available traditional analytical techniques, such as XRF, EMPA and inductively coupled plasma-atomic emission spectroscopy (ICP-AES). In contrast to these techniques, which are mostly laboratory-based and often require complex and time-consuming procedures, sample collection is often not needed in LIBS and little to no sample preparation is required. LIBS also enables the analysis of extremely hard materials that are difficult to digest or dissolve. The chemical analysis is relatively simple and can be conducted in real time with a fast response both in conventional laboratory settings and in the field. LIBS analysis can be considered to be minimally destructive, as it consumes only nanograms of material per laser pulse, and is environmentally friendly because it does not produce waste, noise or pollution.

Broadband LIBS spectrometers have the potential of simultaneous multi-element analysis using a single laser pulse. Furthermore, LIBS is particularly sensitive to light elements such as C, B, Be, H and Li [9], which are often important components of geological samples but are problematic to determine using other analytical techniques. Quantization of N and O is also possible by performing the analysis in non-ambient atmosphere [102]. Portable, remote or stand-off LIBS systems (see section 2.2) allow analysis of samples located in hostile, harsh environments. LIBS can perform local analysis in micro-regions with a spatial resolution power of 10–100 μm, thus allowing *in situ* analysis of individual particles, mineral grains and inclusions [103, 104], and fine-scale compositional mapping of complex samples such as chemically zoned minerals [105]. Stratigraphic, in-depth profiling analysis can also be performed by a relatively simple set-up, as a crater can be obtained which progressively bores down into the sample by successive laser pulses [106]. Finally, LIBS instrumentation is generally less expensive and has lower subsequent operating costs than many other techniques. In the last decade, the possibility of combining LIBS with other complementary spectroscopic techniques, such as Raman spectroscopy [107] and laser-induced fluorescence (LIF) [108, 109] has been widely explored, thus permitting simultaneous, orthogonal, multi-element analysis.

However, similar to all other analytical techniques, LIBS features a number of limitations that need to be carefully taken into account when performing any experiment. LIBS measurements typically show an accuracy higher than 10% RSD (relative standard deviation), and often a precision higher than 5% RSD, whereas the limit of detection (LOD) varies depending on the element, the sample type and the experimental apparatus used, and generally ranges from 1 ng g^{-1} to 100 ng g^{-1} with values of tens of ng/g quite common. Although the LOD and precision of LIBS may often not be as good as those of some other established techniques used for the analysis of geomaterials, they are generally analytically acceptable. However, for some elements showing high LODs, the emission intensities of weaker lines can be enhanced by using an inert atmosphere such as Ar or He [110], whereas problems associated with poor precision can be reduced by ensemble averaging or normalizing the data with respect to the emission intensity of a line associated with a major component.

The main drawbacks of the LIBS technique, which often limit its reproducibility, are the matrix effects and the variations of the laser spark and resultant plasma, which are due to the inherent uneven energy distribution of a nanosecond laser pulse and to the differential coupling of the laser energy to the sample surface from shot-to-shot. Thus, LIBS cannot be regarded as an absolute method for quantitative analysis because its use relies on the comparison of the emission intensity from the unknown sample to that of a standard, which reduces the accuracy due to the difficulty of finding suitable matrix-matched standards [111]. Further, the comparison of LIBS data with those of other bulk analysis techniques, such as ICP-AES, may be meaningless because these methods determine the average concentration of the analyte in a sample mass that is much larger than that sampled by LIBS [112]. However, a number of approaches, such as the use of either internal or external calibration procedures and CF-LIBS methods [88, 90], may avoid the construction of calibration curves (see above, section 2.4). Further, spectra from the same sample collected by different LIBS instruments may not be identical due to the strong dependence of LIBS spectra on the specific system components and configurations used for analysis. This implies that spectral libraries obtained using a certain LIBS instrument may not be reasonably transferable without employing some type of transformation algorithm [110]. Finally, rigorous safety measures are required to avoid ocular damage in using the high-energy laser pulses.

5.2.6 Meteorite characterization by LIBS

More than a decade ago Thompson *et al* [113] were the first scientists to use a remote stand-off LIBS system to analyze at a distance of 5.4 m two Martian basaltic shergottite meteorites (DaG 476 and Zagami) of slightly different composition, texture and grain size. The olivine-phyritic shergottite (DaG 476) could be clearly distinguished from the basaltic shergottite (Zagami) on the basis of MgO and CaO contents. The mean elemental abundances of most of the major elements in the two meteorites agreed within 12% with literature values. The composition of a known andesite standard analyzed by the same LIBS system agreed within 9% with that of the basaltic meteorite sample. Furthermore, information on the weathering process could be obtained and the transition from the weathered surface to the unweathered substrate could be determined by analyzing cut slabs of the meteorites with sawn surfaces that are expected to be different from the heavily weathered, dust-covered surfaces of Mars rocks. In particular, repeated laser pulses on the same spot were able to remove the superficial dust and bore through the thin, weathered outer layers down to a depth of about 1 mm. A number of 10–15 analytical spots for each sample was enough to provide reasonable standard deviations and average estimates of the whole composition for the different grain sizes and spot sizes investigated. The main discrepancy from the literature values found for the Al_2O_3 abundance in DaG 476 probably resulted from the laser not having specifically hit any olivine grains of comparable size on the meteorite. As LIBS analysis spot sizes were fixed, more analytical spots were analyzed with only a few seconds of additional analytical time. The recognition of LIBS as an active remote sensing instrument favored its incorporation in the ChemCam instrumental suite on the Mars Science Laboratory, thus representing the first remote sensing instrument installed onto an extraterrestrial rover or lander.

De Giacomo *et al* [114] applied the calibration-free (CF)-LIBS method to the bench-top analysis of a set of different meteorite samples, i.e. Dhofar 461 (lunar), Chondrite L6 (stony meteorite), Dhofar 019 (Mars meteorite) and Sikhote Alin (iron meteorite). The wt% of the major elements O, Si, Ti, Mg, Fe and Ca and two minor elements, Mn and Cr, were determined successfully on the stony meteorites, and those of Fe, Ni and Co on the iron meteorite. In general, the data obtained agreed well with those in the literature, but the observed accuracy was only 15%. Bench-top CF-LIBS was then used by Dell'Aglio *et al* [115] to measure the major and minor element concentrations and detect qualitatively most trace elements in a large set of different meteorites, including Dhofar 019 (olivine-bearing basaltic shergottite), Dhofar 461 (anorthositic crystalline melt breccia), Sahara 98222 (chondrite L6 with ringwoodite), Sikhote Alin (coarse octahedrite class IIB), Toluca (medium octahedrite, class IAB) and Campo del Cielo (coarse octahedrite, class IAB) (table 5.1 and figure 5.10). All chemical analyses were performed in a vacuum chamber simulating space conditions, which also allowed the measurement of the O concentration in the samples. Furthermore, the concentration profiles of Fe, Ni and Co across the surface (along the Widmanstätten structure) of the iron meteorite Toluca were determined, which allowed the evaluation of the variations of these elements in the taenite and

Table 5.1. Element concentrations of martian, lunar and chondrite meteorites. Reproduced with permission from [115]. Copyright 2010 Elsevier.

	Dhofar 019		Dhofar 461		Sahara 98222
wt%	LIBS[a]	Literature[b]	LIBS[a]	Literature[c]	LIBS[a]
Al	1.4 ± 0.2	3.40	14 ± 1	15.50	–
Ti	0.45 ± 0.06	0.37	0.25 ± 0.04	0.13	0.030 ± 0.004
Mg	9.5 ± 0.9	8.80	1.5 ± 0.2	2.40	18 ± 2
Mn	0.35 ± 0.04	0.38	0.050 ± 0.005	0.05	0.27 ± 0.03
Cr	0.6 ± 0.1	0.40	–	–	–
Ca	3.6 ± 0.6	5.20	12 ± 1	11.90	2.6 ± 0.5
Fe	17 ± 2	14.70	4.9 ± 0.5	3.17	16 ± 2
Si	24 ± 2	22.62	22 ± 2	21.00	21 ± 2
Ni	–	–	–	–	0.41 ± 0.04
O	42 ± 6	44.20	44 ± 5	45.00	41 ± 5

[a] [114].
[b] [116].
[c] [117].

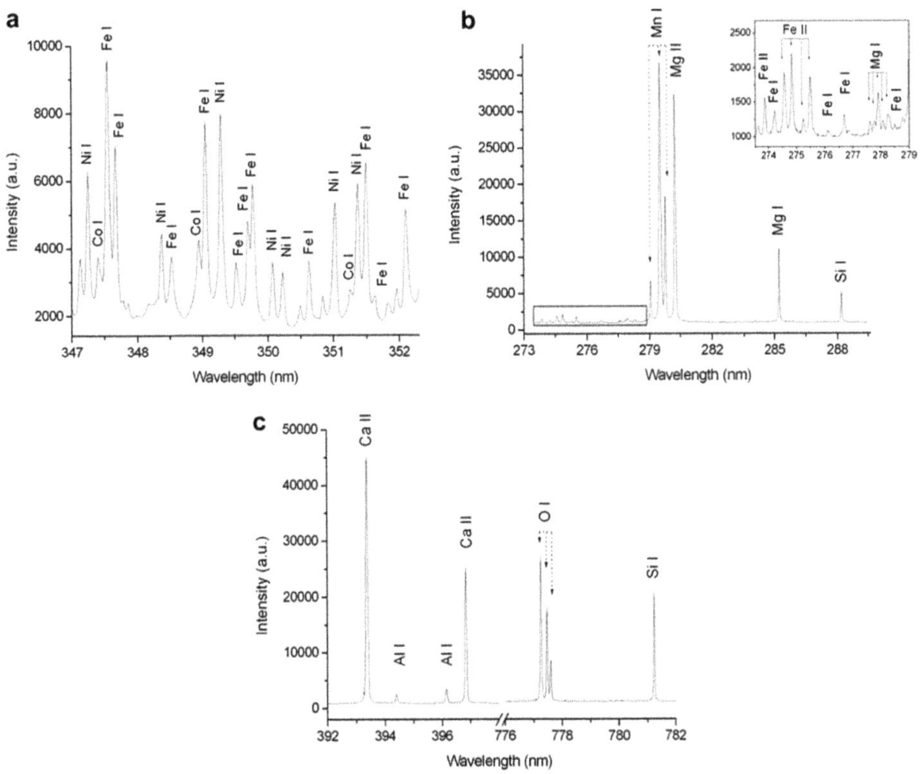

Figure 5.10. Fragments of the emission spectrum of (a) Sikhote Alin, (b) Dhofar 019 and (c) Dhofar 461. The inset in (b) shows an enlargement of a 273–279 nm spectral range. Reproduced with permission from [115]. Copyright 2010 Elsevier.

kamacite lamellae (figure 5.11). In particular, the technique was able to provide a satisfactory spatial resolution thanks to the micrometric laser spot used. The prediction errors of major elements at concentrations above 6% ranged from approximately 0.5 to 15 wt%, whereas the errors for minor elements were as high as 60%. The authors concluded that the proposed CF-LIBS methodology showed unique promising features for remote simultaneous multi-element analysis at distances of up to hundreds of meters in space applications and planetary exploration, and in on-flight measurement of asteroids.

Cousin *et al* [118] successfully applied the LIBS technique associated with principal independent component analysis (PICA) and partial least squares regression (PLSR) methods to evaluate the proportion of different mineral phases, including pyroxene and plagioclase, and identify the distribution of mineral grain sizes in the DAG 476 Martian picritic shergottite meteorite found in the Sahara Desert in Libya, as well as in other terrestrial basaltic rock analogues. The results showed that olivine phenocrystals could easily be distinguished from the groundmass. A good correlation was found between the grain size of the rocks analyzed and the standard deviation of the Ca/Al emission line ratio, which was related to the

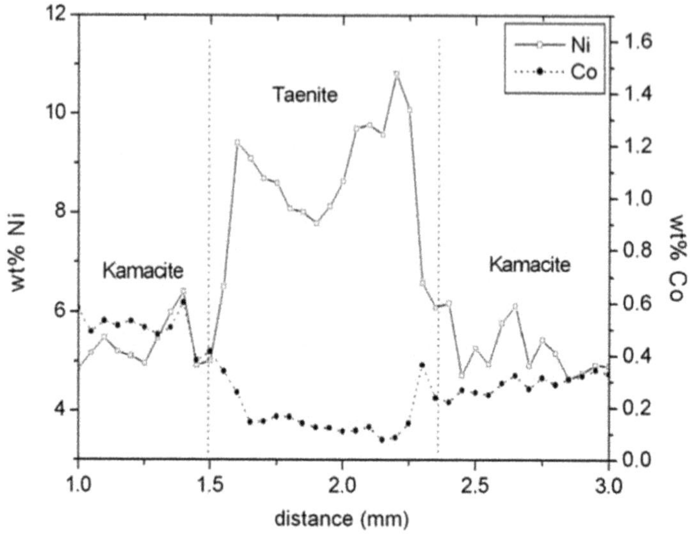

Figure 5.11. Ni and Co concentration profiles across a taenite lamella of Toluca. The typical M-profile of Ni concentration is shown across the taenite lamella. Reproduced with permission from [115]. Copyright 2010 Elsevier.

amount of plagioclase and pyroxene present in the sample, even when the grain size was smaller than the laser spot size. Further, the use of some key elemental ratios (Mg/Si, Ca/Al or Ca/Si) allowed efficient discrimination between olivine, pyroxene and plagioclase crystals in sample matrices. All other conditions being held constant, the RSD values were shown to be good indicators of the crystal size of these rocks. In particular, for high RSD values the rock could be classified as porphyric, whereas for very low RSD values microlitic texture predominated in the matrix. Thus, this approach could be used as a rapid tool to differentiate the degree of crystal basalt fractionation, allowing a preliminary estimate of their formation conditions such as the cooling rate and the inferred depth. The authors [118] concluded that in addition to measuring the quantitative composition and classifying extraterrestrial and terrestrial basaltic rocks, the RSD values of key emission line ratios measured by LIBS would provide important information on their petrogenesis.

Plavčan *et al* [119] confirmed the good performance of CF-LIBS in assessing the composition of several fragments of the meteorite Sikhote Alin, as a suitable alternative to traditional, complex time-consuming analytical methods such as AAS and ICP spectroscopy. The concentrations of major (Fe) and minor (Ni, Co, P, Na, K, Ca, Mn) elements were determined in air at atmospheric pressure, thus the concentrations of O, N and H could not be measured. As some elements such as Na, Ca, K and Mn were not mentioned in the official Sikhote Alin quantitative elemental composition, these elements on the meteorite surface would probably have appeared after the meteorite fall.

Subsequently, Dell'Aglio *et al* [120] measured using CF-LIBS the elemental composition (Fe, Mg, Si, Ti, Al, Cr, Mn, Ca, Fe, Ni, Co) of a chondrite meteorite (Sahara 98222) and its chondrules and the profile of Ni content in an octahedrite iron meteorite (Toluca). Further, a space-resolved analysis of the interface between kamacite and taenite crystals was performed in the Toluca meteorite. In particular, special attention was devoted to exploring the possibilities offered by variants of the basic technique, such as the use of the Fe I Boltzmann distribution as an intensity calibration method for the spectroscopic system and the use of spatially resolved analysis. Finally, the good performance of CF-LIBS for the chemical identification of primordial matter in stone chondrules of a chondrite meteorite, and the estimation of the metallographic cooling rate of iron based on the Ni content distribution in the iron meteorite, confirmed that this technique was mature enough to provide valuable geochemical insight into asteroid matter, and a significant contribution to astrochemistry, in particular to the chemical and geochemical history of the solar system.

Horňáčková *et al* [121] also used the CF-LIBS method to quantitatively determine the elements Al, Ca, Cr, Fe, Mg, Mn, Na, Ni and Si in the crust and inner part of the Košice ordinary H5 chondrite meteorite. In comparison to ICP-MS analysis, the accuracy of CF-LIBS was increased by appropriately selecting the spectral lines to be used and ignoring the lines showing the highest calculated probability of self-absorption. A further study by Odzín *et al* [122] allowed the classification of the Košice chondrite as a monomict breccia of petrological type H5, which suggested that this meteorite probably originated from the same parent body as the H5 chondrite meteorite Morávka from the Czech Republic.

Kovács *et al* [123] applied LIBS supported by optical microscopy, XRD, electron probe devices (EBSD) and micro-Raman spectroscopy to element identification in the Csátalja chondrite meteorite. In addition to Fe, which is the main constituent of this meteorite, the elements Ni, Co, P, K, Na, Ca and Mn were also detected. The mineralogical data (olivine/olivine+pyroxene) and recrystallization temperatures suggested that the parent body of this meteorite was the asteroid Koronis or Agnia and Merxia S-type. The stratigraphic analysis and the degree of terrestrial weathering suggested that the terrestrial resident age of the Csátalja meteorite was <2500 yr. Furthermore, this meteorite represented a very interesting object for further studies of its chondrules, which can be easily separated from the matrix, and due to the presence of highly porous matrix areas and metal veins.

More recently, Senesi *et al* [124] applied the innovative approach of double pulse (DP) micro (μ)-LIBS coupled with optical microscopy (figure 5.12) to the rapid chemical quantitative analysis and characterization in atmospheric air of a fragment of the Agoudal iron meteorite and its petrographic thin section without any treatment of the sample surfaces (figure 5.13). The elements Ca, Co, Fe, Ga, Li and Ni were identified in the thin sections and the whole meteorite. Two different methods, CF-LIBS and one-point calibration LIBS, were used for the quantitative LIBS analysis, which yielded elemental composition data in good agreement with those obtained by the traditional techniques ICP-MS and EDS–SEM (table 5.2). The whole bulk iron meteorite named Dronino was used to validate the analytical results.

Since 2013 the NASA Mars Science Laboratory Curiosity rover has discovered four metallic meteorites along its 17 km exploration of the Gale crater (5°S, 222.5°W),

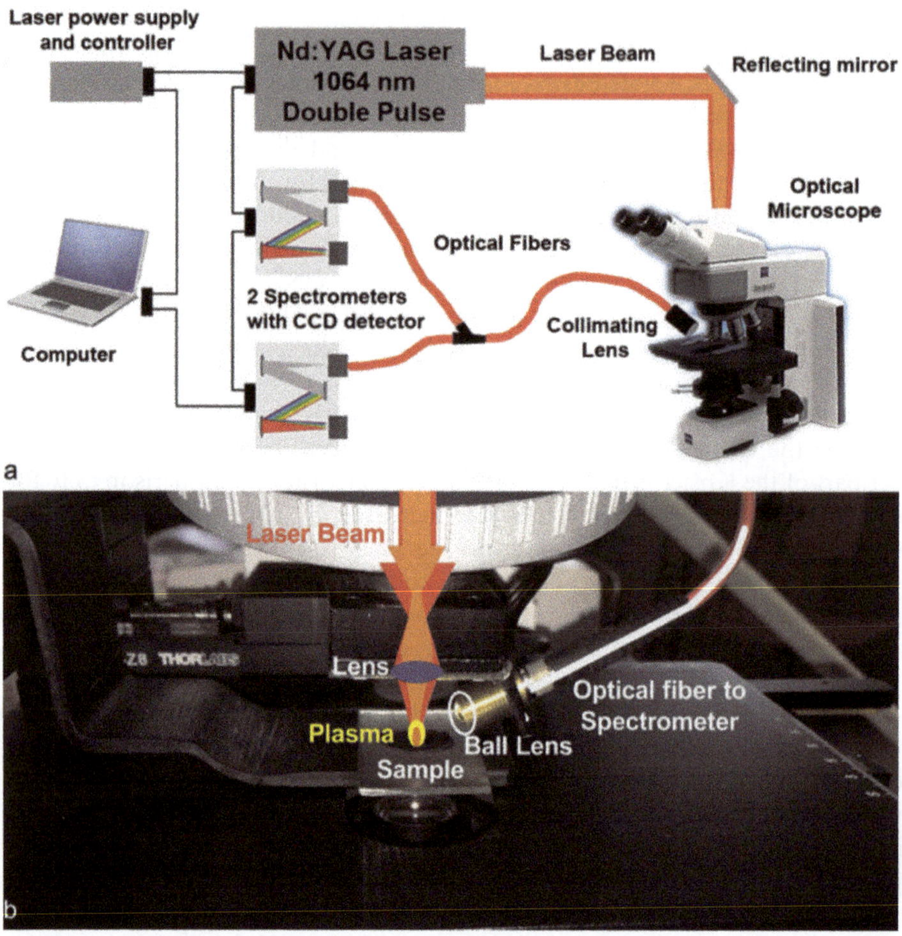

Figure 5.12. Schematic figure of the experimental set-up used for the analysis of the thin section and whole meteorite. Reproduced with permission from [124]. Copyright 2016 Wiley.

two of which were analyzed *in situ* for their elemental composition [125]. The 'Egg Rock' meteorite [126] was imaged using passive VNIR spectroscopy by Mastcam and analyzed by the ChemCam LIBS instrument on the mission day Sol 1505, followed by the 'Ames Knob' on Sol 1577. The Egg Rock meteorite was analyzed by LIBS over a 3 × 3 point raster, while the Ames Knob was probed with a 1 × 3 raster. Each observation point was 0.4 mm in diameter and was hit with 30 laser pulses, each returning a spectrum. The Egg Rock points 1–8 showed very similar spectra dominated by Fe peaks with the presence of Ni peaks. The comparison of these spectra with those of the kamacite phases of several iron meteorites measured previously with a replica of ChemCam [126] indicated the presence of kamacite containing 8 wt% Ni. The three ChemCam observation points of Ames Knob showed a Ni/Fe peak ratio that was nearly identical to that of the Chinga meteorite, an ataxite containing 16.7 wt% Ni, which was analyzed by LIBS for comparison,

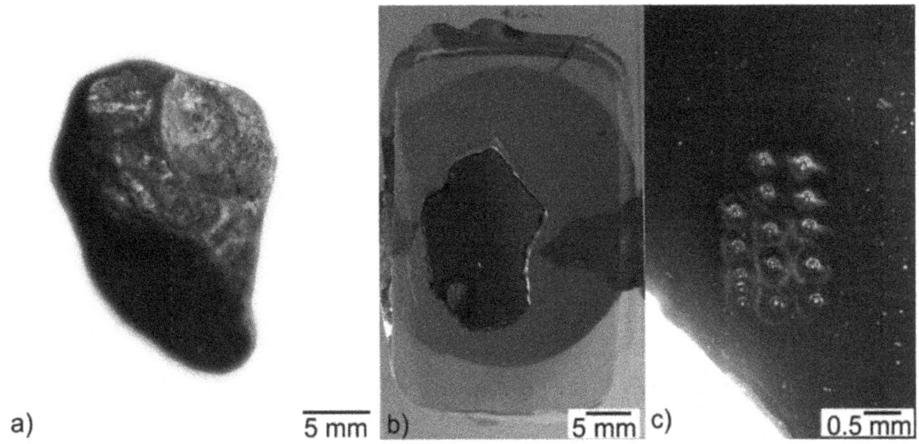

Figure 5.13. Images of (a) a fragment of the whole meteorite Agoudal IIAB and (b) of a thin section cut from it; (c) magnified optical image of LIBS shot area on the thin section. Reproduced with permission from [124]. Copyright 2016 Wiley.

Table 5.2. Comparison of element concentrations data (in % m/m) obtained by different chemical analyses in the literature and this study from whole samples and thin sections of the Agoudal meteorite and from whole samples of the Dronino iron meteorite. Reproduced with permission from [124]. Copyright 2016 Wiley.

	Agoudal meteorite						Dronino meteorite	
Element	ICP-MS[a]	EDS–SEM	CF-LIBS[b]	OPC-LIBS[b]	CF-LIBS[c]	OPC-LIBS[c]	$EPMA^{DR*}$	$OPC\text{-}LIBS^{CDR}$
Fe	94.1	94.41 ± 0.13	94.41 ± 0.31	94.10 ± 0.05	94.94 ± 0.43	94.40 ± 0.28	92.2	92.73 ± 1.64
Ni	5.5	5.57 ± 0.14	4.36 ± 0.44	5.31 ± 0.3	4.72 ± 0.43	5.47 ± 0.21	7.0 ± 0.5	6.26 ± 1.23
Co	0.41	N.D.[d]	1.24 ± 0.14	0.59 ± 0.25	0.58 ± 0.3	0.22 ± 0.17	0.75	1.02 ± 0.72

[a] Reference data from [127].
[b] Thin meteorite section.
[c] Whole meteorite. DR, *Dronino,* $EPMA^{DR*}$. Electron probe micronanalysis reference data from [128].
[d] N.D. not determined.

thus suggesting that the Ames Knob might be the first ataxite observed on Mars. Meteorites observed robotically *in situ* on Mars are of interest, in particular due to their longer preservation time, for understanding the distribution of classes of meteorites on Mars in comparison to Earth and for revealing the past atmospheric conditions of the planet and their weathering history in the Mars environment.

Lanza *et al* [129] compared the Egg Rock and Ames Knob Martian meteorites with three metallic iron meteorites, i.e. Wabar (class IIIA), Canyon Diablo (class IAB) and Calico Rock (class IIA), and two stony meteorite samples, i.e. Thuate (H12) and Tissint (shergottite), obtained from the collection of the Institute of Meteorics (University of New Mexico, NM). The LIBS analysis was performed with a ChemCam laboratory unit in a Mars chamber simulating Martian conditions on the exterior and cut interior sample surfaces in three spots with 50 consecutive laser shots per spot. The exterior surfaces of the three iron meteorites showed a shot-to-shot decrease of Fe, Mg, Ca, Na, K and H, which did not occur on the interior surfaces, thus suggesting terrestrial weathering of the oxidized crust dominated by Fe. These results were consistent with the trends found in the two Martian meteorites, i.e. a decrease in Mg, Ca, Na, K and H with depth, which suggested the presence of a thin layer of dust combined with a thin oxidized layer. In contrast, the stony meteorite Thuate showed an increase of Fe, Mg and K and a decrease of H with depth, whereas Tissint generally did not show evident depth trends in exterior spots.

Recently, Ferus *et al* [130] used the CF method to compare the reference experimental LIBS spectra of three chondritic meteorite samples, i.e. Dhofar 1764 (group CV3), Dhofar 1709 (group LL4) and Porangaba (group L4), to the laboratory simulated plasma of two meteor fireballs, i.e. Perseid and Leonid. The laser-induced ablation of the meteorite samples was conducted under strictly defined laboratory conditions so that the atomic and molecular species were evaporated from a relatively large volume together with the whole matrix, and the plasma was generated in a manner similar to that generated during a meteoroid descent. Thus, the meteorite spectra could be easily and directly compared to those of meteors, allowing LIBS results to be used for the qualitative assignment of meteor spectra and avoiding the computing of artificial spectra with the use of large databases. Furthermore, the LIBS spectra of the meteorites were measured by simultaneously using a high-resolution laboratory echelle spectrograph and a spectral camera for meteor observation. The spectral features of the airglow emission simulated by electric discharges and LIBS in air of the two meteors were qualitatively similar to those of the three meteorites. The CF data evaluation procedure applied to determine the elemental composition of meteorite samples yielded results in good agreement with those obtained for the two meteor plasma, which also agreed well with their chemical composition, previously analyzed independently using model spectra. Further, based on the ratio Fe/Mg = 1, the composition of the two meteoroids was comparable and corresponded to the C-group of chondrites. The authors concluded that the novel method they introduced, based on CF-LIBS comparative analysis, could be applied to evaluate meteor spectral data in real time and determine the chemical composition of meteor plasma which, in the case of the Perseid and Leonid meteors examined, corresponded to that of chondrites of the C-group.

Aramendia *et al* [7] reviewed the features of the most relevant current techniques used in meteorite studies and proposed a combined micro-destructive methodology, which included XRF, Raman-SEM/EDS/RE, LA-ICP-MS and LIBS, for a complete elemental, molecular, structural and mineralogical characterization of meteorites. In particular, LIBS was recommended for element detection and quantification in general, but specifically of the light elements H, Li, N, O and C. Further, the combined use of LIBS and Raman would avoid the use of the gas-chromatography–mass spectrometry (GC–MS) technique.

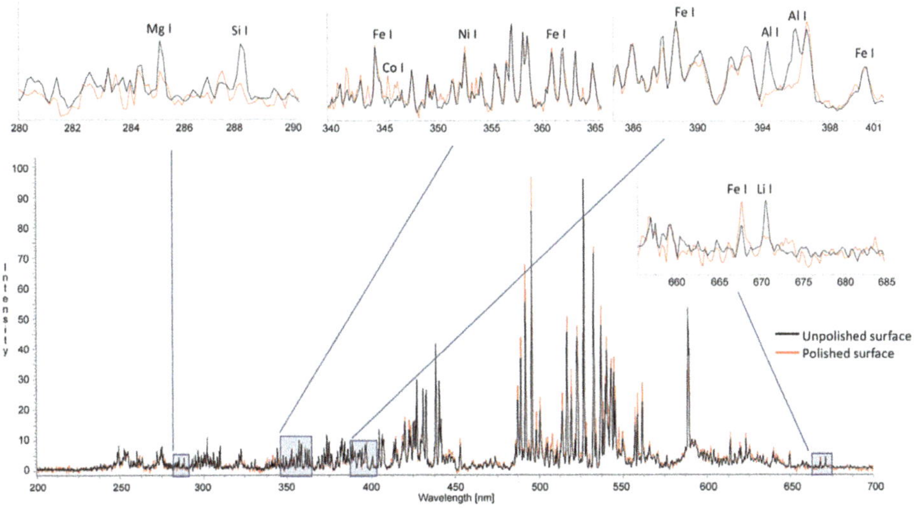

Figure 5.14. DP-μLIBS spectra of the original and polished surfaces of a weathered fragment. Spectral windows: (a) 280–290 nm, (b) 340–365 nm, (c) 385–400 nm and (d) 650–665 nm. Reproduced with permission from [131]. Copyright 2018 Elsevier.

Tempesta *et al* [131] recently studied two fragments of an iron meteorite shower named Dronino by DP-μLIBS combined with optical microscopy, which allowed extensively weathered fragments to be distinguish from poorly weathered fragments, thus providing insights on the geochemistry of the sampling sites. The main difference between the two fragments analyzed was the presence of the elements Al, Ca, Mg, Li and Si only in the extensively weathered fragment (figure 5.14), which was related to the presence of typical minerals of clays and sands. The pervasive presence of Li in the Dronino meteorite (which had not been previously reported, probably due to the lack of detectability of light elements by other routine techniques) was feasibly ascribed to contamination by alkaline-rich fluids occurring in the swamp in which this fragment was found. Quantitative CF-LIBS data showed good agreement with those obtained by traditional methods, i.e. EDS–SEM (this study) and EMPA (literature data). The increase of Ni and decrease of Co determined in the unaltered fragment portion, which exhibited a plessite texture, suggested that this meteorite was subjected to solid state diffusion processes under a slow cooling rate. The different degree of terrestrial alteration between the two analyzed fragments was also confirmed by their different mineralogical composition, i.e. kamacite (low Ni content) as the main component of the extensively weathered fragment and mainly taenite (high Ni content) with some kamacite in the unaltered fragment. The lower Ni content on the surface of the altered fragment was ascribed to the effects of acidic weathering that caused the removal of bivalent Ni. The results of this work confirmed the suitability of DP-μLIBS for analyzing the compositional and mineralogical characteristics of iron meteorites.

Recently, Senesi *et al* [132] proposed an innovative strategy for meteorite classification, which used a handheld LIBS instrument that was able to integrate automated feature selection and supervised learning in a fuzzy-rule-based classifier that could be easily interpreted. This approach was tested successfully to build a model for the automated classification of iron meteorites based on LIBS spectra. Furthermore, the identification of the elements corresponding to the seven wavelengths selected was expected to allow the fuzzy rules to be exploited for further chemometrics studies.

Finally, Senesi *et al* [133] published a study in which a compact handheld LIBS instrument, capable of fast-response multi-element analysis and equipped with a low-resolution spectrometer, was used to identify specific major elements (Fe, Ni and Co) and trace elements (Ga and Ir), so as to discriminate a certified iron meteorite from a suspected meteorite fragment dubbed as a 'meteor-wrong' and a pig iron product. In particular, a CF-LIBS method was used to quantify the main elements Fe, Ni and Co in the iron meteorite and Fe, Mn, Si and Ti in the other two fragments. However, the LIBS quantitative data were only partly satisfactory, requiring adequate improvements of some LIBS instrumental parts to increase the analytical sensitivity. Thus, the handheld LIBS instrument appeared to represent a promising advanced technical tool to obtain a fast, reliable and non-destructive in-field chemical analysis, identification and classification of metallic objects of extraterrestrial origin with respect to human artifacts, enabling discrimination between a true meteorite and a meteor-wrong and allowing one to determine ahead of time any need for further analysis in the laboratory. Furthermore, the preliminary chemical classification of iron meteorites could be used in the near future to identify and recognize mislabelled/unlabeled dubious meteorite specimens in museums and private collections.

In addition to the studies described above on meteorites, a number of studies were conducted in the last decade to test LIBS performance on rocks considered as analogues to meteorites and extraterrestrial geomaterials, often under simulated conditions at reduced or elevated temperatures and pressures in non-air atmosphere at stand-off distances. A comprehensive review of these topics was published recently by Senesi [9]. The advancements made by LIBS thanks to the above studies have benefited and will continue to benefit the LIBS study of meteorites on Earth and in particular in the extreme environmental conditions existing on Mars.

5.3 Conclusions and perspectives

XRD is a well-established analytical technique that in the last few decades has benefited from the production of a large theoretical background, particularly in structural solutions, and from many technical improvements, including that of area detector performance, which allows the performance of a large number of investigations of meteoritic minerals. Despite the new analytical techniques that have emerged in the last years for meteorite characterization, in particular micro-Raman spectroscopy, few of them demonstrate the many intrinsic advantages of XRD. These include its easy sample preparation, micro-destructive nature, high sensitivity and reliability of qualitative and quantitative analysis, and structure solution, from both single crystals or powders. The application of XRD has also led to further advances in meteoritic sciences, such as, for example, in the study of shock conditions.

In conclusion, XRD can be considered to be among the most reliable and complete laboratory techniques for meteorite investigation, and in the portable powder XRD configuration demonstrates a performance comparable to the most advanced techniques. This is confirmed by the successful performance obtained during the mission on the Mars surface in 2012, in which an XRD (CheMin) was included in the Curiosity rover's suite of analytical devices. Finally, an additional advantage of XRD is the possibility of operation coupled to other techniques, including XRF, FTIR and micro-Raman.

In addition to the well-known advantages of LIBS over traditional techniques, which include versatility, minimal destructivity, lack of waste production, low operating costs, rapidity, availability of transportable and portable systems, etc, additional advantages of this technique in the analysis of meteorites are good precision and accuracy, sensitivity to low-atomic-number elements, such as Li, and the capacity to detect and quantify Co contents that cannot be obtained by EDS–SEM.

In general, the results of LIBS studies on meteorites indicate that the higher the concentration of the element, the better the agreement with data obtained using other techniques. As each meteorite group features its own inhomogeneous matrix, the successful application of the CF-LIBS method to meteorite characterization and classification needs further study. In particular, small differences in minor element presence and content may play a fundamental role in meteorite classification. Thus, iron meteorites, which are essentially composed of a few elements, may represent a good basis for building up an LIBS data library to be readily used for studying these kinds of meteorites.

In particular, the LIBS technique shows unique promising features, not only for the laboratory analysis of meteorites, but in particular for their remote, simultaneous and multi-element analysis at distances of up to hundreds of meters. This aspect is very important, in particular for space applications, including planetary exploration and on-flight measurement on asteroids, which allows the calculation of their mass and, in turn, their trajectory and the effect of an impact with a gaseous atmosphere. Furthermore, simulation of the ablation plasma of meteors entering the Earth's atmosphere can be obtained using LIBS analysis of meteorites. Finally, the recently introduced handheld LIBS instrumentation represents a promising advanced technical tool for the fast, reliable and non-destructive in-field chemical analysis and identification of the extraterrestrial origin of metallic objects, allowing them to be distinguished from human artifacts, thus being able to discriminate between a true meteorite and a meteor-wrong.

References

[1] Meteoritical Bulletin Database 2016 http://www.lpi.usra.edu/meteor/index.php

[2] Treiman A H 2005 The nakhlite meteorites: augite-rich igneous rocks from Mars *Chem. Erde-Geochem.* **65** 203–70

[3] Nyquist L E, Bogard D D, Shih C-Y, Greshake A, Stöffler D and Eugster O 2001 Ages and geologic histories of Martian meteorites *Chronology and Evolution of Mars* (Berlin: Springer) pp 105–64

[4] Gooding J L, Wentworth S J and Zolensky M E 1991 Aqueous alteration of the Nakhla meteorite *Meteoritics* **26** 135–43

[5] Bridges J C and Schwenzer S P 2012 The nakhlite hydrothermal brine on Mars *Earth Planet. Sci. Lett.* **359** 117–23

[6] Hicks L J, Bridges J C and Gurman S J 2014 Ferric saponite and serpentine in the nakhlite Martian meteorites *Geochim. Cosmochim. Acta* **136** 194–210

[7] Aramendia J, Gomez-Nubla L, Castro K, de Vallejuelo S F-O, Arana G, Maguregui M, Baonza V G, Medina J, Rull F and Madariaga J M 2018 Overview of the techniques used for the study of non-terrestrial bodies: proposition of novel non-destructive methodology *Trends Anal. Chem.* **98** 36–46

[8] Rubin A E 1997 Mineralogy of meteorite groups *Meteorit. Planet. Sci.* **32** 231–47

[9] Senesi G S 2014 Laser-Induced breakdown spectroscopy (LIBS) applied to terrestrial and extraterrestrial analogue geomaterials with emphasis to minerals and rocks *Earth-Sci. Rev.* **139** 231–67

[10] Brunetti B, Miliani C, Rosi F, Doherty B, Monico L, Romani A and Sgamellotti A 2017 Non-invasive investigations of paintings by portable instrumentation: the MOLAB experience *Analytical Chemistry for Cultural Heritage* (Berlin: Springer) pp 41–75

[11] Hirayama A, Abe Y, van Loon A, De Keyser N, Noble P, Vanmeert F, Janssens K, Tantrakarn K, Taniguchi K and Nakai I 2018 Development of a new portable x-ray powder diffractometer and its demonstration to on-site analysis of two selected old master paintings from the Rijksmuseum *Microchem. J.* **138** 266–72

[12] Burkett D A, Graham I T and Ward C R 2015 The application of portable x-ray diffraction to quantitative mineralogical analysis of hydrothermal systems *Can. Mineral.* **53** 429–54

[13] Hansford G M 2011 Optimization of a simple x-ray diffraction instrument for portable and planetary applications *Nucl. Instrum. Methods Phys. Res.* A **632** 81–8

[14] Bish D, Blake D, Sarrazin P, Treiman A, Hoehler T, Hausrath E M, Midtkandal I and Steele A 2007 Field XRD/XRF mineral analysis by the MSL CheMin instrument *Lunar and Planetary Science Conf.* **vol 1338** p 1163

[15] Gianoncelli A, Castaing J, Ortega L, Dooryhee E, Salomon J, Walter P, Hodeau J-L and Bordet P 2008 A portable instrument for *in situ* determination of the chemical and phase compositions of cultural heritage objects *X-Ray Spectrom.* **37** 418–23

[16] Van de Voorde L, Van Pevenage J, De Langhe K, De Wolf R, Vekemans B, Vincze L, Vandenabeele P and Martens M P 2014 Non-destructive *in situ* study of 'Mad Meg' by Pieter Bruegel the Elder using mobile x-ray fluorescence, x-ray diffraction and Raman spectrometers *Spectrochim. Acta* B **97** 1–6

[17] Bish D *et al* 2014 The first x-ray diffraction measurements on Mars *IUCrJ* **1** 514–22

[18] Whitfield P and Mitchell L 2009 Phase identification and quantitative methods *Principles and Applications of Powder Diffraction* ed A Clearfield, J H Reibenspies and N Bhuvanesh (New York: Wiley) pp 226–60

[19] Waseda Y, Matsubara E and Shinoda K 2011 *X-ray Diffraction Crystallography: Introduction, Examples and Solved Problems* (Berlin: Springer)

[20] Klug H P and Alexander L E 1974 X-ray diffraction procedures: for polycrystalline and amorphous materials *X-Ray Diffraction Procedures: For Polycrystalline and Amorphous Materials* 2nd edn ed H P ed Klug and L E Alexander (New York: Wiley) p 992

[21] Davis B L 1987 Quantitative determination of mineral content of geological samples by x-ray diffraction: discussion *Am. Mineral.* **72** 438–41

[22] Chipera S J and Bish D L 2013 Fitting full x-ray diffraction patterns for quantitative analysis: a method for readily quantifying crystalline and disordered phases *Adv. Mater. Phys. Chem.* **3** 47

[23] Zhou X, Liu D, Bu H, Deng L, Liu H, Yuan P, Du P and Song H 2018 XRD-based quantitative analysis of clay minerals using reference intensity ratios, mineral intensity factors, Rietveld, and full pattern summation methods: a critical review *Solid Earth Sci.* **3** 16–29

[24] Rietveld H M 1969 A profile refinement method for nuclear and magnetic structures *J. Appl. Crystallogr.* **2** 65–71

[25] NASA, Mars Science Laboratory Landing https://www.jpl.nasa.gov/news/press_kits/MSLLanding.pdf

[26] Yavnel A A 1960 Classification of meteorites according to their chemical composition *Int. Geol. Rev.* **2** 380–96

[27] Terho M, Pesonen L J, Kukkonen I T and Bukovanska M 1993 The petrophysical classification of meteorites *Stud. Geophys. Geod.* **37** 65–82

[28] Dementieva A and Ostrogorsky D 2012 *Meteorites and Asteroids: Classification, Geology and Exploration* (Hauppauge, NY: Nova Science)

[29] Griffith R and Hansen S 2017 *Meteorites: Classification, Chemical Composition and Impacts* (Hauppauge, NY: Nova Science)

[30] Krot A N, Keil K, Scott E R D, Goodrich C A and Weisberg M K 2014 Classification of meteorites and their genetic relationships, meteorites and cosmochemical processes *Treatise on Geochemistry* vol 1 (Amsterdam: Elsevier) pp 1–63

[30] Kee T P, Bryant D E, Herschy B, Marriott K E R, Cosgrove N E, Pasek M A, Atlas Z D and Cousins C R 2013 Phosphate activation via reduced oxidation state phosphorus (P). Mild routes to condensed-P energy currency molecules *Life* **3** 386–402

[32] Pasek M A 2017 Schreibersite on the early Earth: scenarios for prebiotic phosphorylation *Geosci. Front.* **8** 329–35

[33] Rubin A E and Ma C 2017 Meteoritic minerals and their origins *Chem. Erde-Geochem.* **77** 325–85

[34] Giacovazzo C (ed) 2008 *Fundamentals of Crystallography* (Oxford: International Union of Crystallography)

[35] Dawson K R, Maxwell J A and Parsons D E 1960 A description of the meteorite which fell near Abee, Alberta, Canada *Geochim. Cosmochim. Acta* **21** 127–44

[36] Fuchs L H 1962 Occurrence of whitlockite in chondritic meteorites *Science* **137** 425–26

[37] Lipschutz M E 1962 Diamonds in the Dyalpur meteorite *Science* **138** 1266–7

[38] Frondel C and Klein C 1965 Ureyite, $NaCrSi_2O_6$: a new meteoritic pyroxene *Science* **149** 742–4

[39] Lipschutz M E and Jaeger R R 1966 X-ray diffraction study of minerals from shocked iron meteorites *Science* **152** 1055–7

[40] Bass M N 1971 Montmorillonite and serpentine in Orgueil meteorite *Geochim. Cosmochim. Acta* **35** 139–47

[41] Putnis A and Price G D 1979 High-pressure $(Mg,Fe)_2SiO_4$ phases in the Tenham chondritic meteorite *Nature* **280** 217

[42] Post J E and Buchwald V F 1991 Crystal structure refinement of akaganéite *Am. Mineral.* **76** 272–7

[43] Britvin S N, Rudashevsky N S, Krivovichev S V, Burns P C and Polekhovsky Y S 2002 Allabogdanite, $(Fe,Ni)_2P$, a new mineral from the Onello meteorite: the occurrence and crystal structure *Am. Mineral.* **87** 1245–9

[44] Zolensky M, Gounelle M, Mikouchi T, Ohsumi K, Le L, Hagiya K and Tachikawa O 2008 Andreyivanovite: a second new phosphide from the Kaidun meteorite *Am. Mineral.* **93** 1295–9

[45] Chukanov N V, Pekov I V, Levitskaya L A and Zadov A E 2009 Droninoite, $Ni_3Fe^{3+}Cl(OH)_8 \cdot 2H_2O$, a new hydrotalcite-group mineral species from the weathered Dronino meteorite *Geol. Ore Deposits* **51** 767–73

[46] International Centre for Diffraction Data, http://www.icdd.com

[47] Ma C and Rossman G R 2009 Tistarite, Ti_2O_3, a new refractory mineral from the Allende meteorite *Am. Mineral.* **94** 841–4

[48] Ma C and Rossman G R 2009 Davisite, $CaScAlSiO_6$, a new pyroxene from the Allende meteorite *Am. Mineral.* **94** 845–8

[49] Ma C and Rossman G R 2009 Grossmanite, $CaTi^{3+}AlSiO_6$, a new pyroxene from the Allende meteorite *Am. Mineral.* **94** 1491–4

[50] Kubo T, Kimura M, Kato T, Nishi M, Tominaga A, Kikegawa T and Funakoshi K-i 2010 Plagioclase breakdown as an indicator for shock conditions of meteorites *Nat. Geosci.* **3** 41

[51] Gismelseed A M, Abdu Y A, Shaddad M H, Verma H C and Jenniskens P 2014 Fe-bearing phases in a ureilite fragment from the asteroid 2008 TC_3 (= Almahata Sitta meteorites): a combined Mössbauer spectroscopy and x-ray diffraction study *Meteorit. Planet. Sci.* **49** 1485–93

[52] Tschauner O, Ma C, Beckett J R, Prescher C, Prakapenka V B and Rossman G R 2014 Discovery of bridgmanite, the most abundant mineral in Earth, in a shocked meteorite *Science* **346** 1100–02

[53] Gabrig Turbay Rangel C V, D'Azeredo Orlando M T, De Morisson Valeriano C and de Oliveira Chaves A 2017 The Varre-Sai chondrite, a Brazilian fall: petrology and geochemistry *Int. Geol. Rev.* **59** 1966–73

[54] Rossi M, Ghiara M R, Chita G and Capitelli F 2011 Crystal-chemical and structural characterization of fluorapatites in ejecta from Somma-Vesuvius volcanic complex *Am. Mineral.* **96** 1828–37

[55] Tait K T, Barkley M C, Thompson R M, Origlieri M J, Evans S H, Prewitt C T and Yang H 2011 Bobdownsite, a new mineral species from Big Fish River, Yukon, Canada, and its structural relationship with whitlockite-type compounds *Can. Mineral.* **49** 1065–78

[56] Jolliff B L, Hughes J M, Freeman J J and Zeigler R A 2006 Crystal chemistry of lunar merrillite and comparison to other meteoritic and planetary suites of whitlockite and merrillite *Am. Mineral.* **91** 1583–95

[57] Cheng H, Chabok R, Guan X, Chawla A, Li Y, Khademhosseini A and Jang H L 2018 Synergistic interplay between the two major bone minerals, hydroxyapatite and whitlockite nanoparticles, for osteogenic differentiation of mesenchymal stem cells *Acta Biomater.* **69** 342–51

[58] Rakovan J F and Pasteris J D 2015 A technological gem: materials, medical, and environmental mineralogy of apatite *Elements* **11** 195
[59] El Khouri A, Elaatmani M, Della Ventura G, Sodo A, Rizzi R, Rossi M and Capitelli F 2017 Synthesis, structure refinement and vibrational spectroscopy of new rare-earth tricalcium phosphates $Ca_9RE(PO_4)_7$ (RE = La, Pr, Nd, Eu, Gd, Dy, Tm, Yb) *Ceram. Int.* **43** 15645–53
[60] Capitelli F, Rossi M, ElKhouri A, Elaatmani M, Corriero N, Sodo A and Della Ventura G 2018 Synthesis, structural model and vibrational spectroscopy of lutetium tricalcium phosphate $Ca_9Lu(PO_4)_7$ *J. Rare Earths* **36** 1162
[61] Hughes J M, Jolliff B L and Gunter M E 2006 The atomic arrangement of merrillite from the Fra Mauro Formation, Apollo 14 lunar mission: the first structure of merrillite from the Moon *Am. Mineral.* **91** 1547–52
[62] Xie X, Minitti M E, Chen M, Mao H-K, Wang D, Shu J and Fei Y 2003 Tuite, γ-$Ca_3(PO_4)_2$: a new mineral from the Suizhou L6 chondrite *Eur. J. Mineral.* **15** 1001–5
[63] Xie X, Zhai S, Chen M and Yang H 2013 Tuite, γ-$Ca_3(PO_4)_2$, formed by chlorapatite decomposition in a shock vein of the Suizhou L6 chondrite *Meteorit. Planet. Sci.* **48** 1515–23
[64] Adcock C T *et al* 2017 Shock-transformation of whitlockite to merrillite and the implications for meteoritic phosphate *Nat. Commun.* **8** 14667
[65] Levine D and Steinhardt P J 1984 Quasicrystals: a new class of ordered structures *Phys. Rev. Lett.* **53** 2477
[66] Bindi L, Steinhardt P J, Yao N and Lu P J 2011 Icosahedrite, $Al_{63}Cu_{24}Fe_{13}$, the first natural quasicrystal *Am. Mineral.* **96** 928–31
[67] Bindi L *et al* 2015 Decagonite, $Al_{71}Ni_{24}Fe_5$, a quasicrystal with decagonal symmetry from the Khatyrka CV3 carbonaceous chondrite *Am. Mineral.* **100** 2340–3
[68] Cremers D A and Radziemski L J 2006 *Handbook of Laser-Induced Breakdown Spectroscopy* 2nd edn (Chichester: Wiley)
[69] Miziolek A W, Palleschi V and Schechter I 2006 *Laser Induced Breakdown Spectroscopy* (Cambridge: Cambridge University Press)
[70] Musazzi S and Perini U 2014 *Laser-induced Breakdown Spectroscopy Springer Series in Optical Sciences* vol 182 (Berlin: Springer)
[71] Fortes F J and Laserna J J 2010 The development of fieldable laser-induced breakdown spectrometer: no limits on the horizon *Spectrochim. Acta* B **65** 975–90
[72] Radziemski L and Cremers A 2012 A brief history of laser-induced breakdown spectroscopy: from the concept of atoms to LIBS *Spectrochim. Acta* B **87** 3–10
[73] Connors B, Somers A and Day D 2016 Application of handheld laser-induced breakdown spectroscopy (LIBS) to geochemical analysis *Appl. Spectrosc.* **70** 810–5
[74] Harmon R S, Hark R R, Throckmorton C S, Rankey E C, Wise M A, Somers A M and Collins L M 2017 Geochemical fingerprinting by handheld laser-induced breakdown spectroscopy *Geostand. Geoanal. Res.* **41** 563–84
[75] Gómez-Nubla L, Aramendia J, de Vallejuelo S F-O and Madariaga J M 2018 Analytical methodology to elemental quantification of weathered terrestrial analogues to meteorites using a portable laser-induced breakdown spectroscopy (LIBS) instrument and partial least squares (PLS) as multivariate calibration technique *Microchem. J.* **137** 392–401
[76] Fortes F J, Moros J, Lucena P, Cabalín L M and Laserna J J 2012 Laser-induced breakdown spectroscopy *Anal. Chem.* **85** 640–69

[77] Bousquet B *et al* 2008 Development of a mobile system based on laser-induced breakdown spectroscopy and dedicated to *in situ* analysis of polluted soils *Spectrochim. Acta* B **63** 1085–90

[78] Laserna J, Reyes R F, González R, Tobaria L and Lucena P 2009 Study on the effect of beam propagation through atmospheric turbulence on standoff nanosecond laser induced breakdown spectroscopy measurements *Opt. Express* **17** 10265–76

[79] Weisberg A, Craparo J, De Saro R and Pawluczyk R 2010 Comparison of a transmission grating spectrometer to a reflective grating spectrometer for standoff laser-induced breakdown spectroscopy measurements *Appl. Opt.* **49** C200–10

[80] Wiens R C *et al* 2012 The ChemCam instrument suite on the Mars Science Laboratory (MSL) rover: body unit and combined system tests *Space Sci. Rev.* **170** 167–227

[81] Tognoni E, Palleschi V, Corsi M, Cristoforetti G, Omenetto N, Gornushkin I, Smith B W and Winefordner J D 2006 From sample to signal in laser-induced breakdown spectroscopy: a complex route to quantitative analysis *Laser Induced Breakdown Spectroscopy (LIBS) Fundamentals and Applications* ed A W Miziolek, V Palleschi and I Schechter (Cambridge: Cambridge University Press) pp 122–70

[82] Harmon R S, De Lucia F C, Miziolek A W, McNesby K L, Walters R A and French P D 2005 Laser-induced breakdown spectroscopy (LIBS)- an emerging field-portable sensor technology for real-time, *in situ* geochemical and environmental analysis *Geochem. Explor. Environ. Anal.* **5** 21–8

[83] Rauschenbach I, Lazic V, Pavlov S, Hübers H-W and Jessberger E K 2008 Laser induced breakdown spectroscopy on soils and rocks: influence of the sample temperature, moisture and roughness *Spectrochim. Acta* B **63** 1205–15

[84] Eppler A S, Cremers D A, Hickmott D D, Ferris M J and Koskelo A C 1996 Matrix effects in the detection of Pb and Ba in soils using laser-induced breakdown spectroscopy *Appl. Spectrosc.* **50** 1175–81

[85] Gornushkin S I, Gornushkin I B, Anzano J M, Smith B W and Winefordner J D 2002 Effective normalization technique for correction of matrix effects in laser-induced breakdown spectroscopy detection of magnesium in powdered samples *Appl. Spectrosc.* **56** 433–6

[86] Anzano J M, Villoria M A, Ruíz-Medina A and Lasheras R J 2006 Laser-induced breakdown spectroscopy for quantitative spectrochemical analysis of geological materials: effects of the matrix and simultaneous determination *Anal. Chim. Acta* **575** 230–5

[87] Michel A P M, Lawrence-Snyder M, Angel S M and Chave A D 2007 Laser-induced breakdown spectroscopy of bulk aqueous solutions at oceanic pressures: evaluation of key measurement parameters *Appl. Opt.* **46** 2507–15

[88] Ciucci A, Corsi M, Palleschi V, Rastelli S, Salvetti A and Tognoni E 1999 New procedure for quantitative elemental analysis by laser-induced plasma spectroscopy *Appl. Spectrosc.* **53** 960–4

[89] Corsi M, Cristoforetti G, Hidalgo M, Legnaioli S, Palleschi V, Salvetti A, Tognoni E and Vallebona C 2006 Double pulse calibration-free laser-induced breakdown spectroscopy: a new technique for *in situ* standard-less analysis of polluted soils *Appl. Geochem.* **21** 748–55

[90] Tognoni E, Cristoforetti G, Legnaioli S and Palleschi V 2010 Calibration-free laser-induced breakdown spectroscopy: state of the art *Spectrochim. Acta* B **65** 1–14

[91] Galbács G 2015 A critical review of recent progress in analytical laser-induced breakdown spectroscopy *Anal. Bioanal. Chem.* **407** 7537–62

[92] Amato G, Cristoforetti G, Legnaioli S, Lorenzetti G, Palleschi V, Sorrentino F and Tognoni E 2010 Progress towards an unassisted element identification from laser induced breakdown spectra with automatic ranking techniques inspired by text retrieval *Spectrochim. Acta* B **65** 664–70

[93] Clegg S M, Sklute E, Dyar M D, Barefield J E and Wiens R C 2009 Multivariate analysis of remote laser-induced breakdown spectroscopy spectra using partial least squares, principal component analysis, and related techniques *Spectrochim. Acta* B **64** 79–88

[94] Death D L, Cunningham A P and Pollard L J 2009 Multi-element and mineralogical analysis of mineral ores using laser induced breakdown spectroscopy and chemometric analysis *Spectrochim. Acta* B **64** 1048–58

[95] Wold S, Sjöström M and Eriksson L 2001 PLS-regression: a basic tool of chemometrics *Chemometr. Intell. Lab. Syst.* **58** 109–30

[96] Naes T, Isaksson T, Fearn T and Davies T 2002 *A User-friendly Guide To multivariate Calibration and Classification* (Chichester: NIR Publications)

[97] Andrade-Garda J M 2013 *Basic Chemometric Techniques in Atomic Spectroscopy* RSC Analytical Spectroscopy Monographs vol 13 (Cambridge: Royal Society of Chemistry)

[98] Tibshirani R 1996 Regression shrinkage and selection via the LASSO *J. R. Stat. Soc.* B **58** 267–88

[99] Rothman A J, Levina E and Zhu J 2010 Sparse multivariate regression with covariance estimation *J. Comput. Graph. Stat.* **19** 947–62

[100] Wiens R *et al* 2013 Pre-flight calibration and initial data processing for the ChemCam laser-induced breakdown spectroscopy instrument on the Mars Science Laboratory rover *Spectrochim. Acta* B **82** 1–27

[101] Anderson R B, Bell J F, Wiens R C, Morris R V and Clegg S M 2012 Clustering and training set selection methods for improving the accuracy of quantitative laser induced breakdown spectroscopy *Spectrochim. Acta* B **70** 24–32

[102] Martin M Z, Wullschleger S D, Garten C T and Palumbo A V 2003 Laser-induced breakdown spectroscopy for the environmental determination of total carbon and nitrogen in soils *Appl. Opt.* **42** 2072–7

[103] Hahn D W, Flower W L and Hencken K R 1997 Discrete particle detection and metal emissions monitoring using laser-induced breakdown spectroscopy *Appl. Spectrosc.* **51** 1836–44

[104] Fabre C, Boiron M-C, Dubessy J and Moissette A 1999 Determination of ions in individual fluid inclusions by laser ablation optical emission spectroscopy: development and applications to natural fluid inclusions *J. Anal. At. Spectrom.* **14** 913–22

[105] Novotný K, Kaiser J, Galiová M, Konečná V, Novotný J, Malina R, Liška M, Kanický V and Otruba V 2008 Mapping of different structures on large area of granite sample using laser-ablation based analytical techniques, an exploratory study *Spectrochim. Acta* B **63** 1139–44

[106] Čtvrtníčková T, Fortes F J, Cabalín L M and Laserna J J 2007 Optical restriction of plasma emission light for nanometric sampling depth and depth profiling of multilayered metal samples *Appl. Spectrosc.* **61** 719–24

[107] Sharma S K, Misra A K, Lucey P G and Lentz R C 2009 A combined remote Raman and LIBS instrument for characterizing minerals with 532 nm laser excitation *Spectrochim. Acta* A **73** 468–76

[108] Hilbk-Kortenbruck F, Noll R, Wintjens P, Falk H and Becker C 2001 Analysis of heavy metals in soils using laser-induced breakdown spectrometry combined with laser-induced fluorescence *Spectrochim. Acta* B **56** 933–45
[109] Lui S L, Godwal Y, Taschuk M T, Tsui Y Y and Fedosejevs R 2008 Detection of lead in water using laser-induced breakdown spectroscopy and laser-induced fluorescence *Anal. Chem.* **80** 1995–2000
[110] Hark R R and Harmon R S 2014 Geochemical fingerprinting using LIBS *Laser-Induced Breakdown Spectroscopy Theory and Applications* (eds) S Musazzi and U Perini (Heidelberg: Springer) pp 309–48
[111] Walid Tawfik Y M 2007 Recent advances in laser induced breakdown spectroscopy as elemental analytical technique for environmental applications and space exploration *Aspects of Optical Sciences and Quantum Information* ed M Abdel-Aty (Kerala: Research Signpost Publishing) pp 1–35
[112] Pasquini C, Cortez J, Silva L M C and Gonzaga F B 2007 Laser induced breakdown spectroscopy *J. Braz. Chem. Soc.* **18** 463–512
[113] Thompson J R, Wiens R C, Barefield J E, Vaniman D T, Newsom H E and Clegg S M 2006 Remote laser-induced breakdown spectroscopy analyses of Dar al Gani 476 and Zagami Martian meteorites *J. Geophys. Res.: Planets* **111** E5
[114] De Giacomo A, Dell'Aglio M, De Pascale O, Longo S and Capitelli M 2007 Laser induced breakdown spectroscopy on meteorites *Spectrochim. Acta* B **62** 1606–11
[115] Dell'Aglio M, De Giacomo A, Gaudiuso R, De Pascale O, Senesi G S and Longo S 2010 Laser induced breakdown spectroscopy applications to meteorites: chemical analysis and composition profiles *Geochim. Cosmochim. Acta* **74** 7329–39
[116] Taylor L A *et al* 2002 Martian meteorite Dhofar 019: A new shergottite *Meteorit. Planet. Sci.* **37** 1107–28
[117] Warren P H, Ulff-Moller F and Kallemeyn G W 2005 New lunar meteorites: impact melt and regolith breccias and large- scale heterogeneities of the upper lunar crust *Meteorit. Planet. Sci.* **40** 1–25
[118] Cousin A, Sautter V, Fabre C, Maurice S and Wiens R C 2012 Textural and modal analyses of picritic basalts with ChemCam laser-induced breakdown spectroscopy *J. Geophys. Res.: Planets* **117** E10002
[119] Plavčan J, Horňáčková M, Grolmusová Z, Kociánová M, Rakovský J and Veis P 2012 Sikhote-Alin meteorite, elemental composition analysis using CF LIBS, *WDS'12 Proceedings of Contributed Papers, Part II* pp 123–7
[120] Dell'Aglio M, De Giacomo A, Gaudiuso R, De Pascale O and Longo S 2014 Laser induced breakdown spectroscopy of meteorites as a probe of the early solar system *Spectrochim. Acta* B **101** 68–75
[121] Horňáčková M, Plavčan J, Rakovský J, Porubčan V, Ozdín D and Veis P 2014 Calibration-free laser induced breakdown spectroscopy as an alternative method for found meteorite fragments analysis *Eur. Phys. J. Appl. Phys.* **66** 10702
[122] Ozdín D, Plavčan J, Horňáčková M, Uher P, Porubčan V, Veis P, Rakovský J, Tóth J, Konečný P and Svoreň J 2015 Mineralogy, petrography, geochemistry, and classification of the Kosice meteorite *Meteorit. Planet. Sci.* **50** 864–79
[123] Kovács J, Sajó I, Márton Z, Jáger V, Hegedüs T, Berecz T, Tóth T, Gyenizse P and Podobni A 2015 Csátalja, the largest H4-5 chondrite from Hungary *Planet. Space Sci.* **105** 94–100

[124] Senesi G S, Tempesta G, Manzari P and Agrosí G 2016 An innovative approach to meteorite analysis by laser-induced breakdown spectroscopy *Geostand. Geoanal. Res.* **40** 533–41

[125] Wiens R C *et al* 2017 Composition and morphology of iron meteorites found in Gale Crater, Mars, *80th Annual Meeting of the Meteoritical Society, LPI Contributions* **vol 1987** p 6168

[126] Meslin P-Y *et al* 2017 Egg Rock encounter: analysis of an iron–nickel meteorite found in Gale crater by Curiosity, *48th Lunar and Planetary Science Conf.* abstract no. 2258

[127] Chennaoui Aoudjehane H, Garvie L A J, Herd C D K, Chen G and Aboulahris M 2013 Agoudal: The most recent iron meteorite from Morocco *76th Meeting of the Meteoritical Society (Edmonton, Canada)* abstract 5026

[128] Grokhovsky V I, Ustyugov V F, Badyukov D D and Nazarov M A 2005 Dronino: An ancient iron meteorite shower in Russia *36th Annual Lunar and Planetary Science Conf., March 14–18, 2005, League City, TX* (Houston, TX: Lunar and Planetary Institute (LIP)) Abstract No.1692

[129] Lanza N L *et al* 2017 Analyzing natural meteorite exteriors with laboratory LIBS for comparison to meteorites encountered by Curiosity in Gale crater, Mars, *80th Annual Meeting of the Meteoritical Society 2017 (LPI Contrib. No. 1987)* p 6402

[130] Ferus M *et al* 2018 Calibration-free quantitative elemental analysis of meteor plasma using reference laser-induced breakdown spectroscopy of meteorite samples *Astron. Astrophys.* **610** A73

[131] Tempesta G, Senesi G S, Manzari P and Agrosí G 2018 New insights on the Dronino iron meteorite by double-pulse micro-laser-induced breakdown spectroscopy *Spectrochim. Acta* B **144** 75–81

[132] Senesi G S, Manzari P, Consiglio A and De Pascale O 2018 A fuzzy logic algorithm for analyzing handheld LIBS spectra of meteorites *49th Lunar and Planetary Science Conf.* Contrib. No. 2083, p 2031

[133] Senesi G S, Manzari P, Tempesta G, Agrosí G, Touchnt A A, Ibhi A and De Pascale O 2018 Handheld laser induced breakdown spectroscopy instrumentation applied to the rapid discrimination between iron meteorites and meteor-wrongs *Geostand. Geoanal. Res.* **42** 607–14

Part II

Hypersonic entry physics

IOP Publishing

Hypersonic Meteoroid Entry Physics

Gianpiero Colonna, Mario Capitelli and Annarita Laricchiuta

Chapter 6

Radiation gas dynamics of centimeter meteoric bodies at an altitude of 80 km

Sergey T Surzhikov

The physical theory of meteor phenomena has been developing for about 100 years [1]. One of the first papers [2] proposed a theory of physical phenomena associated with the flight of meteors, which formed the basis for studying the stratosphere using meteoric astronomy methods, and soon led to the development of a whole new scientific field—the study of a complex of phenomena associated with the motion of cosmic solids in the upper layers of the atmosphere.

At the initial stages of research, it was expected that a closed theory could be created based on the obvious initial data: the size and form of the meteor, its chemical composition, the speed and incidence of the trajectory in the atmosphere, and also on three independent values characterizing the upper layers of the atmosphere: pressure, density and temperature. It was believed that this should be sufficient to determine the height of ignition and extinction of meteors, the spectro-energy characteristics of the radiation of meteors (signatures) and the mass of matter reaching the surface. However, the problem of this formulation remain unresolved, although a considerable number of simplifications are allowed in its solution, which make the mathematical formulation more simple and, in some cases, amenable to an analytical solution.

In [3] the classification of concepts used in the physical theory of meteor phenomena is given. Let us quote it here verbatim [3, p 157]:

> 'Meteors are, in general, small, swiftly moving celestial bodies which, ordinarily invisible, are rendered luminous when they enter the Earth's atmosphere and are heated by impact of the air molecules. They then appear momentarily as 'shooting stars'. Luminous *trails* sometimes persist in the path or track of bright meteors. Very bright meteors are known as *fireballs*, and, if explosive, as *bolides*. Meteors that partially survive the disintegrating effects of the Earth's atmosphere and fall to Earth are known as meteorites. Bodies

doi:10.1088/2053-2563/aae894ch6

> similar to meteors which do not enter the Earth's atmosphere, and are detected by other means, are known as *meteoroids*, and any large-scale distribution of non-self-luminous matter is known as *meteoric matter*.'

Hence it becomes clear that, depending on the objectives of the problem being solved, it is advisable to construct different theories with a specific system of approximations and assumptions.

For the purposes of this work, we will focus on scientific trends that continue to be developed.

The first fundamentally important direction in the physics of meteoric phenomena is the experimental study of the laws governing the emission of cosmic bodies that fly into the Earth's atmosphere. In fact, it was from these observations that this science began. The bulk of the work on the physics of meteor phenomena pertains to the analysis of experimental observational data, from the first observations [1, 4], to the formation of extensive databases of observation results and to the organization of scientific programs for the observation of meteor streams from spacecraft [5–9]. There is no doubt that observational data will continue to be the main source of information for the development of meteor physics.

As our second area of interest, we classify theoretical models that admit analytic solutions. At the initial stage of development of this field, based on the results of observations [4], the first physical theories of meteors were created [2, 10]. These were analytical models based on assumptions about the formation of an air cushion in front of a meteoric body [2] (that is, in fact, on the basis of the model of continuum mechanics) or on the basis of molecular-kinetic theory, according to which a meteoric body heats up, shines and collapses due to numerous impacts of molecules on the meteoroid body [10]. Interesting and instructive is the mutual criticism of the first theoretical models [10, 11]. The results of the development of models in this direction were summarized in the monographs [5, 6, 12, 13], where one can find extensive bibliographies and the results of analyses of observed meteor phenomena.

Two more scientific areas are directly related to the development of meteoric physics. These are the empirical theory of growth curves and high-resolution quantitative spectroscopy [1, 5, 7, 14–17].

A qualitative leap in the development of the theory of meteoric physics was the application of methods of radiation gas dynamics (RadGD) to describe the process of motion of a meteoroid body in the atmosphere and its ablation [18, 19], as well as concomitant studies on the computer simulation of synthetic spectra of meteoric plasma [20]. The fundamental importance of this step in the development of meteoric physics was that all previous theories were based on certain assumptions about the gas-dynamic processes of the formation of a plasma shell around a meteoric body, which is the source of the observed radiation. In computer models of RadGD, it became possible to simulate the gas dynamics and plasmodynamics of the processes accompanying the motion of a meteoroid body in the atmosphere. The paper [18] should be noted, in which many aspects of the transition from approximate theoretical models to computer RadGD models are considered in great detail.

A further important step in the development of theoretical meteoric physics was the incorporation in RadGD models of non-equilibrium physico-chemical processes of the formation of a plasma shell around a meteoroid body and the analysis of the gas-dynamics and electrophysical parameters of its trace [21–23]. We note that the problem of thermalization of the meteoroid plasma shell was formulated in many theoretical works [5], but only the formulation of the non-equilibrium RadGD model allowed the creation of a sufficiently complete computer model, which, however, is so interdisciplinary that today it is not yet possible to talk about its completeness.

The formulation of the problem of non-equilibrium meteor aerophysics is given in [21], and calculations of the radiation gas dynamics of a meteoric body with a characteristic size of 4 cm, moving at a velocity of $V_\infty = 10–40\ \mathrm{km\ s^{-1}}$ at an altitude of 70 km, are performed. The flow regions in the compressed layer near the frontal surface and in the near wake (up to 40 cm behind the body) are considered. The spectral signatures of the plasma air shell are obtained at the observation angles $\theta = 30$ deg and 90 deg with respect to the direction of flight. An important element of this model was the development of a method for solving the spatial problem of transferring selective thermal radiation in the range of wave numbers $\Delta\omega = 10^3–1.2 \times 10^5\ \mathrm{cm^{-1}}$. The role of non-equilibrium relaxation processes in the formation of the plasma shell was shown.

In [22], the flow field around a meteoroid body measuring 4 cm at speeds of 11, 15 and 17 km $\mathrm{s^{-1}}$ at an altitude of 70 km was considered. The products of the destruction of the carbon body are taken into account. The motion of a meteoric body of cometary origin at an altitude of 70 km at a speed of 30 km is considered in [23], with a study of the wake structure up to 1 m long. In the current work, a study of the non-equilibrium aerophysics of a meteoric body 4 cm in size at an altitude of 80 km, flying at a speed of $V_\infty = 10–17\ \mathrm{km\ s^{-1}}$ is carried out. A distinctive feature of the performed calculations is the study of the physicochemical kinetics inside a wake 10 m in length and the regularities of the change in the signature of a meteoric plasma.

6.1 Computer RadGD model

A two-dimensional RadGD model based on a system of self-consistent equations of non-equilibrium physical and chemical mechanics of a partially ionized multicomponent viscous, heat-conducting, selectively radiating and absorbing gas is created and realized. This system includes the equations of continuity and Navier–Stokes (vector function $\mathbf{U}_g$), the equations of continuity of gas mixture components, the equations of conservation of energy of translational motion of particles and conservation of energy accumulated in vibrations of N_2, O_2 and NO molecules in the process of vibrational relaxation (vector function $\mathbf{U}_c$), and also the system of equations of chemical kinetics (6.3) and the transfer of selective thermal radiation (6.4):

$$\frac{\partial \mathbf{U}}{\partial t} + \frac{\partial u\mathbf{U}}{\partial x} + \frac{1}{r}\frac{\partial rv\mathbf{U}}{\partial r} = \mathbf{R}, \quad \mathbf{U} = (\mathbf{U}_g, \mathbf{U}_c), \quad \mathbf{R} = (\mathbf{R}_g, \mathbf{R}_c) \tag{6.1}$$

$$\mathbf{U}_g = (\rho, \rho u, \rho v)^T$$
$$\mathbf{U}_c = (T, \rho_i, E_{V,m})^T \quad i = 1, 2, \ldots, N_s, \quad m = 1, 2, N_V = 3 \tag{6.2}$$

$$\left(\frac{dX_i}{dt}\right)_n = (b_{i,n} - a_{i,n})\left(k_{f,n}\prod_j^{N_s} X_j^{a_{j,n}} - k_{r,n}\prod_j^{N_s} X_j^{b_{j,n}}\right)$$
$$= (b_{i,n} - a_{i,n})(S_{f,n} - S_{r,n}) \quad n = 1, \ldots, N_r \tag{6.3}$$

$$\mathbf{\Omega}\frac{\partial J_\omega(\mathbf{r}, \mathbf{\Omega})}{\partial \mathbf{r}} + \boldsymbol{\kappa}_\omega(\mathbf{r})J_\omega(\mathbf{r}, \mathbf{\Omega}) = j_\omega(\mathbf{r}), \tag{6.4}$$

where

$$\mathbf{R}_g = \left(0; -\frac{\partial p}{\partial x} + S_{\mu,x}; -\frac{\partial p}{\partial r} + S_{\mu,r}\right)^T$$

$$\mathbf{R}_c = \begin{bmatrix} R_T/\rho c_p \\ -\nabla\rho\mathbf{J}_i - \rho_i\nabla\mathbf{V} + \dot{\varpi} \\ \dot{e}_{V,m} - E_{V,m}\nabla\mathbf{V} - \nabla E_{V,m}\mathbf{J}_{i(m)}, \end{bmatrix}$$

$$S_{\mu,x} = -\frac{2}{3}\frac{\partial}{\partial x}(\mu\nabla\mathbf{V}) + \frac{1}{r}\frac{\partial}{\partial r}\left[r\mu\left(\frac{\partial v}{\partial x} + \frac{\partial u}{\partial r}\right)\right] + 2\frac{\partial}{\partial x}\mu\frac{\partial u}{\partial x},$$

$$S_{\mu,r} = -\frac{2}{3}\frac{\partial}{\partial r}(\mu\nabla\mathbf{V}) + \frac{\partial}{\partial x}\left[\mu\left(\frac{\partial v}{\partial x} + \frac{\partial u}{\partial r}\right)\right] + 2\frac{\partial}{\partial r}\mu\frac{\partial v}{\partial r} + 2\mu\frac{\partial}{\partial r}\frac{v}{r},$$

$$R_T = \nabla(\lambda\nabla T) + \nabla\mathbf{q}_R + \frac{\partial p}{\partial t} + \mathbf{V}\cdot\nabla p + \Phi_\mu + Q_V - \sum_{i=1}^{N_s} h_i\dot{\varpi}_i + \sum_{i=1}^{N_s}\rho c_{p,i}D_i(\nabla Y_i \cdot \nabla T),$$

$$\Phi_\mu = \mu\left[2\left(\frac{\partial u}{\partial x}\right)^2 + 2\left(\frac{\partial v}{\partial y}\right)^2 + \left(\frac{\partial v}{\partial x} + \frac{\partial u}{\partial y}\right)^2 - \frac{2}{3}\left(\frac{\partial u}{\partial x} + \frac{\partial v}{\partial y}\right)^2\right],$$

where t is time; x and r are the orthogonal cylindrical coordinates; u, v are the projections of the velocity vector $\mathbf{V}$ on the x- and r-axes; p, ρ are the pressure and density; T is the temperature of the translational movement of particles; μ, λ are the dynamic viscosity coefficient and coefficient of thermal conductivity; c_p is the specific heat of the mixture at constant pressure; $c_p = \sum_i^{N_s} Y_i c_{p,i}$; N_s is the number of chemical components of the gas mixture; $a_{i,n}$, $b_{i,n}$ are the stoichiometric coefficients of the nth chemical reaction; X_i is the volume-molar concentration of the ith component; N_r is the number of chemical reactions; $S_{f,n}$, $S_{r,n}$ are the speeds of direct and reverse reactions; $k_{f,n}$, $k_{r,n}$ are the rate constants of forward and backward reactions;

$c_{p,i}$, h_i, Y_i are the specific heat at constant pressure, associated with the translational and rotational degrees of freedom, enthalpy and mass fraction of the ith component of the mixture; $\dot{\varpi} = M_i \sum_{n=1}^{N_r}(dX_i/dt)_n$, $\mathbf{J}_i = -D_i \nabla Y_i$ and D_i are the mass speed of chemical transformations, the diffusion transport velocity vector and the effective diffusion coefficient of the ith component of the mixture; $Q_V = -\sum_{m=1}^{N_V} \dot{e}_{V,m}$ is the volume power of heat release due to the vibrational relaxation processes in the gas mixture; N_V is the number of vibrational modes ($N_V = 3$; $m = 1$ for vibrational energy N_2, $m = 2$ for O_2, $m = 3$ for NO); $\dot{e}_{V,m}$ is the source of vibrational energy in the mth mode; $e_{V,m}$ is the specific energy of the vibrational motion in the mth vibrational mode of the ith component of the gas mixture $e_{V,m} = R_{I(m)}\theta_m/[\exp(\theta/T_{V,m}) - 1]$; $R_{i(m)} = R_0/M_{i(m)}$ and R_0 are the gas and universal gas constants; $T_{V,m}$ is the vibrational temperature of the mth vibrational mode; $\rho_{i(m)}$, $D_{i(m)}$, $\mathbf{J}_{i(m)}$ and $M_{i(m)}$ are the density, the effective diffusion coefficient in a multicomponent gas mixture, the diffusion transport velocity vector and the molecular weight of the ith component of a gas mixture possessing the mth mode of oscillatory motion; θ_m is the characteristic vibrational temperature; $J_\omega(\mathbf{r}, \mathbf{\Omega})$, $\kappa_\omega(\mathbf{r})$ and $j_\omega(\mathbf{r})$ are the spectral intensity of the radiation, the volume spectral absorption and emission coefficient; $\mathbf{r}$, $\mathbf{\Omega}$ are the radius vector of the point of the computational domain, in which the solution of the radiation transfer equation and the unit radiation propagation vector are sought; $\omega = 10^4/\lambda$ is the wave number of thermal radiation in cm^{-1}, λ is the wavelength in μm; $\mathbf{q}_R = \int_{\Delta\omega_{\text{tot}}} \mathbf{q}_{R,\omega} d\omega = \int_{\Delta\omega_{\text{tot}}} d\omega \int_{4\pi} J_\omega(\mathbf{r}, \mathbf{\Omega})\mathbf{\Omega} d\Omega$ is the density vector of the radiation integral with respect to the spectrum, $\Delta\omega_{\text{tot}} = \omega_{\max} - \omega_{\min} = 150\,000-1000$ cm^{-1}.

The closing relations for the system of equations to be solved include the thermal and caloric equations of the state of an ideal gas,

$$p = \rho R_0 T/M_\Sigma \quad e_i = \int_{T_0}^{T} c_{V,i} dT + e_{i,0},$$

where $M_\Sigma^{-1} = \sum_i^{N_s} Y_i/M_i$, $e = \sum_i^{N_s} Y_i e_i$, $e_{i,0}$ is the internal energy at T_0, $c_{V,i}$ is the specific heat at a constant volume of the ith component and M_Σ is the total molecular weight of the gas.

In the incident flow with a velocity V_∞, the parameters of the atmosphere were set for a given flight altitude of a meteoric body. On its surface, the conditions for aerodynamic braking, absolute catalyticity and recombination for charged particles were set. The boundary conditions in the output section $x = x_{\max}$ were given in the form $\frac{\partial \Psi}{\partial \xi} = 0$, where $\Psi = \{u, v, T, \rho, Y_i, e_{V,m}\}$; ξ is the coordinate line adjusted to the current lines. At the boundary with the streamlined body, a temperature T_w was determined by the model of an equilibrium radiating surface $\varepsilon\sigma T_w^4 = |q_w|$, where ε, σ are the degree of blackness of the surface ($\varepsilon = 0.8$) and the Stefan–Boltzmann constant; $q_w = q_c + q_R$ is the density of the total heat flux that heats the surface. It was assumed $T_V = T_w$. The densities of the convective and integral radiative heat fluxes on a surface with a local normal were determined by the formulas

$$q_c = q_t + q_d = -\lambda_w(\mathbf{n} \cdot \nabla T)_w - \sum_i^{N_s} \rho_i h_i(\mathbf{n} \cdot \mathbf{J}_{iw}) \tag{6.5}$$

$$q_R = -\int_{4\pi} d\Omega \int_{\Delta\omega_{\text{tot}}} J_\omega(\mathbf{r}_w, \mathbf{\Omega})(\mathbf{n} \cdot \mathbf{\Omega})_w d\omega. \tag{6.6}$$

A description of the models of vibrational relaxation, non-equilibrium dissociation, the transport properties of a multicomponent gas and chemical kinetics, as well as a numerical method for integrating the system of equations (6.1)–(6.3), including details of integrating the transfer equation for selective thermal radiation (6.4), taking into account the fine structure of atomic lines and RadGD interactions, are given in [21–23].

6.2 Numerical simulation results

Before analyzing the results of numerical simulation, let us justify the possibility of using the model of a continuous medium for the conditions considered in this paper. The mean free path of molecules at a height $H = 80$ km is $L_\infty = 0.44$ cm. This means that the Knudsen number for the unperturbed atmosphere is $Kn_\infty = 0.1$, so that the conditions in question are related to the flow regime with slip ($0.01 < Kn_\infty < 1$). We note that the results of calculations of the aerothermodynamics of spacecraft in conditions of strong sparseness of the atmosphere have shown that the model of a continuous medium can be used, in particular with regard to the configuration of the perturbed flow region, which is required in this work.

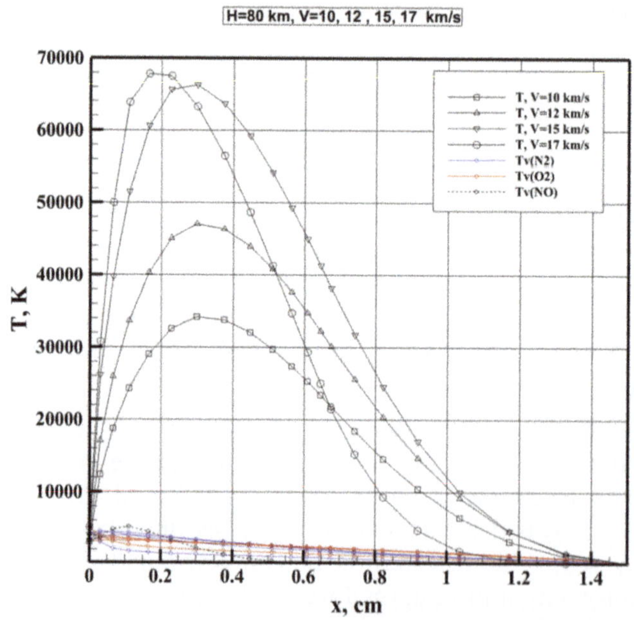

Figure 6.1. Translational and vibrational temperature distributions along the stagnation line.

Figure 6.1 shows the distributions of the translational and vibrational temperatures of N_2, O_2 and NO molecules in a compressed layer near the frontal surface of a meteoric body in the velocity range from $V_\infty = 10$ to $17\ \mathrm{km\ s^{-1}}$. The main feature is that the vibrational temperature does not have time to thermalize. Only at a speed of $V_\infty = 17\ \mathrm{km\ s^{-1}}$, when the translational temperature reaches $T_{tr} = 67\,000$ K, the profile of which is shifting noticeably toward the surface, is a tendency for an increase in the vibrational temperatures observed. From figure 6.2, where the mole

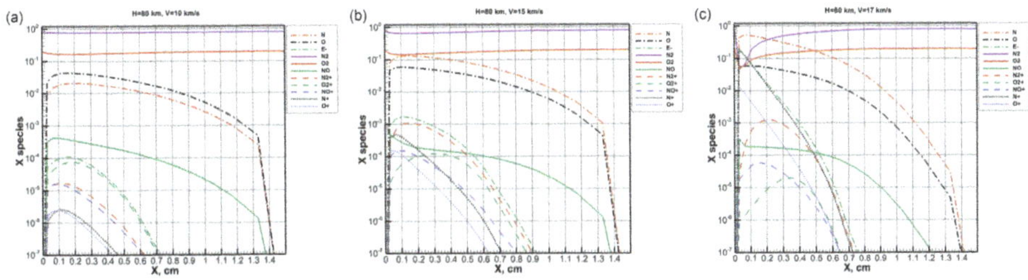

Figure 6.2. Volume fractions of air species along the stagnation line at (a) $V_\infty = 10\ \mathrm{km\ s^{-1}}$, (b) $V_\infty = 15\ \mathrm{km\ s^{-1}}$and (c) $V_\infty = 17\ \mathrm{km\ s^{-1}}$.

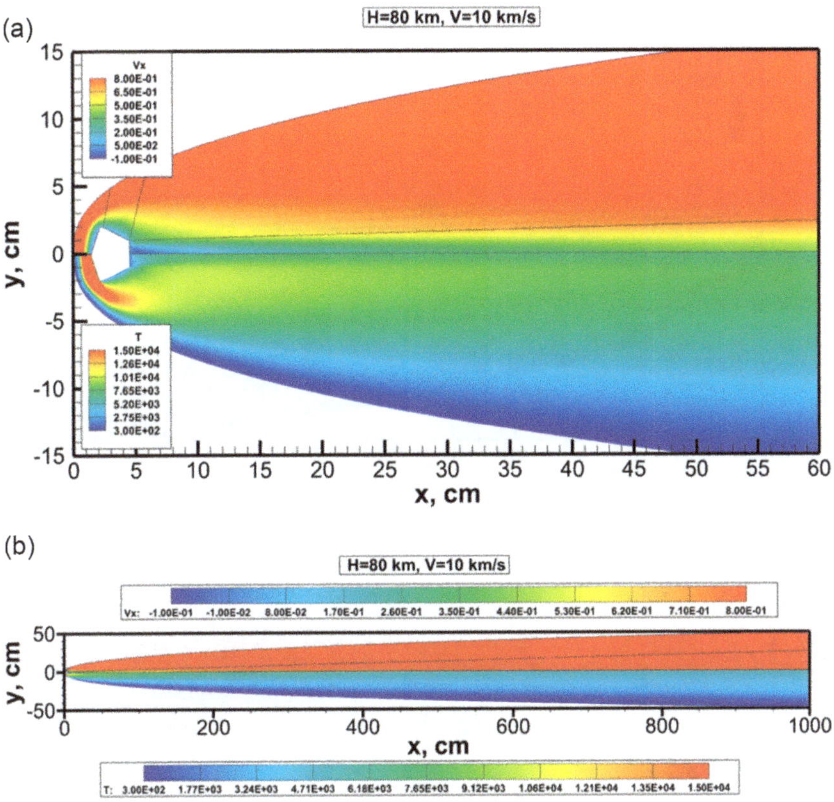

Figure 6.3. Longitudinal velocity $V_x = u/V_\infty$ and translational temperature (in kelvins) at H = 80 km and $V_\infty = 10\ \mathrm{km\ s^{-1}}$.

fraction distributions of high-temperature air components along the critical streamline are shown for velocities of $T = 10$, 15 and 17 km s^{-1}, it is clearly seen that only at the highest velocity $V_\infty = 17$ km s^{-1}, is there a sharp intensification of dissociation of molecular components and ionization of atomic components near the surface.

The following figures compare fields of longitudinal velocity and translational temperature at two speeds of flight $V_\infty = 10$ km s^{-1} (figure 6.3) and $V_\infty = 17$ km s^{-1} (figure 6.4). It is worth noting that if, at a lower velocity at a distance of 10 m from the meteor body, the longitudinal velocity near the axis of symmetry is restored to the value 0.8 from its value in the oncoming stream, then at a higher speed this value is 0.4, that is, at a higher speed, the boundary layer in the wake is significantly longer. In this case, the translational temperature is less than 1000 K.

Figures 6.5 and 6.6 show the fields of the mole fractions of molecular components N_2 and O_2, as well as of the atomic components N and O at a speed of flight of $V_\infty = 17$ km s^{-1}. The degree of dissociation at a distance of 10 m is $\alpha_{N_2} = X_{N_2}/X_{N_2,\infty} = 0.58/0.8 = 0.725$ and $\alpha_{O_2} = X_{O_2}/X_{O_2,\infty} = 0.093/0.197 = 0.47$. At the same distance, the molar fraction of electron concentration is $X_e = 0.045$ (figure 6.7), while the molar fractions of atomic ions $X_{N^+} = 0.03$ and $X_{O^+} = 0.014$ (figure 6.8). The concentration of other ions is much lower.

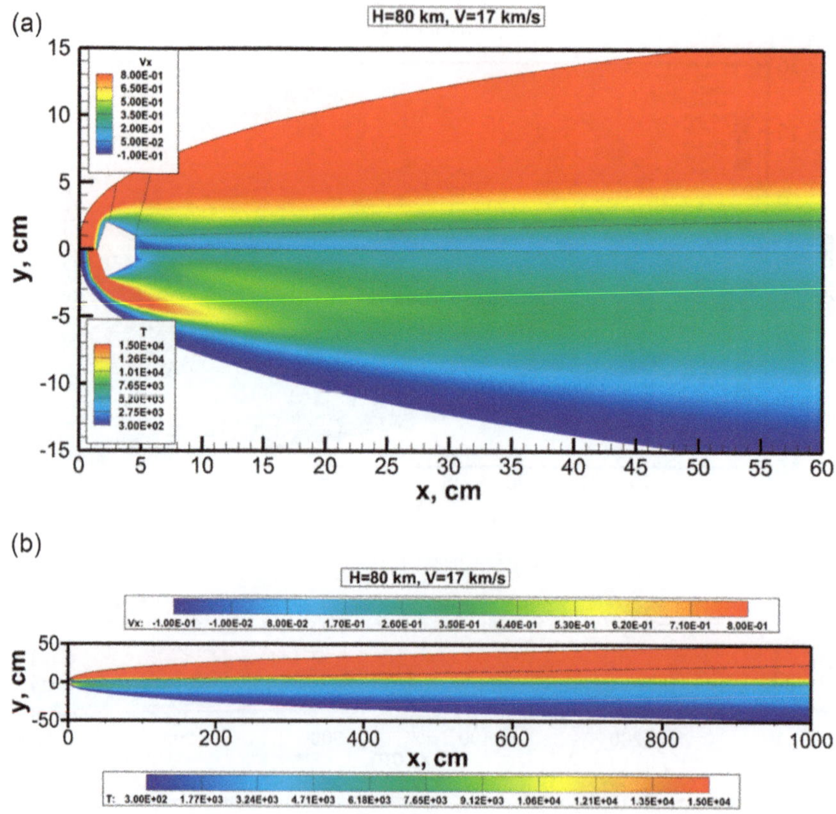

Figure 6.4. Longitudinal velocity $V_x = u/V_\infty$ and translational temperature (in kelvins) at $H = 80$ km and $V_\infty = 17$ km s^{-1}.

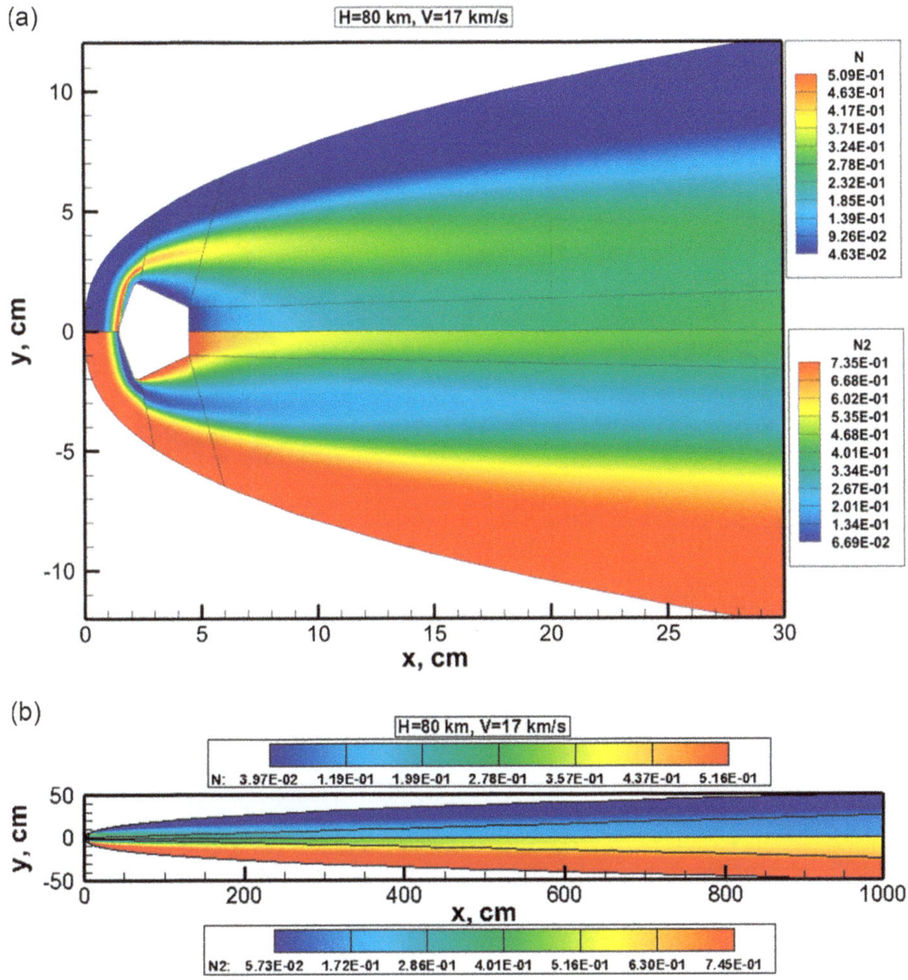

Figure 6.5. Volume fractions of N and N_2 at H = 80 km and V_∞ = 17 km s^{-1}.

The obtained distributions of the gas-dynamics parameters of the compressed layer and the trace behind the meteoroid body made it possible to calculate the spectral signatures of meteoroids at different velocities and different viewing angles with respect to the flight direction (figure 6.9).

It is seen that an increase in the viewing angle from θ = 30° to 90° leads to an increase in the spectral signature by three orders of magnitude. The character of the spectral dependence of the signature indicates that the main radiation mechanism is the radiation from inverse braking processes of the interaction of atomic ions and electrons. When calculating a meteor signature with only the compressed layer and the near trail in mind, the molecular electronic bands are identified in the visible and ultraviolet regions [22]. (In this paper, the emission of atomic lines was not taken into account.)

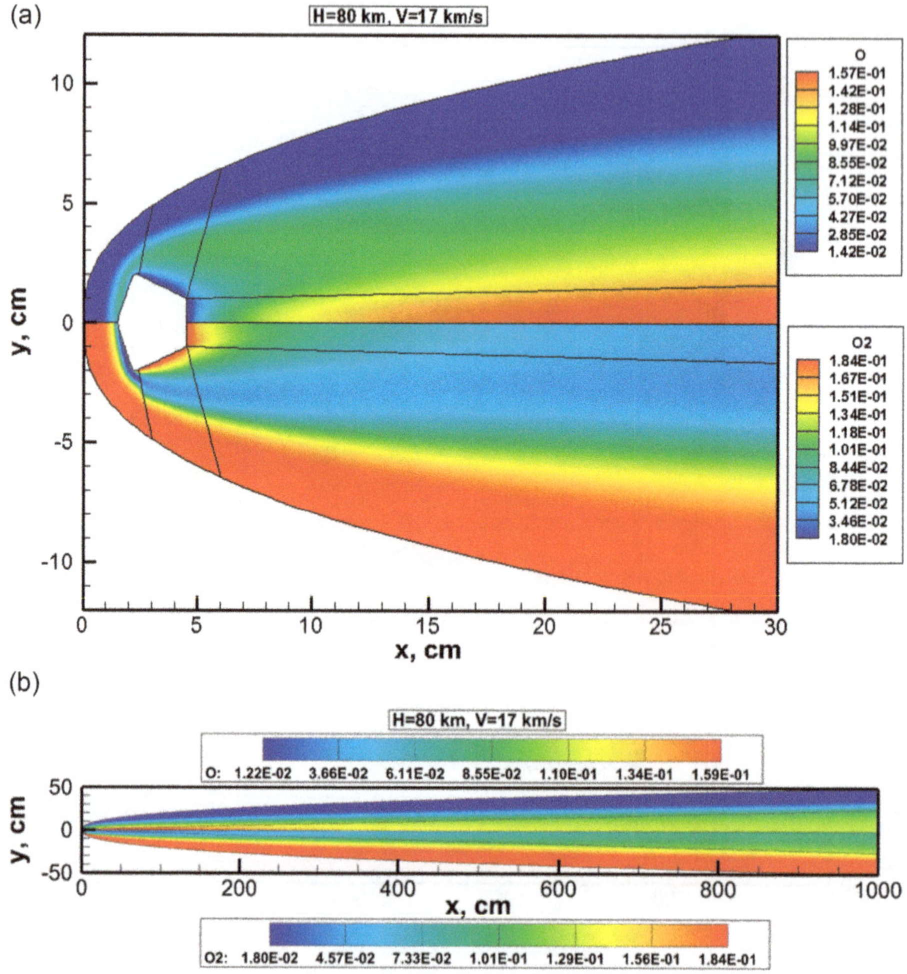

Figure 6.6. Volume fractions of O and O_2 at $H = 80$ km and $V_\infty = 17$ km s^{-1}.

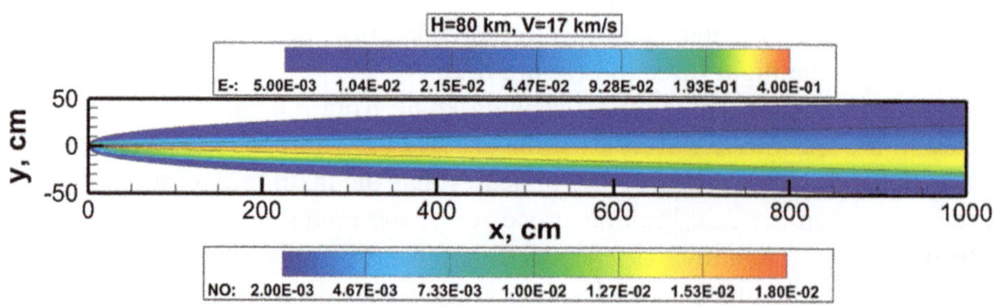

Figure 6.7. Volume fractions of electrons and molecules NO at $H = 80$ km and $V_\infty = 17$ km s^{-1}.

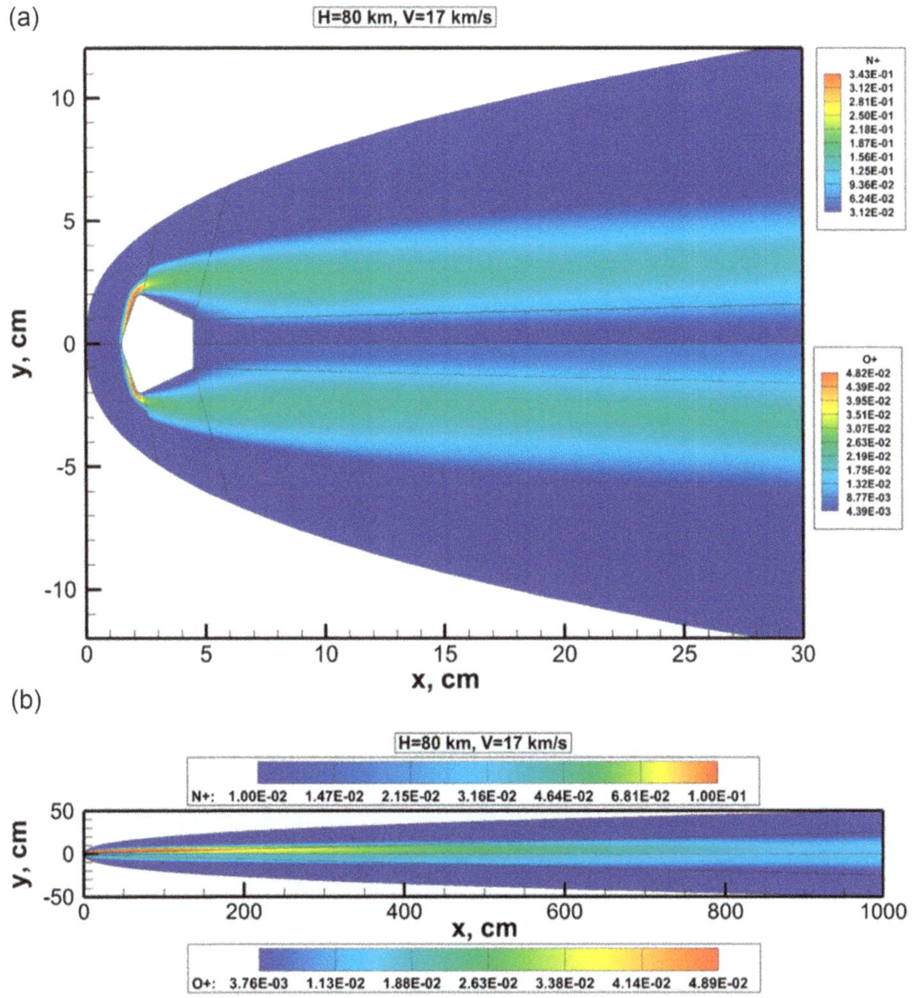

Figure 6.8. Volume fractions of N^+ and O^+ at H = 80 km and V_∞ = 17 km s^{-1}.

As noted in [21], when calculating the spectral signatures of a meteoric body, taking into account the trace, it is expedient to use a special computational procedure, which consists in reformatting the results of numerical simulation of gas-dynamics functions obtained on inhomogeneous grids onto a homogeneous axisymmetric orthogonal grid. The results of such rebuilding are shown in figures 6.10–6.15 for speed V_∞ = 15 km s^{-1}. The fields of translational temperature (figure 6.10) and numerical concentrations of high-temperature air components e^-, N, N^+, O and O^+ are shown here (figures 6.11–6.15, respectively). Note that in the far trace, at a distance of the order of 10 m, the electron concentration is $N_e \sim 10^{11}$cm^{-3}, and the concentrations of nitrogen ions N^+ and oxygen O^+ are 4.3×10^{10} and 4.7×10^{10} cm^{-3}, respectively. Such a concentration of electrons in a meteor track is quite sufficient for studying the patterns of reflection of electromagnetic waves and solving the problem of the

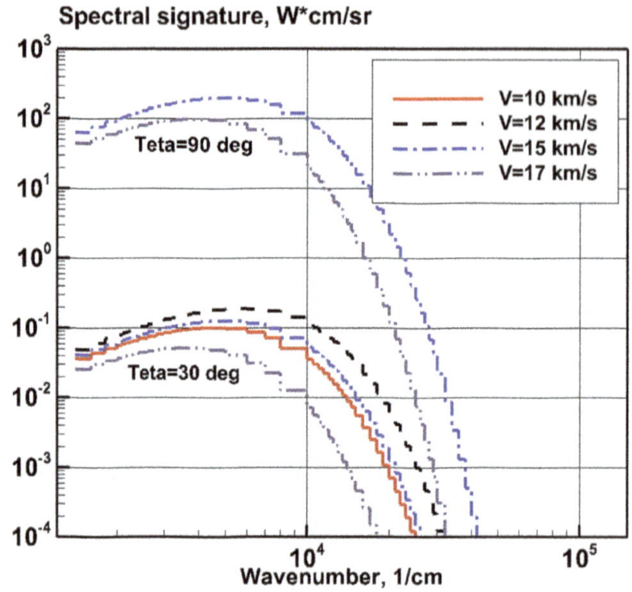

Figure 6.9. Signatures of meteoric bodies of 4 cm at H = 80 km and different velocities.

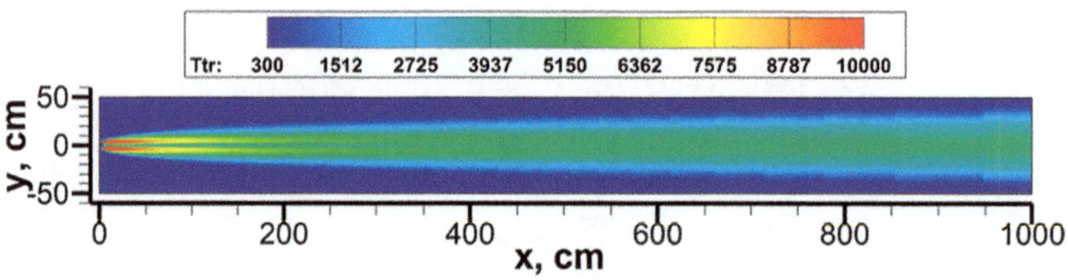

Figure 6.10. Translational temperature in the wake at H = 80 km and V_∞ = 15 km s^{-1}.

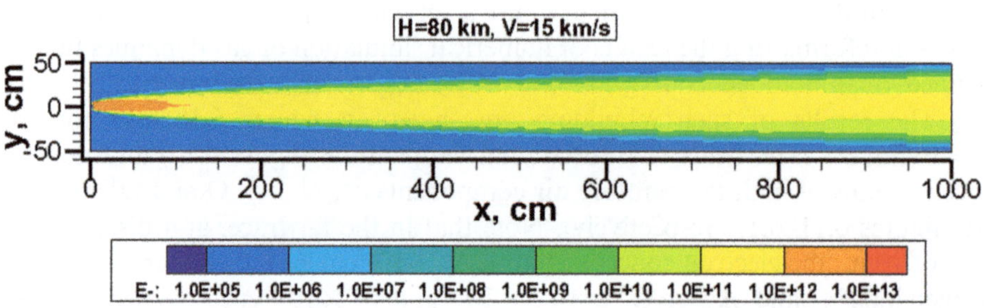

Figure 6.11. Volume concentrations of electrons in the wake at H = 80 km and V_∞ = 15 km s^{-1}.

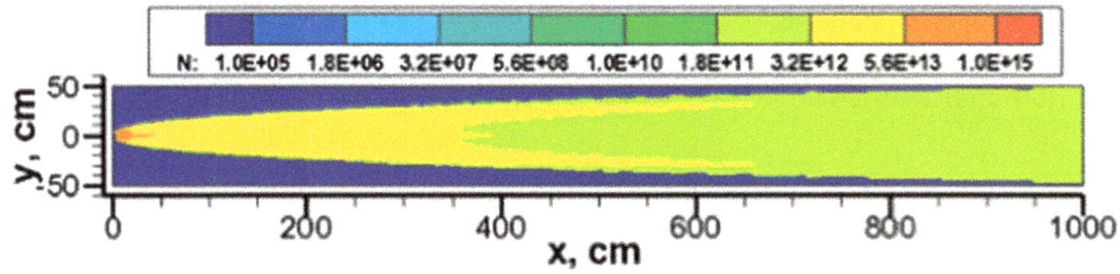

Figure 6.12. Volume concentrations of atoms N in the wake at H = 80 km and V_∞ = 15 km s^{-1}.

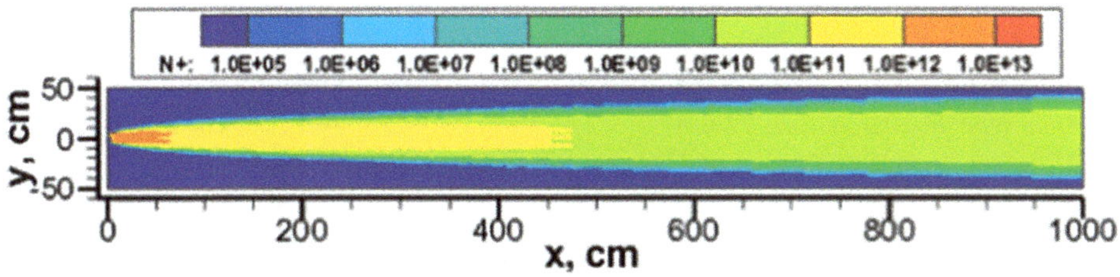

Figure 6.13. Volume concentrations of atoms N^+ in the wake at H = 80 km and V_∞ = 15 km s^{-1}.

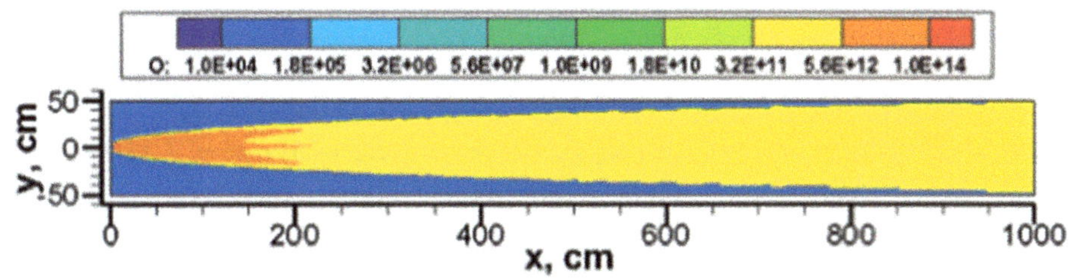

Figure 6.14. Volume concentrations of atoms O in the wake at H = 80 km and V_∞ = 15 km s^{-1}.

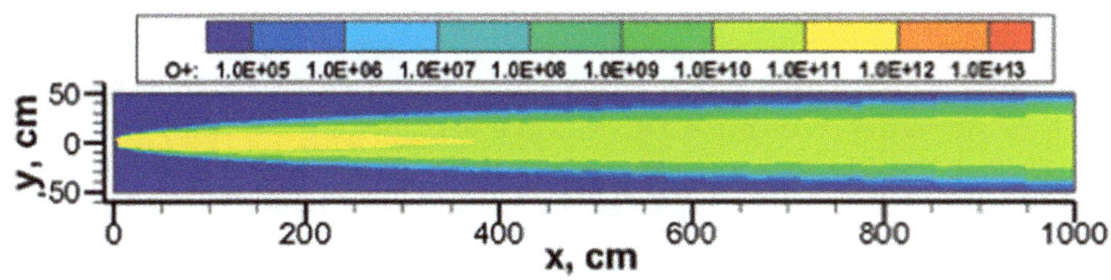

Figure 6.15. Volume concentrations of atoms O^+ in the wake at H = 80 km and V_∞ = 15 km s^{-1}.

evolution of the plasma trace [24], for which the distributions of gas-dynamics functions obtained in the paper can be considered as the initial data.

6.3 Conclusion

A computer model of chemically and thermally non-equilibrium radiating gas was used to predict the parameters of the plasma shell and the long-distance trail behind meteoric bodies of the order of a few centimeters in size at speeds of 10–17 km s^{-1} at an altitude of 80 km. In the calculations, spectral signatures of meteoric bodies and ionization characteristics of tracks with a length of up to 10 m were obtained.

References

[1] Millman P M 1980 One hundred and fifteen years of meteor spectroscopy *Int. Astronomical Union Symp.* (Cambridge: Cambridge University Press) pp 121–8

[2] Lindemann F A and Dobson G M B 1923 A theory of meteors, and the density and temperature of the outer atmosphere to which it leads *Proc. R. Soc. Lond.* A **102** 411–37

[3] Ovenden M W 1947 On the nature and distribution of meteoric matter *J. Br. Interplanet. Soc.* **6** 157

[4] Denning W F 1912 Radiant points of shooting stars observed at Bristol, chiefly in 1899–1911 *Mon. Not. R. Astron. Soc.* **72** 631

[5] Bronshten V A 1981 *Physics of Meteoric Phenomena* (Moscow: Nauka)

[6] Opik E J 2004 *Physics of Meteor Flight in the Atmosphere* (Mineola, NY: Dover)

[7] Smirnov V A 1994 *Spectra of Transient Atmospheric Light Phenomena: Meteors* (Moscow: Fizmatlit)

[8] Brown P, Weryk R J, Wong D K and Jones J 2008 A meteoroid stream survey using the Canadian Meteor Orbit Radar: I Methodology and radiant catalogue *ICARUS* **195** 317–39

[9] Jenniskens P, Tedesco E, Murthy J, Laux C O and Price S 2002 Spaceborne ultraviolet 251–384 nm spectroscopy of a meteor during the 1997 Leonid shower *Meteorit. Planet. Sci.* **37** 1071–8

[10] Sparrow C 1926 Physical theory of meteors *Astrophys. J.* **63** 90

[11] Lindemann F A 1927 Note on the physical theory of meteors *Astrophys. J.* **65** 117

[12] Levin B Y 1956 *The Physical Theory of Meteors, and Meteoric Matter in the Solar System* (Moscow: Russian Academy of Sciences Press)

[13] Astapovich I S 1958 *Meteoric Phenomena in the Earth's Atmosphere* (Moscow: Fizmatgiz) p 640

[14] Ceplecha Z 1964 Study of a bright meteor flare by means of emission curve of growth *Bull. Astron. Inst. Czech.* **15** 102

[15] Borovička J *et al* 1997 Spectral analysis of two Perseid meteors *Planet. Space Sci.* **45** 563

[16] Harvey G A 1973 Spectral analysis of four meteors *Evolutionary and Physical Properties of Meteoroids* vol 319 (Washington, DC: NASA) p 103

[17] Halliday I 1988 The spectra of meteors from Halley's comet *20th ESLAB Symp. on the Exploration of Halley's Comet (Heidelberg, 27–31 October 1986. ESA SP-250)* (Berlin: Springer) pp 921–4

[18] Stulov V P, Mirskii V N and Vislyi A I *Aerodynamics of Bolides* (Moscow: Fizmalit)

[19] Kosarev I B, Loseva T V and Nemchinov I V 1996 Vapor optical properties and ablation of large chondrite and ice bodies in the Earth's atmosphere *Solar Syst. Res.* **30** 265

[20] Kosarev I B 2009 The optical properties of vapors of matter of cosmic bodies invading the Earth atmosphere *High Temp.* **47** 777–87

[21] Surzhikov S 2011 Numerical simulation of spectral signature of bolide entering into Earth atmosphere, *3rd AIAA Atmospheric Space Environments Conf.* AIAA paper 2011-3152

[22] Surzhikov S 2012 Spectral signature of ablating bolide entering into Earth atmosphere, *4th AIAA Atmospheric and Space Environments Conf.* AIAA paper 2012-2943

[23] Surzhikov S 2014 Non-equilibrium radiative gas dynamics of small meteor, *44th AIAA Fluid Dynamics Conf.* AIAA paper 2014-2636

[24] Browne I C, Bullough K, Evans S and Kaiser T R 1956 Characteristics of radio echoes from meteor trails II: the distribution of meteor magnitudes and masses *Proc. Phys. Soc.* B **69** 83

IOP Publishing

Hypersonic Meteoroid Entry Physics

Gianpiero Colonna, Mario Capitelli and Annarita Laricchiuta

Chapter 7

Super-orbital entry of artificial asteroids (Apollo, Hayabusa) and CFD/radiation/thermal analysis of the entry of the Chelyabinsk meteorite

Philippe Reynier, Félix Mouzet, Giulio Seller and Mario Capitelli

Meteoroids are small celestial bodies, originating mostly from the disintegration of an asteroid or a comet core [1]. The former originate from the *asteroid belt*, while the latter are believed to originate from the Oort cloud. Asteroids consist principally of iron and silica, whereas comets can also contain ice. Because of their small size (usually less than one meter [2]), meteroids can be captured by the gravitational field of a planet or its satellites. If there is an atmospheric entry, the meteoroid may be entirely consumed and is called a meteor, otherwise it impacts onto the surface and is called a meteorite.

During atmospheric entry, hypersonic flows are considered: the velocities involved are very high (up to several dozens of km s^{-1} [3]), and the Mach number may reach values as high as 50 or even more. In the shock layer, the air is highly ionized and its temperature can be higher than 10 000 K. In these conditions, several combined phenomena lead to the ablation of the meteoroid. The meteoroid mainly loses material through fusion, oxidation or pyrolysis. These processes depend strongly on the temperature and species present in both the gas and the solid. Spallation can also occur, however, according to [4] it plays a minor role.

Heat dissipation plays a very important role during atmospheric entry, as it prevents the meteoroid from being almost instantly vaporized due to the high plasma temperature. Depending on the size and entry velocity of the object, the heat flux may either be mainly convective or radiative. For slow and small meteors, convection dominates, whereas for larger sizes and velocities, radiative heat transfer can reach very high values (up to several dozens of MW m^{-2}), whereas convection barely ever exceeds 10 MW m^2 and can be mostly compensated by blockage effects [5].

doi:10.1088/2053-2563/aae894ch7

Modeling such phenomena is a complex task, as their influence fluctuates greatly depending on the flow conditions. The meteor's size, entry velocity, shape, chemical composition and the roughness of its surface are parameters that are unknown but have a strong influence. Other phenomena, such as fragmentation (or even explosion) in the atmosphere, particularly common for large meteors [6], are difficult to predict.

The first part of this chapter focuses on a simplified approach describing the phenomenology of meteoroid entry in Earth's atmosphere at a velocity in the range of 10–40 km s^{-1}, an altitude of about 70–100 km and a meteor diameter in the range of 2 mm–20 m. We will try to explore the impact of meteoroids on Earth, distinguishing them in particular by their entry velocity and diameter. To this end a code developed by Seller *et al* [7] for the simulation of the role of electromagnetic fields in the hypersonic entry of spacecraft is used for the small meteoroids.

In parallel, an extensive effort is carried out for large bodies. A survey of meteoroid entries has highlighted the scarcity of available data. As a consequence, two events have been selected, corresponding to recent entries of meteoroids larger than one meter, for which some elements on entry and trajectory are available. The two meteors selected are the famous Chelyabinsk event and a meteor with no official name, which fell into the Atlantic Ocean on 6 February 2016. For convenience, the latter is referred to as the Saint Valentine meteor in the rest of this chapter, based on a press report [8]. Calculations of the trajectory, flow-field, radiation and thermal analysis are undertaken in order to assess the existing state-of-the-art and future necessary developments for improving our knowledge of large meteoroid entry.

7.1 A simplified model for meteoroid entry

7.1.1 Fluid dynamics and chemistry

A CFD scheme is used for hypersonic fluid dynamics with the following features:

- A two-dimensional, non-upwind scheme.
- A finite difference, space-centered stabilized method.
- A Cartesian grid.

The method is a modified Lax–Friedrichs one with a parametric stabilizing term [7]. The scheme is explicit and first order in space and time. Euler equations are used, so viscosity, thermal conductivity and chemical diffusion are not taken into account. In order to have a wave-capturing scheme, the conservative formulation of the Euler equation is used; thus the variables are the density of different chemical species, momentum in the x- and y-directions, and total energy. The stabilizing term is parametric, with a parameter Γ that quantifies the stabilization of the scheme (see [7]). The Γ parameter is uniform and constant over the whole scheme. It is also isotropic, i.e. the method does not take into account the eigenvalues for propagating waves in the scheme. The fluid dynamic equations are modified by introducing chemistry terms; the total density equation is replaced by n density equations for n chemical species. The resulting fluid dynamics equations are:

$$\rho_k^{t+\Delta t} = \rho_k^t - \Delta t \cdot [\nabla \cdot F_k - \Gamma_k - \Sigma_i S_{kl}]^t \tag{7.1}$$

$$(\rho u)^{t+\Delta t} = (\rho u)^t - \Delta t \cdot [\nabla \cdot F_{\rho u} - \Gamma_{\rho u}]^t \tag{7.2}$$

$$(\rho v)^{t+\Delta t} = (\rho v)^t - \Delta t \cdot [\nabla \cdot F_{\rho v} - \Gamma_{\rho v}]^t \tag{7.3}$$

$$(\rho e)^{t+\Delta t} = (\rho e)^t - \Delta t \cdot [\nabla \cdot F_{\rho e} - \Gamma_{\rho e}]^t, \tag{7.4}$$

where F_i ($i \in \{1, \ldots, n, >u, >v, >e\}$) are flux terms, Γ_i are stabilizing terms and S_{kl} are chemical reaction terms (see [7]).The total density is, of course, the sum of the species densities:

$$\rho = \Sigma_k \rho_k. \tag{7.5}$$

To close the system, a relation for the pressure is needed:

$$p = (\gamma - 1)\left(\rho e - \frac{(\rho u)^2 + (\rho v)^2}{2\rho} - \Sigma_k(\rho_k \cdot \varepsilon_k)\right) \tag{7.6}$$

$$\gamma = \Sigma_k \frac{\rho_k \cdot \varepsilon_k}{\rho}. \tag{7.7}$$

The air chemical reactions introduced in our model are reported in table 7.1, including dissociation reactions, ionization of atoms and associative ionization. Eleven species are considered including N_2, N, N_2^+, N^+, N^{++}, O_2, O, O_2^+, O^+, O^{++} and electrons, e^-. The relevant rates are taken from Park's database and introduced in the fluid dynamics as described in [9, 10].

7.1.2 Predictions for small meteoroids

The first results we show are the typical molar fractions of the considered species as a function of distance along the stagnation line for a velocity entry of 30 km s^{-1} and a meteoroid diameter of 2 cm (see figure 7.1). In the same figure we report the two-dimensional density fields including the corresponding gas temperature. Maximum and minimum values of densities and gas temperatures are reported in table 7.2. For

Table 7.1. Chemical reactions in the model. B and M are auxiliary species, representing the total densities of biatomic ($[B] \equiv [N_2] + [N_2^+] + [O_2] + [O_2^+]$) and atomic ($[M] \equiv [N] + [N^+] + [N^{++}] + [O] + [O^+] + [O^{++}]$) species.

$N_2 + B \leftrightarrow N + N + B$	$O_2 + B \leftrightarrow O + O + B$
$N_2 + M \leftrightarrow N + N + M$	$O_2 + M \leftrightarrow O + O + M$
$N + N \leftrightarrow N_2^+ + e^-$	$O + O \leftrightarrow O_2^+ + e^-$
$N + e^- \leftrightarrow N^+ + 2e^-$	$O + e^- \leftrightarrow O^+ + 2e^-$
$N^+ + e^- \leftrightarrow N^{++} + 2e^-$	$O^+ + e^- \leftrightarrow O^{++} + 2e^-$

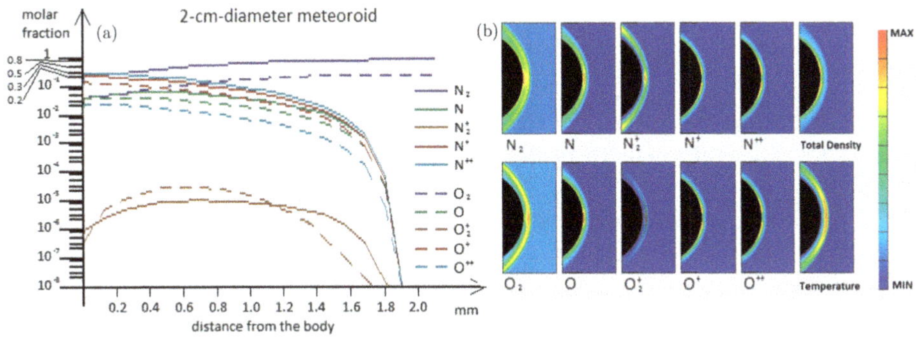

Figure 7.1. (a) Molar fraction of chemical species along the stagnation line for entry of a meteoroid 2 cm of diameter in air at 30 km s^{-1}. (b) Two-dimensional density and temperature fields.

Table 7.2. Maximum and minimum values of densities and gas temperature, for entry of a meteoroid 2 cm in diameter in air at 30 km s^{-1}.

Density	N_2	N	N_2^+	N^+	N^{++}	**Temperature**
Max (g m^{-3})	$4.265\,529 \times 10^{-1}$	$7.252\,904 \times 10^{-2}$	$8.070\,299 \times 10^{-6}$	$7.137\,793 \times 10^{-1}$	$7.802\,693 \times 10^{-1}$	Max (K) 394 596
Min (g m^{-3})	$6.458\,519 \times 10^{-2}$	0	0	0	0	Min (K) 220
Density	O_2	O	O_2^+	O^+	O^{++}	**Total density**
Max (g m^{-3})	$9.862\,295 \times 10^{-2}$	$6.505\,489 \times 10^{-2}$	$5.556\,829 \times 10^{-5}$	$3.827\,251 \times 10^{-1}$	$6.823\,475 \times 10^{-2}$	2.457 353
Min (g m^{-3})	$1.819\,570 \times 10^{-2}$	0	0	0	0	8.278568×10^{-2}

the reported conditions we can see that at 2 mm from the meteoroid we have only the two diatomic species N_2 and O_2, while the atom and ion molar fractions become predominant from 0.4 mm onward. The behavior of the molar fraction of N^{++} is also interesting, as it can become of the same order of magnitude as N^+ ions near the surface.

These results are strongly modified by the meteoroid diameter, as can be appreciated by looking at figure 7.2(a), (b) where two cases (2 mm and 2 m) are reported. One can appreciate (figure 7.2(a)) that the small diameter case presents a small air reactivity as can be appreciated by the predominance of neutral species (diatoms and atoms) on the ionized species in the whole examined distance from the body. Note also that the molar fractions of molecules are larger than the corresponding ones for atoms. The situation completely changes for a meteoroid diameter of 2 m when the reactivity increases with respect to the previous cases (figure 7.2(b)). In particular the molar fractions of N_2 and O_2 strongly decrease from the free-flow condition to the meteoroid surface, while the ion molar fractions strongly increase.

Similar behavior is reported in figure 7.3(a)–(c) for an entry velocity of 18 km s^{-1} and in figure 7.4(a)–(c) for an entry velocity of 10 km s^{-1}.

It is interesting to show the translational temperature profiles along the stagnation line as a function of the reduced distance for different entry velocities (30 to 20 to 10 km s^{-1}) and different meteoroid diameters (see figure 7.5(a)–(c)). The dependence of the gas temperature profile on the meteoroid diameter decreases with decreasing the entry velocity (compare figure 7.5(a) and (c)).

To understand the role of chemistry we compare in figure 7.6 the presented translational temperatures at the relevant maxima with the corresponding results obtained for pure nitrogen. Inspection of the figure shows small variations of the

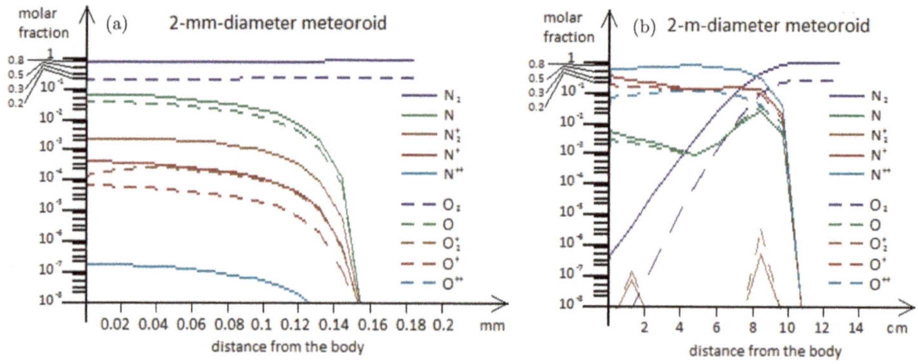

Figure 7.2. Molar fraction of chemical species along the stagnation line for entry of a meteoroid (a) 2 mm or (b) 2 m in diameter in air at 30 km s^{-1}.

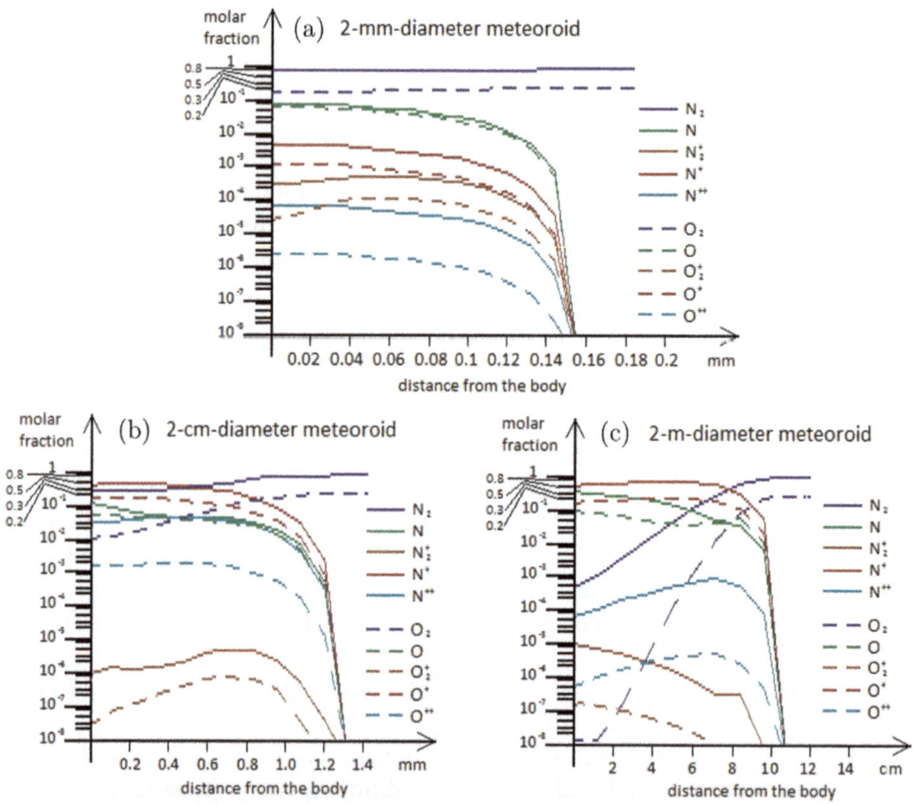

Figure 7.3. Molar fraction of chemical species along the stagnation line for entry of a meteoroid (a) 2 mm, (b) 2 cm or (c) 2 m of diameter in air at 18 km s^{-1}.

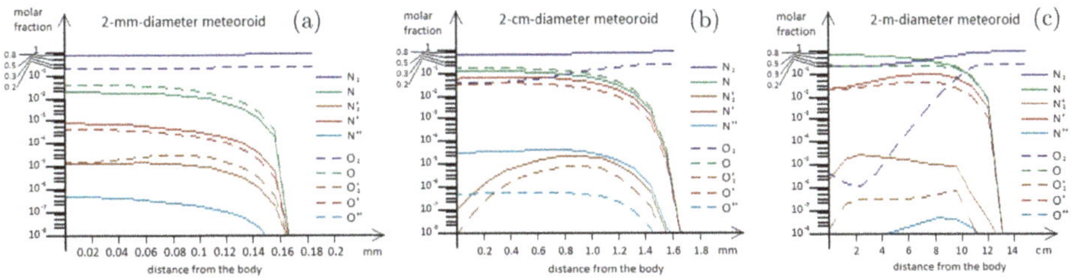

Figure 7.4. Molar fraction of chemical species along the stagnation line for entry of a meteoroid (a) 2 mm, (b) 2 cm or (c) 2 m of diameter in air at 10 km s^{-1}.

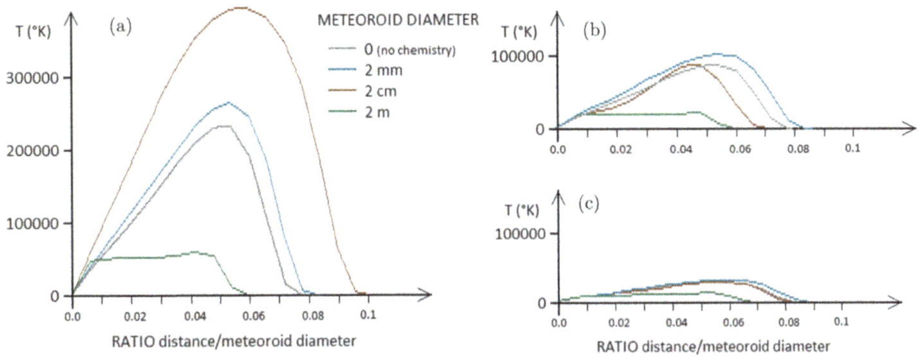

Figure 7.5. Temperature profiles as a function of the reduced distance for meteoroid entry in air at (a) 30 km s^{-1}, (b) 18 km s^{-1} and (c) 10 km s^{-1}.

temperature in the two cases up to 18 km s^{-1} while appreciable differences appear at 30 km s^{-1}, emphasizing the low dependence of this quantity on the chemistry used, i.e. air and nitrogen give similar results.

A comparison of the gas temperature in the present contribution to the corresponding temperature calculated with the more sophisticated model of Surzhikov [11], that includes a more complex chemistry, ablation and radiation, is reported in figure 7.7(a) for a meteoroid diameter of 2 cm. Differences appear in the two calculations either in the magnitude of observed maxima or in the displacement of the profiles as a function of the distance along the stagnation line. The more complex model in fact presents profiles in the x range 0–0.7 cm, while our profiles span the range 0–0.2 cm. A different behavior is obtained by comparing the present gas temperature profile to the corresponding one presented by Stern *et al* [12] (figure 7.7(b)).

In conclusion the presented results can be considered a useful tool to predict the gas temperature in the impact meteoroid–Earth atmosphere qualitatively following the results which can be obtained by using the more sophisticated methods described in this book.

7.2 Entry of large meteoroids

The Chelyabinsk meteor exploded in the Earth's atmosphere on 15 February 2013, above the city of Chelyabinsk. This event, which was the most energetic atmospheric

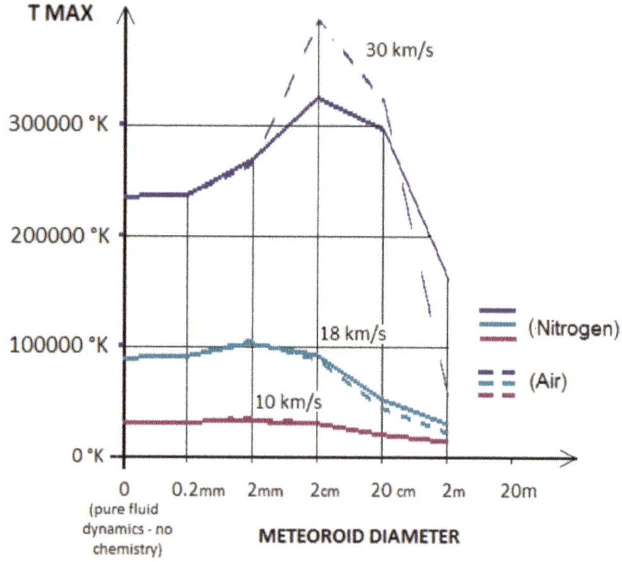

Figure 7.6. Maxima of translational temperature as a function of the meteoroid diameter, at different values of the entering velocity, for air (dashed lines) and pure nitrogen (solid lines).

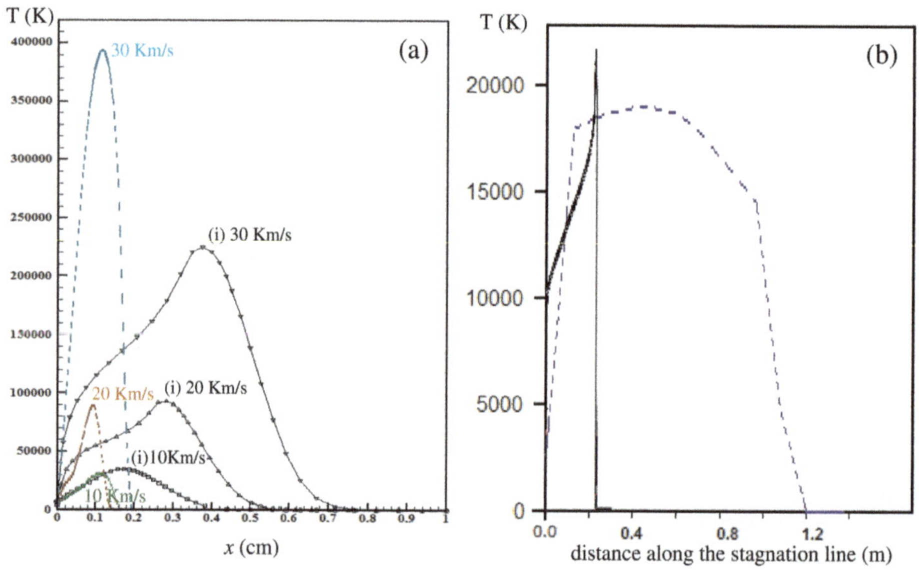

Figure 7.7. Temperature profiles along the stagnation line at different entry velocities (dashed lines), compared with results (a) from [11] for a meteoroid diameter of 2 cm ((i)-curves, solid lines with markers) and (b) from [12] at 20 km s^{-1} for a meteoroid diameter of 20 m (solid line).

entry since the Tunguska meteor, was equivalent to the explosion of 440 kt of TNT, and has been selected because it is well documented. Indeed, numerous observations allowed a fairly precise determination of the entry data [13], as well as the estimation of the mass and diameter of the meteoroid. The Saint Valentine meteor was selected for practical reasons of mesh size because, despite its reduced size, it was still sufficient to be detected. The data are more sparse [14], but they still give an estimation of the meteoroid mass, from the entry velocity and the emitted energy; it can be assumed that all of the meteoroid kinetic energy was converted into heat during the explosion. In the following, the most likely trajectories are calculated for both meteoroids using a three-degrees-of-freedom trajectory code. This includes the sensitivity study the ballistic coefficient, lift and stagnation point correlation. Then, CFD and radiation calculations are performed in order to have a thermal analysis of meteoroid behavior during entry.

7.2.1 The Chelyabinsk meteor

From available flight data, it was determined [13] that the diameter and mass of the Chelyabinsk meteoroid were, respectively, between 15 and 19 m, and 7 and 13 kt. From these values, four limit cases are considered to determine the influence of the parameters. Table 7.3 shows the densities and ballistic coefficient (mass divided by the drag force: $\beta = \frac{M}{C_D S}$, with C_D the drag coefficient, and S the cross sectional surface) for different configurations. The rock fragments that have been recovered provide an estimation of the density that is between 2000 and 4000 kg m^{-3}. The second and third cases are thus not relevant, but they could be realistic for a meteoroid made of iron or ice.

Figure 7.8 shows the evolution of velocity and stagnation point radiative heating during entry for the four cases in table 7.3. In figure 7.8(a), the dashed zone represents the explosion area and in figure 7.8(b) the plain line shows the altitude of the observed radiation peak.

The ballistic coefficient appears to have some influence at low altitudes but only after the body explosion. The radiative heat flux is proportional to the ballistic coefficient, as shown in figure 7.9. Moreover, these results have to be considered

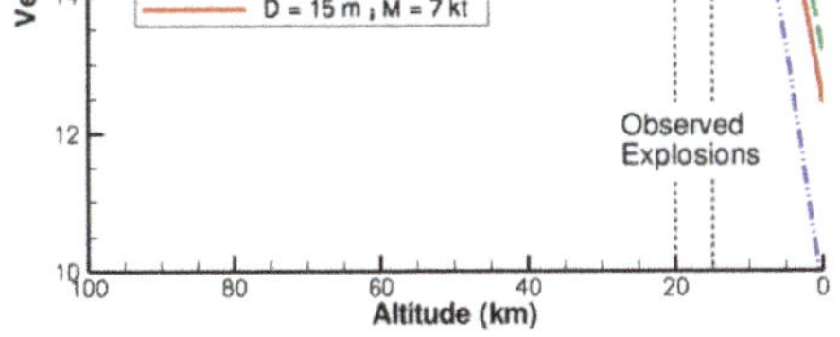

(a) Velocity

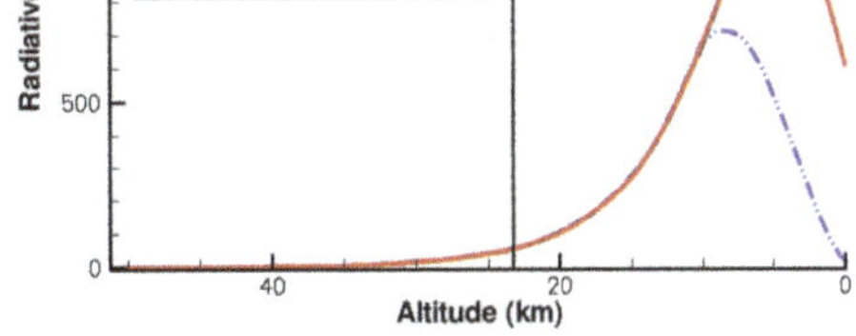

(b) Radiative heat flux at stagnation point

Figure 7.8. Velocity and radiative heat flux for different diameters and masses of meteor.

Table 7.3. Density and ballistic coefficients for the Chelyabinsk event.

Mass	7×10^6 kg		13×10^6 kg	
Diameter	15 m	19 m	15 m	19 m
Density (kg m^{-3})	3963	1950	7360	3622
Ballistic coefficient	42.1	26.2	78.3	48.8

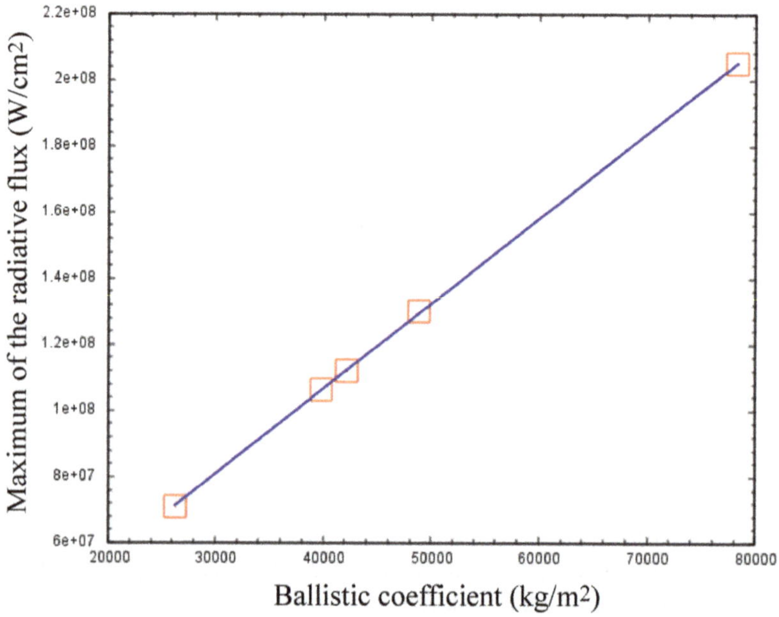

Figure 7.9. Correlation between the ballistic coefficient and radiation maximum.

carefully since the radiative heating is estimated using correlations that were not developed for high-pressure conditions.

The geometry of the meteoroids is unknown, so a spherical shape is assumed for estimating the drag coefficient. A sensitivity study of lift (and a non-spherical shape) is also conducted. The ratio of the lift and drag forces is considered between 0 and 0.32. The last value corresponds to the limit after which the lift is too high and the meteoroid does not reach the Earth's surface but has an elliptical trajectory around the planet. The results are plotted in figure 7.10, where the dashed and solid lines have the same meaning as in figure 7.8(b). The lift has a very minor influence on velocity, and it is only observable after the explosion. The radiative heat flux is slightly reduced (less than 10%), with the addition of lift, since the meteoroid is slowed down slightly.

The radiative heating computed with the correlations proposed by Brandis and Johnston [15], Tauber and Sutton [16], and Detra and Hidalgo [17] (extended to higher velocity in [18]) are plotted in figure 7.11. The predicted radiative heat fluxes are not consistent with the observations, since the peaks occur at a lower altitude and are far less energetic (by several orders of magnitude). This shows that the usual engineering correlations for stagnation point heating are not adequate. They were developed for entries of blunt body vehicles, slow-down at high altitude (around 80–100 km) and low pressure. They are adequate for velocities below 16 km s^{-1} (17 km s^{-1}) for [15] and for pressures around one dozen pascals (~10^3 Pa here). The problem is highlighted in figure 7.11, where the radiation peak occurs at an altitude

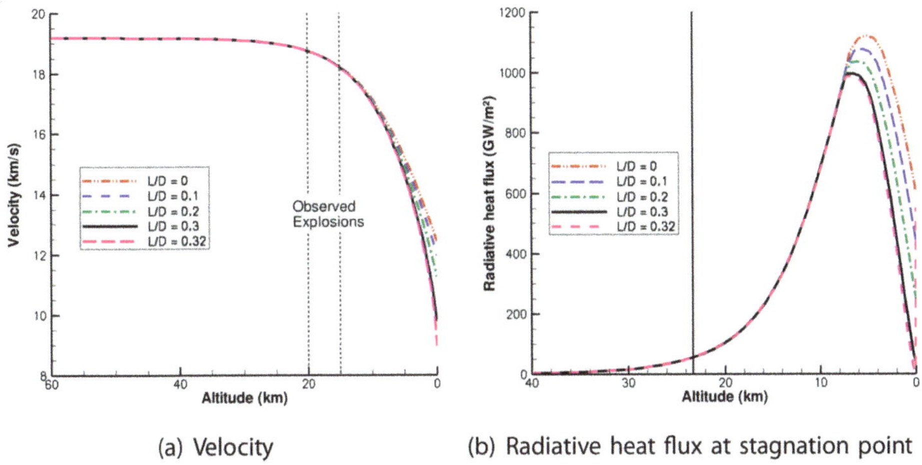

Figure 7.10. Sensitivity study of lift on velocity and radiative heat flux.

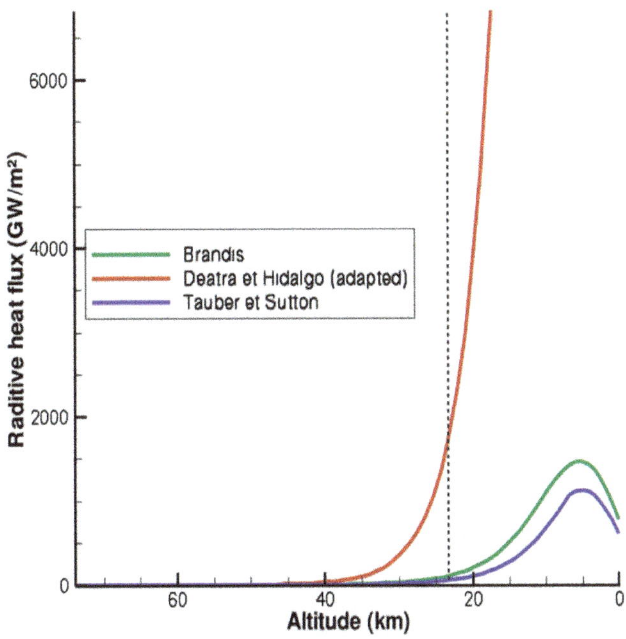

Figure 7.11. Correlation comparison for radiative heating.

Table 7.4. Possible densities for the Saint Valentine meteor.

Diameter (m)	Density	Material
2	53.6	
3	15.9	
4	6.70	Metal
5	3.43	Rock
6	1.98	Dirty ice
7	1.25	
8	0.84	

20 km lower than the observation, for the three correlations. For the Detra and Hidalgo [17] correlation, the peak happens at the same altitude as the others, but reaches much higher values (more than 10^4 GW m^{-2}). Thus, none of these correlations is suited for such a study, and the reliability of the results obtained is questionable. However, this conclusion will be revised if the meteoroid has fragmented before its final explosion.

7.2.2 The Saint Valentine meteor

During its atmospheric entry, the Saint Valentine meteor released energy estimated to be around 13 kt TNT equivalent. By assuming that its kinetic energy was entirely converted during the explosion and, given its entry velocity of 15.5 km s^{-1}, its mass can be approximated to 224 t. Its diameter has been estimated to be between 2 and 8 m, which corresponds to densities between 0.84 and 53.6, as reported in table 7.4. The extremal values can be eliminated as being unrealistic. Moreover, the altitude of the explosion of the meteor implies that it was most likely made of rock, which means a diameter around 5 m. As a consequence, this value is retained in our study.

Figure 7.12 represents the evolution of the velocity and radiative heat flux during the Saint Valentine meteor entry. The dashed line is the altitude of the explosion and the solid line is the altitude of the radiation peak. For this velocity, convection is negligible before radiation. The discontinuity of the radiative heat flux near an altitude of 10 km is due to the change of correlation, from Brandis and Johnson [15] to Detra and Hidalgo [17] when the velocity reaches a threshold of 9 km s^{-1}.

7.2.3 CFD computations

For the CFD calculations, several points of the trajectories have to be selected. The relevant points are the maxima of the radiative and convective heat flux. However, the calculated maxima occur only after the observed explosions and therefore are not relevant. Hence, for both meteoroids, the most relevant points to perform the calculations are the observed radiation peak and the explosion. For this study, computations are carried out at the radiation maxima, with the data reported in table 7.5.

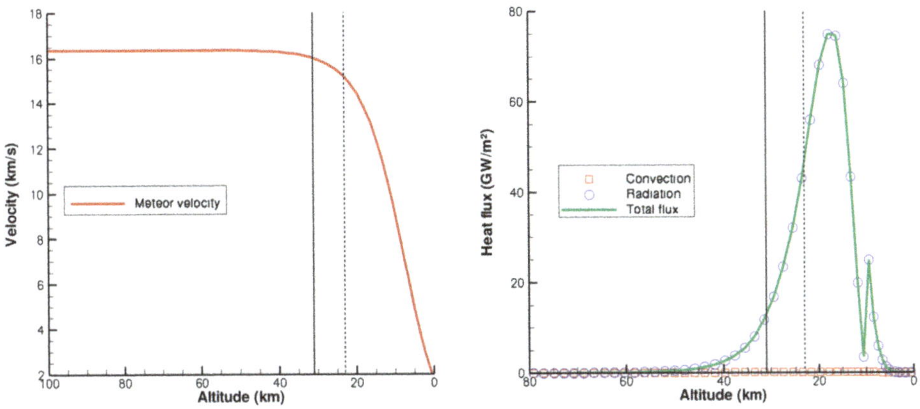

(a) Velocity (b) Heat flux at stagnation point

Figure 7.12. Velocity and heat flux at the stagnation point for the Saint Valentine meteor.

Table 7.5. Calculation matrix for the two meteors.

Meteor	Velocity (km s^{-1})	Altitude (km)	Temperature (K)	Pressure (Pa)
Saint Valentine	16.01	31	227.5	1.039×10^3
Chelyabinsk	18.90	23.3	219.6	3.468×10^3

The Saint Valentine meteor

The mesh is constructed in two steps: a first rough mesh to determine the position of the shock, and a second mesh refined near the meteoroid surface and the shock to help capture those two sensitive areas. The latter is presented in figure 7.13, with 120×160 square cells; the grid dependence of the results was checked [19] and the level of refinement was found to be sufficient. The cells near the meteoroid surface need to be no more than 10 μm thick to ensure the grid independence of the results.

The flow around the meteoroid is computed using the non-equilibrium CFD code TINA [20], developed by Fluid Gravity Engineering. Park's model [9] with 11 species (N_2, O_2, NO, N, O, N_2^+, O_2^+, NO^+, N^+, O^+ and e^-) and 16 reactions is retained to model the mixture composition around the object. Ablation is not accounted for, since the chemical reactions involved are unknown due to the lack of data on meteoroid composition. A strict Courant–Friedrichs–Lewy condition has to be implemented first to avoid shock layer detachment during its formation. Since the mesh is made of quadrangles, an instability may develop spontaneously around 40° from the stagnation line for Courant numbers above 0.01. This problem is no longer present when the shock layer is well established. Calculations are carried out over 300 000 iterations, and are stopped when a residual of 8×10^{-3} is obtained. The final flow-field with the Mach number distribution around the body and the bow shock is shown in figure 7.14.

The translational and vibrational temperature distributions along the stagnation line are plotted in figure 7.15(a). The two lines are practically coincident, which

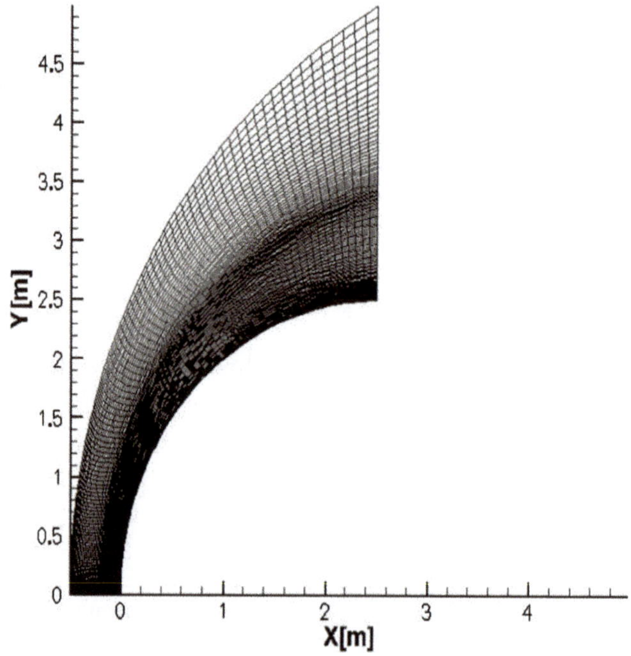

Figure 7.13. Mesh for the Saint Valentine meteor.

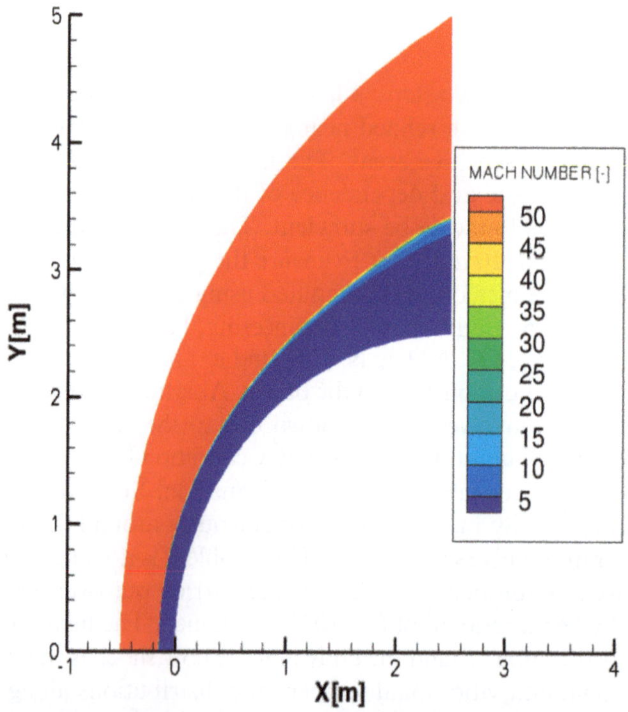

Figure 7.14. Mach number distribution around the Saint Valentine meteor.

means that thermal equilibrium is reached. The temperature inside the shock layer reaches a maximum of the order of 20 000 K.

The molar fraction distributions along the stagnation line are presented in figure 7.15(b) and the flow-field distributions in figures 7.16–7.18. N_2 and O_2 are almost entirely dissociated inside the shock layer; the gas is mostly composed of atomic nitrogen and oxygen, as well as their ionized equivalents (N^+ and O^+), and electrons. Ionized species are dominant near the stagnation point, where the mole fraction of electrons reaches a level of 0.3 (with a local maximum of 0.35), while atomic species are more present near the corner of the meteoroid since less energy is available there for the ionization process. The other species from the model (such as NO, N_2^+, O_2^+ and NO^+) are very sparsely present in a very thin band along the shock.

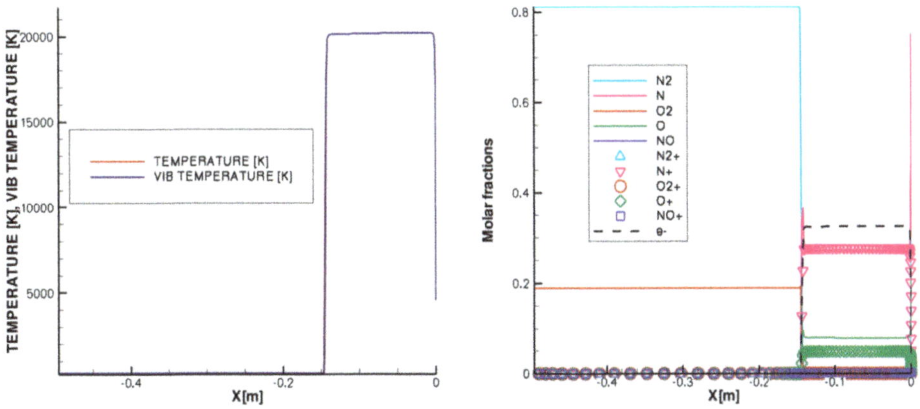

(a) Temperatures (b) Molar fractions

Figure 7.15. (a) Temperatures and (b) molar fractions along the stagnation line for the Saint Valentine meteor.

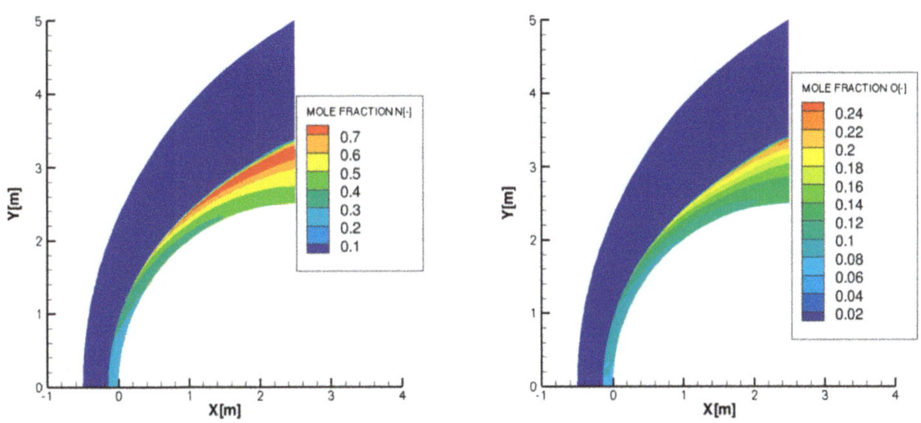

(a) N (b) O

Figure 7.16. Atomic nitrogen and oxygen molar fraction distributions around the Saint Valentine meteor.

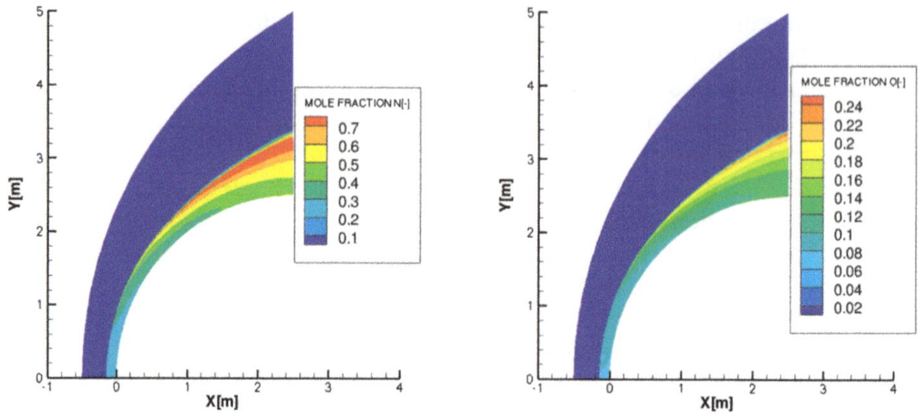

(a) N^+ (b) O^+

Figure 7.17. N^+ and O^+ molar fraction distributions around the Saint Valentine meteor.

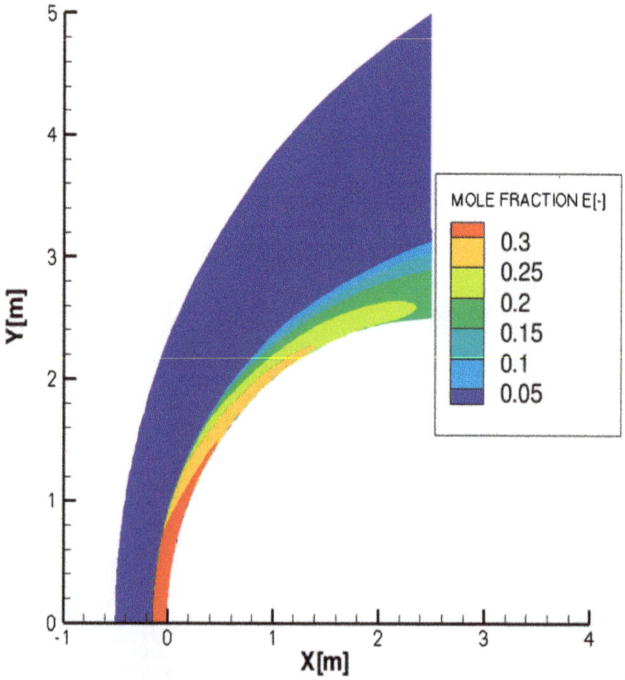

Figure 7.18. Electron molar fraction distribution around the Saint Valentine meteor.

Th Chelyabinsk meteor

The same calculations were performed for the Chelyabinsk meteor using a similar mesh, with 100×100 mesh cells and similar results in term of flow-field predictions. The calculations were converged up to a residual of 3×10^{-3} and 300 000 iterations.

The Mach number and pressure distributions are presented in figure 7.19. Similar distributions as for the Saint Valentine case are obtained for the species molar fraction distributions. The only noteworthy difference compared to the Saint Valentine meteor is a higher level of ionization; the molar fraction of electrons reaches a maximum of 0.4 near the stagnation point against 0.35 for the Saint Valentine meteor. This is highlighted by the electron molar fraction distribution shown in figure 7.20.

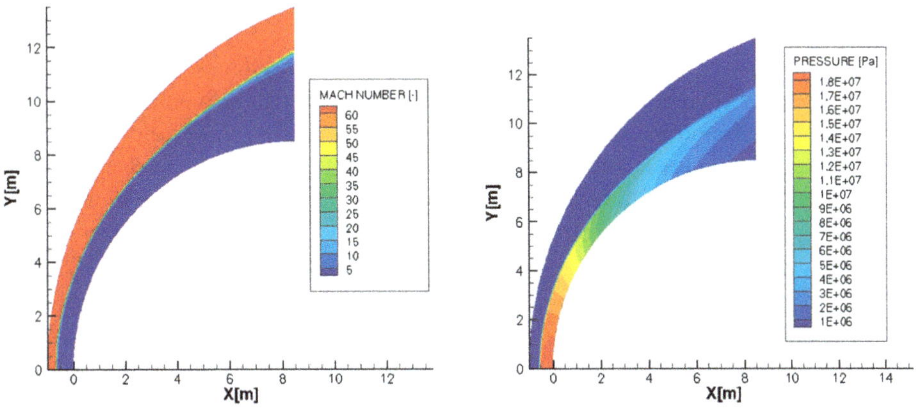

(a) Mach number (b) Pressure

Figure 7.19. Mach number and pressure distributions around the Chelyabinsk meteor.

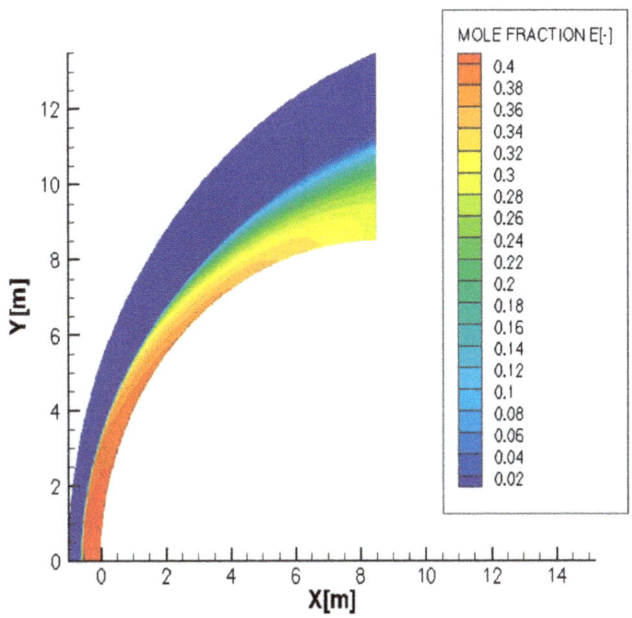

Figure 7.20. Electron molar fraction distributions around the Chelyabinsk meteor.

7.3 Heating

7.3.1 Convective heating

Convective heat fluxes for both large meteoroids obtained from the numerical simulations are analyzed. Figure 7.21 shows the total convective heat flux along the meteoroid surface obtained with several meshes for the Saint Vantentine case. For the first two meshes (green and red), a singularity can be observed between 40° and 50° from the stagnation line. This problem is due to meshes that are not refined enough in this area, which leads to a detachment of the shock layer. This is not visible in the pressure distribution, but has an important effect on the surface heat flux distribution. This problem is eliminated by using more refined meshes.

The value predicted at the stagnation point for the Saint Valentine meteoroid is 111 MW m^{-2}, which is notably higher than that given by the trajectory analysis, 72 MW m^{-2}. However, due to the severity of the entry the convective blockage can be expected to counter-balance the convective heating [5] and, as a consequence, in future studies the focus will be put on the radiative heating.

7.3.2 Meteoroid composition

The composition of the Chelyabinsk meteoroid is provided by the study of some of its fragments [21]. For the Saint Valentine meteor no data are available so the general composition of L-type meteorites [22] is used. For both meteoroids, the chemical species present are reported in table 7.6 and their radiation intensities are specified from their observation in meteor spectra [23].

For the atomic metals, the Roman numerals correspond to the oxidation number of the atom. A question mark means that the emission is mixed with those from

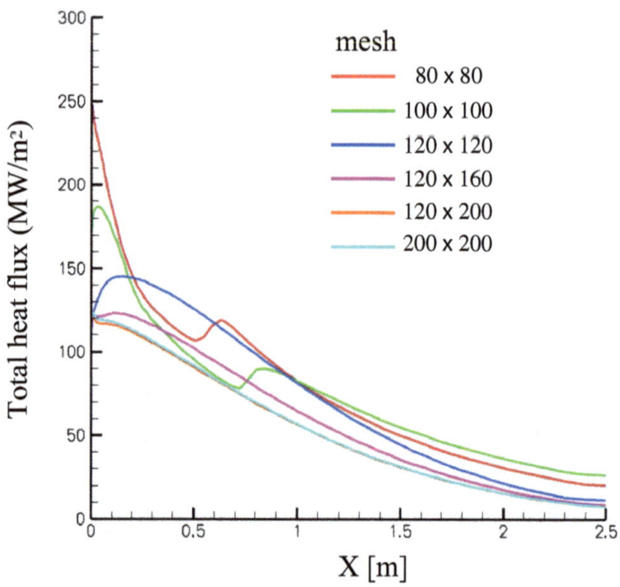

Figure 7.21. Total heat flux along the Saint Valentine meteoroid surface.

other species and may not actually be visible. Species whose radiation frequencies are not included in PARADE [24] are in blue and those whose influence on the meteor spectrum is unknown are in red. Complementary data can be found in databases such as CHIANTI [25] or NIST [26], particularly for metallic atoms and ions.

7.3.3 Radiation computations

The mixture composition along the stagnation line as well as the density and temperatures are then used to compute the radiation with the code PARADE [24]. PARADE (PlasmA RAdiation DatabasE) is a line-by-line radiation code that performs radiation calculations (emissions and absorption coefficients) for non-equilibrium gases. The spectral emission and absorption are determined as functions of transition level (from the upper level to the lower level) and emitting populations of this level. The radiative computations are performed with the Boltzmann assumption for the determination of excited state populations. Phenomena such as precursor effects and photo-ionization cannot be accounted for in PARADE. Self-absorption by molecules is considered.

Table 7.6. Chemical species present in a meteor emission spectrum. See the text for details.

Species present in the meteoroid		Strongly visible	Slightly visible	No data
In high quantities (>10%)	Fe	I	II	
	Si	I / II		
	FeO	X		
	MgO	?		
	SiO_2			X
In medium quantities (∼1 %)	Al	I	II	
	Ca	I/ II		
	Ni	I	II	
	S		I/ II	
	CaO	X		
In low quantities (<1 %)	Ti		I/ II	
	C		I/ II	
	Cr	I	II	
	Mn	I	II	
	Na	X		
	K	I	II	
	Co		I/ II	
	Mg	I/ II		
	C_2	X		
	Al_2O_3			X
	Na_2O			X
	K_2O			X
	P_2O_5			X
Possible creation during the entry	CN	X		
	SiO			X

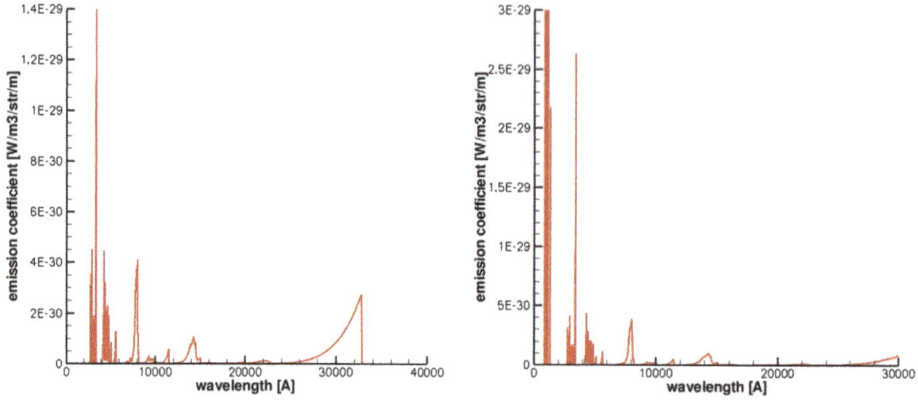

(a) Cheliabinsk (b) Saint Valentine

Figure 7.22. Emission spectra computed for the Chelyabinsk (without VUV) and Saint Valentine (with VUV) meteors.

The species bands and systems accounted for in the calculations are N_2 (first positive, second positive, Birge–Hopfield 2, Birge–Hopfield, Lyman–Birge–Hopfield, Carol–Yoshino, Worley–Jenkins, Worley and e–X Rydberg transition), O_2 (Schuman–Runge), NO (Beta, Gamma, Delta, Epsilon) and N_2^+ (first negative and Meinel).

Figure 7.22 shows the emission spectrum for the Saint Valentine and Chelyabinsk meteors. The spectra are calculated for wavelength between 1000 Å and 40 000 Å using 10 000 points over the whole range. The computed spectrum highlights the strong emission in the VUV and UV ranges. Calculations start at 2000 Å when the VUV is not accounted for. The results for the calculations of the radiation for the Chelyabinsk case indicate a radiation heating of 9.18 GW m^{-2} at the stagnation point when the VUV is not considered, while a value of 11 GW m^{-2} is obtained when accounting for VUV; this indicates that the VUV contribution to radiative heating is around 16%. However, the reliability of these values cannot be estimated due to the lack of available data for a cross-check. Moreover, the meteoroid species are not considered and a quite important radiative blockage can be expected [5].

7.4 Thermal analysis

The heat flux at the meteoroid surface is then used as an input to compute the internal heating for the Saint Valentine meteoroid. To perform these calculations, the multiphysics code ELMER [27], developed by Computer Science Corporation (CSC) in collaboration with several Finnish universities, is retained. The composition of both meteoroids is considered to be similar and correspond to L-type chondrites. A thermal capacity of 800 J kg^{-1} K^{-1}, a density of 3000 kg m^{-3} and a thermal diffusion coefficient of 2 W m^{-1} K^{-1} are assumed. Since the meteoroid came from space, its initial temperature was very low. Its atmospheric entry did not last long enough to warm it up significantly, so its initial temperature is arbitrarily chosen to be 100 K.

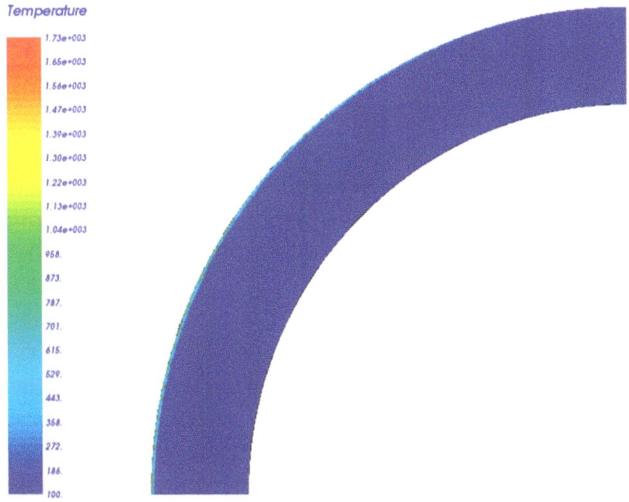

Figure 7.23. Temperature distribution inside the Saint Valentine meteoroid.

To reduce the calculation times, the meshed area is composed of a quarter of annulus 500 mm thick. The previously calculated convective heat flux is used as an input for the external border, whereas for the other borders the heat flux is taken as null. This choice is justified by the symmetry of the geometry. The input convective heat flux is considered constant over 10 s, with the value calculated at the observed radiation peak. The results of the calculations are shown in figure 7.23.

Although the convective heat flux is very high (120 MW m^{-2} at the stagnation point), the duration of the atmospheric fall is less than 10 s, and the meteoroid temperature does not increase significantly. Only the first centimeters are affected by the temperature increase, up to 1700 K. These results highlight the key issue, which is radiative heating for meteoroid demise. Much better modeling and simulation accounting for this phenomenon will be necessary to improve the current state-of-the-art in this domain. A first point is that, certainly due to convective blockage, convective heating might be negligible and the total heating would be from radiation. Another point is that the material's opacity needs to be considered as the heat flux is radiative, so it could not only ablate the external layer, but also penetrate very deep inside the meteoroid. In this case, the demise of the entry object could be due to heating of its core. It has to be noted that silica materials possess an excellent transparency to radiation from VUV to IR [28].

7.5 Conclusions

Different models and methods were applied to the entry of small and large meteoroids. First, a simplified method was developed for the entry of small meteoroids at very high velocities. The results demonstrate the potential of the method for predicting the temperature environment of the meteoroid. Due to the high velocities, such a method would be very attractive when coupled to state-to-state approaches.

In parallel, an extensive effort was made to apply the existing state-of-the-art in terms of computational tools to the simulation of large meteoroid entry. The other objective was to identify existing gaps and efforts required to improve the modeling.

The trajectory analysis allowed the assessment of the influence of different parameters and highlighted the lack of reliability of the current stagnation point heating correlations for such entries. CFD calculations were performed for the Chelyabinsk and Saint Valentine meteoroids to predict the heating distribution, temperatures and mixture composition. Radiation calculations were performed and, in parallel, missing species in the database used were identified. The results demonstrate that the key issue is VUV radiation, since its emission is important in the computed spectra. Another aspect to consider is the material's opacity, to determine if the radiative heating is absorbed at the surface or in the internal layers of the meteoroids. A deeper investigation of radiation accounting for ablated material species is, however, needed to confirm the current results and to try to cover some of the identified gaps. Finally, we used multiphysics software to perform a thermal analysis of a meteoroid during entry. The results show the need to account for radiative heating and to have good knowledge of the material's opacity to improve the level of accuracy of the simulations.

References

[1] Gladman B and Coffey J 2009 Mercurian impact ejecta: meteorites and mantle *Meteorit. Planet. Sci.* **44** 285–91

[2] Rubin A E and Grossman J N 2010 Meteorite and meteoroid: new comprehensive definitions *Meteorit. Planet. Sci.* **45** 114–22

[3] Egorova L 2011 Effect of thermal explosion, *Int. Meteor Conf., Sibiu, Romania Sept. 15–18, 2011*

[4] Rogers L A, Hill K A and Hawkes R L 2005 Mass loss due to sputtering and thermal processes in meteoroid ablation *Planet. Space Sci.* **53** 1341–54

[5] Reynier P 2013 Survey of convective blockage for planetary entries *Acta Astronaut.* **83** 175–95

[6] Longo G 2007 The Tunguska Event *Comet/Asteroid Impact on Human Society* ed P T Bobrowsky and H Rickman (Berlin: Springer) ch 18

[7] Seller G, Capitelli M, Longo S and Armenise I 2005 Numerical MHD of aircraft re-entry; fluid dynamic, electromagnetic and chemical effects, *36th AIAA Plasmadynamics and Lasers Conf., Toronto 6–9 June* AIAA paper no. 2005-5048

[8] MeteoWeb, Le meteore di San Valentineo e quella esplosa sull'oceano Atlantico il 6 febbraio, 2016 (http://www.meteoweb.eu/2016/02/le-meteore-di-san-Valentineo-e-quella-esplosa-sulloceano-atlantico-il-6-febbraio/639239/).

[9] Park C, Jaffe R L and Partridge H 2001 Chemical-kinetic parameters of hyperbolic Earth entry *J. Thermophys. Heat Transf.* **15** 76–90

[10] Park C, Howe J T, Jaffe R L and Candler G V 1994 Review of chemical-kinetic problems of future NASA missions. II—Mars entries *J. Thermophys. Heat Transf.* **8** 9–23

[11] Surzhikov S 2017 private communication (results presented at the *61st Course 'Hypersonic Meteoroid Entry Physics' of the Int. School of Quantum Electronics of the Ettore Majorana Foundation and Centre for Scientific Culture, Erice October 3–8, 2017)*

[12] Stern E 2017 private communication (results presented at the *61st Course 'Hypersonic Meteoroid Entry Physics' of the Int. School of Quantum Electronics of the Ettore Majorana Foundation and Centre for Scientific Culture, Erice October 3–8, 2017)* http://users.ba.cnr.it/imip/cscpal38/HYMEP/lecturers.php
[13] Popova O P *et al* 2013 Chelyabinsk airburst, damage assessment, meteorite recovery, and characterization *Science* **342** 1069–73
[14] NASA, Fireball and Bolide Reports https://cneos.jpl.nasa.gov/fireballs/.
[15] Brandis A M and Johnston C O 2014 Characterisation of stagnation-point heat flux for Earth entry, *45th AIAA Plasmadynamics and Lasers Conf., Atlanta 16–20 June* AIAA paper no. 2014-2374
[16] Tauber M E and Sutton K 1991 Stagnation-point radiative heating relations for Earth and Mars entries *J. Spacecraft Rockets* **28** 40–2
[17] Detra R W and Hidalgo H 1961 Generalized heat transfer graphs for nose cone re-entry into the atmosphere *Am. Rocket Soc. J.* **31** 318–21
[18] Smith A J and Parnaby G 2012 TRAJ3D V2.6: Three degrees of freedom planetary entry and descent trajectory. *Simulation User Manual* (Emsworth: Fluid Gravity Engineering)
[19] Mouzet F and Reynier P 2016 Numerical analysis of large meteoroid atmospheric entries, *7th Int. Workshop on Radiation of High Temperature Gases in Atmospheric Entry, Stuttgart 21–25 Nov 2016*
[20] Fluid Gravity Engineering 2008 *TINA Version 4: Theory and User Manual* (Emsworth: Fluid Gravity Engineering) TN89/96
[21] The Meteoritical Society Entry for Chelyabinsk Meteoritical Bulletin https://www.lpi.usra.edu/meteor/metbull.php?code=57165.
[22] Permanent Meteorite Classifications and Compositions http://www.permanent.com/meteorite-compositions.html.
[23] Borovićka J 1994 Line identifications in a fireball spectrum *Astron. Astrophys.* **103** 83–96
[24] Smith A J, Beck J, Fertig M, Liebhart H and Marraffa L 2013 *PARADE v3.2: Plasma Radiation Database* (Emsworth: Fluid Gravity Engineering) TN28/96
[25] CHIANTI, An Atomic Database for Spectroscopic Diagnostics of Astrophysical Plasmas http://www.chiantidatabase.org/chianti_direct_data.html.
[26] NIST, Atomic Spectra Database Lines Form http://physics.nist.gov/PhysRefData/ASD/lines_form.html.
[27] Råback P and Malinen M 2014 Overview of Elmer *Technical Report* CSC, IT Center for Science, University of Helsinki, Finland
[28] Kajihara K 2008 Vacuum–ultraviolet transparency of silica glass and its relation to processes involving mobile interstitial species, *Winter School of New Functionalities in Glass Tokyo Metropolitan University*

Chapter 8

High-enthalpy ionized flows

Gianpiero Colonna, Mario Capitelli, Lucia Daniela Pietanza, Alessandro Munafò and Marco Panesi

When an orbital space vehicle enters the Earth's atmosphere, the temperature inside the shock layer is large enough to promote molecular dissociation. Under these conditions the chemical model for air, considered to be formed by N_2 (80%) and O_2 (20%), includes five species (N_2, O_2, NO, N, O). It is known that to characterize the flowfield within the shock layer, chemical and thermal non-equilibrium effects must be accounted for, in particular, when kinetic processes (such as dissociation) are enhanced by vibrational excitation. The most conventional and common methodology to tackle this modeling problem is the multi-temperature approach (see chapter 6), which assigns to each molecule an independent vibrational temperature (described by a related continuity equation) which, in turn, affects the chemical rate coefficients. An alternative, the state-to-state approach (see chapters 12 and 13), describing the evolution of the vibrational distributions of the molecular ground electronic state, is being investigated by a growing number of researchers.

In the case of super-orbital atmospheric entry, conditions encountered by a space vehicle returning from outer planets or by a meteoroid, the kinetic energy to be dissipated is very large, leading to temperatures which are large enough to cause ionization. At the same time, the electronically excited states of atoms and molecules become relevant in affecting the flow properties and the radiative signature of the plasma in the shock layer. In this context, the state-to-state approach also becomes a fundamental tool to model high-enthalpy flows. At these energies, almost all the molecules are dissociated and the level kinetics of atomic species must be considered. The state-to-state approach for atomic systems is known in the literature as the collisional-radiative model, where the level distribution is determined by the balance among ion recombination, excitation and ionization by electron impact and radiative decay.

The computational cost of a collisional-radiative model simulation is typically 1–2 orders of magnitude larger than that required by multi-temperature models. As a

doi:10.1088/2053-2563/aae894ch8

result the use of state-to-state models is not popular among the hypersonic fluid-dynamics community, and often remains limited to 0D/1D calculations. To overcome this issue, reduced-order models are being developed, such as coarse-grained models based on energy level grouping, described in section 8.1.

Collisional-radiative models often assume a Maxwellian distribution for electron energy, and the related electron temperature is a parameter determining the rate coefficients of electron induced processes such as excitation and ionization. In fluid-dynamic calculations, electron temperature is described by an additional continuity equation. However, non-equilibrium electron energy distribution functions (EEDFs) can play an important role in the level and chemical kinetics. The solution of a suitable Boltzmann equation is the answer to this problem. Moreover, the mutual interaction between the level population and EEDF must also be considered, requiring the simultaneous solution of the chemical kinetics and the Boltzmann equation. This approach, known as self-consistent state-to-state kinetics, is described in section 8.2.

8.1 Modeling of non-local thermodynamic equilibrium plasmas

Accurate prediction of high-enthalpy plasma flows requires, in general, accounting for non-local thermodynamic equilibrium (NLTE) consequences of collisional and radiative processes. Under conditions where the length and time scales of the problem allow for adopting a hydrodynamic description (e.g. Euler/Navier–Stokes), the most accurate formulation is provided by a state-to-state fluid (StS) approach [1–26]. In the StS approach, the internal energy states, s_i, of a chemical component s are treated as separate *pseudo-species* and, for one of these, a balance equation is written

$$\frac{\partial \rho_{s_i}}{\partial t} + \nabla \cdot (\rho_{s_i}\mathbf{v} + \mathbf{J}^{\mathrm{m}}_{s_i}) = \omega_{s_i}, \tag{8.1}$$

where the symbols ρ_{s_i} and $\mathbf{v}$ denote, respectively, the partial density of the state s_i and the bulk velocity of the gas mixture. The constitutive relations for the mass diffusion fluxes, $\mathbf{J}^{\mathrm{m}}_{s_i}$, and the production terms due to kinetic processes, ω_{s_i}, should be obtained by applying a multi-scale Chapman–Enskog expansion to the Boltzmann equation of the Kinetic Theory of gases [27–29]. The complete StS fluid description of a plasma is provided by equation (8.1) coupled with the global momentum and energy governing equations, and the balance equation for the energy of free electrons [30–33],

$$\frac{\partial}{\partial t}\left(\frac{3}{2}p_{\mathrm{e}}\right) + \nabla \cdot \left(\frac{3}{2}p_{\mathrm{e}}\mathbf{v} + \mathbf{J}^{\mathrm{e}}_{\mathrm{e}}\right) = -p_{\mathrm{e}}\nabla \cdot \mathbf{v} + \Omega_{\mathrm{e}}, \tag{8.2}$$

where the free-electron pressure is $p_{\mathrm{e}} = n_{\mathrm{e}}k_{\mathrm{B}}T_{\mathrm{e}}$, with quantities n_{e} and T_{e} denoting, respectively, the free-electron number density and temperature, whereas k_{B} is the Boltzmann constant. The introduction of a separate temperature for free electrons, T_{e}, is necessary in light of the inefficient energy transfer in electron–heavy-particle collisions (due to the large mass disparity between the colliding partners) [31]. The occurrence of radiative transitions (i.e. photo-ionization, line emission/absorption) complicates the modeling as the flow governing equations (8.1)–(8.2) must be coupled to the transport equation for the radiation field [34–36]. This equation can be obtained from the kinetic equation for the photon distribution function and, under conditions of (i) steady-state radiation field and (ii) the absence of scattering and refraction, reads [35]

$$\mathbf{\Omega} \cdot \nabla I_\lambda = \varepsilon_\lambda - \kappa_\lambda I_\lambda, \tag{8.3}$$

where $I_\lambda = I_\lambda(\mathbf{r}, \mathbf{\Omega})$ denotes the monochromatic intensity, whereas quantities ε_λ and κ_λ are, respectively, the (phenomenological) emission and absorption coefficients [35]. The former depend on the local thermodynamic state of the material gas. The intrinsic coupling between fluid and radiation can be readily appreciated by considering, for instance, bound-free (BF) transitions occurring in photo-ionization (PI) reactions, $s_i + h_P v \rightarrow s_j^+ + e^-$ (where h_P is the Planck constant, $v = c/\lambda$ the photon frequency and c the vacuum speed of light). The mass production term and the monochromatic absorption coefficient for this case are [37]

$$\omega_{s_i}^{\mathrm{BF}} = m_s\, n_{s_i} \left(\frac{4\,\pi}{h_P c}\right) \int_0^{\lambda_{s_{ij}}} \sigma_{s_{ij}}^{\mathrm{PI}}(\lambda)\, J_\lambda\, \lambda\, d\lambda, \quad \alpha_\lambda^{\mathrm{BF}} = n_{s_i}\, \sigma_{s_{ij}}^{\mathrm{PI}}(\lambda), \tag{8.4}$$

where m_s is the mass of the chemical component s, whereas σ^{PI} denotes the total photo-ionization cross section. The directionally averaged monochromatic intensity (or, simply, average intensity) is $J_\lambda = (1/4\pi) \oint I_\lambda(\mathbf{\Omega})\, d\mathbf{\Omega}$. The upper limit, $\lambda_{s_{ij}}$, in the integral in equation (8.4) corresponds to the threshold photon energy for the process under consideration, $h_P c/\lambda_{s_{ij}}$. Equation (8.4) shows that the number of photo-ionization events per unit time depends on the local radiation field. This is in turn influenced by the local state of the gas through the dependence of the BF absorption coefficient, $\alpha_\lambda^{\mathrm{BF}}$, on the population of the bound state s_i. The same conclusions also hold for other types of radiative transitions such as bound–bound and free–free [35].

The self-consistent and coupled solution of the equations of radiation hydrodynamics (8.1)–(8.3) in unsteady and multi-dimensional configurations is impractical for the following reasons:

- The adoption of a StS approach leads to large sets of species and kinetic processes compared to conventional (and less accurate) multi-temperature models [38–40]. To give a practical example, the *ab initio* database for the $N_2(^1\Sigma_g^+) - N(^4S_u)$ system developed during the last ten years by the Computational Quantum Chemistry Group at NASA Ames Research Center provides data for more than 20×10^6 million transitions involving the 9390 ro-vibrational states of the ground electronic state of the N_2 molecule [41–45].
- Obtaining cross sections and thermal rate coefficients for elementary processes (e.g. line emission/absorption, electron impact excitation) usually requires complex quantum chemistry calculations [46, 47].
- The inclusion of radiative processes introduces a global coupling between all points of the material gas. Moreover, resolving the spectral and directional dependence of the radiation field requires one to *attach* a tangent space

$(\lambda, \mathbf{\Omega})$ to each point of the position space, $\mathbf{r}$, leading to even more costly calculations [37].

The above considerations motivate the development of reduced-order models with the following desired features: (i) retaining, as much as possible, the accuracy of StS formulations and (ii) overcoming the *empiricism* and lack of theoretical foundations of conventional multi-temperature models. This is the subject of section 8.1.1 which discusses the construction of reduced-order kinetic models based on the maximum entropy principle. The accuracy of these models is assessed in section 8.1.2 by comparison with StS predictions in the study of radiative shock waves in a hydrogen plasma.

8.1.1 Development of reduced-order models

The development of reduced-order kinetic models is performed within the framework of the principle of maximum entropy [48–51]. The procedure consists of two steps as follows:

- Lump the discrete energy states into groups.
- Prescribe a local (and approximate) representation of the population within each group subjected to a series of moment constraints (e.g. mass, energy).

Following the principle of maximum entropy, the logarithm of the normalized population within the group k is written as a polynomial in internal energy,

$$\ln\left(\frac{n_i}{g_i}\right) = \alpha_k + \beta_k E_i + \gamma_k E_i^2 + \text{H. O. T.}, \tag{8.5}$$

where quantities E_i and g_i are, respectively, the energy and the statistical weight of the state i (the chemical component index s has been dropped in order not to burden the notation). The group properties α_k, β_k and γ_k are obtained by imposing a series of moment constraints. In the case of a first-order approximation, these constraints correspond to group mass and energy conservation,

$$\tilde{n}_k = \langle n_i \rangle_k^0, \quad \tilde{n}_k \tilde{E}_k = \langle n_i \rangle_k^1, \tag{8.6}$$

where the notation ~ denotes a group-averaged property, whereas the group moment operator of order r is defined as

$$\langle f_i \rangle_k^r = \sum_{i \in \mathcal{I}_k} f_i E_i^r, \tag{8.7}$$

where the set I_k stores the states contained within the group k. The substitution of equation (8.5) in the moment constraints (8.6) leads, after some algebraic manipulation, to the following result [20, 24, 51],

$$\frac{n_i}{g_i} = \frac{\tilde{n}_k}{\tilde{Z}_k} \exp\left(-\frac{E_i}{k_B T_k}\right), \tag{8.8}$$

where the group *internal* temperature and partition function are $T_k = -1/k_B \beta_k$ and $\tilde{Z}_k = \sum_{i \in \mathcal{I}_k} g_i \exp(-E_i/k_B T_k)$, respectively. Equation (8.8) shows that truncating the polynomial expansion (8.5) at first-order is equivalent to assuming a Maxwell–Boltzmann distribution, at its own *internal* temperature, within each group (maximum entropy linear (MEL) model). The zeroth-order approximation may be obtained by taking the infinite *internal* temperature limit of the MEL model (i.e. $\beta_k = 0$). This gives a uniform distribution on a Boltzmann plot as sketched in figure 8.1 (maximum entropy uniform (MEU) model). The intrinsic problem of the MEU model is the impossibility of retrieving equilibrium (i.e. the Maxwell–Boltzmann distribution).

The governing equations for the reduced-order models are obtained by applying the group moment operator (8.7) to the mass balance equations (8.1):

$$\left\langle \frac{\partial \rho_i}{\partial t} + \nabla \cdot (\rho_i \mathbf{v} + \mathbf{J}_i^{\mathrm{m}}) \right\rangle_k^r = \langle \omega_i \rangle_k^r. \tag{8.9}$$

For the sake of consistency, the group moment operator (8.7) must also be introduced in all quantities (e.g. emission and absorption coefficients) involving states being grouped. For more details, the interested reader may consult [24, 51]. Reduced-order models developed based on the maximum entropy principle have been successfully applied to studying excitation, dissociation and ionization in atomic and molecular gases [14, 20, 24, 51].

Before concluding this section and moving on to applications, it is worth discussing the following points:

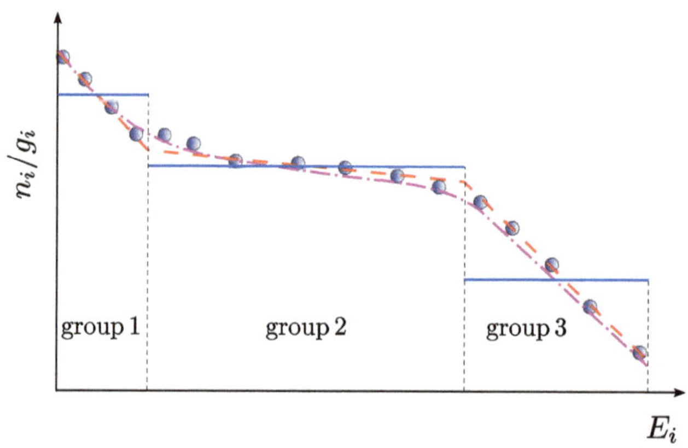

Figure 8.1. Example of application of the maximum entropy principle for the construction of a reduced-order model in the case of three energy groups. In the zeroth-order approximation, the actual NLTE distribution (circles) is approximated by three energy groups having infinite internal temperature (unbroken line). The first-order approximation introduces finite internal temperatures and is equivalent to assuming a Maxwell–Boltzmann distribution within each group (dashed line). The introduction of second-order terms (dashed-dotted line) allows for a more accurate representation at the cost, however, of a more complex formulation.

- The moment equations (8.9) assume the availability of kinetic data (e.g. cross sections and rate coefficients) for elementary processes such as excitation and ionization. This may be feasible for atomic plasmas, but soon becomes intractable for ro-vibrationally resolved molecular systems. One of the proposed methodologies to circumvent this issue consists in applying the averaging operator (8.9) while performing quasi-classical trajectory (QCT) calculations to obtain group mass and energy production terms. This QCT-based grouping approach (called the coarse-grained quasi-classical trajectory (CG-QCT) method) leads to large savings of both CPU time and storage, and has been successfully applied to the $N_2(^1\Sigma_g^+)$ – $N_2(^1\Sigma_g^+)$ system [52, 53].
- The general polynomial expansion (8.5) may also be applied to situations that require going beyond a hydrodynamic description (e.g. flow across a normal shock wave, discharges, lasers) [7, 54]. To this aim, one may replace the discrete internal energy, E_i, with the continuous translational energy, ε, and apply the moment operator (8.6) (where the sum now becomes an integral) to the Boltzmann equation for gas particles. Preliminary results making use of this methodology have been obtained in [55] for the model kinetic equation proposed by Bathnagar, Gross and Krook (the BGK model) [56].
- The radiation field is still treated in a *line-by-line* fashion with no simplifying assumption on the directional dependence of the monochromatic intensity. To reduce the computational cost, one may consider the use of opacity binning methods in analogy with the grouping approach employed for gas particles [57]. However, this possibility is not explored in the current work.

8.1.2 Application: radiative shock waves in a hydrogen plasma

This section assesses the accuracy of the reduced-order MEU and MEL models discussed in section 8.1.1. The application considers the structure of one-dimensional radiative shock waves in a hydrogen plasma. Details on the physical model (e.g. rate coefficients, energy levels) and the numerical method can be found in [24]. Radiative shock waves occurring in laboratory and astrophysical plasmas were among the first NLTE radiation hydrodynamics problems to be investigated. Examples are the early works (this list is not complete) of Whitney and Skalafuris [58], Skalafuris [59–61], Murty [62, 63], Clarke and Ferrari [64], and the most recent studies by Fadeyev and Gillet [65–68], Kapper and Cambier [22, 23], Panesi *et al* [69, 70], Colonna *et al* [25, 26] and Munafò *et al* [24]. Under the assumption of inviscid flow (e.g. no electron heat conduction), the structure of a radiative shock wave in an atomic plasma can be, in general, subdivided into four regions (see figure 8.2): (i) a far precursor, (ii) a near precursor ahead of the gas-dynamic jump, (iii) an internal relaxation region behind the shock dominated by collisional excitation and ionization, and (iv) a radiative relaxation (or cooling) region where the temperature drops due to the escape of optically thin radiation produced by radiative recombination. In the far precursor, atoms may undergo excitation by absorbing resonant radiation in atomic line wings [71]. These excited states may be then photo-ionized by absorbing low-energy photons. On the other hand, in the near precursor,

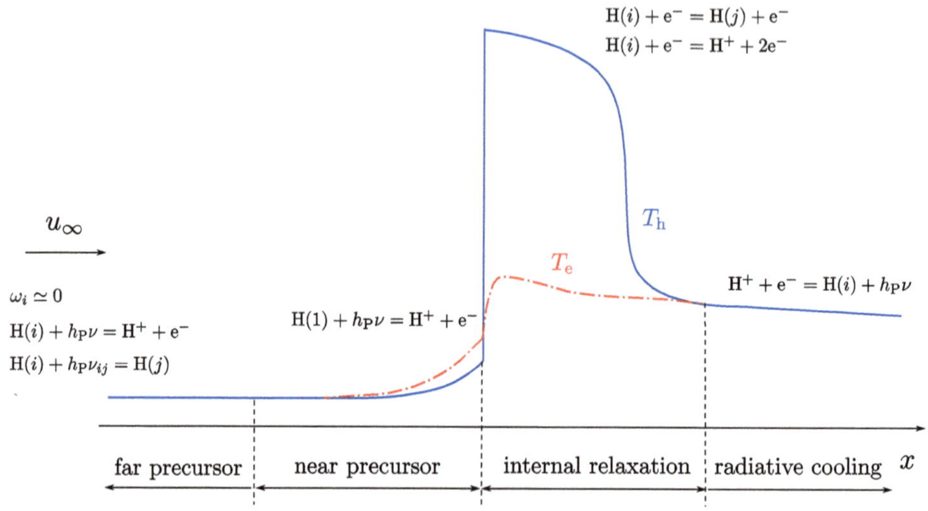

Figure 8.2. Structure of a radiative shock wave in hydrogen (the length-scales assigned to the various zones, such as the radiative cooling region, do not refer to any specific conditions and have been chosen only for illustrative purposes).

photo-ionization occurs from the ground state due to absorption of Lyman *continuum* photons. The extent (and also the existence) of the aforementioned zones strongly depends on the shock velocity, and the free-stream values of pressure and temperature [72].

In the current work, the free-stream pressure, temperature and velocity are set to 5 Pa, 5000 K and 40 km s^{-1}, respectively. These values correspond to conditions found in the outer layers of pulsating stars [65–68]. It is important to mention that, due to radiation, the chemical composition of the gas far upstream from the shock cannot be set to LTE. Instead, the free-stream composition must be obtained from the solution of the problem by imposing the condition of statistical equilibrium, $\omega_i = 0$, (or quasi-steady-state (QSS)) for the population of hydrogen bound-states [37].

Figure 8.3 shows the evolution of the free-electron mole fraction, and the temperatures of heavy particles and free electrons for the StS model. To underline the impact of radiation on the flowfield, predictions obtained accounting only for collisional processes are also reported. The extent of the near precursor is two orders of magnitude larger than the internal relaxation region. The structure of the latter results from the *avalanche/cascade* ionization process where the energy lost by free electrons in excitation and ionization is replenished by elastic collisions with heavy particles, until an equilibrium state is reached where ionization is balanced by three-body recombination [34].

The inclusion of radiative transitions has the effect of shrinking the length of the internal relaxation region, due to the larger degree of ionization ahead of the shock caused by photo-ionization. In the radiative cooling region, the degree of ionization drops due to radiative recombination. In this zone, the plasma is in thermal equilibrium, $T_h \simeq T_e$, and its temperature is lower than the post-shock LTE value for the non-radiating case due to the escape of optically thin radiation. This can also

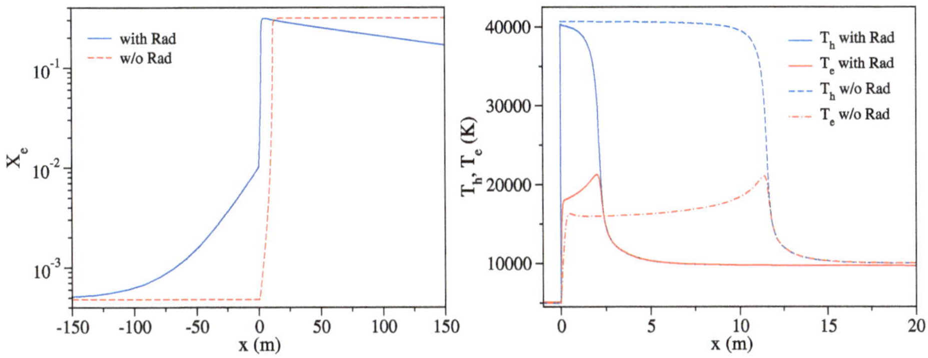

Figure 8.3. Free-electron mole fraction (left), and heavy-particle and free-electron temperature (right) evolution with and without radiation (StS model). Note that the temperature evolution refers to the internal relaxation region only.

be observed from the average intensity 10 m behind the shock (see figure 8.4) which shows that the first three lines of Balmer and Paschen series are in the emission. On the other hand, the Lyman *continuum* part of the spectrum is in equilibrium, as confirmed by the comparison with the Planck function evaluated at the local free-electron temperature.

The StS predictions discussed above are compared with those obtained using the MEU and MEL reduced-order models to assess their accuracy. To this aim, the grouping illustrated in figure 8.5 is adopted. This scheme is developed by taking care not to include, within the same group, those states coupled by strong radiative transitions (e.g. the Hα line). In the present case, the number of groups is equal to 3 (MEU(3) and MEL(3) models). The first contains only the ground electronic state. The second contains the states with $n = 2$, where n denotes the principal quantum number [73]. The third contains all the remaining excited states (i.e. $n > 2$), where the population of bound-states is approximated by a local uniform or Boltzmann distribution depending on the reduced-order model in use.

Figure 8.6 compares the StS and MEU(3)/MEL(3) predictions in terms of free-electron mole fraction and temperature. The solution obtained using the MEL(3) model is essentially indistinguishable from the StS prediction. This is an important result because it means that, for the conditions adopted in this work, the StS flowfield can be reproduced using only four variables: (i) the number densities of the three groups and (ii) the *internal* temperature of the third/last group. The MEU(3) model shows significant deviations from the StS results, in particular in the internal and radiative relaxation regions. The reason behind this lack of accuracy is the use of a uniform distribution for the population of high-lying states. This is demonstrated in figure 8.7, which displays the normalized population predicted by the three models 2.5 m behind the shock (internal relaxation region). It is worth recalling that, for the MEU(3) and MEL(3) models, the population distribution is obtained *a posteriori* by substituting the computed group properties in equation (8.5).

The uniform distribution poorly approximates the under-populated high-energy tail of the actual distribution, with dramatic consequences on macroscopic quantities

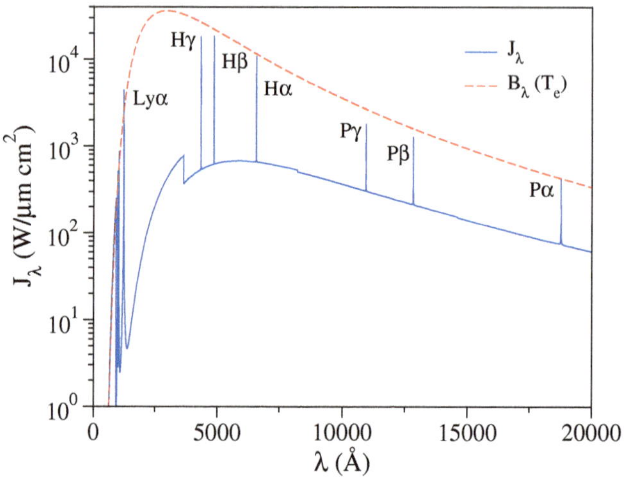

Figure 8.4. Average monochromatic intensity at $x = 10$ m behind the shock front (radiative relaxation region; StS model). The dashed line represents the Planck function, B_λ, evaluated at the local value of the free-electron temperature.

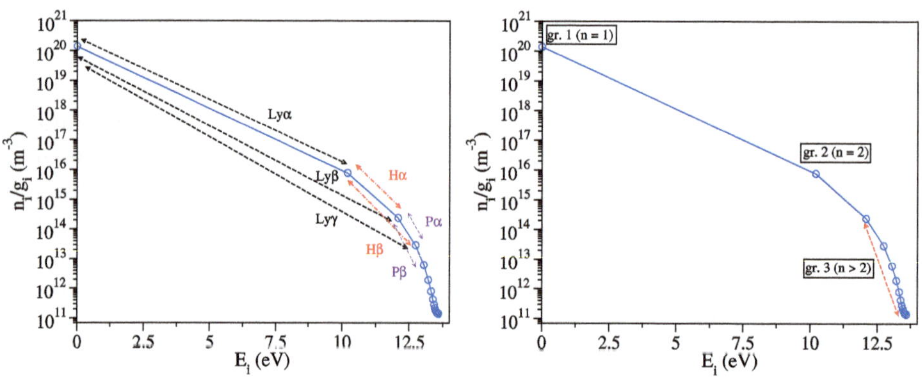

Figure 8.5. Lumping scheme adopted for the electronic states of the hydrogen atom. The panel on the left shows a typical NLTE distribution during ionization highlighting the main radiative bound–bound transitions. The panel on the right illustrates the three energy group lumping schemes. The first group contains only the ground state. The second group contains the states having $n = 2$. The third group contains all the other states (i.e. $n > 2$).

such as chemical composition and temperature (see figure 8.6). The introduction of an *internal* temperature (MEL model) leads to a more accurate representation. To further improve the agreement with the StS distribution, one may consider splitting the third group into two or three parts (MEL(4) and MEL(5) models, respectively) or using a second-order local representation for the third group, bearing in mind that this gain in accuracy comes at the cost of a more complex and expensive formulation.

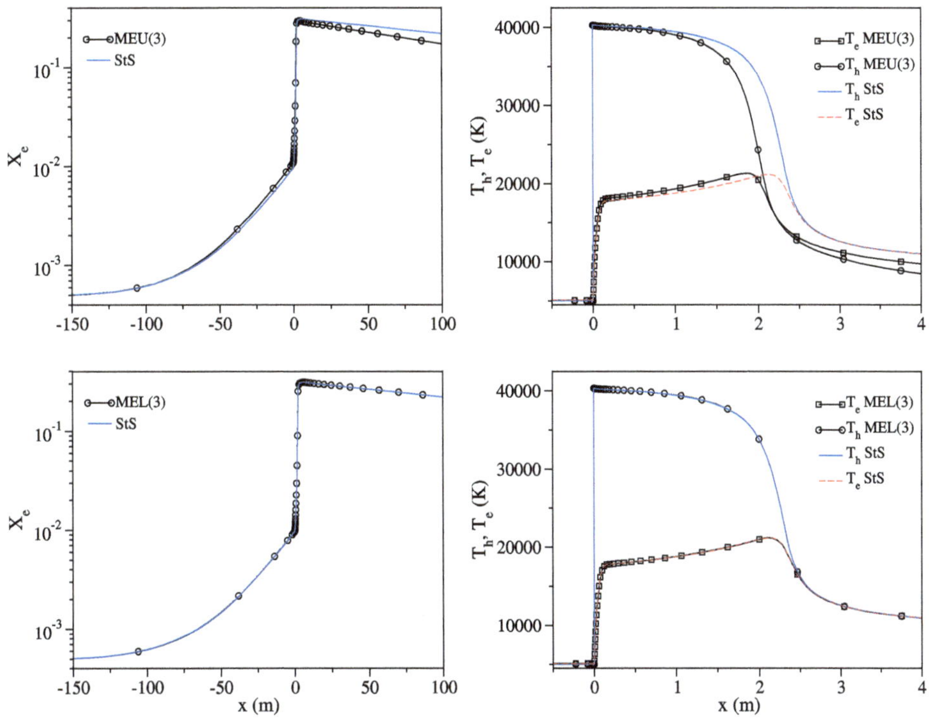

Figure 8.6. Comparison between StS, MEU(3) and MEL(3) models for the free-electron mole fraction (left), and heavy-particle and free-electron temperatures (right).

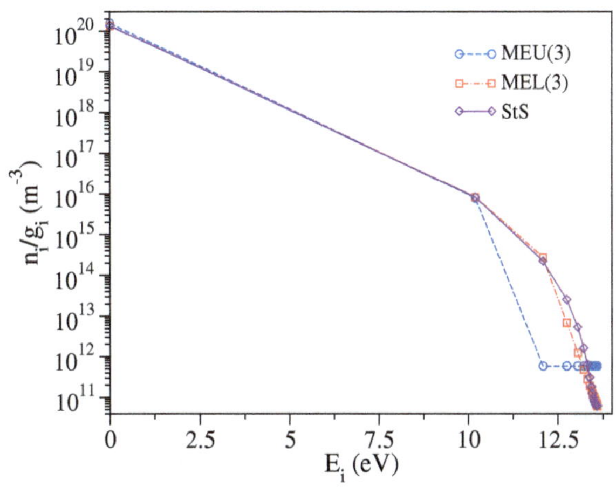

Figure 8.7. Population distribution predicted by the StS, MEU(3) and MEL(3) models at $x = 2.5$ m behind the shock front (internal relaxation region).

8.2 Self-consistent state-to-state approach

The state-to-state model described in the previous sections determines the electron energy from a continuity equation (see (8.2)). This macroscopic approach assumes a Maxwellian EEDF, resulting in Arrhenius type rates expressed as a function of the free-electron temperature T_e [5, 38, 74]. Alternatively, in the cold plasma community, the Boltzmann equation [75–78] is solved to account for the EEDF departing from the Maxwell distribution [79]. In a common approach, assuming that the relaxation time of the EEDF is much smaller than that of the chemical processes, due to the small electron–molecule mass ratio, a stationary Boltzmann equation is solved [80, 81]. Therefore, the EEDF, and as a consequence also the rate coefficients of electron induced processes, depends only on the reduced electric field (E/N). This approach is known as the *mean local field approximation* (LFA) [82]. Moreover, under the same assumptions and the mean electron energy ($\bar{\varepsilon}_e$) being a growing function of E/N, the rate coefficients are tabulated against $\bar{\varepsilon}_e$. This approach is known as the *mean local energy approximation* (LEA) [83]. The stationary EEDF is also considered in a high-frequency field [84], correcting the mean electric field by a factor depending on the elastic collision frequency, the rapid oscillations of the EEDF around the mean distribution being ineffective.

Both the LFA and LEA neglect the contribution of excited states to the EEDF, for this reason they are also known as *ground-state approaches*. The LEA, having as an independent variable the electron temperature or the electron mean energy, presents a clear advantage in fluid dynamics, when a continuity equation for the electron energy is included in the model (see equation (8.2) for example). These approaches are commonly applied to discharge modeling, because, creating *a priori* look-up tables of the electron properties, they avoid the solution of the Boltzmann equation in each time step. The main advantage of ground state approaches is the low computational effort required for determining the rates of electron induced processes. Moreover, the Boltzmann solver *Bolsig+* [85, 86] is freely available through the *LXCat* website [87] for this purpose. However, distributions and related quantities are also functions of the population of excited levels, as shown in [88] for helium plasma.

To overcome the limits of the ground-state models, a more complex approach must be used, the so-called self-consistent state-to-state (SC-StS) kinetics [89]. It consists in solving, at the same time, the Boltzmann equation for free electrons and the StS master equation, allowing one to account for the coupling between the EEDF and the internal level distributions [90–93]. In particular, for each time step, the self-consistent loop (see figure 8.8) is solved: starting from the Boltzmann equation, the EEDF is calculated, and is needed to determine the rate coefficients of electron impact induced processes, the quantities entering in the master equation. Solving the master equation gives as a result the gas composition and level distributions, input data for the Boltzmann equation. To have a fully non-linear solver, the self-consistent loop can be repeated more than once for each time step [94, 95].

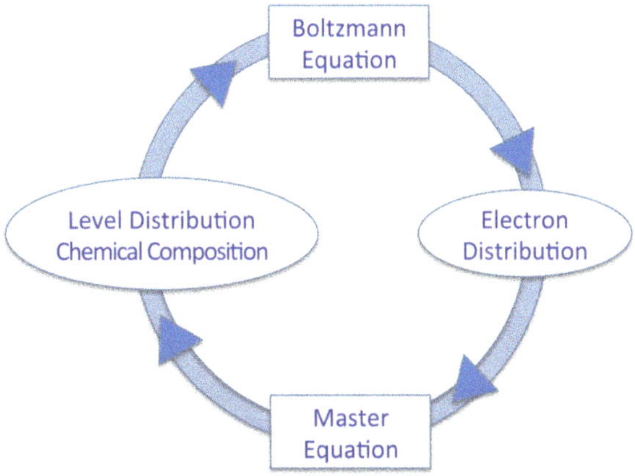

Figure 8.8. Scheme of the self-consistent loop.

The self-consistent approach is more accurate than the ground-state approximation, in accounting for the effects of excited states in the electron kinetics. Due to superelastic collisions [96, 97], the EEDF shows relevant non-equilibrium structures. In figure 8.9 a schematic view of the effects of inelastic and superelastic processes on the electron energy distribution is sketched. Inelastic processes, included in the ground models, show a deflection of the EEDF, decreasing its slope for energy larger that the transition threshold $\varepsilon^{\star}$ with respect to the Maxwell distribution. On the other hand, superelastic collisions create peaks around the threshold energy, or, if an electron is subject to another superelastic collision, two peaks appear, resulting in a substantial growing of the EEDF with respect to the Maxwell distribution, at energies higher than the threshold. As a consequence, the rate coefficients of processes with energy above $\varepsilon^{\star}$ increases correspondingly. This is a simplified view of the effects of superelastic collisions, because the synergy among different processes makes the shape more complex [80].

It must be pointed out that the SC-StS approach can give a more detailed description at the expense of computational time and memory requirements. Moreover, a large number of electron impact cross sections, including transitions starting from excited states [98, 99], are necessary for consistency. As a consequence, only simplified fluid-dynamics configurations (1D) are suitable to include in SC-StS models, even if in the last few years the use of GPUs seems a promising strategy to speed up the kinetic calculations, allowing the application of the StS approach in multi-dimensional configurations [100] (see also chapter 13).

8.3 The self-consistent model in hypersonic flows

As an extension of the state-to-state approach, the self-consistent model has been applied to high-enthalpy flows. The first system investigated using the SC-StS was the boundary layer [101] of a blunt body entering a nitrogen atmosphere. This model included only vibrational kinetics and the stationary electron Boltzmann equation was solved locally.

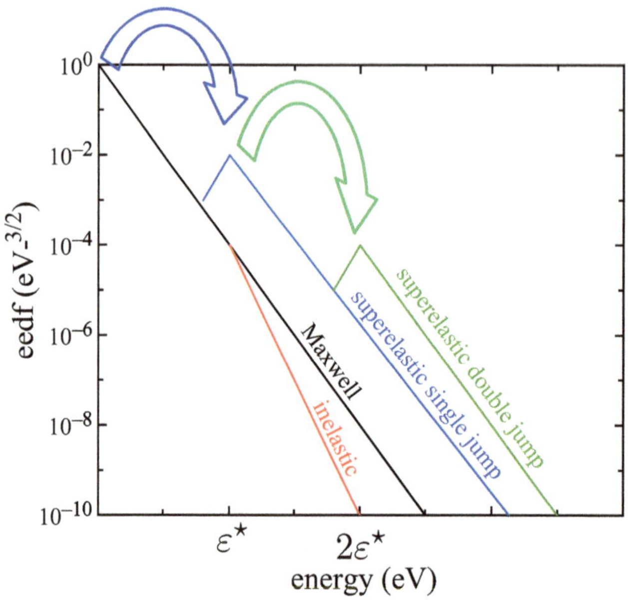

Figure 8.9. Sketch of the effects of different processes on the EEDF.

The kinetic model was fully developed for supersonic nozzle expansion in nitrogen including molecular [3, 102] and atomic [103] electronically excited states and for air [104, 105]. In a further improvement, the model was used to investigate argon supersonic flow in the presence of an electric and magnetic field [106, 107], extending the theory of the two-term expansion of the Boltzmann equation to magnetized plasmas. This model was used to rationalize experimental results in supersonic wind tunnels [108–112].

In a further evolution, the self-consistent model was coupled with the radiation transport equation in a shock tube and applied to hydrogen [25, 113, 114] and a hydrogen/helium mixture [115–117]. These papers present an alternative approach to the collisional-radiative model of atoms [118].

For non-Boltzmann and non-Maxwellian distributions, the temperature loses its meaning, becoming open to different definitions, all converging to the same value in equilibrium, but leading to different values in non-equilibrium. In particular, two definitions are commonly used. The first one assumes as temperature that of the Boltzmann (Maxwell) distribution with the same mean energy as the effective distribution. This is used to calculate the electron temperature, related to the electron mean energy as

$$T_e = \frac{2}{3}\bar{\varepsilon}_e. \tag{8.10}$$

Alternatively, it is possible to define the *two-level temperature* from the Boltzmann distribution passing from two levels, i.e.

$$T_{i,j} = \frac{\varepsilon_i - \varepsilon_j}{k_b} \ln \frac{g_i n_j}{g_j n_i}, \tag{8.11}$$

where ε_i, g_i and n_i are the energy, statistical weight and population of the ith level, respectively. In the rest of this chapter the electron temperature is calculated as in equation (8.10), while the internal temperature of heavy species is given by equation (8.11), where j and i refer to the ground and first excited states.

8.3.1 Boundary layer

The first attempt to use the self-consistent kinetics in high-enthalpy flows was focused on the boundary layer of a supersonic body entering a pure nitrogen atmosphere. The fluid-dynamic model Si based on the Lees–Dorodnitsyn transformation [119], which—considering the flow around the stagnation point, where the heat flux is maximum, and assuming the flow is self-similar along the body—calculates the profile of temperature and chemical species as a function of the coordinate η perpendicular to the surface ($\eta = 0$ on the vehicle surface and $\eta = 5$ at the limit of the boundary layer). This model assumes a fixed temperature $T_w = 1000$ K at the surface, considered to be non-catalytic, i.e. neither reactions nor vibration deactivation happen on the surface. The vibrational kinetics implemented in previous works for non-ionized gases [120–122] was improved by adding electron–molecule collisions by solving the Boltzmann equation for free electrons. In this case, the coupling was not fully self-consistent because the stationary Boltzmann equation was solved under local conditions, neglecting electron diffusion and ionization kinetics. Nevertheless, the solution was recalculated at each iteration, updating the composition and vibrational distributions, in order to include superelastic and electron–electron collisions. The cross section dataset was not complete, based on the Phelps report [123], extended with data taken from Cacciatore [124]. At the limit of the boundary layer ($\eta = 5$), local equilibrium was assumed, fixing the temperature T_∞, with the exception of the ionization degree (α), varied as a parameter and kept constant along the boundary layer.

The results showed that the electron kinetics is governed by the vibrational distribution function (VDF), attributing a small effect of electron–atom collisions on the EEDF. Moreover, the coupling between electron and vibrational kinetics grows in importance only for ionization degree[1] α above 10^{-4}, but only for $\alpha > 10^{-3}$ this contribution becomes really effective. In any case, EEDFs are far from equilibrium, as can be observed in figure 8.10(left). Moreover, the EEDF and VDF, figure 8.10(right), at the surface strongly depend on the ionization degree. Remembering that, at the edge of the boundary layer, the VDF were in equilibrium with the gas temperature, the differences observed on the surface are due to the coupling between electrons and molecular vibration. Accounting for e–M coupling results in the faster cooling of EEDF and VDF.

[1] $\alpha = N_e/N_h$, the ratio between electron, e, and heavy-particle, h, densities.

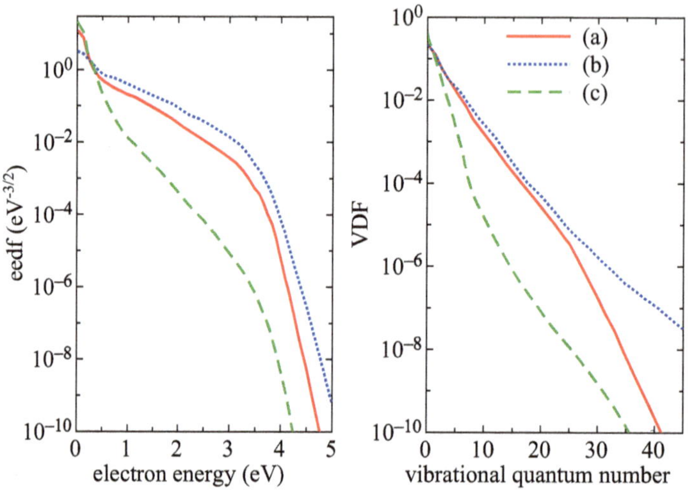

Figure 8.10. Electron (left) and N_2 vibrational (right) distributions in the boundary layer of an hypersonic body entering a pure nitrogen atmosphere. (a) $\alpha = 10^{-7}$, $\eta = 0$; (b) $\alpha = 10^{-7}$, $\eta = 5$; and (c) $\alpha = 10^{-3}$, $\eta = 0$.

These behaviors were obtained in the absence of atom–molecule (A–M) processes, namely vibration–translation (VTa) and dissociation–recombination (DRa), whose presence makes the e–M collision less effective, being the distribution tails affected by DRa and the low-energy VDF by VTa. The importance of A–M processes is proportional to the atom density and, as a consequence, to T_∞. Therefore, the largest effects of e–M processes are reached for the lowest value of atom density and for the largest value of ionization degree, i.e. $T_\infty = 5000$ K and $\alpha = 10^{-3}$. The vibrational distributions with two different values of T_∞, with and without A–M processes, are compared in figure 8.11. The plateaux in the distribution tails, absent when A–M processes have been neglected, are due to three-body recombination. The recombination model considered here assumes that the formed molecules can be in the states $v = 25$, corresponding to the peak in figure 8.11, and $v = 45$, the last vibrational level. The height of the plateaux in cases (b) and (d) is influenced by the atom molar fraction at the edge of the boundary layer. When A–M are neglected, i.e. without recombination, a colder distribution tail is present for the lower temperature case (c), a consequence of the larger molecular density resulting in faster vibrational–translational relaxation by molecular collisions (VTm).

To better understand the role of VTa and e–M vibrational excitation, figure 8.12 compares the corresponding rate coefficients at the edge of the boundary layer (BL) ($\eta = 5$) and on the surface ($\eta = 0$). The rate coefficients of e–M are reported as multiplied by α, considering that their contribution to the source terms is proportional to the electron density. At the edge of the boundary layer, both rates are of the same order of magnitude, even if the e–M are higher at $T_\infty = 7000$ K and lower at $T_\infty = 5000$ K, a consequence of the peak structure of the resonant cross sections [125, 126]. On the other hand, at the surface, e–M rates are an order of magnitude higher than the VTa, due to the non-Maxwellian EEDF, heated by superelastic collisions.

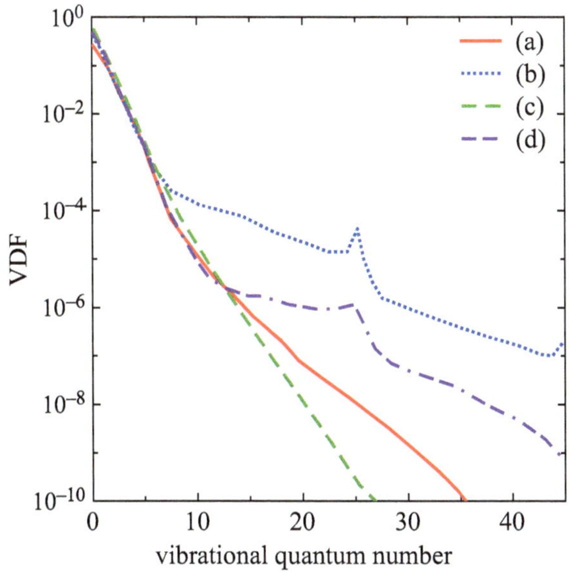

Figure 8.11. N_2 vibrational distributions adjacent to the body surface for $\alpha = 10^{-3}$, in the following conditions (a) $T_\infty = 7000$ K, without A–M; (b) $T_\infty = 7000$ K, with A–M; (c) $T_\infty = 5000$ K, without A–M; and (d) $T_\infty = 5000$ K, with A–M.

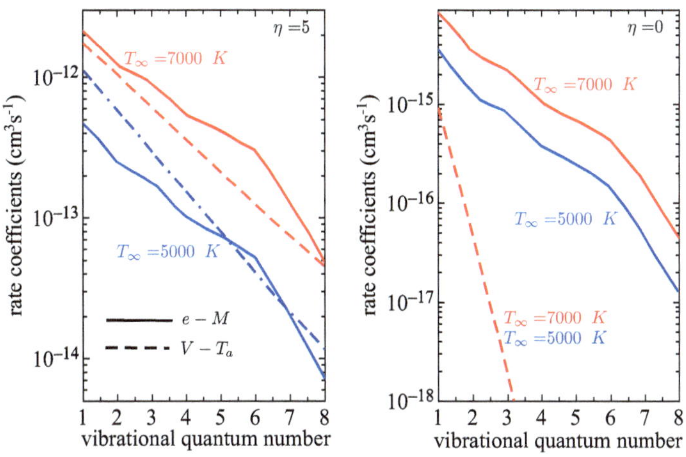

Figure 8.12. Comparison of e–M and VTa rate coefficients of the $0 \rightarrow v$ transition as a function of the vibrational quantum number for $\alpha = 10^{-3}$ at $\eta = 5$ (left) and $\eta = 0$ (right). e–M rates have been multiplied by α.

The non-equilibrium character of the distributions also affects macroscopic quantities such as the temperature profiles. As the e–M rates grow with α, the translation temperature along the BL decreases correspondingly, as can be observed in figure 8.13(left), where the profile of translational, vibrational and electron temperatures calculated for $\alpha = 10^{-3}$ and $T_\infty = 7000$ K are reported, together with

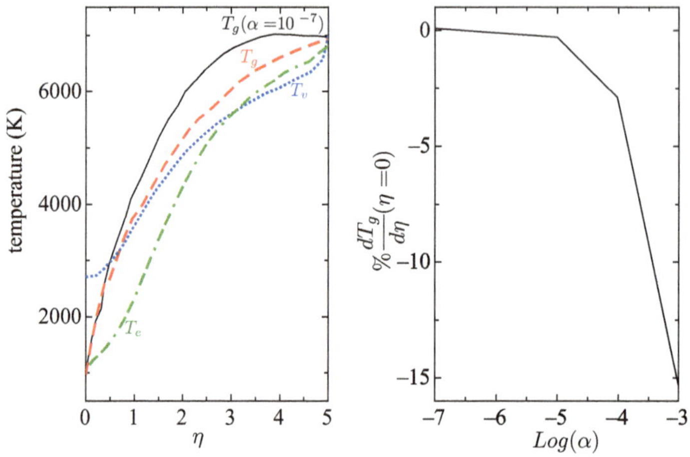

Figure 8.13. Comparison of gas (T_g), electron (T_e) and vibrational (T_v) temperature profiles for $\alpha = 10^{-3}$ with the gas temperature for $\alpha = 10^{-7}$ ($T_g(\alpha = 10^{-7})$) (right). On the left is shown the percent variation, with respect to the value for $\alpha = 10^{-7}$, of the temperature gradient at the surface as a function of the logarithm of α; $T_\infty = 7000$ K.

the T_g for $\alpha = 10^{-7}$, where the contribution of e–M collisions is practically null. All the temperatures show different profiles. Electron temperature is determined by the interplay among elastic, inelastic and superelastic collisions, the vibrational temperature by VT and e–M processes, with a small contribution of VV, and the gas temperature is the balance between translational, internal and chemical energy. The difference in the profiles of the gas temperature for $\alpha = 10^{-3}$ and $\alpha = 10^{-7}$ shows that the processes induced by electron collisions change the global temperature. As a consequence, the temperature gradient is also influenced by e–M collisions. In particular, $\partial T_g / \partial \eta$ at the surface, proportional to the conductive heat flux to the body, is a function of α (see figure 8.13(right)). In [101], this quantity was compared in the cases (a)–(d) considered in figure 8.11, showing that the largest effects are obtained in case (b), where A–M processes are considered and $T_\infty = 7000$ K, contrary to what was obtained for the VDF, where the largest effects are observed in the absence of A–M and at lower BL edge temperatures. To conclude this section, it must be pointed out that, in the presence of ionized species, charge accumulation on the surface can create an electric field, capable of affecting the distributions [101] and as a consequence, also the temperature profiles.

8.3.2 Nozzle flow

In supersonic nozzle expansion, non-equilibrium is more pronounced than in the boundary layer, due to the very low temperature and pressure and high velocity in the divergent section. The pure vibrational kinetics [2, 127] is extended by including the kinetics of molecular and atomic electronically excited states of nitrogen [3, 103] and oxygen [105]. The stationary nozzle equations are solved in the quasi-1D approximation [128], starting from the reservoir temperature, T_0, pressure, P_0, and composition. As a result, the profile of the flow properties along the nozzle axis is calculated, taking as the reference position the throat. Usually, equilibrium compositions are reached in the reservoir, with few exceptions [110]. As a consequence, electron density, species concentrations and excited state populations in the reservoir, and therefore also at the nozzle exit, depend only on T_0 and P_0.

As discussed above (see figure 8.9), electronically superelastic collisions considerably affect the EEDF, and as a consequence the whole kinetics. Vibrational superelastic collisions, due to the small energy threshold, give a secondary contribution, slightly increasing the electron temperature, mimicking an increase of the electric field [80].

The effects of superelastic collisions in nozzle expansion have been discussed in different papers [3, 103, 105] using the SC-StS approach. In particular, in [103] the synergy between electron–electron (e–e) and superelastic collisions by atomic nitrogen are investigated for different reservoir temperatures and for the test cases reported in table 8.1, in the 4 m long F4 nozzle. The EEDFs at the nozzle exit (see figure 8.14) are strongly influenced by the atomic superelastic collisions in the whole temperature range considered, while e–e collisions are relevant only at very high reservoir temperatures, where the ionization degree α is very large ($>10^{-2}$).

There is no process involving $N^\star$ and $N_2^\star$, nevertheless, if the electron molar fraction is sufficiently high, the action of superelastic collisions with atomic excited states on the EEDF is reflected in the molecular level distribution. For example the VDF (figure 8.15) shows relevant sensitivity to the e–e collisions and $N^\star$ only for $T_0 = 10\,000$ K, while at $T_0 = 7000$ K the VDF is practically the same in all the considered simulations. For electronically excited states of nitrogen molecules (figure 8.16) some effects are observed at the nozzle exit for $N_2(A^3\Sigma)$ also when $T_0 = 7000$ K, while the other levels are practically insensitive.

As a further improvement, the role of electronically excited states was investigated in an air mixture [105], with a more complex chemistry, flowing through a short conic nozzle. In reference [105] the roles of nitrogen and oxygen electronically excited states are analyzed separately, both including and neglecting the electron Boltzmann equation. Electronically excited states have a large influence on the flow properties, but the more pronounced differences are obtained in combination with electron kinetics. To this purpose, in figure 8.17(a) and (b), the results obtained with the full model 'f', including the kinetics of atomic and molecular electronically excited states and the electron Boltzmann equation, are compared to those of the pure vibrational kinetics 'v' showing large differences in the NO and O_2 molar

Table 8.1. The processes included in different test cases for the F4 nozzle [103] and molar fractions of electrons and atoms in the ^2D state. e–e: electron–electron collisions; $N^\star$: atomic nitrogen metastable kinetics.

Cases	a	b	c	d	Molar fractions	$T_0 = 5000$ K	$T_0 = 7000$ K	$T_0 = 10\,000$ K
e–e	×	0	×	0	Electrons	8.9×10^{-7}	3.8×10^{-4}	2.4×10^{-2}
$N^\star$	×	×	0	0	N (^2D)	3.7×10^{-4}	2.9×10^{-2}	1.3×10^{-1}

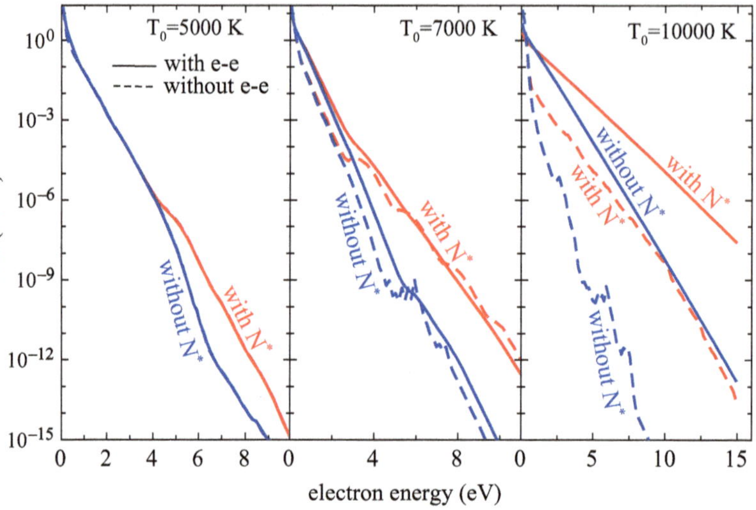

Figure 8.14. EEDF at the exit of the F4 nozzle [103] for three different reservoir temperatures T_0 and for pure nitrogen mixture at $P_0 = 1$ atm. Comparison of the four models in table 8.1.

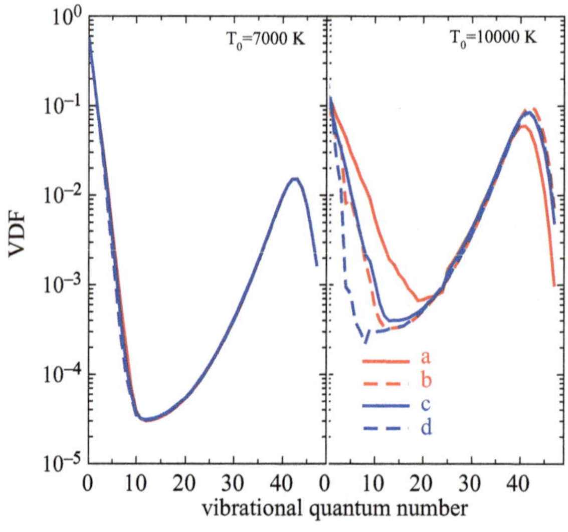

Figure 8.15. Vibrational distributions at the exit of the F4 nozzle [103] for two different reservoir temperatures T_0 and for a pure nitrogen mixture at $P_0 = 1$ atm. Comparison of the four models in table 8.1.

fractions (a) and in the vibrational distributions at the nozzle exit (b). The solution of the pure vibrational kinetics predicts the decreasing of NO and O_2, while the full model shows a more complex behavior. The NO molar fraction presents a maximum in the throat ($x = 0$), reaching a value at the exit much higher than that in the v case. The oxygen molar fraction grows up to a small maximum in the throat, followed by a small minimum and then increasing up to the exit. This

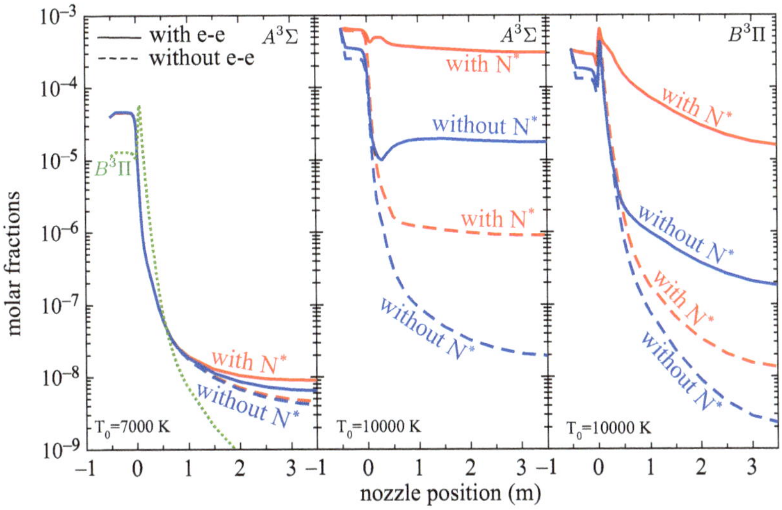

Figure 8.16. Molar fractions of electronically excited states of N_2 along the F4 nozzle [103] for two different reservoir temperatures T_0 and for a pure nitrogen mixture at P_0 = 1 atm. Comparison of the four models in table 8.1.

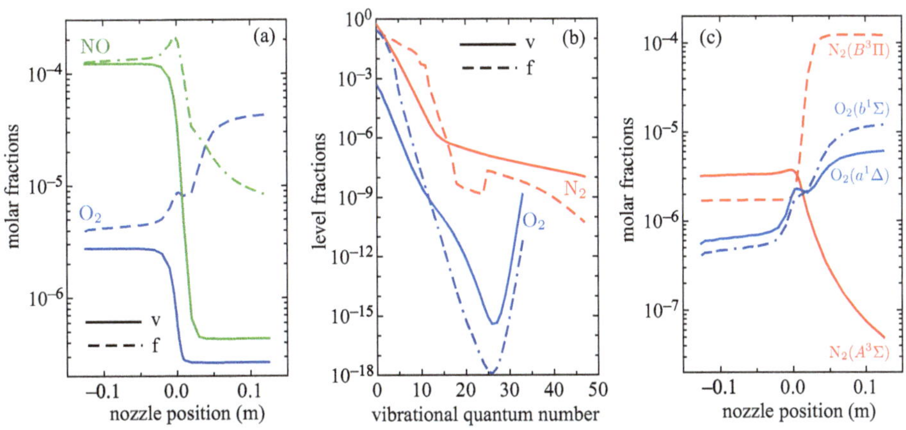

Figure 8.17. Comparison of the results of the pure vibrational kinetics 'v' and that with the full kinetic scheme 'f' including electronically excited states of atoms and molecules in the conic nozzle [105] for air at T_0 = 10 000 K and P_0 = 1 atm. (a) Molar fractions along the nozzle, (b) fraction of the vibrational levels at the nozzle exit and (c) electronically excited state molar fractions in the case of the full model.

behavior, a consequence of atom recombination in electronically excited states, is amplified by considering the electron kinetics (see [105]) for the O_2 molar fraction, while it is mitigated for NO.

Another important feature, depicted in figure 8.17(c), is the population inversion between the first two electronically excited states of nitrogen and oxygen molecules in the expansion region, an effect due to the balance between atom recombination and level quenching. This behavior was also observed in figure 8.16 for the F4 nozzle, but only in a small region close to the throat, at a distance comparable to that of the exit in the conic nozzle. For longer distances, electronically excited states cannot be produced by recombination due to the low pressure and high velocity, and radiative decay, independent of the pressure, makes the molar fraction of $B^3\Pi$ decrease rapidly.

8.3.3 Shock tubes

For a meteoroid entering a planetary atmosphere, the surface heat flux is determined by the conditions inside the shock layer. Also in this case, strong non-equilibrium has been observed, and multi-temperature models fail in predicting relevant quantities, i.e. the distance between the shock front and the body. For low-speed entry, when the ionization in the shock is negligible, the state-to-state vibrational kinetics have shown good agreement with experimental results [100]. When ionization becomes sufficiently high, radiation gives a relevant contribution to the total heat transfer [129], as a consequence of the higher dissociation degree and large concentration of electronically excited states. In section 8.1.2, the effects of a state-to-state collisional-radiative model are discussed in affecting the flow properties of the shock tube. However, in this case, the electron temperature is determined from the energy continuity equation. To overcome this limit, the SC-StS model for hydrogen [25] and hydrogen/helium [116] mixtures, also coupled self-consistently with the radiation transport equation, has been developed and applied to shock tube configurations at different free-stream flow speeds.

The above papers focus on the role of radiation in affecting the flow properties when coupled with the SC-StS model, considering also the contribution of atom–atom (A–A) collisions. The latter are fundamental processes when the atomic molar fraction becomes very large due to dissociation, activating atomic excitation and ionization in the region where electron concentration is still small. Electron concentration (figure 8.18(a)) starts to grow at $x = 0.4$ mm, x being the distance from the shock when atomic molar fraction is large enough ($>10^{-4}$) to have efficient ionization by A–A collisions. This space interval is limited to a very small region (0.4–1 mm from the shock front) and the ionization peak is reached at 3 mm. On the other hand, in the absence of A–A processes, the ionization peak is obtained at 3 m [130].

The total molar fractions (figure 8.18(a)) calculated with the self-consistent radiation transport coupling (SRC) are very close to the thick case (see also [25, 116]) while, in the thin case, variations are observed in minor species for $x > 3$ mm. The differences are due to superelastic collisions with excited atomic hydrogen. In this region, the plasma is in the recombination regime, and the e + H^+ produces excited state atoms, that, in the optically thin plasma approximation, decay radiatively, reducing their population with respect to the thick and SRC cases. As a consequence, fewer superelastic collisions occur, reducing the EEDF tail. As the electron molar fraction is quite high, this behavior is reflected on the composition. A transition region is located in the interval 0.4–1 mm where the

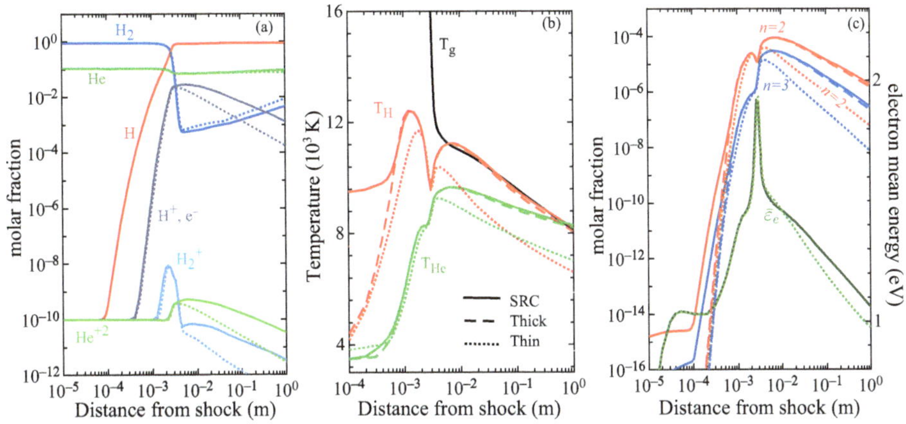

Figure 8.18. Spatial distribution of (a) species molar fractions, (b) internal temperatures and (c) the population of H in the first two excited states in the self-consistent radiation transport coupling (SRC) compared to the optically thin and thick approximations. Free-stream conditions: $T_0 = 160$ K, $P_0 = 700$ Pa and $M \approx 31$. Reproduced with permission from [116]. Copyright 2014 IOP.

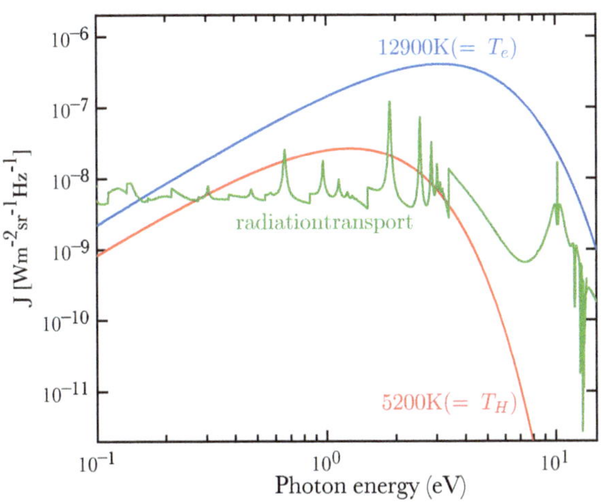

Figure 8.19. Monochromatic average J_ν (solid lines) and the local Planck distribution $B_\nu(T_e)$ (dotted lines) as a function of the photon energy just behind the shock wave (0.1 nm from the shock) for pure atomic hydrogen plasma. Free-stream conditions: $T_0 = 300$ K, $\rho_0 = 8.1 \times 10^5$ g/cm^3 and $M = 30$. Reproduced with permission from [89]. Copyright 2016 IOP.

plasma becomes thick. The T_H profile presents two peaks (see figure 8.18(b)), the first due to thermalization of internal and electron temperatures, the second due to electron–ion recombination. At a very short distance from the shock front, T_{H} obtained from the SRC model is much higher than the approximated cases, while T_{He} gives a higher temperature in the thin case. At a short distance, the atomic hydrogen temperature is determined from absorption of radiation emitted far from the shock front, where collisions did not have the time to induce excitation. This effect is a consequence of the level molar fraction profile (figure 8.18(c)). The radiation absorption in the proximity of the shock wave can be detailed by looking at the spectral average intensity in a pure hydrogen mixture. The spectral signature (figure 8.19) shows emission in the low-energy region, while a negative contribution comes from Lyman and Balmer series.

On the other hand, the first excited state of helium is metastable, i.e. radiative decay is prohibited, and therefore in a thin plasma, higher triplet levels decay radiatively on the metastable state, increasing its molar fraction. The electron mean energy (figure 8.18(c)) is weakly affected by the radiation model except at long distances, where the mean energy is influenced by superelastic collisions.

References

[1] Capitelli M, Ferreira C M, Gordiets B F and Osipov A I 2000 *Plasma Kinetics in Atmospheric Gases* (Berlin: Springer)

[2] Colonna G, Tuttafesta M, Capitelli M and Giordano D 1999 Non-Arrhenius NO formation rate in one-dimensional nozzle airflow *J. Thermophys. Heat Transfer* **13** 372–5

[3] Colonna G and Capitelli M 2000 Self-consistent model of chemical, vibrational, electron kinetics in nozzle expansion *J. Thermophys. Heat Transfer* **15** 308–16

[4] Bultel A, van Ootegem B, Bourdon A and Vervisch P 2002 Influence of Ar_2^+ in an argon collisional-radiative model *Phys. Rev.* E **65** 046406

[5] Bultel A, Chéron B G, Bourdon A, Motapon O and Schneider I F 2006 Collisional-radiative model in air for Earth re-entry problems *Phys. Plasmas* **13** 043502

[6] Colonna G, Armenise I, Bruno D and Capitelli M 2006 Reduction of state-to-state kinetic to macroscopic models in hypersonic flows *J. Thermophys. Heat Transfer* **20** 477–86

[7] Colonna G, Pietanza L D and Capitelli M 2008 Recombination-assisted nitrogen dissociation rates under nonequilibrium conditions *J. Thermophys. Heat Transfer* **22** 399–406

[8] Panesi M, Magin T E, Bourdon A, Bultel A and Chazot O 2009 FIRE II flight experiment analysis by means of a collisional-radiative model *J. Thermophys. Heat Transfer* **23** 236–48

[9] Panesi M, Magin T E, Bourdon A, Bultel A and Chazot O 2011 Electronic excitation of atoms and molecules for the FIRE II flight experiment *J. Thermophys. Heat Transfer* **25** 361–74

[10] Munafò A, Panesi M, Jaffe R L, Colonna G, Bourdon A and Magin T E 2012 QCT-based vibrational collisional models applied to nonequilibrium nozzle flows Eur *Phys. J.* D **66** 188

[11] Magin T E, Panesi M, Bourdon A, Jaffe R L and Schwenke D W 2012 Coarse-grain model for internal energy excitation and dissociation of molecular nitrogen *Chem. Phys.* **398** 90–5

[12] Munafò A, Lani A, Bultel A and Panesi M 2013 Modeling of non-equilibrium phenomena in expanding flows by means of a collisional-radiative model *Phys. Plasmas* **20** 073501

[13] Panesi M, Jaffe R L, Schwenke D W and Magin T E 2013 Rovibrational internal energy transfer and dissociation of $\mathrm{N}(^4\mathrm{S}_\mathrm{u}) + \mathrm{N}_2(^1\Sigma_\mathrm{g}^+)$ system in hypersonic flows *J. Chem. Phys.* **138** 044312

[14] Panesi M and Lani A 2013 Collisional radiative coarse-grain model for ionization in air *Phys. Fluids* **25** 057101

[15] Munafò A and Magin T E 2014 Modeling of stagnation-line nonequilibrium flows by means of quantum based collisional models *Phys. Fluids* **26** 097102

[16] Munafò A, Panesi M and Magin T E 2014 Boltzmann rovibrational collisional coarse-grained model for internal energy excitation and dissociation in hypersonic flows *Phys. Rev.* E **89** 023001
[17] Panesi M, Munafò A, Magin T E and Jaffe R L 2014 Study of the non-equilibrium shock heated nitrogen flows using a rovibrational state-to-state method *Phys. Rev.* E **90** 013009
[18] Laporta V and Bruno D 2013 Electron-vibration energy exchange models in nitrogen-containing plasma flows *J. Chem. Phys.* **138** 104319
[19] Heritier K L, Jaffe R L, Laporta V and Panesi M 2014 Energy transfer models in nitrogen plasmas: analysis of $N_2(^1\Sigma_g^+)$–$N(^4S_u)$–e interaction *J. Chem. Phys.* **141** 184302
[20] Munafò A, Liu Y and Panesi M 2015 Physical models for dissociation and energy transfer in shock-heated nitrogen flows *Phys. Fluids* **27** 127101
[21] Le H P, Karagozian A P and Cambier J-L 2013 Complexity reduction of collisional-radiative kinetics for atomic plasma *Phys. Plasmas* **20** 123304
[22] Kapper M G and Cambier J-L 2011 Ionizing shocks in argon. Part I: collisional-radiative model and steady-state structure *J. Appl. Phys.* **109** 113308
[23] Kapper M G and Cambier J-L 2011 Ionizing shocks in argon. Part II: transient and multi-dimensional effects *J. Appl. Phys.* **109** 113309
[24] Munafò A, Mansour N N and Panesi M 2017 A reduced-order NLTE kinetic model for radiating plasmas of outer envelopes of stellar atmospheres *Astrophys. J.* **838** 126
[25] Colonna G, Pietanza L D and D'Ammando G 2012 Coupling of radiation, excited states and electron distribution function in non equilibrium hydrogen plasma *Chem. Phys.* **398** 37–45
[26] Capitelli M, Colonna G, Pietanza L D and D'Ammando G 2013 Coupling of radiation, excited states and electron distribution function in non equilibrium hydrogen plasma *Spectrochim. Acta Part* B **83–84** 1–13
[27] Ferziger J H and Kaper H G 1972 *Mathematical Theory of Transport Processes in Gases* (Amsterdam: North-Holland)
[28] Giovangigli V 1999 *Multicomponent Flow Modeling* (Berlin: Springer)
[29] Nagnibeda E and Kustova E 2009 *Non-Equilibrium Reacting Gas Flows* (Berlin: Springer)
[30] Appleton J P and Bray K N C 1964 The conservation equations for a non-equilibrium plasma *J. Fluid Mech.* **20** 659–72
[31] Mitchner M and Kruger C H 1973 *Partially Ionized Gases* (New York: Wiley)
[32] Park C 1990 *Nonequilibrium Hypersonic Aerothermodynamics* (New York: Wiley)
[33] Gnoffo P A, Gupta R N and Shinn J L 1989 Conservation equations and physical models for hypersonic air flows in thermal and chemical nonequilibrium *NASA Technical Paper* 2867
[34] Zel'dovich Y B and Raizer Y P 1967 *Physics of Shock Waves and High-Temperature Hydrodynamic Phenomena* (New York: Academic)
[35] Oxenius J 1986 *Kinetic Theory of Particles and Photons Springer Series in Electronics and Photonics* vol 20 (Berlin: Springer)
[36] Mihalas D and Mihalas B W 1999 *Foundations of Radiation Hydrodynamics* Dover Books on Physics (Mineola, NY: Dover)
[37] Mihalas D 1978 *Stellar Atmospheres* 2nd edn (San Francisco, CA: Freeman)
[38] Park C 1989 Assessment of two-temperature kinetic model for ionizing air *J. Thermophys. Heat Transfer* **3** 233–44

[39] Park C 1993 Review of chemical–kinetic problems of future NASA missions, I: Earth entries *J. Thermophys. Heat Transfer* **7** 385–98
[40] Park C, Howe J T, Jaffe R L and Candler G V 1994 Review of chemical-kinetic problems of future NASA missions, II: Mars entries *J. Thermophys. Heat Transfer* **8** 9–23
[41] Schwenke D W 2008 Dissociation cross-sections and rates for nitrogen *Non-Equilibrium Gas Dynamics—From Physical Models to Hypersonic Flights, Lecture Series* (Sint-Genesius-Rode: von Karman Institute for Fluid Dynamics)
[42] Jaffe R L, Schwenke D W, Chaban G and Huo W 2008 Vibrational and rotational excitation and relaxation of nitrogen from accurate theoretical calculations, *46th AIAA Aerospace Sciences Meeting and Exhibit, Reno, NV* AIAA Paper 2008–1208
[43] Chaban G, Jaffe R L, Schwenke D W and Huo W 2008 Dissociation cross-sections and rate coefficients for nitrogen from accurate theoretical calculations, *46th AIAA Aerospace Sciences Meeting and Exhibition, Reno, NV* AIAA Paper 2008–1209
[44] Jaffe R L, Schwenke D W and Chaban G 2009 Theoretical analysis of N_2 collisional dissociation and rotation-vibration energy transfer, *47th AIAA Aerospace Sciences Meeting and Exhibit, Orlando, FL* AIAA Paper 2009–1569
[45] Jaffe R L, Schwenke D W and Chaban G 2010 Vibration–rotation excitation and dissociation in N_2-N_2 collisions from accurate theoretical calculations, *10th AIAA/ASME Joint Thermophysics and Heat Transfer Conf., Chicago, IL* AIAA Paper 2010–4517
[46] Huo W M and Green S 1996 Quantum calculations for rotational energy transfer in nitrogen molecule collisions *J. Chem. Phys.* **104** 7572–89
[47] Esposito F, Armenise I and Capitelli M 2006 $N - N_2$ state-to-state vibrational relaxation and dissociation rate coefficients based on quasi-classical calculations *Chem. Phys.* **331** 1–8
[48] Liu Y, Vinokur M, Panesi M and Magin T E 2010 A multi-group maximum entropy model for thermo-chemical nonequilibrium, *10th AIAA/ASME Joint Thermophysics and Heat Transfer Conference, Chicago, IL* AIAA Paper 2010–4332
[49] Liu Y, Panesi M, Vinokur M and Clarke P 2013 Microscopic simulation and macroscopic modeling of thermal and chemical non-equilibrium gases, *44th AIAA Thermophysics Conf., San Diego, CA* AIAA Paper 2013–3146
[50] Liu Y, Panesi M, Sahai A and Vinokur M 2014 General multi-group macroscopic modeling for thermo-chemical non-equilibrium gas mixtures, *7th AIAA Theoretical Fluid Mechanics Conf., Atlanta, GA* AIAA Paper 2014–3205
[51] Liu Y, Panesi M, Sahai A and Vinokur M 2015 General multi-group macroscopic modeling for thermo-chemical non-equilibrium gas mixtures *J. Chem. Phys.* **142** 134109
[52] MacDonald R L, Jaffe R L, Scwhenke D L and Panesi M 2018 Construction of a coarse-grain quasi-classical trajectory method. I. Theory and application to N_2–N_2 system *J. Chem. Phys.* **148** 054309
[53] MacDonald R L, Grover M S, Schwartzentruber T E and Panesi M 2018 Construction of a coarse-grain quasi-classical trajectory method. II. Comparison against the direct molecular simulation method *J. Chem. Phys.* **148** 054310
[54] Munafò A, Haack J R, Gamba I M and Magin T E 2014 A spectral-Lagrangian Boltzmann solver for a multi-energy level gas *J. Comput. Phys.* **264** 152–76
[55] Jayaraman V, Liu Y and Panesi M 2017 Multi-group maximum entropy model for translational non-equilibrium, *47th AIAA Thermophysics Conf.* AIAA Paper 2017–4024

[56] Bhatnagar P L, Gross E P and Krook M 1954 A model for collision processes in gases. I. Small amplitude processes in charged and neutral one-component systems *Phys. Rev.* **94** 511–25
[57] Johnston C O, Sahai A and Panesi M 2018 Extension of multiband opacity-binning to molecular, non-Boltzmann shock layer radiation *J. Thermophys. Heat Transfer* **32** 816–21
[58] Whitney C A and Skalafuris A J 1963 The structure of a shock front in atomic hydrogen. I. The effects of precursor radiation in the Lyman continuum *Astrophys. J.* **138** 199–215
[59] Skalafuris A J 1965 The structure of a shock front in atomic hydrogen. II. The region of internal relaxation *Astrophys. J.* **142** 351–68
[60] Skalafuris A J 1968 The structure of a shock front in atomic hydrogen. III. The region of radiative relaxation *Astrophys. Space Sci.* **2** 258–78
[61] Skalafuris A J 1968 The structure of a shock front in atomic hydrogen. IV. Stability dissipation and propagation *Astrophys. Space Sci.* **3** 234–57
[62] Murty S S R 1968 Effect of line radiation on precursor ionization *J. Quant. Spectrosc. Radiat. Transfer* **8** 531–54
[63] Murty S S R 1971 Electron energy equation for an atomic radiating plasma *J. Quant. Spectrosc. Radiat. Transfer* **11** 1681–90
[64] Farnsworth A V and Clarke J H 1971 Radiatively and collisionally structured shock wave exhibiting large emission-conversion ratio *Phys. Fluids* **14** 1352–60
[65] Fadeyev Y A and Gillet D 1998 The structure of radiative shock waves. I The method of global iterations *Astron. Astrophys.* **333** 687–701
[66] Fadeyev Y A and Gillet D 2000 The structure of radiative shock waves. II The multilevel hydrogen atom *Astron. Astrophys.* **354** 349–64
[67] Fadeyev Y A and Gillet D 2001 The structure of radiative shock waves. III The model grid for partially ionized hydrogen gas *Astron. Astrophys.* **368** 901–11
[68] Fadeyev Y A, Coroller H L and Gillet D 2002 The structure of radiative shock waves IV. Effects of electron thermal conduction *Astron. Astrophys.* **392** 735–40
[69] Panesi M and Huo W 2011 Non-equilibrium ionization phenomena behind shock waves, *11th AIAA/ASME Joint Thermophysics and Heat Transfer Conf., Honolulu, HW* AIAA Paper 2011–3629
[70] Huo M W, Panesi M and Magin E T 2011 Ionization phenomena behind shock waves *Physical Phenomena in Shock Waves* ed R Brun Shockwaves vol 7 (Berlin: Springer)
[71] Libermann M A and Velikovich A L 1986 *Physics of Shock Waves in Gases and Plasmas* Springer Series in Electronics and Photonics vol 19 (Berlin: Springer)
[72] Foley W H and Clarke J H 1973 Shock waves structured by nonequilibrium ionizing and thermal phenomena *Phys. Fluids* **16** 375–83
[73] Pauling L and Wilson E B Jr 1985 *Introduction to Quantum Mechanics with Applications to Chemistry* (Mineola, NY: Dover)
[74] Bourdon A, Térésiak Y and Vervisch P 1998 Ionization and recombination rates of atomic oxygen in high-temperature air plasma flows *Phys. Rev.* E **57** 4684–92
[75] Colonna G 2016 Boltzmann and Vlasov equations in plasma physics *Plasma Modeling: Methods and Applications* ed G Colonna and A D'Angola (Bristol: Institute of Physics Publishing) ch 1
[76] Colonna G and D'Angola A 2016 The two-term Boltzmann equation *Plasma Modeling: Methods and Applications* ed G Colonna and A D'Angola (Bristol: Institute of Physics Publishing) ch 2

[77] Loffhagen D 2016 Multi-term and non-local electron Boltzmann equation *Plasma Modeling: Methods and Applications* ed G Colonna and A D'Angola (Bristol: Institute of Physics Publishing) ch 3

[78] Capitelli M, Celiberto R, Colonna G, Esposito F, Hassouni C, Laricchiuta A and Longo S 2016 *Fundamental Aspects of Plasma Chemical Physics* Springer Series on Atomic, Optical and Plasma Physics vol 85 (Berlin: Springer)

[79] Capitelli M, Colonna G, Gicquel A, Gorse C, Hassouni K and Longo S 1996 Maxwell and non-Maxwell behavior of electron energy distribution function under expanding plasma jet conditions: the role of electron–electron, electron–ion, and superelastic electronic collisions under stationary and time-dependent conditions *Phys. Rev.* E **54** 1843

[80] Colonna G, Capitelli M, De Benedictis S, Gorse C and Paniccia F 1991 Electron-energy distribution-functions in CO_2-laser mixture—the effects of 2nd kind collisions from metastable electronic states *Contrib Plasma Phys.* **31** 575–9

[81] Alves L, Bogaerts A, Guerra V and Turner M 2018 Foundations of modelling of nonequilibrium low-temperature plasmas *Plasma Sources Sci. Technol.* **27** 023002

[82] Rassou S, Packan D and Labaune J 2017 Numerical modeling of repetitive nanosecond discharge in air *Phys. Plasmas* **24** 100704

[83] Chen X, Lan L, Lu H, Wang Y, Wen X, Du X and He W 2017 Numerical simulation of Trichel pulses of negative DC corona discharge based on a plasma chemical model *J. Phys. D: Appl. Phys.* **50** 395202

[84] Ridenti M A, Guerra V and Amorim J 2016 Atmospheric pressure plasmas operating in high-frequency fields *Plasma Modeling: Methods and Applications* ed G Colonna and A D'Angola (Bristol: Institute of Physics Publishing) ch 11

[85] Hagelaar G and Pitchford L 2005 BOLSIG+: Electron Boltzmann equation solver http://www.bolsig.laplace.univ-tlse.fr/

[86] Hagelaar G and Pitchford L 2005 Solving the Boltzmann equation to obtain electron transport coefficients and rate coefficients for fluid models *Plasma Sources Sci. Technol.* **14** 722

[87] LXcat 2016 http://fr.lxcat.net/home/Plasma Data Exchange Project database http://fr.lxcat.net/home/

[88] Capriati G, Colonna G, Gorse C and Capitelli M 1992 A parametric study of electron energy distribution functions and rate and transport coefficients in nonequilibrium helium plasmas *Plasma Chem. Plasma Process.* **12** 237–60

[89] Colonna G, Pietanza L D and D'Ammando G 2016 Self-consistent kinetics *Plasma Modeling: Methods and Applications* ed G Colonna and A D'Angola (Bristol: Institute of Physics Publishing) ch 8

[90] Pietanza L D, Colonna G, D'Ammando G, Laricchiuta A and Capitelli M 2016 Electron energy distribution functions and fractional power transfer in 'cold' and excited CO_2 discharge and post discharge conditions *Phys. Plasmas* **23** 013515

[91] Capitelli M, Colonna G, D'Ammando G and Pietanza L 2017 Self-consistent time dependent vibrational and free electron kinetics for CO_2 dissociation and ionization in cold plasmas *Plasma Sources Sci. Technol.* **26** 055009

[92] Pietanza L D, Colonna G and Capitelli M 2018 Non-equilibrium electron and vibrational distributions under nanosecond repetitively pulsed CO discharges and afterglows: I. optically thick plasmas *Plasma Sources Sci. Technol.* **27** 095004

[93] Pietanza L D, Colonna G and Capitelli M 2018 Non-equilibrium electron and vibrational distributions under nanosecond repetitively pulsed CO discharges and afterglows: II. the role of radiation and quenching processes *Plasma Sources Sci. Technol.* **27** 095003

[94] Formaggia L and Scotti A 2011 Positivity and conservation properties of some integration schemes for mass action kinetics *SIAM J. Numer. Anal.* **49** 1267–88

[95] Verwer J G 1994 Gauss–Seidel iteration for stiff ODEs from chemical kinetics *SIAM J. Sci. Comput.* **15** 1243–50

[96] Capitelli M, Colonna G, Hassouni K and Gicquel A 1994 Electron energy distribution functions in non-equilibrium H_2 discharges. The role of superelastic collisions from electronically excited states *Chem. Phys. Lett.* **228** 687–94

[97] D'Ammando G, Colonna G, Capitelli M and Laricchiuta A 2015 Superelastic collisions under low temperature plasma and afterglow conditions: a golden rule to estimate their quantitative effects *Phys. Plasmas* **22** 034501

[98] Capitelli M, Colonna G, D'Ammando G, Laporta V and Laricchiuta A 2013 The role of electron scattering with vibrationally excited nitrogen molecules on non-equilibrium plasma kinetics *Phys. Plasmas* **20** 101609

[99] Colonna G, Laporta V, Celiberto R, Capitelli M and Tennyson J 2015 Non-equilibrium vibrational and electron energy distributions functions in atmospheric nitrogen ns pulsed discharges and μs post-discharges: the role of electron molecule vibrational excitation scaling-laws *Plasma Sources Sci. Technol.* **24** 035004

[100] Bonelli F, Tuttafesta M, Colonna G, Cutrone L and Pascazio G 2017 An MPI-CUDA approach for hypersonic flows with detailed state-to-state air kinetics using a GPU cluster *Comput. Phys. Commun.* **219** 178–95

[101] Colonna G and Capitelli M 1996 Electron and vibrational kinetics in the boundary layer of hypersonic flow *J. Thermophys. Heat Transfer* **10** 406–12

[102] Shizgal B D and Lordet F 1996 Vibrational nonequilibrium in a supersonic expansion with reaction: application to O_2–O *J. Chem. Phys.* **104** 3579–97

[103] Colonna G and Capitelli M 2001 The influence of atomic and molecular metastable states in high-enthalpy nozzle expansion nitrogen flows *J. Phys. D: Appl. Phys.* **34** 1812

[104] Colonna G and Capitelli M 2001 Coupled solution of electron and vibrational kinetics in supersonic nozzle expansion, *32nd AIAA Plasmadynamics and Lasers Conf.* AIAA paper 2001–2729

[105] Colonna G, Capitelli M and Giordano D 2002 State to state electron and vibrational kinetics in supersonic nozzle air expansion: an improved model, *33rd Plasmadynamics and Lasers Conf.* AIAA paper 2002–2163

[106] Colonna G and Capitelli M 2003 The effects of electric and magnetic fields on high enthalpy plasma flows, *34th AIAA Plasmadynamics and Lasers Conf.* AIAA paper 2003–4036

[107] Colonna G and Capitelli M 2008 Boltzmann and master equations for magnetohydrodynamics in weakly ionized gases *J. Thermophys. Heat Transfer* **22** 414–23

[108] Borghi C, Carraro M, Cristofolini A, Biagioni L, Fantoni G and Passaro A 2005 Experimental investigation on the magneto-hydrodynamic interaction in the shock layer on a hypersonic body, *43rd AIAA Aerospace Sciences Meeting and Exhibit* AIAA paper 2005–1181

[109] Borghi C A, Carraro M R, Cristofolini A, Veefkind A, Biagioni L, Fantoni G, Passaro A, Capitelli M and Colonna G 2006 Magnetohydrodynamic interaction in the shock layer of a wedge in a hypersonic flow *IEEE Trans. Plasma Sci.* **34** 2450–63
[110] Borghi C, Cristofolini A, Carraro M, Gorse C, Colonna G, Passaro A and Paganucci F 2006 Non-intrusive characterization of the ionized flow produced by nozzle of an hypersonic wind tunnel, *14th AIAA/AHI Space Planes and Hypersonic Systems and Technologies Conf.* AIAA paper 2006–8050
[111] Cristofolini A, Borghi C and Colonna G 2007 Numerical analysis of the experimental results of the MHD interaction around a sharp cone, *38th AIAA Plasmadynamics and Lasers Conf. in Conjunction with the 16th Int. Conf. on MHD Energy Conversion* AIAA paper 2006–4252
[112] Cristofolini A, Borghi C, Neretti G, De Filippis F, Purpura C and Colonna G 2012 Non-intrusive characterization and numerical investigation on a Mach 10 ionized air flow, *43rd AIAA Plasmadynamics and Lasers Conf.* AIAA paper 2012–3179
[113] D'Ammando G, Pietanza L, Colonna G, Longo S and Capitelli M 2010 Modelling spectral properties of non-equilibrium atomic hydrogen plasma *Spectrochim. Acta* B **65** 120–9
[114] D'Ammando G, Pietanza L, Colonna G, Longo S and Capitelli M 2011 Study of radiation transfer, level and electron kinetics in a 1D steady shock in atomic hydrogen, *Proc. of 4th Int. Workshop Radiation of High Temperature Gases in Atmospheric Entry* **vol 689**, id.26
[115] Colonna G, Pietanza L D, D'Ammando G and Capitelli M 2012 Self-consistent coupling of chemical, electron and radiation models for shock wave in Jupiter atmosphere *AIP Conf. Proc.* **1501** 1400–7
[116] Colonna G, D'Ammando G, Pietanza L and Capitelli M 2014 Excited-state kinetics and radiation transport in low-temperature plasmas *Plasma Phys. Control. Fusion* **57** 014009
[117] D'Ammando G, Pietanza L, Colonna G and Capitelli M 2012 Coupling of radiation transfer, level and electron kinetics under high-speed shock wave conditions in an H_2/He mixture *ESA Spec. Publ.* **vol 714**, id.8
[118] Colonna G, Pietanza L and Capitelli M 2001 Coupled solution of a time-dependent collisional-radiative model and Boltzmann equation for atomic hydrogen plasmas: possible implications with LIBS plasmas *Spectrochim. Acta* B **56** 587–98
[119] Brun R 2009 *Introduction to Reactive Gas Dynamics* (Oxford: Oxford University Press)
[120] Armenise I, Capitelli M, Colonna G, Koudriavtsev N and Smetanin V 1995 Nonequilibrium vibrational kinetics during hypersonic flow of a solid body in nitrogen and its influence on the surface heat-flux *Plasma Chem. Plasma Process.* **15** 501–28
[121] Armenise I, Capitelli M, Colonna G and Gorse C 1996 Nonequilibrium vibrational kinetics in the boundary layer of re-entering bodies *J. Thermophys. Heat Transfer* **10** 397–405
[122] Capitelli M, Armenise I and Gorse C 1997 State-to-state approach in the kinetics of air components under re-entry conditions *J. Thermophys. Heat Transfer* **11** 570–8
[123] Phelps A and Pitchford L 1985 Anisotropic scattering of electrons by N_2 and its effect on electron transport: tabulation of cross sections and results *Technical report* 26 JILA Information Center https://jila.colorado.edu/sites/default/files/assets/files/publications/JILA.ICR_.26.pdf
[124] Cacciatore M, Capitelli M and Gorse C 1982 Non-equilibrium dissociation and ionization of nitrogen in electrical discharges: the role of electronic collisions from vibrationally excited molecules *Chem. Phys.* **66** 141–51

[125] Laporta V, Celiberto R and Wadehra J 2012 Theoretical vibrational-excitation cross sections and rate coefficients for electron-impact resonant collisions involving rovibrationally excited N_2 and NO molecules *Plasma Sources Sci. Technol.* **21** 055018

[126] Laporta V, Little D, Celiberto R and Tennyson J 2014 Electron-impact resonant vibrational excitation and dissociation processes involving vibrationally excited N_2 molecules *Plasma Sources Sci. Technol.* **23** 065002

[127] Capitelli M, Colonna G and Esposito F 2004 On the coupling of vibrational relaxation with the dissociation-recombination kinetics: from dynamic to aerospace applications *J. Phys. Chem.* A **108** 8930–34

[128] Colonna G, Tuttafesta M and Giordano D 2001 Numerical methods to solve Euler equations in one-dimensional steady nozzle flow *Comput. Phys. Commun.* **138** 213–21

[129] Surzhikov S 2016 High enthalpy radiating flows in Aerophysics *Plasma Modeling, Methods and Applicaitions* ed G Colonna and A D'Angola (Bristol: Institute of Physics Publishing) ch 12

[130] Colonna G, D'Ammando G, Pietanza L D and Capitelli M 2011 Radiation transfer, level and free-electron kinetics in non-equilibrium atomic hydrogen plasma *AIP Conf. Proc.* **1333** 1100–5

IOP Publishing

Hypersonic Meteoroid Entry Physics

Gianpiero Colonna, Mario Capitelli and Annarita Laricchiuta

Chapter 9

Precursor ionization during high-speed Earth entry

Adrien Lemal, Satoshi Nomura and Kazuhisa Fujita

The upcoming sample return missions being considered by JAXA [1–4] and planetary defense [5] activities have motivated the characterization of air radiation and its underlining mechanisms (excitation, ionization) to determine the radiative heating than can be withstood by a re-entry body. During entry into the Earth's atmosphere, the main components of the air mixture undergo significant dissociation and ionization and the resulting plasma radiates significantly, a substantial part of which impinges on the body. To characterize the radiation magnitude and spectral distribution, shock-tube experiments have been carried out [6, 7]. Electron measurements in the non-equilibrium region behind a shock wave traveling in air were used in [8, 9] to assess the performances of the Park [10] ionization model. Very good agreement was obtained between simulations and measurements for shock speeds up to 11 km s^{-1}, demonstrating that the underlying mechanism consisting of associative ionization followed by avalanche ionization was possible. For higher speeds, Fujita *et al* [11] showed that the ionization rate predicted with the Park [10] ionization model was slower than the measurement in the shock tube. It was deemed that electrons were already present ahead of the shock wave and would prompt collisional processes and ionization reactions. This phenomenon is called precursor ionization and is the subject of the present work. Subsequently, a new model was developed [11]. As precursor phenomenon modeling is complex, various processes were assumed to be negligible, chief among them: electron diffusion, conduction and recombination, the work of the electrical field, shock layer photo-absorption, shock layer non-equilibrium, precursor emission and the molecules' internal excitation. The precursor region was assumed to be 1D and stationary and was bounded by the shock front, the temperature of which was set to T_{eq}, characterizing the equilibrium temperature of the shock layer. The photoionization cross-section, σ_{ph} was assumed to be independent of the spectral range. Good agreement with experimental data was obtained by adjusting these parameters from the nominal values of T_{eq} (= 12 000 K) and σ_{ph} (= 10^{-16} cm^2),

doi:10.1088/2053-2563/aae894ch9

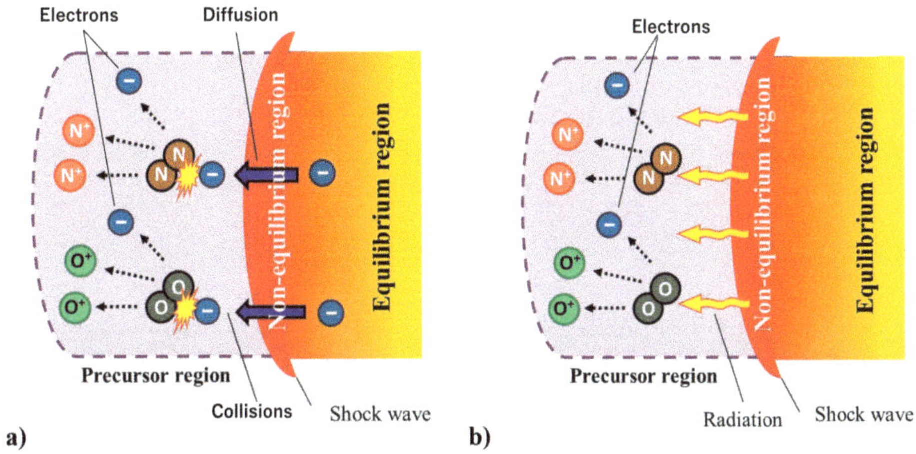

Figure 9.1. Creation of electrons ahead of a shock wave in air by (a) diffusion and (b) photoionization.

suggesting the model had taken into account the main features. Further modeling improvements were implemented and are discussed in [12].

Currently, the presence of electrons ahead of the shock wave is explained by two competing processes, illustrated in figure 9.1.

Precursor electrons could diffuse from the post-shock region, as first suggested in [13, 14], or can be due to the photoionization of molecules ahead of the shock wave, as mentioned in [15–18]. The modeling of the precursor phenomenon in air has received significant attention [19–31] and was shown to enhance the heating withstood by a re-entry body at speeds exceeding 12 km s^{-1}, highlighting the crucial role played by this phenomenon for the missions currently being considered at JAXA and for activities related to meteor threat assessment. Over the last few decades, very few measurements of precursor electrons have been undertaken [32–35] and thus the measurement of precursor ionization has gained interest at JAXA [36] and is the subject of this chapter. Specifically, the latter aims to address the measurement of the electron properties ahead of the shock wave and provide some insight on the ionization mechanisms. Section 9.1 presents the basics of Langmuir probe theories. Section 9.2 describes the experimental facility, the flow and Langmuir probe diagnostics. Section 9.3 introduces the test conditions selected to measure precursor electron densities and temperatures. Section 9.4 presents the electron properties inferred from the electrical measurements carried out ahead of the shock wave and discusses the ionization mechanisms.

9.1 Langmuir probe analysis

9.1.1 Basics

The design of the probe as well the analysis of the current–voltage curves (hereafter the I–V curves) depends on the magnitudes of the probe geometrical dimensions (length L and radius a), with respect to the ion and electron mean free paths λ_i and λ_e, the sheath length λ_s which is of the order of the Debye length λ_D, the ion diffusion velocity u_i, the flow velocity u_f, the sweep frequency of the electrical circuit f_{sw} with respect to the plasma frequency f_p and the flow frequency f_f ($f_f = u_f/a$). The theoretical expressions of the plasma parameters are recalled in equations (9.1)–(9.4) and are plotted in figure 9.2 for convenience.

$$f_j = \frac{q_e}{2\pi}\sqrt{\frac{n_j}{m_j\varepsilon_0}} \tag{9.1}$$

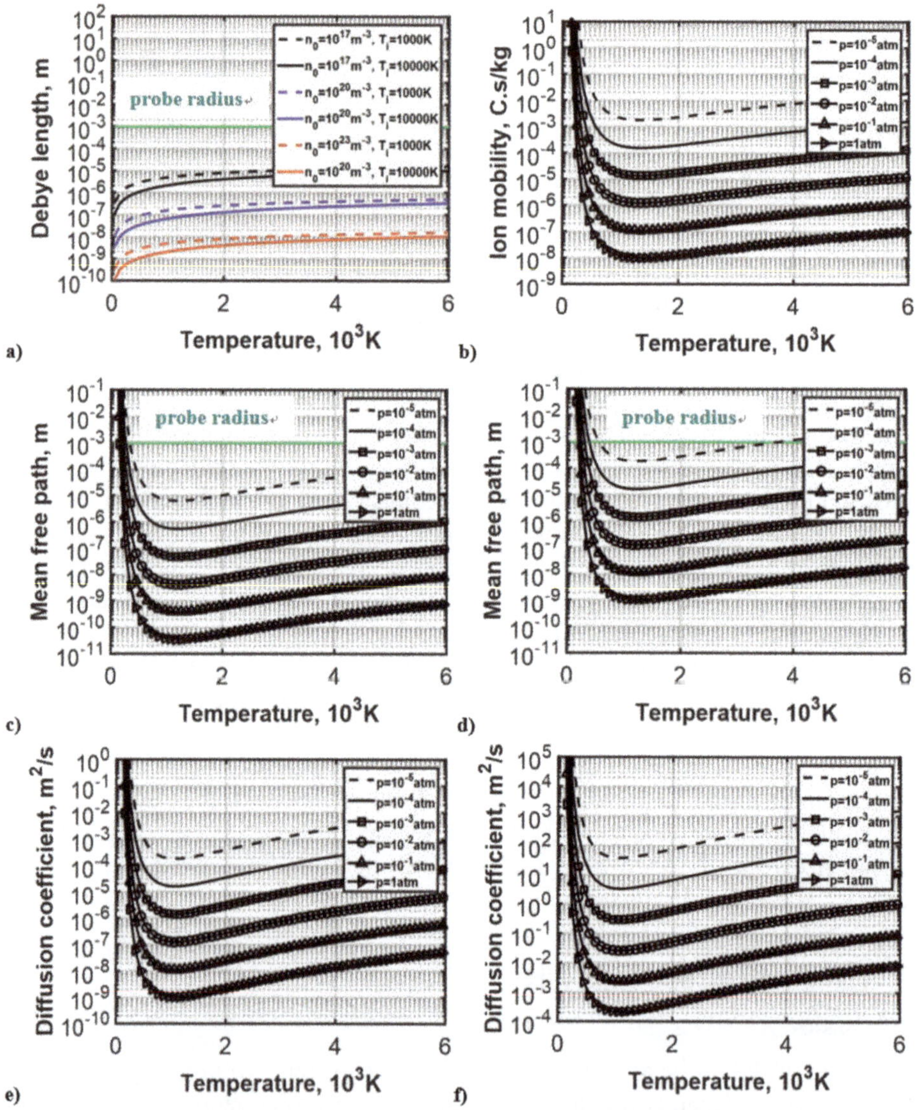

Figure 9.2. (a) Debye length, (b) ion mobility, mean free path of the (c) ion and (d) electron, and diffusion coefficient of the (e) ion and (f) electron.

$$\lambda_j = \frac{\eta_j(T_j)}{p}\sqrt{\frac{\pi k_B T_j}{2m_j}} \tag{9.2}$$

$$\lambda_D = \frac{1}{q_e}\sqrt{\frac{\varepsilon_0 k_B T_e}{n_e(1+t)}} \tag{9.3}$$

$$u_j = \frac{\mu_j k_B T_j}{a q_e} = \frac{D_j(T_j)}{a} = \frac{\eta_j(T_j)}{p}\frac{R_g T_j}{a M_j}\frac{6\pi\Omega_{jj}^{2,2}}{5\pi\Omega_{jj}^{1,1}}, \tag{9.4}$$

where T_j is the temperature of the jth ion, T_e the electron temperature and $t = T_j/T_e$, p the pressure, n, m, M, the number density, the mass and the molar mass of the sth species, η the viscosity, μ the mobility, D the diffusion coefficient, λ_j the mean free path, u the velocity, $k_B = 1.380\,648\,52 \times 10^{-23}$ J K^{-1} the Boltzmann constant, $R_g = 8.314\,463\,17$ J K^{-1} mol^{-1} the ideal gas constant, $q_e = 1.602\,176\,62 \times 10^{-19}$ C the electron charge, $\varepsilon_0 = 8.854\,187\,82 \times 10^{-12}$ F m^{-1} the vacuum permittivity, $\Omega_{jj}^{l,l}$ the collision integrals and $a = 10^{-3}$ m the radius of the probe used in the current work, computed according to [37]. Low density effects were neglected. The equations governing the motion of the charge carriers and their collection on the probe encompass the conservation of mass, momentum and Poisson's equations. The resolution of these equations is highly dependent on the key dimensionless numbers, chief among them the electrical Reynolds number $R_e = u_f/u_i$, the Knudsen numbers $K_{n,i} = \lambda_i/a$, $K_{n,e} = \lambda_e/a$, $K_{n,D} = \lambda_D/a$ and the frequency ratios f_f/f_p and f_{sw}/f_p. The analysis of the I–V curve is complicated when the plasma is flowing ($R_e > 1$), continuum, i.e. driven by collisional processes ($K_{n,i}, K_{n,e} \ll 1$), and unsteady ($f_f/f_p > 1$) and therefore the current work was achieved by probing the electron properties ahead of the shock where the flow is quiescent and stationary ($R_e < 1$), by designing the probe and the electrical circuit to enable the plasma be collisionless ($K_{n,i}, K_{n,e} \gg 1$) and steady ($f_f/f_p < 1$). The magnitude of the ratio $K_{n,D}$ determined the theory to be employed to analyze the I–V curve: for $K_{n,D} \ll 1$, the sheath is thin and the theory initially derived by Langmuir, Mott–Schmidt and Bohm [38] can be employed, while for $K_{n,D} > 1$, more sophisticated theories are required. Under the former assumption, the sweep frequency was finally adjusted following the works of [39]. Subsequent theories were developed to address continuum flowing regimes. When $K_{n,D} > K_{n,i}, K_{n,e}$, the theoretical framework is complicated and it is necessary to resort to asymptotic theories. For further details, the reader is referred to the comprehensive textbooks [40–42].

9.1.2 Measurement of electron properties

Langmuir probe analysis has been the subject of extensive works which highlighted the role played by the plasma spatial, charge and time scales. Under the assumptions of a stationary, steady, collisionless, thin-sheath plasma and a Maxwell–Boltzmann distribution of the charge carriers, the electron number density can be determined from the electron saturation current $I_{e,\mathrm{sat}}$ or from the ion saturation current $I_{i,\mathrm{sat}}$ as derived in equations (9.5*a*) and (9.5*b*), respectively,

$$I_{i,\mathrm{sat}} = \frac{\chi}{K} q_e A_0 n_e \sqrt{\frac{k_B T_e}{m_i}} \tag{9.5a}$$

$$I_{e,\mathrm{sat}} = \frac{1}{4} q_e A_0 n_e \sqrt{\frac{8 k_B T_e}{\pi m_e}} \tag{9.5b}$$

where A_0 is the collecting area of the probe ($A_0 = \pi a^2$ for a plane probe with radius a), χ is the Bohm constant [38] and K is a corrective parameter [40, 41] which depends on the ratios t and λ_j/a, respectively. When $t \ll 1, \chi = 0.61$, while for $t \approx 1, \chi$ is undetermined. When $\lambda_j/a \gg 1$ $K \approx 1$, while for $\lambda_j/a \ll 1$ $K \approx 0.5$; thus χ/K ranges from 0.61 to 1.22. Under the thin-sheath approximation, the increase of the ion saturation current with the potential, as is sometimes observed in measurements, is explained by the growth of the sheath and thus the collecting area. This process can be modeled by equations (9.6*a*) and (9.6*b*). The increase of the electron saturation current with respect to the voltage is modeled by equation (9.7*b*). Under the thick-sheath approximation, the equations are more complicated, and the reader is referred to the works referred to previously.

$$A = A_0\left(1 + \frac{x_s}{a}\right) \tag{9.6a}$$

$$x_s = \frac{2\lambda_D}{3}\sqrt[4]{2e}\left(\sqrt{\frac{q_e|V|}{T_j}} + \sqrt{2}\right)\sqrt{\sqrt{\frac{q_e|V|}{T_j}} + \sqrt{\frac{1}{2}}} \tag{9.6b}$$

$$I_{e,\mathrm{sat}} = \frac{A_0}{9\pi} q_e A_0 n_e \sqrt{\frac{2q_e}{m_e}\frac{V^{3/2}}{x_s^2}} \quad \text{for } V \gg \frac{k_B T_e}{q_e} \tag{9.7a}$$

$$I_{e,\mathrm{sat}} = \frac{A_0}{9\pi} q_e A_0 n_e \sqrt{\frac{2q_e}{m_e}\frac{V^{3/2}}{x_s^2}}\left(1 + 2.66\sqrt{\frac{k_B T_e}{q_e V}}\right) \text{for } V \ll \frac{k_B T_e}{q_e}, \tag{9.7b}$$

where $e = 2.718\,281\,82$ is the Napier constant. When collisions are not negligible ($\lambda_i/a \approx 1$), the saturation currents are given by [40–42]

$$I_{i,\mathrm{sat}} = \frac{1}{2} q_e A_0 n_i \sqrt{\frac{k_B T_e}{m_i}} \tag{9.8a}$$

$$I_{e,\text{sat}} = \frac{1}{2} q_e A_0 n_e \sqrt{\frac{k_B T_e}{\pi m_e}} \tag{9.8b}$$

$$n_i = n_e = \frac{n_\infty}{\zeta} \tag{9.8c}$$

$$\zeta = 1 + \frac{a}{\lambda_e} \frac{\lambda_i + t\lambda_e}{1 + t}. \tag{9.8d}$$

The mean free paths were computed and it was observed that the ratio $\lambda_e/\lambda_i \approx 10$. The ratio t ranges from 0 to 1; thus, the collected plasma density was reduced by a factor ζ ranging from $1 + 0.50/K_{n,e}$ to $1 + 11/K_{n,e}$, where $K_{n,e}$ is the electron Knudsen number. When assuming the electron energy distribution is Maxwellian, the current in the transition zone is given by equation (9.9) and therefore the electron temperature can be inferred by determining the slope of the $\ln(I) - V$ function.

$$I = \frac{1}{4} q_e A_0 n_e \sqrt{\frac{8 k_B T_e}{\pi m_e}} \exp\left(\frac{q_e V}{k_B T_e}\right). \tag{9.9}$$

9.2 Experimental set-up

The hypervelocity shock-tube (HVST) facility was commissioned at JAXA Chofu Aerospace Centre to support research into the aerothermochemistry of hypervelocity flight through Earth's and other planets' atmospheres, and to simulate the high enthalpy non-equilibrium phenomena encountered by spacecraft during atmosphere entry. This section describes the HVST facility, its flow and plasma diagnostics.

9.2.1 The HVST facility

The HVST facility [43] consists of a 16 m free-piston driven shock tube, with a reservoir tank, a compression tube, a high-pressure tube, a low-pressure tube, a vacuum tank and a free piston. The 70 mm square cross-section in the low-pressure tube is made of aluminum alloy to reduce the radiation emitted from impurities. High- and low-pressure tubes are evacuated down to 10^{-2} Pa with a turbomolecular pump (ULVAC Inc, UTM1001FW) and a dry pump (ULVAC Inc, LR60) before being filled with gases. Two pressure transducers (PCB, HM109C11) and (PCD, HM113A24) are mounted near the first and second diaphragms to measure the pressure evolution. The signal of the pressure transducer was monitored by digital oscilloscopes (Yokogawa, DL-1640L and DL-6154).

9.2.2 Velocity measurements

The shock velocity is measured using two methods: pressure piezoelectric sensors and a Schlieren laser system. The Schlieren system consists of two He–Ne laser beams, whose width is about 0.48 mm, aligned along the low-pressure axis and separated from each other by $d_L = 31.6$ mm, as sketched in figure 9.3. The laser beam passes through the test section and arrives at the avalanche photodiodes (Hamamatsu Photonics K.K., C5331-02, time of rise 1 ns) through the flat mirrors. In order to improve the sensitivity, lasers were restricted by a spatial filter. The time elapsed between the two signal changes of the photodiodes, Δt_L, is monitored using digital oscilloscopes (Yokogawa, DL-1640L and DL-6154) and is used to determine the shock speed. The pressure sensors (PCB Piezotronics Inc, 113A28, rise time of 1 μs) are mounted along the shock-tube axis and are separated by $d_P = 160$ mm. The rise time, Δt_P, of the sensors is monitored using the digital oscilloscopes (Yokogawa, DL-1640L and DL-6154) and is used to determine the shock speed. Figure 9.4 shows an example of the data acquired by the sensors.

9.2.3 Shock location measurements

The shock front location is of paramount importance as it enables discrimination between the electrons created ahead and behind the shock front, and determine the ionization length. In previous works [6, 9], the shock front location was determined by studying the inflection points of the post-shock intensity and electron number density profiles. Here, the shock front location is determined more accurately with the combination of laser beams and piezoelectric sensors.

9.2.4 Electron measurements

The electron energy distribution, temperature and number density were inferred from the analysis of I–V curve, of a single plane Langmuir probe. The probe

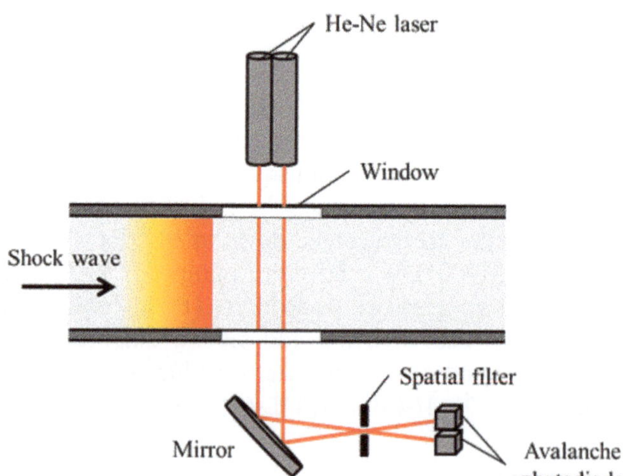

Figure 9.3. Sketch of the He–Ne laser system.

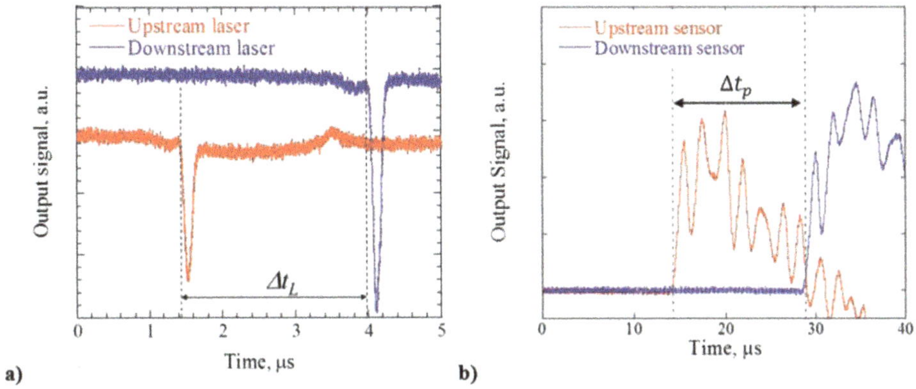

Figure 9.4. Output signal from (a) the He–Ne laser beams and (b) the piezoelectric sensors for $u_{sh} = 10$ km s^{-1}.

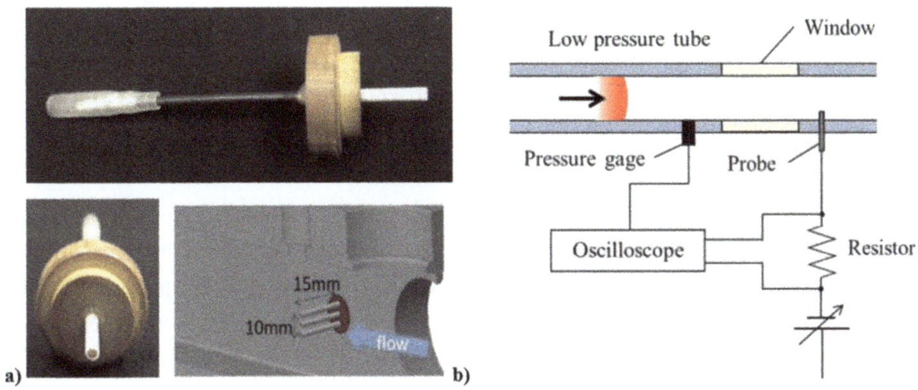

Figure 9.5. Langmuir probe (a) in the shock tube and (b) electrical circuits.

voltage, V, was varied with respect to the plasma potential V_p (which is initially unknown) and the probe current was monitored. The collection of electrons on the probe is governed by electrostatic effects and is significantly influenced by the probe geometry, the temporal resolution of the electrical circuit, the flow regimes and the various plasma scales. The choice of the resistor value, R, is the result of an optimization procedure to lower the voltage drop and increase the signal-to-noise ratio (S/N). A value of $R = 240\ \Omega$ was chosen in the current work. High time resolution measurements were obtained by limiting the sweeping frequency, f_{sw}, to characteristic frequencies of various phenomena, as discussed in section 9.1. Under the current probe regimes, a sweeping frequency $f_{sw} = 1.3$ MHz was chosen for the following conditions. A 1 mm-radius 15 mm-length probe was made of Al_2O_3 and inserted in the shock tube, perpendicular to the direction of the flow, as sketched in figure 9.5.

9.3 Test conditions

The current work aims to reproduce on the ground the shock layer radiation and ionization occurring in front of a body during its entry into Earth's atmosphere. Over the last few decades, various objects, encompassing spacecraft and meteors, have entered Earth's atmosphere and are reviewed in table 9.1. Their heating profiles are depicted in figure 9.6. Their trajectories are overlapped in figure 9.7 with the current performance envelope of HVST. Preliminary computations of the heating than can be withstood by the spacecraft currently being considered by JAXA [1–4] indicate that the peak heating will occur at a free-stream velocity exceeding 12 km s^{-1}. As stated in the introduction, the ionization processes are poorly understood with respect to governing the heating. Therefore, the current work aims to measure the electron number density ahead of the shock wave for the representative conditions. The flight conditions are taken from the JAXA mission trajectory and are scaled to ground operating conditions by keeping the free-stream density constant, and the resulting operating conditions are listed in table 9.2. The convective Q_C and radiative Q_R fluxes are usually correlated with respect to the spacecraft nose radius R_n, the free-stream density ρ_∞ and velocity u_∞ and are computed with the recent work in [63], together with representative ground shock speed u_{sh} and pressure p_{lab}.

9.4 Results

9.4.1 Electrical measurements

Ideally, the I–V curves are composed of two saturation regions and a transition region. Some of the underlying mechanisms are well understood and are recalled here for convenience. In the first saturation region ($V \ll V_p$, the plasma potential), the electric field produced in the space surrounding the probe prevents the electrons from being collected on the probe. The current I is only due to the ion current I_i. When the potential is further increased ($V < V_p$), the faster electrons are able to overcome the electrical field so the electron current I_e is no longer negligible. There is thus a rapid increase of I in the transition region until the potential reaches the plasma potential V_p. Finally, in the second saturation region ($V \gg V_p$), the electrons are accelerated and collected by the probe. Figures 9.8–9.11 display the measured I–V curves for conditions 1–4, respectively. For some locations ahead of the shock wave, the positive and negative sweeps yield different curves, the reason for which is the phenomenon of hysteresis, which is due to the contamination of the probe and the alteration of its electrical properties [54]. To mitigate this phenomenon, the probes were first cleaned with sand paper, then inserted into the shock tube and cleaned with a 500 V/5 mA discharge for 15 min.

9.4.2 Electron properties

The ion saturation $I_{i,\text{sat}}$ current was determined at negative probe potential and subtracted from the total current I. Subsequently, the $(I - I_{i,\text{sat}})$–V curves were fitted

Table 9.1. Peak heating operating conditions of various spacecraft.

#	Re-entry body	R_n [m]	u_∞ [km s^{-1}]	max(Q_C) [MW m^{-2}]	max(Q_R) [kW m^{-2}]	u_{sh} [km s^{-1}]	p_{lab} [Pa]	Ref.
1	BSUV 1	0.1	3.49	3.59	1.28	3.24	492	[44]
2	BSUV 2	0.1	5.11	1.83	0.12	5.11	10	[45]
3	EXPERT	0.3	5.04	4.04	42.49	4.08	621	[46]
4	US Shuttle	1.0	7.30	2.15	8.78	5.80	19	[47]
5	Orion[a]	4.0	7.46	0.47	43.02	6.37	6	[48]
6	MIRKA 1	0.5	7.49	2.15	20.18	6.13	6	[49]
7	MIRKA 2	0.025	7.53	4.20	0.03	5.69	42	[50]
8	QARMAN	0.23	7.53	2.48	3.58	6.89	12	[51]
9	ARD	3.36	7.57	0.63	68.37	6.10	22	[52]
10	IXV	1.0	7.59	0.55	1.35	6.57	2	[53]
11	Soyuz	1.67	7.61	1.22	88.35	6.69	30	[54]
12	PTV[a]	5.0	7.62	0.52	113.98	7.03	12	[48]
13	HTV-R capsule	0.94	7.88	0.88	5.57	7.37	3	[55]
14	OREX	1.35	7.91	0.91	17.87	6.37	43	[56]
15	RAMC 2	0.152	7.65	22.78	2.15	6.31	1000	[57]
16	Apollo 9	4.7	7.73	1.36	2.39	6.69	132	[47]
17	Apollo 11	4.7	8.31	1.76	6.74	7.76	10	[47]
18	Apollo 20	4.7	9.56	1.31	2.06	8.65	29	[47]
19	Genesis	0.43	10.54	5.02	0.22	9.01	25	[47]
20	Orion[b]	4.0	10.59	1.53	2.19	10.0	14	[48]
21	PTV[b]	5.0	10.59	1.76	6.74	9.69	32	[48]
22	Orion[c]	4.0	10.63	1.92	4.75	9.00	47	[48]
23	Apollo 17	4.7	10.72	1.88	6.98	9.78	33	[48]
24	Apollo 4	4.7	10.76	1.83	6.37	9.92	28	[48]
25	FIRE 1	0.9(9)	11.36	7.36	6.81	9.91	99	[58]
26	FIRE 2	0.9(9)	11.59	7.58	7.49	9.92	159	[59]
27	Hayabusa 1	0.2	11.78	12.97	0.64	9.96	52	[60]

28	Hayabusa 2[d]	0.2	n/a	n/a	n/a	n/a	n/a	[1]
29	MMX[d]	0.6	n/a	n/a	n/a	n/a	n/a	[2]
30	Phoebus	0.255	12.41	12.41	1.12	8.97	110	[61]
31	Rastas Spear[e]	0.275	12.35	8.58	0.38	10.1	44	[62]
32	Rastas Spear[g]	0.275	12.39	10.71	0.82	10.2	24	[62]
33	Stardust	0.23	12.55	9.20	0.29	10.4	20	[5]
34	Trojan[g]	0.2	14.46	17.99	1.90	11.9	39	[3]
35	Trojan[h]	0.2	14.58	18.91	2.26	12.5	31	[4]
36	HTV-R[f]	n/a	n/a	n/a	n/a	n/a	n/a	n/a
37	Marco-Polo[i]	n/a	n/a	n/a	n/a	n/a	n/a	n/a
38	Osirix Rex[d]	n/a	n/a	n/a	n/a	n/a	n/a	n/a
39	Orion EFT 1[d]	n/a	n/a	n/a	n/a	n/a	n/a	n/a
M1	MORP 138	10[j]	16.92	7.32	6	15.4	129	[5]
M2	MORP 18	10[j]	18.51	12.72	40	14.7	432	[5]
M3	Chelyabinsk	10[j]	19.06	46.15	3120	18.1	1961	[5]
M4	Grande Prairie	10[j]	26.45	25.77	410	22.6	216	[5]

[a] ISS return.
[b] Guided Moon return.
[c] Ballistic Moon return.
[d] Confidential.
[e] Flight path angle = −12.5°.
[f] Flight path angle = −8.5°.
[g] Flight path angle = −11.0°.
[h] Flight path angle = −11.2° and modified front shield.
[i] Average value over the trajectory (see the references for further information).
[j] Assumed initial value before fragmentation (see the references for further information).

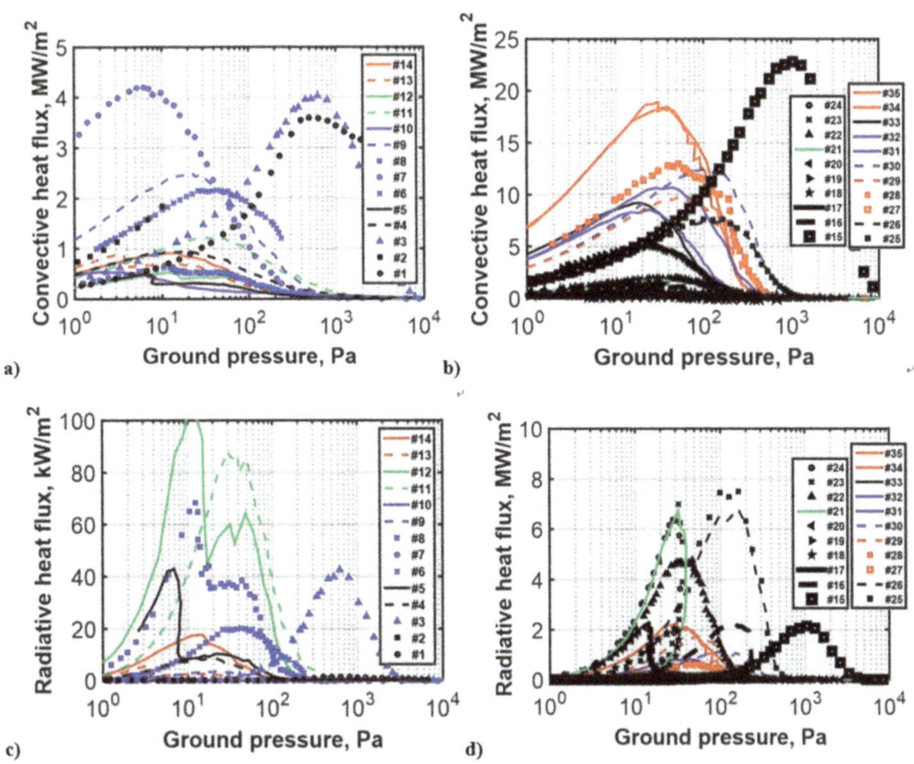

Figure 9.6. High-speed Earth re-entries with convective (a),(b) and radiative (c),(d) heat fluxes.

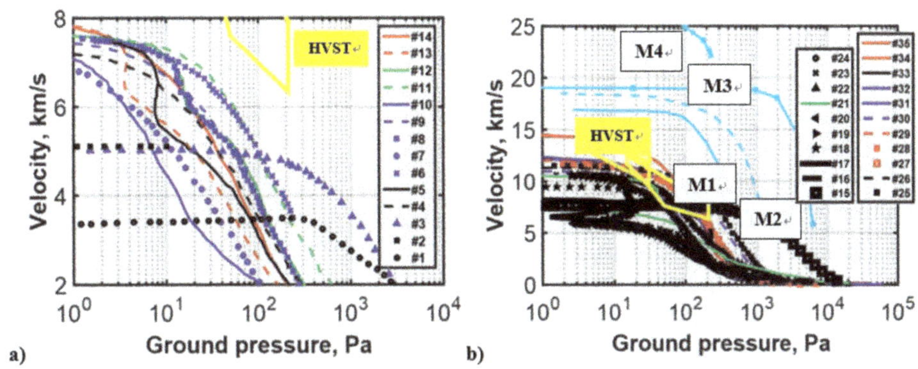

Figure 9.7. Ground-testing strategy for (a) low- and (b) high-speed Earth re-entry missions.

according to $I - I_{i,\text{sat}} = I_e = B \times \exp[C(V - A)]$ with $C = q_e/k_B T_e$. At some locations, a low S/N hindered the fit of the $\ln(I_e)$–V curve and yielded larger uncertainty in the determination of the electron temperature, which propagated in the determination of the electron number density. Additional uncertainty may remain as the electrons may not be Maxwellian [64, 65] for the low-pressure conditions (3 and 4 in

Table 9.2. HVST operating conditions.

Condition	u_{sh} [km s^{-1}]	p_{lab} [Pa]
1	11.5	30
2	12.5	30
3	13.8	5
4	14.5	5

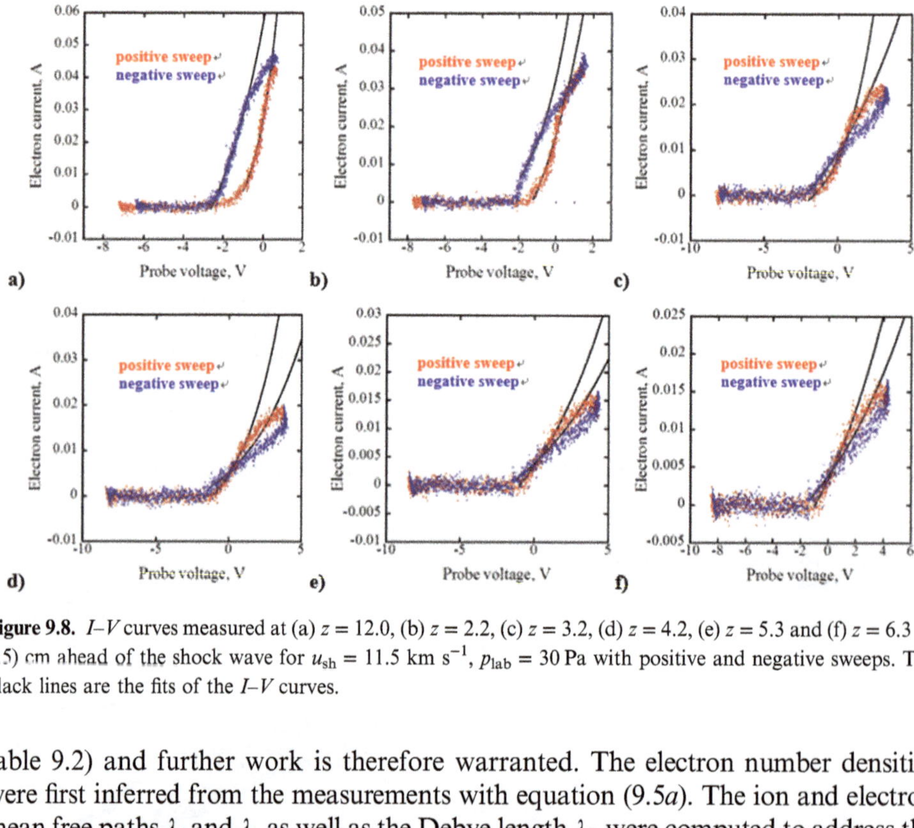

Figure 9.8. *I–V* curves measured at (a) $z = 12.0$, (b) $z = 2.2$, (c) $z = 3.2$, (d) $z = 4.2$, (e) $z = 5.3$ and (f) $z = 6.3$ (± 0.5) cm ahead of the shock wave for $u_{sh} = 11.5$ km s^{-1}, $p_{lab} = 30$ Pa with positive and negative sweeps. The black lines are the fits of the *I–V* curves.

table 9.2) and further work is therefore warranted. The electron number densities were first inferred from the measurements with equation (9.5*a*). The ion and electron mean free paths λ_i and λ_e as well as the Debye length λ_D were computed to address the validity of the aforementioned assumptions and the magnitude of the collisional effects. The computed Debye length was found to be negligible with respect to the probe radius, which prevented the sheath from growing significantly and confirmed the saturation of the ion current $I_{i,\mathrm{sat}}$. The ion and electron mean free paths were of the same order of magnitude as the probe radius; which requires the electron number densities be corrected with equation (9.8) to consider the collisions between particles. The comparison between the results from the various approaches therefore gives an estimation of the uncertainties. Figure 9.12 displays the electron temperature and number densities inferred from the *I–V* curve measurements and with equations (9.5) and (9.8) for the conditions 1–4 in table 9.2, respectively. It is observed that condition

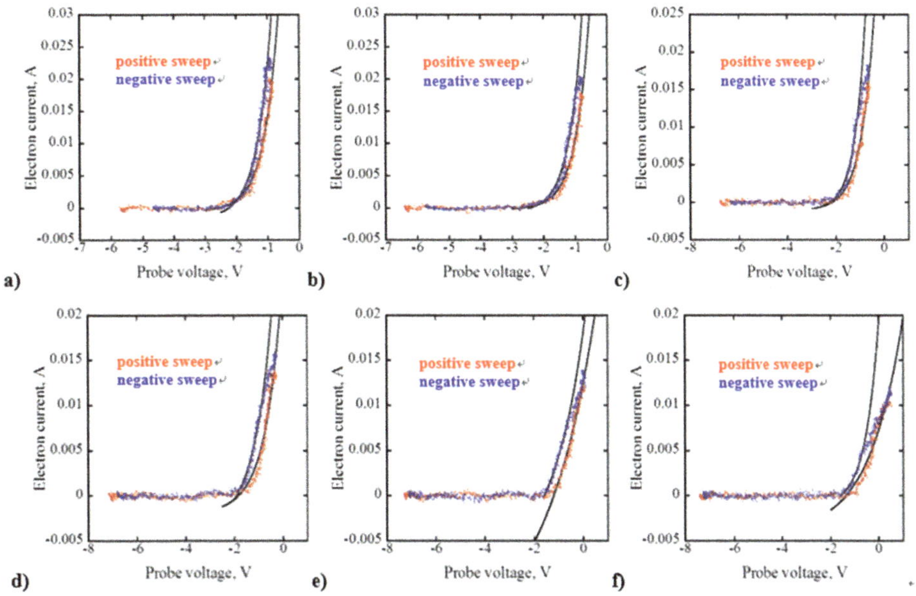

Figure 9.9. *I–V* curves measured at (a) z = 2.8, (b) z = 3.9, (c) z = 5.0, (d) z = 6.1, (e) z = 7.2 and (f) z = 8.3 (± 0.6) cm ahead of the shock wave for u_{sh} = 12.5 km s^{-1}, p_{lab} = 30 Pa with positive and negative sweeps. The black lines are the fits of the *I–V* curves.

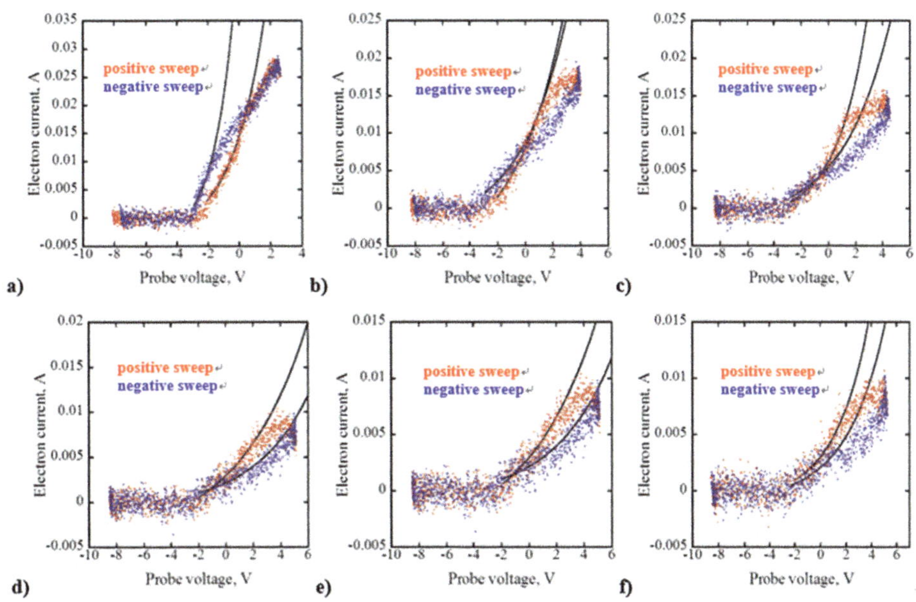

Figure 9.10. *I–V* curves measured at (a) z = 2.5, (b) z = 3.7, (c) z = 4.9, (d) z = 6.1, (e) z = 7.3 and (f) z = 8.5 (± 0.6) cm ahead of the shock wave for u_{sh} = 13.8 km s^{-1}, p_{lab} = 5 Pa with positive and negative sweeps. The black lines are the fits of the *I–V* curves.

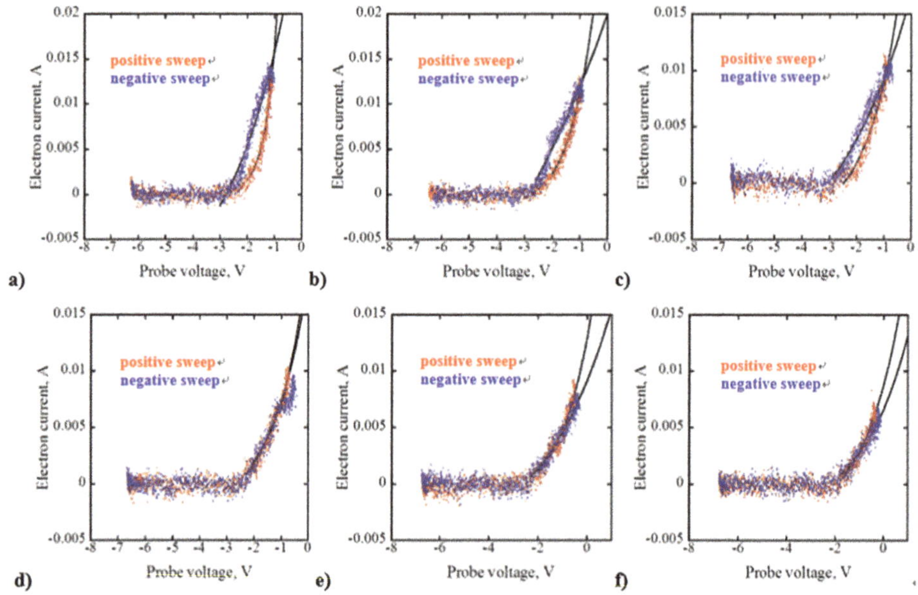

Figure 9.11. *I–V* curves measured at (a) $z = 2.7$, (b) $z = 3.9$, (c) $z = 5.2$, (d) $z = 6.4$, (e) $z = 7.6$ and (f) $z = 8.8$ ($\pm$ 0.6) cm ahead of the shock wave for $u_{sh} = 14.5$ km s^{-1}, $p_{lab} = 5$ Pa with positive and negative sweeps. The black lines are the fits of the *I–V* curves.

3 ($u_{sh} = 12.5$ km s^{-1}, $p_{lab} = 30$ Pa), which corresponds to the peak heating condition of a mission currently considered at JAXA, yields the maximum value of the electron number density ahead of the shock. Further insight was achieved by computing the integrated spectral radiance of the shock layer, as in [10–12]. The shock layer was assumed to be in equilibrium and the thermophysical properties were computed using the Code Equilibrium with Applications (CEA) [66] as a first approximation. Note that the conditions investigated in the current work yield a significant degree of ionization, which affects the thermodynamic properties of the flow when its temperature exceeds 10 000 K. This effect, which was quantified in [7] was, however, neglected in the current analysis. The integrated radiances were then computed with the JAXA in-house structured package for radiation analysis (SPRADIAN) [67] and are listed in table 9.3. It is shown that condition 3 yields to the highest radiance transferred to the precursor region of the shock wave.

9.4.3 Ionization mechanism

To understand the mechanism responsible for the creation of the electrons and discriminate between photoionization and diffusion processes, it is interesting to compare the characteristic length of the electron number density profiles with the typical lengths of diffusion and radiation processes [68]. The electron number density profiles $n_e(z)$ were modeled according to $[n_e(z) - n_{e,\infty}]/[n_{e,sh} - n_{e,\infty}] = \exp(z/L)$ and the characteristic length L was determined by taking the slope of the $\ln(n_e(z))$ profile.

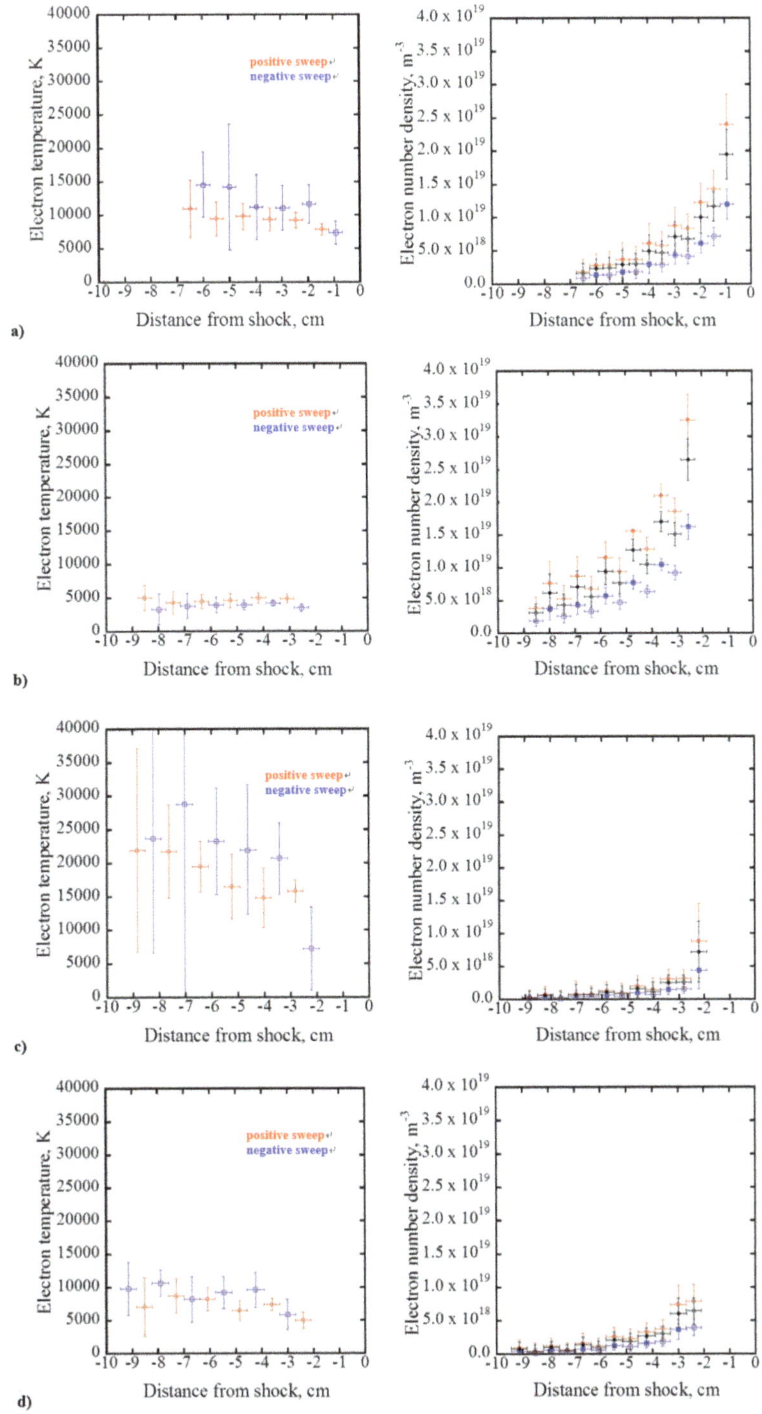

Figure 9.12. Precursor electron temperature (left) and number density (right) profiles for (a) condition 1, (b) condition 2, (c) condition 3 and (d) condition 4 in table 9.2. The red, blue and black points were inferred from the *I–V* curves with equation (9.5*a*) ($\chi/K = 0.61$), equation (9.5*a*) ($\chi/K = 1.22$) and equation (9.8*c*), respectively. Solid and empty diamonds were obtained from negative and positive sweeps, respectively.

Table 9.3. Characteristic lengths (T_i = 300 K, T_e = 5000 ± 2000 K, σ_{ph} = 1.5 × 10^{-21} m^2).

Condition	u_{sh} [km s^{-1}]	p_{lab} [Pa]	I_R [MW m^{-2} sr^{-1}]	L_{exp} [cm]	L_{ph} [cm]	D_e [m^2 s^{-1}]	$L_{d,e}$ [cm]
1	11.5	30	0.47	2.46	9.20	40	0.34
2	12.5	30	0.99	3.46	9.20	40	0.32
3	13.8	5	0.30	2.44	55.24	276	2.00
4	14.5	5	0.38	2.66	55.24	276	1.90

The typical length of the diffusion process $L_{d,j}$ was obtained by solving the 1D drift-diffusion equation and is given by $L_{d,j} = D_j/u_{sh}$. The solution of the 1D radiation transport equation in the precursor [68] yields the characteristic length of photoionization $L_{ph} = 1/(\sigma_{ph} n_i) = k_B T_i/(\sigma_{ph} p)$, where σ_{ph} is the cross-section of air photoionization. The various lengths are tabulated in table 9.3. It is observed that the characteristic length inferred from the electron number density profiles is bounded by the diffusion and photoionization lengths.

9.5 Conclusions

This chapter presents the first efforts in measuring the electron number densities and temperatures ahead of a shock wave traveling in an air mixture at speeds ranging from 11 to 14 km s^{-1} to characterize the precursor phenomenon and subsequent heating, and thus support the design of the TPS for the upcoming sample return mission currently being considered at JAXA. Flight conditions were replicated in the HVST facility at JAXA Chofu Aerospace Centre. A Langmuir probe was used to measure the *I–V* characteristics and the electron properties were inferred using appropriate theoretical methods. The electron number densities and temperatures were shown to increase up to 10^{19} m^{-3} and 5000 K, respectively. The spatial lengths were determined from the electron number density profiles and were compared to the diffusion and photoionization lengths. The diffusion lengths were found to be smaller than the experimental lengths, suggesting that diffusion processes, when solely considered, fail to account for the creation of electrons ahead of shock waves. Photoionization processes can therefore be considered a valuable, additional and competitive mechanism in the formation of precursor electrons in a shock wave in air when its speed exceeds 12 km s^{-1}. Ongoing work includes the mitigation of the error bars in the temperature measurements and the effect of the low-pressure regime to ascertain the range of electron diffusion coefficients and corresponding diffusion lengths, the analysis of the electrical measurement to infer the energy distribution of the electrons and the characterization of the radiative transfer within the shock-tube walls. Future work will assess the physico-chemical models currently in use in JAXA solvers and the significance of precursor electrons on the radiative heating that can be withstood by a spacecraft during its re-entry into Earth's atmosphere, to support JAXA high-speed sample return missions and planetary defense activities.

Acknowledgements

Experimental devices were provided by JAXA under the internal competition funding scheme and are gratefully acknowledged. Thanks are due to T Kawakami (University of Shizuoka, Japan) for operating the shock-tube facility under a tight schedule and for his analysis of the voltage measurements. The authors are grateful to the editors; their suggestions greatly improved this work and are much appreciated.

References

[1] Yamada T, Yoshikawa K, Yamada K and Kawahara K 2017 Development of Hayabusa 2 capsule and its return preparation (In Japanese) *61st Space Science and Technology Alliance Lecture (Niigata, Japan)*

[2] Sugimoto R, Yamada K, Shimoda T, Maru Y and Sato Y 2017 Sample return capsule concept study of MMX (in Japanese) *61st Space Science and Technology Alliance Lecture (Niigata, Japan)*

[3] Fujita K, Takayanagi H, Matsuyama S, Yamada K and Abe T 2014 Assessment of convective and radiative heating for Jupiter Trojan sample return capsule *11th AIAA/ASME Joint Thermophysics and Heat Transfer Conf., Atlanta, GA* AIAA paper 2014-2673

[4] Takahashi Y and Yamada K 2018 Aerodynamic-heating analysis of sample-return capsule in future Trojan-asteroid exploration *J. Thermophys. Heat Transfer* **32** 547–59

[5] Prabhu D K *et al* 2016 Thermophysics issues related to entries of large meteoroids, *AIAA SciTech 2016 Conf., Special Session: Aerothermodynamics of Meteor Entries*, *San Diego, CA*

[6] Brandis A M and Cruden B A 2017 Benchmark shock tube experiments of radiative heating relevant to Earth re-entry *55th AIAA Aerospace Sciences Meeting, Grapevine, TX* AIAA paper 2017-1145

[7] Lemal A, Matsuyama S, Nomura S, Takayanagi H and Fujita K 2018 Calculation of intensity profiles behind a shock wave travelling in air at speeds exceeding 12 km s^{-1} *Shock Wave* **32** 1–7

[8] Fujita K, Sato S, Abe T and Otsu H 2003 Electron density measurements behind strong shock waves by H_β profile matching *J. Thermophys. Heat Transfer* **17** 210–16

[9] Lemal A, Jacobs C, Perrin M-Y, Laux C, Tran P and Raynaud E 2015 Prediction of nonequilibrium air plasma radiation behind a shock wave *J. Thermophys. Heat Transfer* **30** 197–210

[10] Park C 1993 Review of chemical–kinetic problems of future NASA missions. I—Earth entries *J. Thermophys. Heat Transfer* **7** 385–98

[11] Fujita K, Sato S, Abe T and Matsuda A 2001 Electron temperature and density measurement ahead of strong shock waves *35th AIAA Thermophysics Conf., Anaheim, CA* AIAA paper 2001-2765

[12] Passeron A, Dantec E, Lemal A, Nomura S and Fujita K 2018 Efforts in modeling precursor phenomena in a shocked air mixture *2018 AIAA Aerospace Sciences Meeting, Kissimmee, FL* AIAA Paper 2018-742

[13] Weymann H D 1969 Precursors ahead of shock waves: I. Electron diffusion *Phys. Fluids* **12** 1193–99

[14] Kim M, Gülhan A and Boyd I D 2012 Modeling of electron energy phenomena in hypersonic flows *J. Thermophys. Heat Transfer* **26** 244–57

[15] Zel'Dovich Y B and Raizer Y P 2012 Physics of shock waves and high-temperature hydrodynamic phenomena (Mineola, NY: Courier)

[16] Holmes L B and Weymann H D 1969 Precursors ahead of shock waves: II. photoionization *Phys. Fluids* **12** 1200–10

[17] Zhelezniak M B and Mnatsakanian A K 1969 Photoionization ahead a strong shock in air *J. Tech. Phys.* **12** 1200–10

[18] Drake R P 2007 Theory of radiative shocks in optically thick media *Phys. Plasmas* **14** 043301

[19] Wetzel L 1964 Far flow approximations of precursor profiles *AIAA J.* **2** 1209–14

[20] Yoshikawa K K 1967 Analysis of radiative heat transfer for large objects at meteoric speeds *NASA Technical note* D-4051 NASA

[21] Lasher L E and Wilson K H 1968 Effect of shock precursor absorption on superorbital entry heating *AIAA J.* **6** 2419–20

[22] Smith P C 1968 Radiative effects in the precursor region of high-speed wedge flows *AIAA J.* **6** 937–38

[23] Edwards K 1969 Precursor plasma formation for blunt reentry vehicles *2nd Fluid and Plasma Dynamics Conf., San Francisco, CA* AIAA paper 1969-718

[24] Vertushkin V K and Romishevskii E A 1970 Influence of the precursor effect on hypersonic blunt body flows *Mekh. Zhidkosti Caza* **6** 40–7

[25] Zinn J and Anderson R 1973 Structure and luminosity of strong shock waves in air *Phys. Fluids* **16** 1639–44

[26] Nemtchinov I, Popova O, Shuvalov V and Svetsov V 1994 Radiation emitted during the flight of asteroids and comets through the atmosphere *Planet. Space Sci.* **42** 491–506

[27] Yoshizawa R, Fujita K, Ogawa H and Inatani Y 2007 Numerical analysis of shock wave with precursor heating *45th AIAA Aerospace Sciences Meeting and Exhibit, Reno, NV* AIAA paper 2007-809

[28] Johnston C O, Gnoffo P A and Mazaheri A 2013 Influence of coupled radiation and ablation on the aerothermodynamic environment of planetary entry vehicles *Technical report* STO-AVT-218 NATO

[29] Golub A, Kosarev I, Nemchinov I and Shuvalov V 1996 Emission and ablation of a large meteoroid in the course of its motion through the Earth's atmosphere *Solar Syst. Res.* **30** 183

[30] Johnston C O, Samareh J and Brandis A M 2015 Aerothermodynamic characteristics of 16-22 km s^{-1} Earth entry *45th AIAA Thermophysics Conf., Dallas, TX* AIAA paper 2015-3110

[31] Park C 2016 Radiation phenomenon for large meteoroids, *AIAA SciTech 2016 Conf., Special Session: Aerothermodynamics of Meteor Entries, San Diego, CA*

[32] Zivanovic S 1963 Investigation of precursor ionization in front of the shock waves of hypersonic projectiles, *Conf. on Physics of Entry into Planetary Atmospheres* p 458

[33] Dunn M and Lordi J 1969 Measurement of electron temperature and number density in shock-tunnel flows. I—Development of free-molecular Langmuir probes *AIAA J.* **7** 1458–65

[34] Lederman S and Wilson D S 1967 Microwave resonant cavity measurement of shock produced electron precursors *AIAA J.* **5** 70–7

[35] Omura M and Presley L L 1969 Electron density measurements ahead of shock waves in air *AIAA J.* **7** 2363–65

[36] Kawakami T, Nomura S, Lemal A, Fujita K and Matsui M 2017 Measurement of electron temperature and electron number density ahead of a shock wave using fast sweep Langmuir probes, *Space Dynamics and Navigation Symp.*, *(Matsuyama, Japan)*

[37] Gupta R N, Yos J M, Thompson R A and Lee K-P 1990 A review of reaction rates and thermodynamic and transport properties for an 11-species air model for chemical and thermal nonequilibrium calculations to 30 000 K *NASA Reference Publication* 1232
[38] Bohm D 1949 *Minimum Ionic Kinetic Energy for a Stable Sheath* (New York: McGraw-Hill) pp 77–86
[39] Lobbia R B and Gallimore A D 2010 Temporal limits of a rapidly swept Langmuir probe *Phys. Plasmas* **17** 073502
[40] Swift J D and Schwar M 1969 *Electrical Probes for Plasma Diagnostics* (Amsterdam: Elsevier) appendices 4, 5, pp 317–22
[41] Huddleston R H and Leonard S L 1965 *Plasma Diagnostic Techniques* (New York: Academic) ch 4, pp 113–200
[42] Xu K and Doyle S 2016 Measurement of atmospheric pressure microplasma jet with Langmuir probes *J. Vac. Sci. Technol.* A **34** 1–10
[43] Takayanagi H, Lemal A, Nomura S and Fujita K 2018 Measurements of carbon dioxide nonequilibrium infrared radiation in shocked and expanded flows *J. Thermophys. Heat Transfer* **32** 483–94
[44] Erdman P W, Zipf E C, Espy P, Howlett C, Levin D A, Loda R, Collins R J and Candler G V 1993 Flight measurements of low-velocity bow shock ultraviolet radiation *J. Thermophys. Heat Transfer* **7** 37–41
[45] Erdman P W, Zipf E C, Espy P, Howlett C, Christou C, Levin D A, Collins R J and Candler G V 1993 *In situ* plume radiance measurements from the bow shock ultraviolet 2 rocket flight *J. Thermophys. Heat Transfer* **7** 704–08
[46] Muylaert J, Walpot L, Ottens H and Cipollini F 2007 Aerothermodynamic re-entry flight experiment—EXPERT *NATO Technical Report* RTO-EN-AVT-130
[47] Grinstead J H, Wilder M C, Reda D C, Cornelison C J, Cruden B A and Bogdanoff D W 2010 Shocktube and ballistic range facilities at NASA Ames research center *NATO Technical Report* RTO-EN-AVT-186
[48] Surzhikov S 2015 Radiative gasdynamics of re-entry space vehicle of large size with superorbital velocity *53rd AIAA Aerospace Sciences Meeting, Kissimmee, FL* AIAA paper 2015-0980
[49] Fertig M and Frühauf H-H 2006 Detailed computation of the aerothermodynamic loads of the MIRKA capsule, *3rd European Symp. of Aerothermodynamics for Space Vehicles, (Noordwijk, The Netherlands)*
[50] Ehresmann M *et al* 2015 CubeSat-sized re-entry capsule MIRKA 2, *10th IAA Symp. on Small Satellites for Earth Observation, (Berlin, Germany)*
[51] Chazot O, Sakraker I, Turchi A, Trifoni E and van der Haegen V 2017 Ground testing of QARMAN reentrySat in SCIROCCO plasma wind tunnel under flight relevant conditions, *14th Int. Planetary Probe Workshop, (The Hague, The Netherlands)*
[52] Tran P, Paulat J and Boukhobza P 2007 Re-entry flight experiments lessons learned—the atmospheric reentry demonstrator ARD *NATO Technical Report* RTO-EN-AVT-130-10
[53] Sakraker I 2016 Aerothermodynamics of pre-flight and in-flight testing methodologies for atmospheric entry probes *PhD thesis* Université de Liège, Belgium
[54] Plastinin Y, Karabadzhak G, Khmelinin B, Zemliansky B, Gorshkov A and Zalogin G 2007 Measurements of the UV radiation generated by the Soyuz spacecraft transport capsule during reentry *45th AIAA Aerospace Sciences Meeting and Exhibit, Reno, NV* AIAA paper 2007-815

[55] Fujita K, Suzuki T, Matsuyama S, Mizuno M, Aoki T, Ogasawara T and Takizawa N 2015 Derogation of aeroshell aerodynamic performance due to recession of lightweight ablator along a reentry trajectory, *8th Int. Symp. on Aerothermodynamics for Space Vehicles, (Lisbon, Portugal)*
[56] Inouye Y 1995 OREX flight-quick report and lessons learned *Aerothermodynam. Space Vehicles* **367** 271
[57] Jones W L Jr and Cross A E 1973 Electrostatic-probe measurements of plasma parameters for two reentry flight experiments at 25000 feet per second NASA Technical Report TM-D-6617
[58] Cauchon D L 1996 Project FIRE flight 1 radiative heating experiment *NASA Technical Report* TM X-1222
[59] Cauchon D 1967 Radiative heating results from the FIRE II flight experiment at a reentry velocity of 11.4 km s^{-1}, 1967 *NASA Technical Report* TM-X-1402
[60] Suzuki T, Fujita K, Yamada T, Inatani Y and Ishii N 2013 Postflight thermal protection system analysis of Hayabusa reentry capsule *J. Spacecr. Rockets* **51** 96–105
[61] Storelli A 2011 Assessment of phoebus reentry blackout *Master's thesis* European Space Agency
[62] Bouilly J, Pisseloup A, Chazot O, Vekinis G, Bourgoing A, Chanetz B and Sladek O 2011 RASTAS SPEAR, *9th Int. Planetary Probe Workshop, (Portsmouth, VA)*
[63] Brandis A M and Johnston C O 2014 Characterization of stagnation-point heat flux for Earth entry *45th AIAA Plasmadynamics and Lasers Conf., Atlanta, GA* AIAA paper 2014-2374
[64] Colonna G and Capitelli M 1996 Electron and vibrational kinetics in the boundary layer of hypersonic flow *J. Thermophys. Heat Transfer* **10** 406–12
[65] Colonna G, D'Ammando G, Pietanza L and Capitelli M 2014 Excited-state kinetics and radiation transport in low-temperature plasmas *Plasma Phys. Controlled Fusion* **57** 014009
[66] Gordon S and McBride B J 1994 *Computer Program for Calculation of Complex Chemical Equilibrium Compositions and Applications* vol 1 (Cleveland, OH: National Aeronautics and Space Administration, Office of Management, Scientific and Technical Information Program)
[67] Fujita K and Abe T 1997 SPRADIAN, structured package for radiation analysis: theory and application *JAXA Technical Report* 669
[68] Cruden B A and Bogdanoff D W 2017 Shock radiation tests for Saturn and Uranus entry probes *J. Thermophys. Heat Transfer* **54** 1246–57

IOP Publishing

Hypersonic Meteoroid Entry Physics

Gianpiero Colonna, Mario Capitelli and Annarita Laricchiuta

Chapter 10

Response of the meteoroid/meteorite to aerodynamic forces and ablation

Georges Duffa, James Beck and Kelly Stephani

Aerodynamic forces and heating of meteoroids/meteorites by convective and radiative effects results in superficial ablation, volume and surface constraints. The high velocities encountered upon atmospheric entry make radiation the dominant contribution to heating of the surface. Rarefaction effects are also important owing to low-density (high-altitude) effects as well as cracks and roughness at the surface. Most meteorites, such as chondrites, develop a liquid layer flowing at the surface. An overview of these phenomena is given in the first part of this chapter, including surface pattern creation. In the second part of this chapter, a more detailed description of physico-chemical phenomena is presented for liquid-free surfaces, including rarefaction effects, using the DSMC tool.

Fragmentation is a complicated phenomenon due to the heterogeneity of meteorites, making a detailed science-based approach difficult. A description of phenomenological approaches is given for breakup and subsequent phenomena.

10.1 Ablation models

The description of ablation and its coupling with the boundary layer is not trivial. The first direct modeling including solid and gaseous phases is only recent [1]. In a classical approach one treats the conservation of elemental mass fractions and energy and includes a correlation for the influence of blowing [2].

10.1.1 Mass conservation

The elemental mass conservation is written, using a mass Stanton number C_M,

$$\tilde{J}_k = -\dot{m}\left(\tilde{c}_{k_s} - \tilde{c}_{k_w}\right) = -\rho_e u_e C_M\left(\tilde{c}_{k_e} - \tilde{c}_{k_w}\right), \tag{10.1}$$

or, introducing the blowing coefficient $B' = \frac{\dot{m}}{\rho_e u_e C_M}$,

doi:10.1088/2053-2563/aae894ch10

$$(B' + 1)\tilde{c}_{k_w} = B'\tilde{c}_{k_s} + \tilde{c}_{k_e}. \tag{10.2}$$

The blowing equation is a correlation giving the effect of gas injection on mass exchange (the same expression exists for energy fluxes)

$$C_M = f(C_{M_0}, B'). \tag{10.3}$$

This kind of expression is relatively accurate for moderate values of B', but can produce important errors for massive injection $B' > 2.7$ where drastic alteration of the flow takes place with possible destabilization as seen in experiments made using the Giant Planet Facility (Galileo program).

In general this massive blowing tends to create a region near the surface where the species contained in the external flow cannot enter in significant amounts [3, 4], in particular oxygen.

10.1.2 Energy conservation

The equation of energy is

$$F_{\text{conv}} + F_{\text{rad}} = F_{\text{cond}} + F_{\text{rerad}} + \dot{m}(h_G - h_S), \tag{10.4}$$

where $F = \mathbf{F} \cdot \mathbf{n}$.

All the terms except for F_{rad} depend on wall temperature T_w.

- *Steady-state approximation.* If we assume a local steady-state we have (taking into account $T_w \gg T_0$)

$$F_{\text{cond}} = \dot{m}[h_S(T_w) - h_S(T_0)] \simeq \dot{m}h_S(T_w). \tag{10.5}$$

  Then equation (10.4) can be rewritten as

$$F_{\text{conv}} + F_{\text{rad}} = F_{\text{rerad}} + \dot{m}h_G. \tag{10.6}$$

  An important consequence is that in-depth coupled thermal calculations are not necessary for ablation estimates. This approximation is no longer valid when important surface temperature gradients are present on the surface. 'Important' means of the same order of magnitude as gradients normal to the surface.

- *Gas in thermodynamic equilibrium.*

  For the calculation of $\dot{m}$, the chemical reactions of gaseous species are assumed to be in thermodynamic equilibrium, a reasonable hypothesis for these media at high pressures. It is then possible to calculate the gas composition through knowing the elemental composition

$$c_i = f(\tilde{c}_k, T_w, p). \tag{10.7}$$

  Then the resolution of the problem is as sketched in figure 10.1.

- *Solids and gases in thermodynamic equilibrium.*

  Another approximation is often used which supposes thermodynamic equilibrium between the solid and gas phases. This results in the possibility of constructing *a priori* functions $B' = f(T_w, p)$, ignoring the influence of

$\rho_e u_e C_M$. This approximation is incoherent since equilibrium leads to $\dot{m} = 0$. This approximation must be seen as the limit

$$B'_{\text{lim}} = \lim_{\rho_e u_e C_M \to 0} \left(\frac{\dot{m}}{\rho_e u_e C_M} \right).$$

The error induced by using this expression is not negligible, as can be seen in figure 10.2. But the effect on ablation is limited by the fact that this phenomenon is basically based on energy conservation: a displacement of

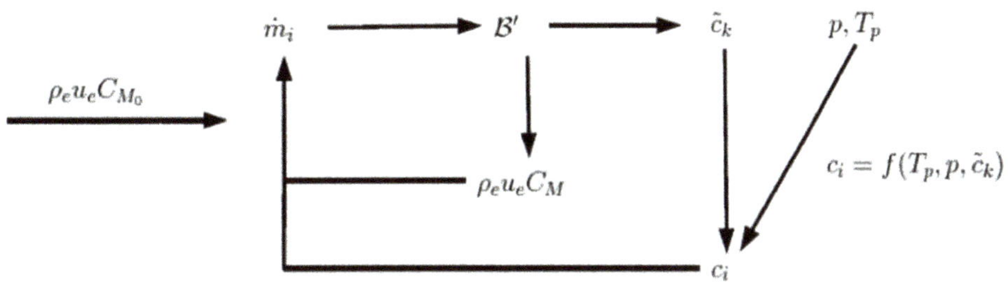

Figure 10.1. Calculation scheme.

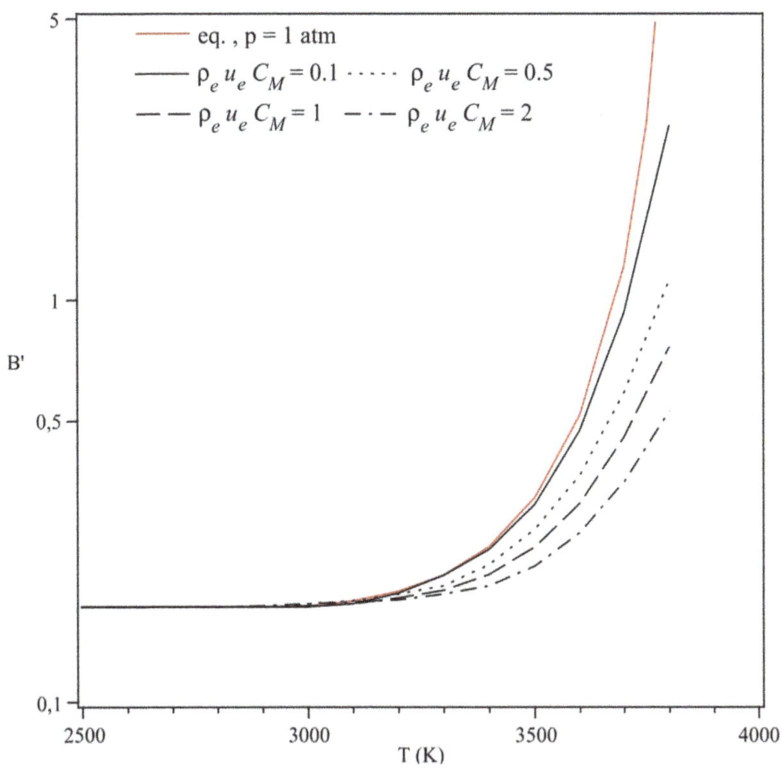

Figure 10.2. Carbon B′ models.

say 300 K on the curves corresponds to an error of less than 10% for the temperature and about the same value for the ablation velocity given by equation (10.4) or (10.6).

Note that this approximation hides the influence of boundary layer exchanges, and also hides the difference between a laminar and a turbulent flow for given heat fluxes.

- *Equivalent (apparent) ablation enthalpy.*

  A quantity based on the linearization of the blowing law has been commonly used in the literature [5]. This will be excluded here where gas injection is important.

10.2 An example

An example is given in [7, 8] for a carbonaceous chondrite (figure 10.3). The elemental composition, measured from the Orgueil meteorite (bulk density 1580 kg m^{-3}, open porosity 35% [9]), is 35.2% O, 19.4% Si, 15.9% Mg, 27.2% Fe and 2.2% S. Assuming thermodynamic equilibrium and no mixing with external flow the composition is as given in figure 10.4. We see that a liquid phase (SiO_2, MgO) is present for low temperatures, roughly between 2000 K and 3000 K (figure 10.5). For higher temperatures the diffusion of oxygen is necessary for the presence of liquids. For large objects and a re-entry velocity of 20 km s^{-1}, rough calculations (no fluid phase) give a

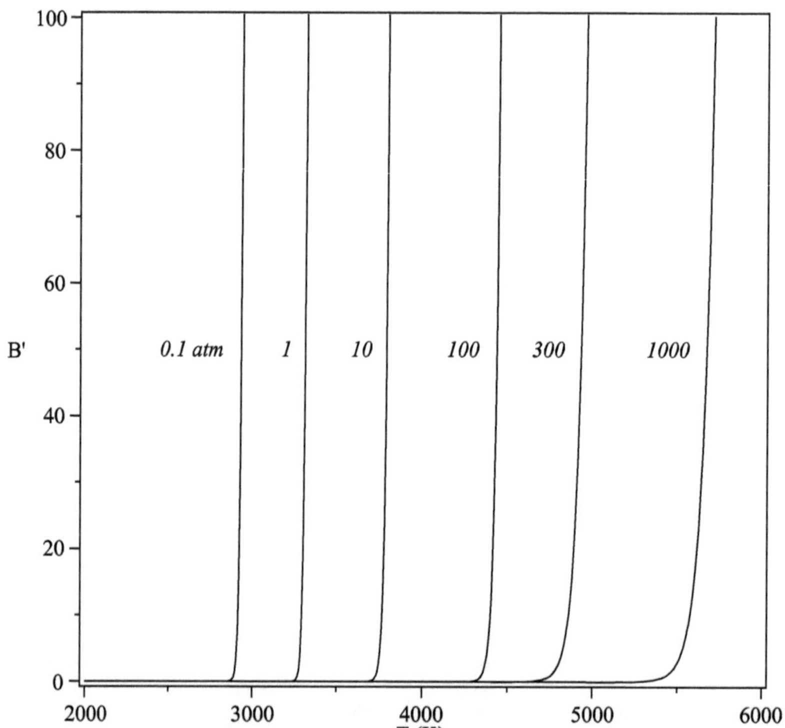

Figure 10.3. The approximate model for chondrite B′.

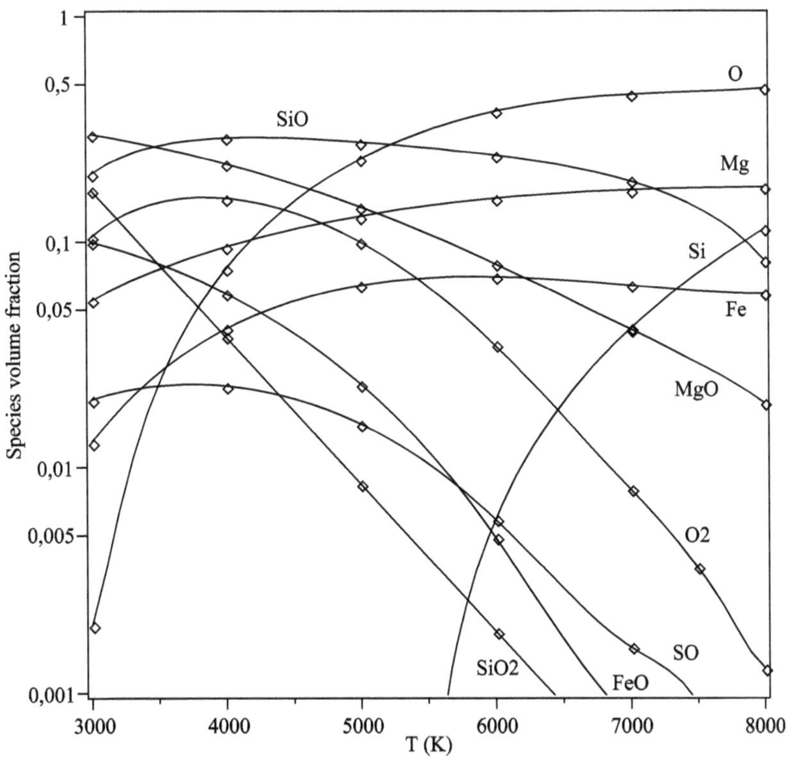

Figure 10.4. Composition of a gaseous chondrite versus temperature for $p = 100$ bars.

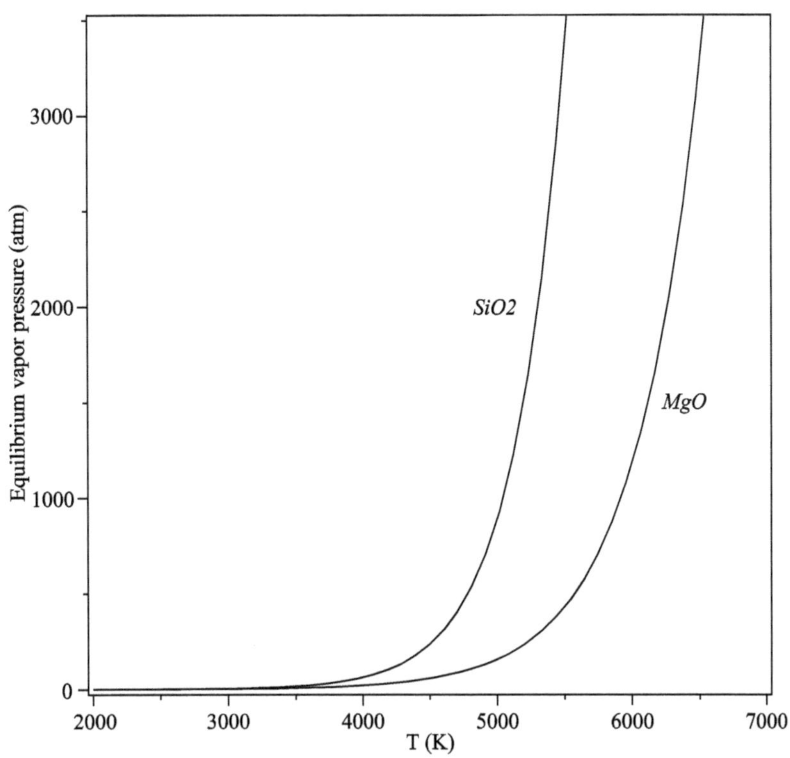

Figure 10.5. Evaporation of SiO_2 and MgO versus temperature [6].

temperature near 4000 K and an ablation velocity of about 10 cm s^{-1} [10]. These values are obtained in a large domain of altitudes, 0–60 km.

10.3 Porosity

An open porosity of 35% as measured for the Orgueil meteorite can lead to internal flow, depending on the associated permeability [11]. This problem is negligible for meteorites of important dimensions. Moreover, other measured permeabilities on meteorites are lower [9].

10.4 The presence of a fluid phase

Most meteoroids are composed of metals (iron, nickel, ...) or silicates. In oxygen-rich media the components are rapidly oxidated and the formed species melt, directly or indirectly (for example in the case of the formation of SiO as an intermediate). Carbon is a counter-example because its tendency to form a stable gas phase (CO) or sublime, giving many possible carbons [2].

The viscosities of saturated oxides and silica are highly dependent on the temperature. These oxides are very viscous and flow on the surface and can be ejected in the form of droplets [12] or freeze in colder regions in the case of small stable objects.

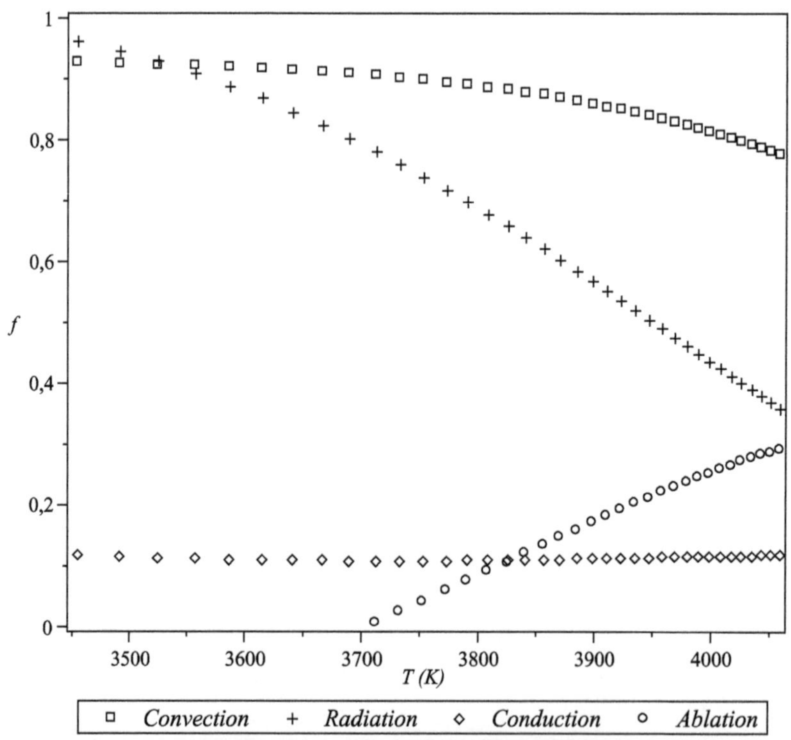

Figure 10.6. Energy partition on a carbon wall. $f = 1$ corresponds to the flux on an inert cold wall.

The presence of a liquid layer constitutes an important difficulty in modeling, not only for the induced liquid flowing on the surface, making the problem non-local, but also for its influence on the emissive properties of the surface, which play a fundamental role in ablation; in capsule re-entries the re-radiation of surface is the dominant aspect [2] (figure 10.6). Then the radiative properties of the liquid influence the ablation [6]. For very high temperatures obtained in meteorite entries, this effect is less important.

The models used [2, 6] calculate the profile of the liquid layer $V_l(s, y)$ supposing a Stokes flow. The gaseous species are assumed to be in thermodynamic equilibrium, and the mass exchanges between liquid and gaseous species are determined by the Hertz–Knudsen law. In a pure quartz material at moderate temperatures the ablation is divided in two equivalent parts for liquid and gaseous flowing [6].

The vaporization coefficients in these laws are generally unknown and a rough hypothesis (see section 10.1.2) of liquid–gas thermodynamic equilibrium is used, as shown in figure 10.3.

This simplification also has the advantage allowing us to ignore the composition of the liquid layer. Due to the different-temperature evaporation of species (see figure 10.5) an inhomogeneity can be anticipated. A simple mixing approximation is not sufficient and the problem is to model the diffusion of species in the layer.

10.5 Creation of surface patterns

Ablation generally creates patterns on the surface of any material. Different mechanisms contribute to this effect.

10.5.1 Diffusion–reaction mechanisms

In this kind of mechanism the creation of the pattern is due to heterogeneities in the chemical composition at submillimetric scales. Examples are the presence of silicate chondrules in chondrite meteorites (figure 10.9) or Widmanstatten heterogeneities in nickel–iron meteorites (figure 10.10).

The patterns are created at the scales of surface heterogeneities, but the three-dimensional patterns display a length scale depending on both the reactivity and diffusion. In the case of figure 10.7 this scale is present in the curvature of the ogival geometry. This mechanism is now well known [13].

These patterns result in a modification of the apparent reactivity due to the perturbation of the shear flow. This effect can be assumed to be low in our case, where the component of velocity parallel to the mean surface is low itself.

These patterns play an important role in laminar–turbulent transitions in the case of low blowing [2].

10.5.2 Convection–reaction mechanisms

At higher scales (millimetric to centimetric) the typical geometries encountered are called scallops or thumb-prints. This kind of pattern, visible in figure 10.8, is encountered in numerous domains, such as sub-glacial caves, snow fields and solid–liquid interfaces (frozen rivers, pipelines).

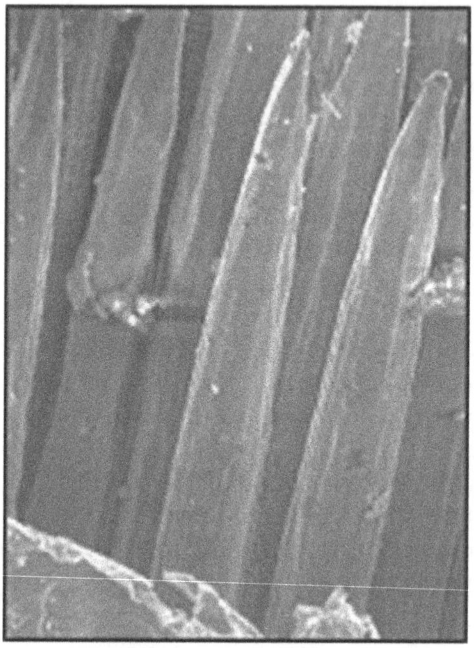

Figure 10.7. Ablated carbon fiber.

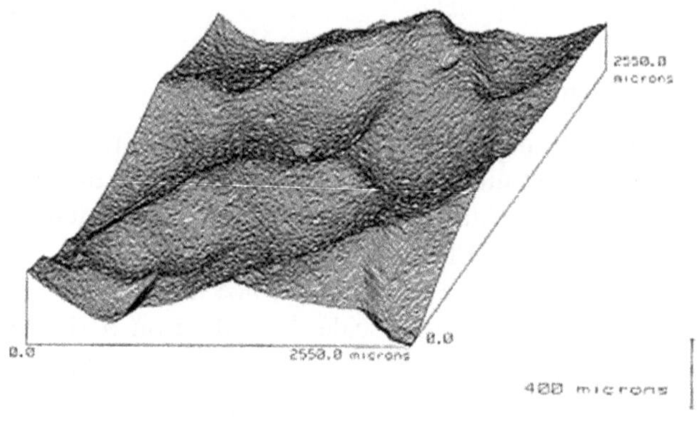

Figure 10.8. Turbulent roughness on polycrystalline graphite.

A physical description of the associated mechanism is lacking. Experimental studies [14] show that the underlying mechanism is convective in origin and is associated with a turbulent environment. The modifications of mass and energy exchanges are important in these experiments. We can, however, infer a limited effect in massive blowing cases.

10.5.3 Radiative-reaction mechanisms

On cannot exclude the role of radiation in the creation of patterns, an example of such a mechanism is the ice elongated geometries called penitentes [15].

Figure 10.9. Chondrules in the chondrite Grassland meteor. The scale is in millimeters. (Reproduced from Wikipedia https://en.wikipedia.org/wiki/Chondrule).

Figure 10.10. Hoba iron–nickel meteorite. (Reproduced from Wikipedia).

10.6 Fragmentation processes

The fragmentation of a meteoroid/meteorite during atmospheric entry is a complex process, and is dependent on how the body itself was formed. Some meteoroids are monolithic, whereas others are formed through successive collision and cohesion of smaller bodies. These compound objects are termed 'rubble-pile' meteoroids or 'dustballs', and their internal structure of a collection of weakly bound parts has a significant impact on their re-entry behavior. During the entry process they fragment into smaller objects. For larger meteoroids, this is generally considered the main mass loss process, due to there being insufficient time for substantial heat penetration into a large object in the limited re-entry time. The resulting smaller fragments have a larger surface area-to-mass ratio, and thus they transfer more energy to the atmosphere through drag and ablate more readily.

10.6.1 Physical processes

The majority of researchers agree that little ablation occurs for large objects above an altitude of 30 km. Once this altitude is reached, the exponentially increasing density results in a rapid increase in thermal stress and aerodynamic shear. These higher thermal and aerodynamic loads result in an initial fragmentation event after which the body undergoes a continuous fragmentation process. This continuous fragmentation was first understood to be due to meteor decelerations from photographic evidence, being greater than that theoretically calculated for an ablating solid body. The fragmentation process is enhanced by the phenomenon that neighboring supersonic objects produce a repulsive lateral force between them due to shock wave interactions.

If the object is sufficiently massive that it does not fragment or ablate completely in the upper atmosphere, then it can undergo an airburst event. This is where the combination of the high pressures and the weak fragmenting meteoroid result in a rapid cascade fragmentation of the remaining object mass, producing small debris which ablate very quickly in the high-density, high-heat-flux environment. The kinetic energy of the meteoroid is transferred to the atmosphere, causing a hypersonic blast wave. This is analogous to an explosion, although the energy release time-scales are longer. In these cases, such as the Tunguska event of 1908, the blast wave can cause significant damage even though very little of the meteoroid mass reaches the surface. Indeed, the destruction can be as much as twice as severe due to the increased energy transfer to the atmosphere resulting from fragmentation.

Not all large objects will airburst. Very large objects, say 100 m or more in diameter, will not be sufficiently slowed down, even in the lower atmosphere, and can reach the ground, forming an impact crater before the catastrophic fragmentation event is able to occur. It is difficult to predict a threshold for airburst, as this is dependent upon the unknown internal structure of the specific meteoroid/meteorite. Extremely large objects will reach landfall without significant reduction in velocity, where the impactor mass is much larger than the mass of the atmosphere displaced in the trajectory. Certainly, any object over 1 km in diameter would not be expected to experience significant drag or ablation in the atmosphere.

10.6.2 Modeling

The standard models for the breakup of meteoroids are the gross fragmentation model of Hills and Goda [16] and the pancake model of Lyne *et al* [17]. These models are based on the concept that the meteoroid body remains essentially intact until a given flight condition is reached. At this point, it is allowed to deform and/or fragment. More recently, Celepcha and Revelle [18] and others have modeled the breakup as a series of events that split the body into separate pieces, including differentiating between events which produce a set of large fragments of similar size to the original body, and events which produce a cluster of small fragments.

It is worth noting that observations suggest a number of cascades of breakup, which is thought to be due to the meteoroid being made up of a range of material components. However, the breaking strengths inferred from observations of meteoroids are often a factor of ten greater than the measured crushing strengths of specimens in ground tests [19], demonstrating that the subject of meteoroid strength is still poorly understood.

The basic dynamics of the body are similar in all the modeling approaches. Standard methods for the calculation of a ballistic trajectory through an atmosphere under Earth's gravity are used. Whilst the body remains intact, the drag acts in the opposite direction to the motion, and is calculated simply as

$$\frac{dv}{dt} = -\frac{\rho v^2 C_D A}{2m}, \tag{10.8}$$

where ρ is the local atmospheric density, v is the velocity, A is the reference area of the body, m is the body mass and C_D is the drag coefficient of the body, which for a hypersonic sphere in continuum air will be approximately 0.92. As the body penetrates deeper into the atmosphere, the density increases and the pressure at the stagnation point of the impactor rises until the point at which the body begins to break up.

There are a number of different thresholds used to predict the onset of breakup of a meteoroid. For example, Register *et al* [20] use a stagnation pressure correlation for the yield strength, Y, which is a function of the body mass, and Collins *et al* [21] use a function of the body density, which is

$$\log_{10} Y = 2.107 + 0.0624\sqrt{\rho_s} \tag{10.9}$$

and is considered to be valid for body densities, ρ_s, between 1000 and 8000 kg m^{-3}. From this the altitude at which the meteoroid is predicted to break up can be calculated, and it is at this point where the models diverge. Note that this correlation suggests a minimum fragmentation stagnation pressure of about 12 000 Pa. This is relevant for the majority of meteoroids, but it is observed that the fragmentation of a weak 'dustball' meteoroid can begin at dynamic pressures as low as 2000 Pa.

The pancake model

The basic idea of the pancake model is that the fractured body can be deformed. Therefore, under the high stagnation pressures, the front of the body is compressed towards the rear, resulting in the lateral expansion of the body. It is generally assumed [22] that the increase in the diameter of the body, now assumed to be ellipsoidal rather than spherical, is analogous to the behavior of an inviscid fluid. The expansion of the impactor diameter, L, is then given by

$$\frac{d^2L}{dt^2} = \frac{C_D P_s}{\rho_s L}, \tag{10.10}$$

where P_s is the stagnation pressure, which is approximately twice the dynamic pressure for hypersonic flows. This increase in the diameter, with an assumed constant drag coefficient, results in increased drag forces. The body is allowed to spread to a limiting diameter, L_{crit}, which is usually defined as a multiple of the original diameter, L_0. Good agreement with observational data has been obtained [22] with the value of the pancake factor, L_{crit}/L_0, between 5 and 10, although it would be expected that the fragmenting body would demonstrate discrete fragment behavior once the pancake factor has reached 2–4. This model differs from the gross fragmentation model only in that the lateral expansion in that model is empirically fitted to observational data.

In the case that the pancake factor is reached prior to the body impacting the surface, the body is considered to undergo an airburst event. The energy of the body is then dissipated into the atmosphere, and the speed of the blast wave is assumed to be the velocity at airburst. If an airburst is not predicted, then the pancake will impact the surface, creating a cluster of craters, as observed at the Henbury formation in Australia [23]. This is shown schematically in figure 10.11.

Discrete fragmentation model

The discrete fragmentation model has a number of variants, depending on how the wake of the meteoroid is treated. Essentially, a collective wake model is similar to the pancake model in that it treats the fragmenting body as having a single bow shock, whereas an independent wake model considers each fragment to be completely separate with its own interaction with the airflow. These concepts are shown schematically in figure 10.12.

The collective wake model assumes a simple splitting of the body into two equal parts in both size and volume in each fragmentation event. These fragments continue their flight alongside one another with a common bow shock, thereby

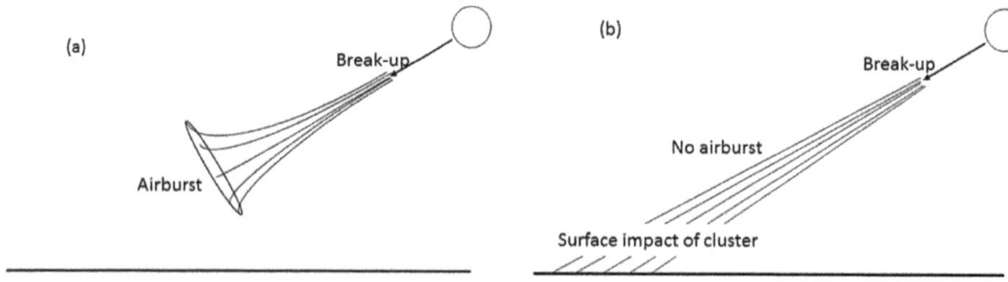

Figure 10.11. Pancake model schematics (a) with and (b) without an airburst.

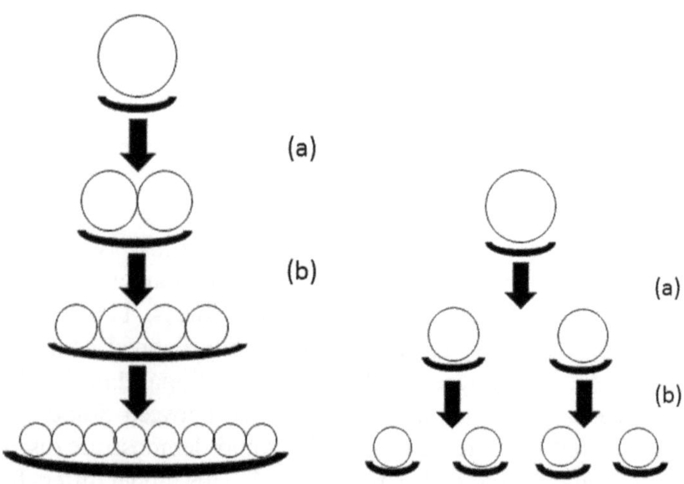

Figure 10.12. Left: Collective wake model. The ballistic coefficient reduces, as in the pancake model, as the bow shock spreads. Right: independent wake model. (a) First fragmentation and (b) second fragmentation in both figures.

reducing the ballistic coefficient of the overall body. It is further assumed that the drag area is doubled in the fragmentation, although it is acknowledged that this is not strictly consistent with the spherical object model. In order to prevent an immediate cascade of fragmentation, a model where the smaller fragments have greater strength is required. This is physically consistent, and Register *et al* [20] use a relationship where the parent (subscript p) and child (subscript c) critical fragmentation stagnation pressures are related by

$$P_{sc} = P_{sp}\left(\frac{m_p}{m_c}\right)^{\alpha}, \tag{10.11}$$

where α is an exponential strength scaling parameter, often tuned to a particular observation. The net effect of this model is that it essentially forms a pancake in discrete steps, rather than in a continuous manner.

The independent fragmentation model considers the individual fragments as completely separate from the moment of fragmentation, which is essentially the same approach as is used in the modeling of the destructive re-entry of spacecraft. In this case, the relative sizes of the resultant fragments can be determined stochastically. The net result is that the decrease of the ballistic coefficient is less than in the other models, and thus the energy deposition into the atmosphere is increased by a smaller amount at each fragmentation relative to the other models.

10.7 Chemically reacting surfaces

Ablation processes contribute to the overall aerodynamic forces of a meteoroid through chemical and mechanical erosion, which transform the size, shape and roughness of the meteoroid. In this section, we present recent studies which address the modeling of gas–surface interactions at chemically reactive surfaces, particularly for cases in which surface roughness and porous length-scales approach the molecular mean free path. Although a number of processes, including ablation, sublimation, melting and evaporation, work to chemically and mechanically erode the meteoroid surface, the discussion here is limited to chemically reactive surface processes. We also discuss the construction of surface chemistry models from molecular beam experiments as an attractive alternative to previous approaches based on macroscopic experimental data.

Accurate high-temperature aerothermodynamics computational tools provide an important means for unraveling the complex thermophysical processes intrinsic to meteoroid entry into Earth's atmosphere. During the high-altitude phase of Earth atmospheric entry, typically above 80 km, the atmosphere is considered to be rarefied, and the aerodynamic forces which develop on the meteoroid are most appropriately characterized by a kinetic description of the gas. At lower altitudes, the extreme temperatures established within the shock layer surrounding the meteoroid result in rapid dissociation of the molecular species, providing a source of atomic nitrogen and oxygen which interact with the surface of the meteoroid. The high pressures and densities which develop within the shock layer around the meteoroid are often described using a continuum description of the flowfield.

The accuracy of a continuum flowfield description near a meteoroid surface, however, is questionable, owing to the roughness and porosity length-scales that may be approaching or comparable to the molecular mean free path of the gas. Furthermore, it has also been demonstrated that strong non-continuum behavior can be established through surface chemical reaction processes, even in cases involving a smooth surface [24].

The development of predictive surface chemistry models for high-speed atmospheric entry applications has been a long-standing challenge owing to the complexity of gas–surface interactions. Finite-rate surface chemistry models are often developed from macroscopic experimental data, such as composition, heat and mass flux, and radiative signatures. These quantities, however, are often a result of gas–surface interactions coupled with collisional gas-phase processes, making it nearly impossible to isolate surface processes for model construction. Recent experimental efforts involving molecular beam–surface scattering measurements provide a characterization of (non-)reactively scattered products in a near-vacuum environment. This kind of approach allows for the isolation of gas–surface interactions from gas-phase collisional processes, and enables the study of surface reaction kinetics with a controlled source of gas particles.

The current study is motivated by ongoing efforts to develop a DSMC-based simulation tool for carbon-based ablative materials. The simulation tool incorporates a surface recession model for a realistic porous carbon-fiber based substrate (FiberForm®) into DSMC, enabling a fully coupled simulation capability involving surface recession, gas–surface interactions and gas-phase reactions for an ablation environment [25, 26]. Kinetic based methods such as DSMC employ a molecular description of the gas, and are necessary to obtain accurate solutions where rarefaction effects are important [24, 27–29]. This work focuses on the development and validation of a DSMC gas–surface interaction model for carbon oxidation within this simulation tool [30]. Although the final objective of this effort is to simulate the oxidation and recession of FiberForm®, we first aim to establish an appropriate gas–surface interaction model for the carbon surface (i.e. the individual fibers) without the added complexity introduced by simulating the full FiberForm® microstructure. Vitreous carbon was chosen as the template for developing the surface reaction model, owing to the similarity of carbon and FiberForm® [31] in terms of their characteristic chemical structure and sp^2 bonding. This model is then used to simulate scattering from a FiberForm® sample, and time-of-flight data from experiments and simulations are compared. It is noted that these finite-rate models have been developed specifically for carbon-based ablative thermal protection system (TPS) materials, but this approach is readily extended to meteoroid materials as well.

We first present the time-of-flight (TOF) and angular distribution data obtained from experiments of Murray *et al* [31], which are used to construct and validate the DSMC model. The details of the experiment and analysis of the scattered products have been discussed at length by Murray *et al* [31]. Here, we describe some key observations for completeness. A hyperthermal O/O_2 beam with a nominal velocity of $7760\,\mathrm{m\,s^{-1}}$, and a mole ratio of approximately 93% $O(^3P)$ (~5 eV) and 7%

$O_2(^3\Sigma_g^-)$ (~10 eV), was directed at a vitreous carbon surface with an incidence angle (θ_i) of 45°. The surface temperature of the sample was controlled by resistively heating the material, and was determined by fitting the TOF data to a Maxwell–Boltzmann distribution characterized by the surface temperature. First, the number densities of the species are collected as functions of the flight time (TOF distribution) and scattering angle in the plane defined by the surface normal and the beam. The TOF distributions are then converted to energy distributions ($P(E_T)$), which may be integrated to obtain the angular flux distributions. These distributions were collected over the full range of final angles (θ_f) at 5° increments for surface temperatures of 800 and 1875 K only. At the intermediate temperatures, the TOF distributions are obtained at a single angle θ_f of 45°.

A total of four chemical species are detected as scattered products, including O and O_2, which are supplied by the beam, and CO and CO_2, which are formed through chemical reactions at the surface. Representative TOF distributions for O and CO at a temperature of 1700 K are shown in figure 10.13. The O and O_2 TOF distributions have a peak at very short times corresponding to impulsively scattered (IS) products. The interaction time of the IS products with the surface is very brief, such that this population does not achieve thermal equilibrium with the surface. These products scatter with energies slightly below that of the incident beam energy, and show up at early times in the TOF distribution.

A second population of scattered products is observed at later times in the TOF distributions, corresponding (for example) to the second peak in the O and the first peak in the CO TOF distributions (see also figures 8, 11 and 12 of Murray *et al* [31]). A considerable amount of information may be gained from analysis of these products, and detailed discussions may be found in [30, 31]. First, it is noted that these products can be fit to a Maxwell–Boltzmann (MB) distribution based on the surface temperature, and include O, CO and CO_2 products. This implies that these

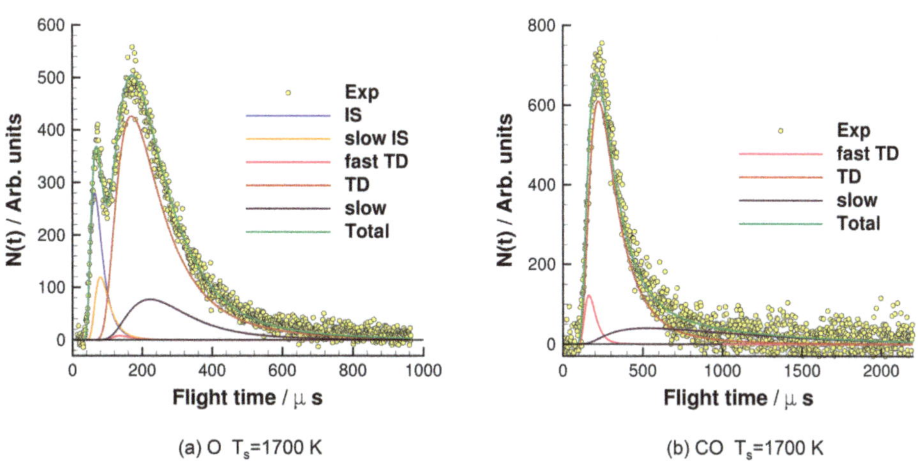

Figure 10.13. Decomposition of the experimental TOF distribution of (a) O and (b) CO at 1700 K into individual components, based on flux contributions from (slow) IS, (fast) TD and slow processes. Reproduced with permission from [30]. Copyright 2018 Elsevier.

products are formed via thermal mechanisms and are completely accommodated to the surface temperature. Second, these products may desorb from the surface in a number of different ways. Products that desorb immediately after the beam pulse (at $t = 0$ ms) and that follow a MB distribution are referred to as thermally desorbed (TD) products; again, these include O, CO and CO_2, and make up the majority of the detected TOF distribution following the IS products. Products may also desorb from the surface at later times, resulting in a distribution component that is not captured by a TD description. For example, if the surface reaction products are formed by a relatively slow process whose rate is governed by the surface temperature, then the products will desorb from the surface based on the decay rate equation of the formation reaction since the beam is the only source of oxygen on the surface. In this case, the final observed TOF will be a convolution of the MB distribution and the specific decay rate equation (exponential for a first order reaction, linear inverse for a second order reaction, etc). As explained in [30, 31], this behavior is observed for O and CO products, which make up the majority of the long tail in the detected TOF distribution.

The reaction mechanisms considered in the current surface interaction model include adsorption, desorption and Langmuir–Hinshelwood (LH) mechanisms. Identification of these mechanisms is based on careful analysis from both the TOF and angular distribution data of the detected products, as well as a mass balance analysis to account for long-desorption time products that are not detected in the experiments at low temperatures. In general, the LH mechanism is a thermal surface mechanism consisting of three major steps, including adsorption, formation and desorption. The designation of an LH model mechanism refers to the reaction of a gas-phase adsorbate reacting with a surface species that is part of the bulk. The LH mechanisms considered for this model are further classified into four different types (hereafter referred to as LH1–4) and are distinguished based on the time-scales of the formation and desorption processes:

LH1. Rapid formation and desorption: prompt LH mechanism.
LH2. Slow formation with rapid desorption (formation limited).
LH3. Rapid formation with slow desorption (desorption limited).
LH4. Slow formation and desorption.

Table 10.1 summarizes the nine major reaction mechanisms that are elucidated from the experimental TOF distributions. The reaction mechanisms include: adsorption of O; LH3 formation of CO$\{a\}$ and CO$\{b\}$; LH1 formation of O, CO, and CO_2; desorption of O; and LH3 desorption of CO$\{a\}$ and CO$\{b\}$. The first six reactions are gas–surface reactions, which involve the atoms from the incident beam pulse as reactants. The last three reactions are pure-surface reactions, independent of the beam pulse.

The LH1 or prompt LH mechanisms are responsible for the observed TD products in O, CO and CO_2. In addition, a small component of these products has super-thermal energies and corresponds to non-thermal products in the TOF distributions. The LH1 formation of CO and CO_2 involves additional adsorbed atoms, denoted by O′(ads). The adsorption/desorption and LH3 mechanisms are

Table 10.1. Set of reactions modeled in DSMC for an O/O_2 hyperthermal beam. Reproduced with permission from [30]. Copyright 2018 Elsevier.

	Gas–surface (GS) reactions
A1. LH3 O{*a*} formation	$O(g) + (s) \longrightarrow O\{a\}(s)$
A2. LH3 CO{*a*} formation	$O(g) + (s) + C(b) + O'(s) \longrightarrow CO\{a\}(s) + O'(s)$
A3. LH3 CO{*b*} formation	$O(g) + (s) + C(b) + O'(s) \longrightarrow CO\{b\}(s) + O'(s)$
A4. LH1 O formation	$O(g)(IS) + (s) \longrightarrow O(g)(TD) + (s)$
A5. LH1 CO formation	$O(g) + C(b) + O'(s) \longrightarrow CO(g) + O'(s)$
A6. LH1 CO_2 formation	$O(g) + (s) + O(s) + C(b) + 4O'(s) \longrightarrow CO_2(g) + 2(s) + 4O'(s)$
	Pure-surface (PS) reactions
B1. LH3 O{*a*} desorption	$O\{a\}(s) \longrightarrow O(g) + (s)$
B2. LH3 CO{*a*} desorption	$CO\{a\}(s) \longrightarrow CO(g) + (s)$
B3. LH3 CO{*b*} desorption	$CO\{b\}(s) \longrightarrow CO(g) + (s)$

responsible for the slow products observed in the TOF distributions of O and CO. These products are formed immediately after they strike the surface (adsorption and LH3 formation), followed by delayed desorption of these species. Thus these products are represented as a two-step process: adsorption/LH3 formation and desorption/LH3 desorption. The two types of CO that are formed through LH3 mechanisms are indicated as CO{*a*} and CO{*b*}, and the CO{*b*} comprises the missing flux in the TOF distributions at lower temperatures, as discussed in Swaminathan-Gopalan *et al* [30].

In order to compute the reaction rates corresponding to the mechanisms outlined in table 10.1, the TOF distributions must first be decomposed into contributions from each reaction mechanism. The decomposition of the O and CO scattered products' TOF distributions are shown in figure 10.13, including impulsively scattered and thermally desorbed populations. The total product fluxes may then be computed as a function of temperature from experimental data at a 45° angle, and finally the surface reaction rate constants constrained by experimental flux values may be determined.

The final reaction mechanisms of the finite-rate model and their rate constant expressions and values are summarized in table 10.2. The reactions are generally classified as adsorption, adsorption-mediated, gas–surface (GS) or pure-surface (PS). The quantity Φ appearing in the expressions for the rate constants in table 10.2 is the surface site density. A value of 3.5×10^{19} m^{-2} was used for the surface site density of carbon in this work [32, 33].

This surface chemistry model was employed in DSMC to simulate the O/O_2 molecular beam scattered from a vitreous carbon surface. The DSMC TOF distributions are shown in figure 10.14 for O and CO products at 1000 K, alongside the experimental TOF data. The current model shows excellent agreement with the experimental data, and IS, TD and tail components in the TOF distributions are properly captured by the DSMC simulation using the current model.

Table 10.2. Reaction rate constants in the finite-rate carbon oxidation model. Reproduced with permission from [30]. Copyright 2018 Elsevier.

Type	Mechanisms	Reaction	Rate constant (k)	Units
Adsorption	Adsorption	$O(g) + (s) \longrightarrow O(ads)$	$\frac{1}{\Phi}*\frac{1}{4}\sqrt{\frac{8k_bT_g}{\pi m}}*0.85$	$m^3\ mol^{-1}\ s^{-1}$
	LH3 O{*a*} formation	$O(ads) \longrightarrow O\{a\}(s)$	1	s^{-1}
	LH3 CO{*a*} formation	$O(ads) + C(b) + O'(ads) \longrightarrow CO\{a\}(s) + O'(ads)$	$\frac{1}{\Phi}*153.0\ \exp(-\frac{4172.8}{T_s})$	$m^2\ mol^{-1}\ s^{-1}$
Adsorption-mediated	LH3 CO{*b*} formation	$O(ads) + C(b) + O'(ads) \longrightarrow CO\{b\}(s) + O'(ads)$	$\frac{1}{\Phi}*71.2\ \exp(-\frac{1161.2}{T_s})$	$m^2\ mol^{-1}\ s^{-1}$
GS reactions	LH1 O formation	$O(ads) \longrightarrow O(TD)(g) + (s)$	$20.9\ \exp(-\frac{2449.3}{T_s})$	s^{-1}
	LH1 CO formation	$O(ads) + C(b) + O'(ads) \longrightarrow CO(g) + (s) + O'(ads)$	$\frac{1}{\Phi}*1574.9\ \exp(-\frac{6240.0}{T_s})$	$m^2\ mol^{-1}\ s^{-1}$
	LH1 CO_2 formation	$O(ads) + O(s) + C(b) + 4O'(ads) \longrightarrow CO_2(g) + 2(s) + 4O'(ads)$	$\frac{1}{\Phi^5}*536.3\ \exp(-\frac{655.6}{T_s})$	$m^{10}\ mol^{-1}\ s^{-1}$
PS reactions	LH3 O{*a*} desorption	$O\{a\}(s) \longrightarrow O(g) + (s)$	$0.05T^2\ \exp(-\frac{3177.2}{T_s})$	s^{-1}
	LH3 CO{*a*} desorption	$CO\{a\}(s) \longrightarrow CO(g) + (s)$	$4485.5\ \exp(-\frac{1581.4}{T_s})$	s^{-1}
	LH3 CO{*b*} desorption	$CO\{b\}(s) \longrightarrow CO(g) + (s)$	$1.2\ \exp(-\frac{2251.6}{T_s})$	s^{-1}

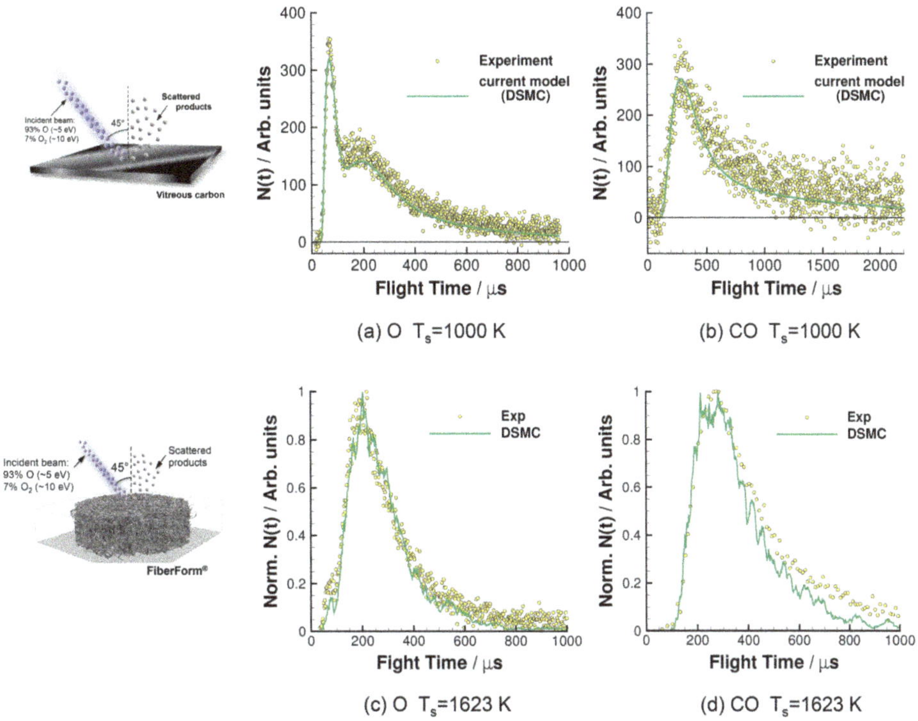

Figure 10.14. Comparison of TOF distributions at $\theta_f = 45°$ of O (left) and CO (right) obtained from DSMC with the current finite-rate model and the experiment [31, 34] following bombardment with the O/O_2 beam at $\theta_i = 45°$. Results of beam scattering from vitreous carbon are shown in (a) and (b), and scattering from FiberForm® are shown in (c) and (d). The corresponding surface temperatures as determined from the experiments are indicated.

The vitreous carbon oxidation model is next employed to simulate molecular beam scattering on FiberForm®, which is a carbon preform material commonly used as a precursor in thermal protection systems (TPSs). The detailed microstructure of the FiberForm® surface is obtained from x-ray microtomography to capture the complex microstructure that is characteristic of the preform material. The surface chemistry framework outlined above is again employed in DSMC to describe the gas–surface interactions of the beam with the FiberForm® material, represented schematically in figure 10.14. The carbon oxidation model is applied to each fiber of the FiberForm® material, and the results of these simulations are compared to the hyperthermal beam experiments performed on FiberForm®. Figure 10.14 shows the experimental and simulated TOF distributions for O and CO, respectively, at 1623 K and a final angle of 45°. Comparison between the experimental and DSMC TOF distributions of both O and CO show good agreement. It is found that a significantly higher amount of CO is generated when the beam scatters from FiberForm®, when compared to the vitreous carbon. This is postulated to primarily be a result of multiple collisions of oxygen with the fibers, resulting in a higher rate of CO production. The occurrence of multiple collisions with the network of fibers was also found to thermalize the O atoms, in addition to the adsorption/desorption process. The effect of microstructure was concluded to be very significant in determining the final composition and energy distributions of the products.

References

[1] Dubroca B, Duffa G and Leroy B 2002 High temperature mass and heat transfer fluid–solid coupling, *11th Meeting AIAA/AAAF Space Planes and Hypersonic Systems and Technologies, Orléans, France* AIAA paper 2002-5180

[2] Duffa G 2013 *Ablative Thermal Protection Systems Modeling* (Reston, VA: AIAA)

[3] Libby P A 1962 The homogeneous boundary layer at an axisymmetric stagnation point with large rates of injection *J. Aerospace Sci.* **29** 48–60

[4] Wilson K H 1970 Massive blowing effects on viscous, radiating, stagnation-point flow *AIAA 8th Aerospace Science Meeting, New York, January 19-21, 1970* AIAA paper 70-203

[5] Wercinski P, Wright M, Prabhu D K, Tauber M E, Laub B and Curry D M 2004 Tutorial on ablative TPS, *2nd Planetary Probe Workshop, NASA Ames Conf. Center, Moffett Field, CA*

[6] Chen Y-K 2016 Thermal ablation modeling for silicate materials, *54th AIAA Aerospace Science Meeting* AIAA Paper 2016-1414

[7] Park C 2013 Rosseland mean opacities of air and H-chondrite vapor in meteor entry problems *J. Quant. Spectrosc. Radiat. Transfer* **127** 158–64

[8] Park C 2015 Rosseland mean opacities for the flow around cometary meteoroids *J. Quant. Spectrosc. Radiat. Transfer* **154** 44–54

[9] Consolmagno G J and Britt D T 1998 The density and porosity of meteorites from the Vatican collection *Meteorit. Planet. Sci.* **33** 1231–41

[10] Park C 2016 Inviscid-flow approximation of radiative ablation of asteroidal meteoroids by line-by-line method, *54th AIAA Aerospace Science Meeting*, *San Diego, CA* AIAA Paper 2016-0506

[11] Duffa G 2015 Porous media interaction with high temperature and high speed flows *STO-AVT-216 Lecture Series* (Belgium: Von Karman Institute)

[12] Brownlee D E, Bates B and Bauchamp R H 1983 Meteor ablation spherules as chondrule analogs *Lunar and Planetary Institute Technical Report*

[13] Vignoles G L, Lachaud J and Aspa Y 2014 Environmental effects: ablation of C/C materials. Surface dynamics and effective reactivity *Ceramic Matrix Composites: Materials, Modeling and Technology* (New York: Wiley)

[14] Wool M R 1975 Summary of experimental and analytical results *Technical report* SAMSO-TR-74-86 Acurex/Aerotherm

[15] Claudin P, Jarry H, Vignoles G, Plapp M and Andreotti B 2015 Physical processes causing the formation of penitentes *Phys. Rev.* E **92** 033015

[16] Hills J G and Goda M P 1993 The fragmentation of small asteroids in the atmosphere *Astron. J.* **105** 1114–44

[17] Lyne J E, Tauber M E and Fought R M 1996 An analytical model of the atmospheric entry of large meteors and its application to the Tunguska event *J. Geophys. Res.* **101** 23207–12

[18] Celepcha Z and Revelle D O 2005 Fragmentation model of meteoroid motion, mass loss, and radiation in the atmosphere *Meteorit. Plant. Sci.* **40** 35–54

[19] Svetsov V V, Nemtchinov L V and Teterev A V 1995 Disintegration of large meteoroids in Earth's atmosphere: theoretical models *Icarus* **116** 131–53

[20] Register P J, Mathias D L and Wheeler L F 2017 Asteroid fragmentation approaches for modelling atmospheric energy deposition *Icarus* **284** 157–66
[21] Collins G S, Melosh H J and Marcus R A 2005 Earth impact effects program: a web based computer program for calculating the regional environmental consequences of a meteoroid impact on Earth *Meteorit. Planet. Sci.* **40** 817–40
[22] Chyba C F, Thomas P J and Zahnle K J 1908 Tunguska explosion: atmospheric disruption of a stony asteroid *Nature* **361** 40–4
[23] Hodge P W 1965 The Henbury meteorite craters *Smithsonian Contrib. Astrophys.* **8** 199–213
[24] Swaminathan-Gopalan K, Subramaniam S and Stephani K A 2016 Generalized Chapman–Enskog continuum breakdown parameters for chemically reacting flows *Phys. Rev. Fluids* **1** 083402
[25] Borner A, Panerai F and Mansour N N 2017 High temperature permeability of fibrous materials using direct simulation Monte Carlo *Int. J. Heat Mass Transfer* **106** 1318–26
[26] White C, Scanlon T J and Brown R E 2015 Permeability of ablative materials under rarefied gas conditions *J. Spacecr. Rockets* **53** 134–42
[27] Stephani K A, Goldstein D B and Varghese P L 2013 A non-equilibrium surface reservoir approach for hybrid DSMC/Navier–Stokes particle generation *J. Comput. Phys.* **232** 468–81
[28] Deschenes T R, Holman T D and Boyd I D 2011 Effects of rotational energy relaxation in a modular particle-continuum method *J. Thermophys. Heat Transfer* **25** 218–27
[29] Subramaniam S, Swaminathan Gopalan K and Stephani K A 2016 Assessment of continuum breakdown for high-speed chemically reacting wake flows, *46th AIAA Thermophysics Conf.* AIAA Paper 2016-4434
[30] Swaminathan-Gopalan K, Borner A, Murray V J, Poovathingal S, Minton T K, Mansour N N and Stephani K A 2018 Development and validation of a finite-rate model for carbon oxidation by atomic oxygen *Carbon* **137** 313–32
[31] Murray V J, Marshall B C, Woodburn P J and Minton T K 2015 Inelastic and reactive scattering dynamics of hyperthermal O and O_2 on hot vitreous carbon surfaces *J. Phys. Chem.* C **119** 14780–96
[32] Zhluktov S V and Abe T 1999 Viscous shock-layer simulation of airflow past ablating blunt body with carbon surface *J. Thermophys. Heat Transfer* **13** 50–9
[33] Blyholder G and Eyring H 1957 Kinetics of graphite oxidation *J. Phys. Chem.* **61** 682–8
[34] Poovathingal S J, Qian M, Murray V J and Minton T K 2018 Reactive and scattering dynamics of hyperthermal O and O_2 from a carbon fiber network *J. Phys. Chem.* C **119** 14780–96

Chapter 11

Experimental investigation of meteorites: ground test facilities

Christophe O Laux, Megan E MacDonald, Carolyn Jacobs, Pierre Mariotto, Augustin Tibère-Inglesse, Sean D McGuire, Fabian Zander, Richard Morgan, Luigi Savino, Mario De Cesare, Giuseppe Ceglia, Antonio Del Vecchio, Fabrizio Paganucci and Massimo D'Orazio

The capability to reproduce atmospheric entry phenomena in ground-based facilities is crucial for studying the entry of space missions and of any objects de-orbiting or entering the Earth's atmosphere. During entry, these objects are exposed to intense mechanical and thermal stresses due to interactions with the high-temperature plasma that forms in front of them. Complex physical and chemical phenomena occur at hypersonic conditions. The characterization of plasma–meteoroid interactions requires the ability to understand the thermochemical state of the plasma flowfield, both in the free stream and in the boundary layer surrounding the meteoroid, as well as the determination of the evolution of the meteoroid under the effect of the plasma. The latter requires monitoring the temporal evolution of the shape and surface temperature of the material during melting and ablation processes and the ability to characterize the evolution of the structure and composition of the material. Several complementary ground-based facilities worldwide can be used for that purpose. A few of them are presented in this chapter. They include the SCIROCCO and GHIBLI facilities of CIRA in Italy, the HEAT facility at SITAEL (Italy) and the CP50 Plasma Torch facility at CentraleSupélec (CNRS, France). These ground-based facilities operate under a wide variety of conditions, with test times ranging from less than 1 s (HEAT) to continuous (SCIROCCO, GHIBLI, CP50), specific enthalpies up to 30 MJ kg^{-1} and heat fluxes up to 30 MW m^{-2}. The following sections present these various facilities and illustrate their measurement capabilities. Illustrative results from experiments on various materials, including meteorites, are presented. Additional experimental facilities and results may be found in [1–3].

doi:10.1088/2053-2563/aae894ch11

11.1 The CP50 plasma torch facility at CentraleSupélec

The CP50 plasma torch facility was installed in 2014 at Laboratoire EM2C (CNRS/CentraleSupélec, France) to study a variety of fundamental issues associated with high-enthalpy flows, in particular plasma–material interactions. The torch operates in continuous mode at atmospheric pressure and temperatures up to 8000 K in air, with specific enthalpies of 15–25 MJ kg^{-1}. Materials placed in the air plasma stream are exposed to high heat fluxes ranging from 0.5 to about 30 MW m^{-2}. The torch can be fitted with several nozzles from 1 to 7 cm in diameter, thus allowing the testing of samples of dimensions up to a few centimeters. The facility is equipped with various optical diagnostic instruments allowing the investigation of gas and material temperatures, and the detection of chemical species present in the gas phase. The ability to characterize the evolution of the structure and composition of the material is also desirable, but this has not yet been implemented in our facility.

In the following sections, we describe the plasma torch facility and the associated suite of implemented experimental diagnostics. We then illustrate the capabilities of the facility with the results of studies of the interactions between a high-enthalpy air plasma flow and a phenolic-impregnated carbon ablative material used for atmospheric re-entry heat shields. These investigations can be extended to the study of plasma flow interactions with any objects (such as meteorites) less than a few centimeters in diameter.

11.1.1 The 50 kW CP50 plasma torch facility

The facility is centered around an inductively coupled plasma (ICP) torch (TAFA Model 66) powered by a radio frequency (4 MHz) power supply (LEPEL Model T-50-3) (figure 11.1). This power supply can deliver up to 120 kVA of line power to the oscillator plates with a maximum of 12 kV DC and 7.5 A. The measured maximum RF power output from the oscillator plates is 64 kW. The torch schematic (figure 11.1) shows the five-turn copper induction coil (mean diameter 8.6 cm) encapsulated

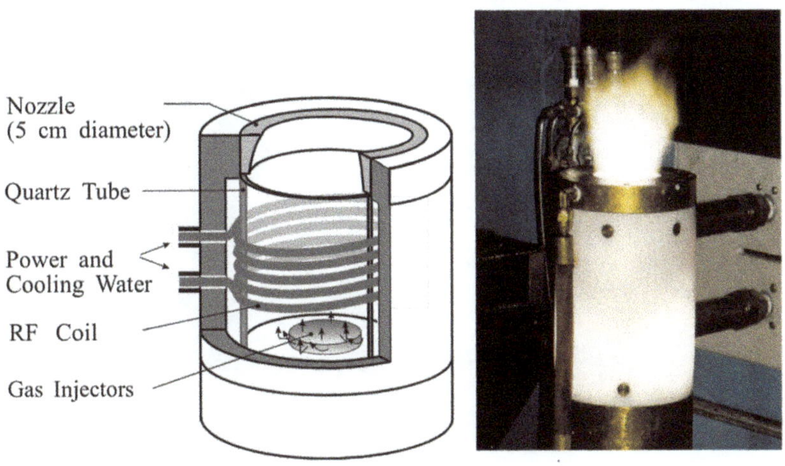

Figure 11.1. Plasma torch schematic.

between a confinement quartz tube (inner diameter 7.6 cm, thickness 3 mm) and a Teflon body. The region between the quartz tube and the Teflon body is cooled with deionized water to prevent arcing between the turns of the coil. The plasma torch operates at atmospheric pressure with a variety of gases (argon, air, nitrogen, methane, carbon dioxide, hydrogen, …). For the current experiments, the torch was operated with air. More details about the facility can be found in [4].

The torch exit can be equipped with various nozzles, ranging in diameter from 1 to 7 cm. Here three different nozzles were employed: a straight nozzle with a diameter of 7 cm and two converging nozzles with 2 cm and 5 cm exit diameters. The typical velocity at the exit of the nozzle is about 10 m s^{-1} for the 7 cm nozzle, 20 m s^{-1} for the 5 cm nozzle, and 140 m s^{-1} for the 2 cm nozzle. The ablative samples to be tested are placed a few centimeters downstream of the nozzle exit where the flow remains laminar and is not affected by entrainment of ambient air.

11.1.2 Diagnostics

The torch facility is equipped with two spectrometers, an Acton Research SpectraPro SP2750i, and an Ocean Optics Maya 2000Pro spectrometer. The characteristics of these spectrometers are described below and a photograph of the facility is shown in figure 11.2.

The first spectrometer is a broadband Ocean Optics Maya 2000Pro spectrometer with a spectral resolution of 1.16 nm. This instrument is used to record emission spectra over a wide spectral range extending from the UV to the near infrared (250–1100 nm). The typical spatial resolution can be as low as 60 μm. The spectrometer is useful to collect the gray body emission of the material, from which surface temperature measurements can be made, and to record optical emission of the plasma in the free flow and around the ablating material. The location of the viewing region is shown as a red dot in figure 11.3.

The second instrument is a high-resolution spectrometer (Acton SpectraPro SP2750i) fitted with an intensified CCD camera (Princeton Instruments PIMAX:1024-UV-18 mm with 1024 × 256 pixels, 26 μm pitch) with response

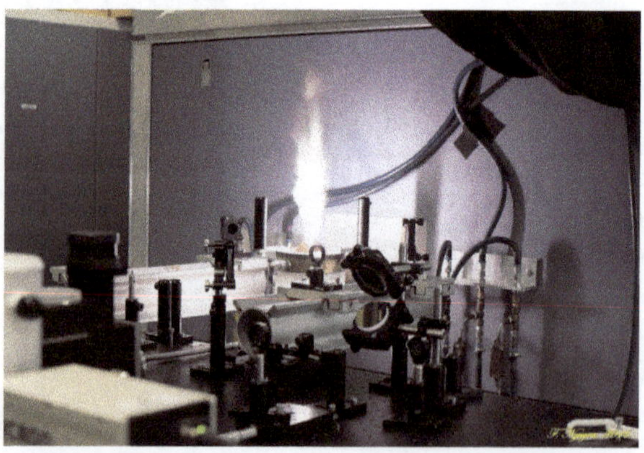

Figure 11.2. The 50 kW RF torch facility and associated optical diagnostics.

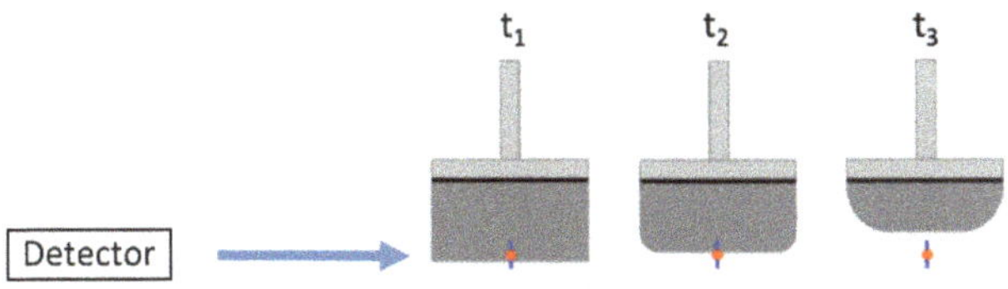

Figure 11.3. Set-up for optical emission spectroscopy measurements across the boundary layer.

from the VUV to the near-IR spectral range (140–900 nm). Six interchangeable gratings provide coverage over this entire spectral range. With gratings of 1200 grooves/mm, the typical spectral resolution is 0.05 nm and the covered spectral range is 15 nm.

Used in spectroscopic mode, the CCD provides 2D images, with wavelength in the horizontal direction and spatial position in the vertical direction. With a typical magnification of 1 and a CCD height of 6.6 mm, spectra can be recorded over a vertical distance of 6.6 mm.

The high-resolution VUV–visible–IR spectrometer can be used to determine the radial temperature profiles in the plasma flow (using for instance the O triplet at 777 nm), to identify emitting species when high resolution is required, and to interrogate the VUV and infrared spectral ranges that are not accessible with the Maya instrument.

The measured spectra can be calibrated in absolute intensity using an Optronics Laboratory tungsten ribbon lamp for measurements in the UV/visible and near infrared (350–6000 nm), and an argon mini-arc for calibrations in the range 140–400 nm.

The facility is also equipped with two digital SLR cameras (Canon 450D and 1100D) in order to obtain time resolved images of the ablator placed in the plasma flow. One of these two cameras captures images of the evolution of the shape of the material, and the other is used for a two-color ratio pyrometry (TCRP) method developed at the University of Queensland [5, 6]. Photographs for the TCRP are typically recorded every 1–30 s during the experiments.

Finally, an optical pyrometer (Minolta Cyclops Model 153A) is used to provide point-measurements of the surface temperature of the ablating material. The pyrometer measures the radiation emitted by the material in the spectral range 800–1100 nm. If the emissivity of the observed material is known in this wavelength range, the radiation measured by the pyrometer can be converted into a temperature. The measurements provided by the pyrometer have been found to be in excellent agreement with the two-dimensional temperature maps obtained with the TCRP method, as will be shown below.

11.1.3 Characterization of the plasma free stream

The plasma flows produced by the torch are in local thermodynamic equilibrium at atmospheric pressure. Therefore, the thermochemical state of the plasma can be entirely characterized with knowledge of the temperature profile. In air plasma, this profile is conveniently obtained from Abel-inverted profiles of the absolute intensity of the atomic oxygen triplet at 777 nm. A typical temperature profile in the plasma

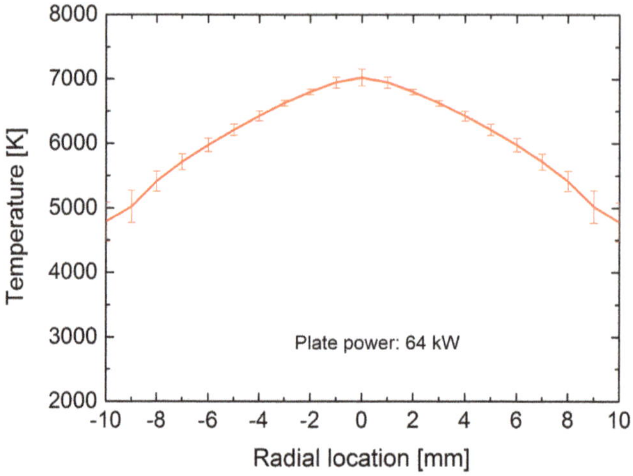

Figure 11.4. Measured temperature profile in the free plasma flow at the exit of the 2 cm nozzle. Plate power: 64 kW [7, 15].

free-stream is shown in figure 11.4. Depending on the torch power conditions, the maximum temperatures in air range from 6000 to 8000 K with the current facility. Because the initial thermochemical state of the incoming plasma state is completely determined, the initial conditions of the flow impinging on the ablator are well known, which is a great advantage for comparison with numerical simulations of plasma–ablator interactions.

11.1.4 Characterization of plasma–material interactions

Samples of ablative material can be placed in the plasma free stream a few centimeters above the exit of the torch nozzle. The samples are glued to a water-cooled disk held by a water-cooled sting, as shown in figure 11.5. In this case, the sample is an ASTERM sample (ablative material made of a phenolic-impregnated carbon fiber matrix (provided by Airbus Defence and Space)) of density around 350 kg m^{-3}.

Figure 11.6 shows a sample of ASTERM placed in the air plasma flow at the exit of the 5 cm nozzle. The plate power is 41 kW and the velocity of the incoming plasma flow is 19.3 m s^{-1}. For these conditions, the maximum temperature of the flow impinging on the material is 6550 ± 100 K and the specific centerline enthalpy is 19.9 MJ kg^{-1}. The heat flux on the sample is mainly convective and its value is estimated to be 1.4 ± 0.2 MW m^{-2} [8, 9]. As shown in figure 11.6, the material undergoes various ablative processes under the effect of the air plasma. The surface temperature is measured with the TRCP method, which provides a two-dimensional temperature map as the sample ablates. In this case, the surface temperature is around 2200 K. These measurements are confirmed with the single-color pyrometer, and also with the gray body emission measurements, as shown in figure 11.7.

Figure 11.5. Photograph of ASTERM ablative sample glued on a water-cooled copper holder. Torch nozzle diameter: 5 cm. Distance of sample above nozzle exit: 2 cm.

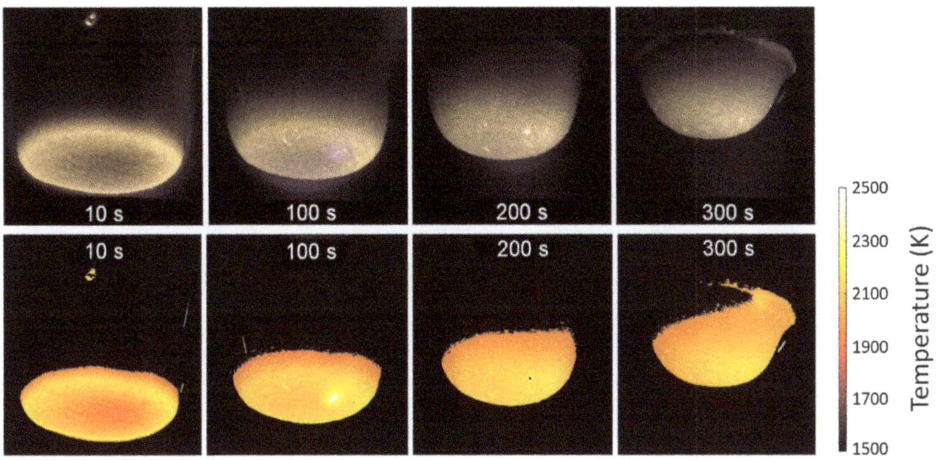

Figure 11.6. Top row: temporal evolution of the ASTERM ablative sample placed in an air plasma flow produced with 41 kW plate power. Nozzle diameter: 5 cm. Bottom row: two-dimensional images of the temperature evolution obtained with the TCRP technique [5–8].

The evolution of the shape of the material is obtained by the digital camera. The recession profiles are shown in figure 11.8. From these profiles, the recession rate is determined to be 2.8 mm min^{-1} under the conditions investigated.

11.1.5 Characterization of chemical species in the gas phase

When exposed to a plasma flow, the material undergoes pyrolysis and ablation, and various chemical species enter the gas phase where they may combine with the species present in the plasma. The products in the gas phase can be detected by non-intrusive optical diagnostic techniques, as described in this section. In the work presented below, the gas phase was characterized by optical emission spectroscopy (OES) only, but additional diagnostic techniques are also available in the facility: laser Raman scattering [10], TALIF of O atoms [11], CRDS of N_2^+ [12], NO^+ [13] and N_2 $A^3\Sigma_u^+$ [11].

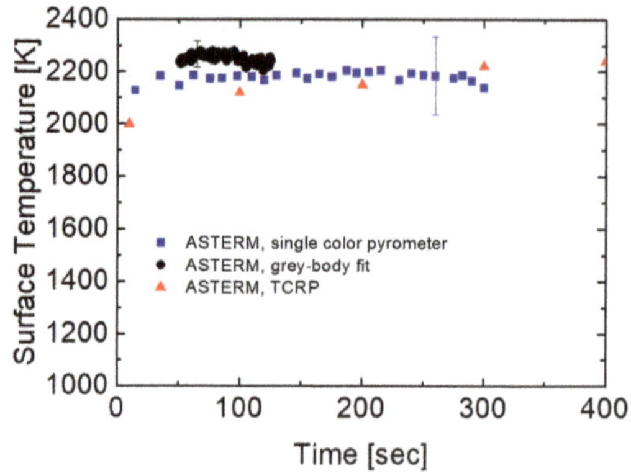

Figure 11.7. Comparison of the surface temperature measurements obtained using three different methods. Plate power: 41 kW. Nozzle diameter: 5 cm [7, 8].

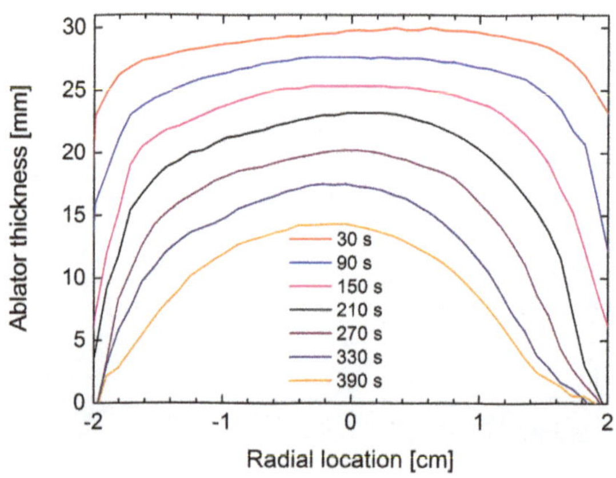

Figure 11.8. Temporal evolution of the shape of the ablating material. 41 kW plate power. Nozzle diameter: 5 cm [7, 8].

For OES measurements, emission is collected sideways of the sample material and plasma flow as shown in figure 11.3. Typically, the optical train looks at a fixed location as the material recedes. This allows one to measure the emission of gaseous species across the boundary layer in front of the material.

The density of excited species present in the boundary layer is determined using the SPECAIR code [14, 15] from the absolute intensity emission measurements. An example of spatial profiles across the boundary layer is shown in figure 11.9. The figure shows that sodium, detected from its emission at 588 nm, is ejected from the material. We also see CN emission bands; this CN is produced by combination of ablated carbon atoms with the dissociated nitrogen present in the incoming plasma flow. In another study at higher heat flux [7], we also detected calcium and barium lines. Figure 11.10 shows another configuration with a 2 cm diameter ASTERM sample exposed to a high heat flux. The nozzle diameter is 2 cm, and the maximum

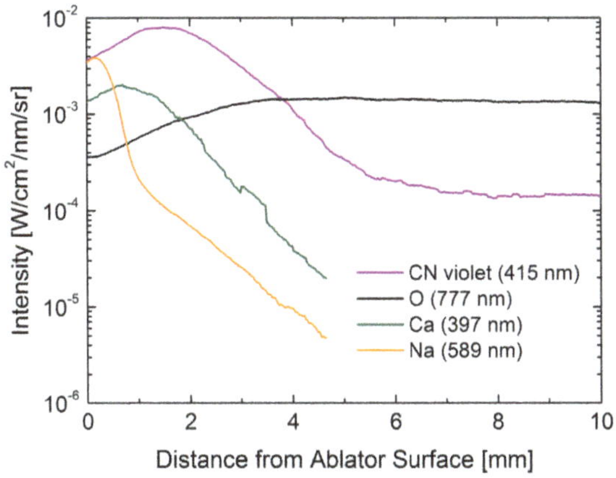

Figure 11.9. Measured excited species density profiles in the boundary layer of an ASTERM material exposed to an air plasma flow. Plate power: 40 kW. Nozzle diameter: 5 cm [7, 8].

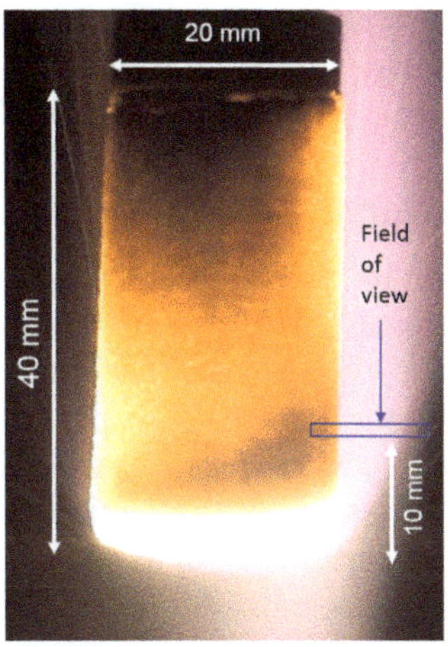

Figure 11.10. Photograph of ASTERM sample placed in the plasma flow. The field of view of the spectrometer is highlighted.

centerline temperature and velocity of the air plasma flow are 7000 K and 140 m s^{-1}, respectively [7, 16].

The heat flux to the surface is about 7.4 MW m^{-2} at the stagnation point [7]. The measured surface temperature of the material at the stagnation point is up to 3300 K [7]. It is expected that carbon sublimates from the material and forms carbonaceous molecules such as CN and CO. CN is actually observed in the stagnation region [7] but CO is not detected in that region. However, CO is detected in the side boundary layer of the material. Infrared emission spectra were acquired in the range 4.58–4.62 mm to record the emission of several ro-vibrational lines of CO. The spectra were Abel-inverted and the radial profiles of temperature and CO density were determined, as discussed in [16]. Figure 11.11 presents these profiles. The density of CO molecules

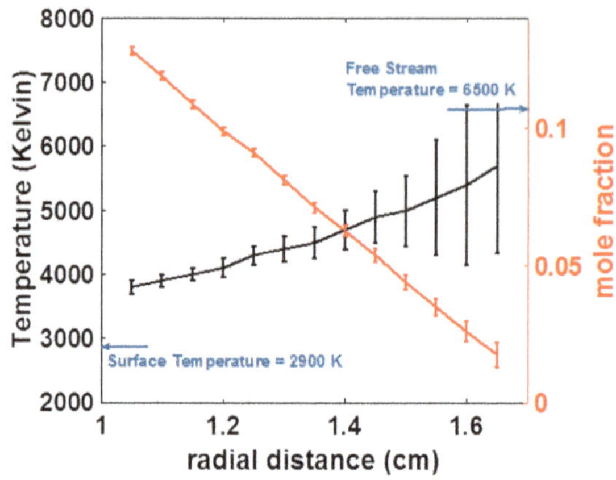

Figure 11.11. Measured temperature (dashed red line) and CO mole fraction (solid blue line) profiles in the side boundary layer of the sample ASTERM material placed in air plasma flow. 1 cm corresponds to the edge of the sample ASTERM material. Nozzle diameter: 2 cm. Plate power: 64 kW [16].

reaches high values, with mole fractions up to 12%. This CO is produced by combining the carbon ablated from the material at the stagnation point with the incoming air plasma. These data can be used to validate ablation models.

11.2 The PWT facility for testing meteorites at CIRA

The capability to experimentally reproduce the atmospheric re-entry in terms of mechanical and thermal stresses is a crucial point for manned or unmanned aerospace missions as well as for the study of the re-entry phenomena of any objects de-orbiting or entering the Earth's atmosphere, due to the complex physical and chemical phenomena that occur under hypersonic conditions [17]. In particular there is increasing interest in the analysis of potentially hazardous asteroids and their entry scenarios, which has led aerothermodynamic engineers to investigate meteor entries by applying the engineering tools developed for the entry of space capsules [18, 19].

One type of facility used to reproduce atmospheric re-entry consists of plasma wind tunnels (PWT). The arcjet wind tunnels such as SCIROCCO and GHIBLI located at CIRA allow one to generate a hypersonic continuous plasma that strikes the tested models [20–23] allowing, at the same time, the verification of the behavior of the materials subjected to the stresses induced by the hypersonic jet and to provide, through the experimental characterization of the plasma as well as of the tested materials and components, experimental data in order to support and improve the numerical models that describe the hypersonic conditions and the plasma–material interactions.

The flow and material characterization is based on both standard intrusive experimental measurements (temperature, pressure and heat flux sensors) and innovative diagnostic techniques such as:

- OES—optical emission spectroscopy.
- LIF—laser induced fluorescence.

- IBA—ion beam analysis.
- IR–infrared thermography.

In this section, a brief description of the plasma wind tunnel as a test facility and of the most innovative advanced diagnostic techniques is presented.

11.2.1 Hypersonic plasma wind tunnels

For this kind of experimental analysis (atmospheric entry) two kinds of plasma wind tunnels can be used: impulse tunnels [20–24] and continuous tunnels. In the following applications the continuous tunnels are described. Continuous arc-heated tunnels are necessary to study radiative heating, a very important aspect at high enthalpies and in particular during interplanetary missions, and to study the ablation phenomena typical of meteors and atmospheric entry vehicles, that cannot be obtained with impulse facilities. The SCIROCCO plasma wind tunnel [21–23] is an arc-heated facility, capable of producing a low-pressure high-enthalpy hypersonic flow of large dimensions on samples in scale up to 1:1 with a test duration up to 30 min. This facility is operational in Capua (Italy) at CIRA (Centro Italiano Ricerche Aerospaziali). A scheme of the facility is shown in figure 11.12. The arc-heater of the PWT-SCIROCCO is a segmented constricted type with a bore diameter of 0.11 m and a length of 5.5 m. At the extremities of this tube are the cathode and the anode, each made of nine disks, through which a confined electrical discharge is generated. Process air at a mass flow rate in the range of 0.3–3.5 kg s^{-1} at a high pressure of 87 bar is supplied to the arc-heater, inside which temperatures in the range 2000–10 000 K are achieved [20, 21].

The heated compressed air is accelerated through a convergent–divergent conical nozzle. There are seven different nozzle configurations (from 187 mm up to 1950 mm of exit diameter) in order to achieve the desired flow conditions and to match any model size. A model support system inserts the test article in the plasma flow. The test chamber has a cylindrical vertical shape with an overall height of 9.2 m and an inner diameter of 5.17 m. Finally, the hypersonic jet is sent by a pick-up line to a very long diffuser (50 m length and average diameter of 3 m). The unique intrusive

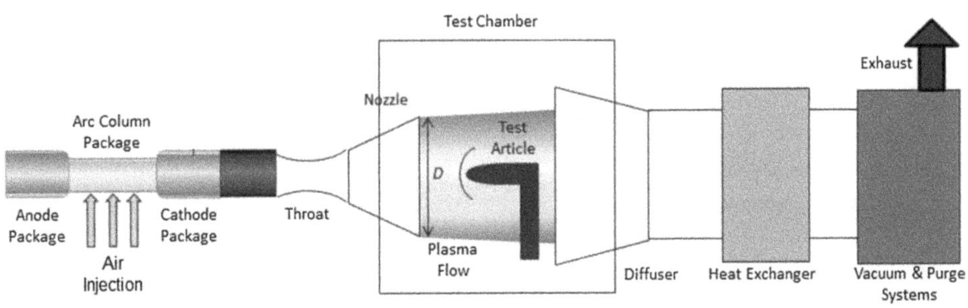

Figure 11.12. Schematic of the main components of the hypersonic arc-heater plasma wind tunnel SCIROCCO [25].

investigations of the plasma flow characteristics acting on the test model are performed using two calibration probes. They are two water-cooled copper arms with a spherical nose 100 mm in diameter. They support and locate instruments for stagnation heat flux and pressure measurements.

One of the most sensitive aspects in using a hypersonic plasma wind tunnel is the experimental characterization of the hypersonic jet. The high voltage and current at the arc-heater (it can absorb up to 70 MW of electrical power) [21–23] necessary for the proper working of the facility, allows one to obtain information about total enthalpy through a proper energy balance technique between the energy provided to the arc-heater and the amount absorbed by the plasma [25], while knowledge of plasma features in terms of temperature and concentration of the plasma species can be obtained through non-intrusive techniques such as OES and laser induced fluorescence (LIF). The other facility located at CIRA is called GHIBLI. It is a hypersonic plasma wind tunnel, with the hot flow generated by a 2 MW segmented arc-heater similar to that of PWT-SCIROCCO, but scaled down. The facility has equivalent technology to the latter, but with its smaller sized components, it is able to perform test campaigns on models up to 80 mm in diameter [26].

11.3 Optical emission spectroscopy (OES)

11.3.1 Basic principles of OES

Emission spectroscopy deals with transitions between different energy levels in a system composed of atoms and molecules. It is known from quantum mechanics that the energy released by an atom, given the jump of an electron from an excited to unexcited state, occurs as photon emission, observable as an emission line. The emission (or absorption) of a molecule is related to the transition from upper to lower electronic, vibrational and rotational states. The investigation of the spectrum requires the knowledge of the energy levels associated with a species. Referring to the simple concept of the diatomic molecule 'dumbbell' (rigid rotor or harmonic oscillator) it is possible to identify different forms of energy [17].

- *Translational energy* $\varepsilon'_{\mathrm{trans}}$ is associated with the kinetic energy of the center of mass of a molecule. The speed can be decomposed along the three Cartesian coordinates (in this case three geometric degrees of freedom).
- *Rotational energy* $\varepsilon'_{\mathrm{rot}}$ is due to the rotation of the molecule around three coordinated axes. The source is the rotational kinetic energy connected to the rotational velocity and to the moment of inertia. For a diatomic molecule the moment of inertia associated with the intermolecular axis is equal to zero.
- *Vibrational energy* $\varepsilon'_{\mathrm{vib}}$ is due to the vibration of the molecule with respect to the equilibrium state constituted by two sources: kinetic energy connected to the motion and potential energy due to intermolecular forces.
- *Electronic energy* $\varepsilon'_{\mathrm{el}}$ is due to the motion of the electrons around the nucleus. There are two energy sources associated with each electron: the kinetic energy due to its translational motion on the orbit around the nucleus and the

potential energy due to its position in the field of electromagnetic motion, generated mainly by the nucleus.

Total energy can be written as the sum of the energies in the different degrees of freedom, in particular for a molecule

$$\varepsilon'_{\text{mol}} = \varepsilon'_{\text{trans}} + \varepsilon'_{\text{rot}} + \varepsilon'_{\text{vib}} + \varepsilon'_{\text{el}}, \tag{11.1}$$

and for an atom

$$\varepsilon'_{\text{atom}} = \varepsilon'_{\text{trans}} + \varepsilon'_{\text{el}}. \tag{11.2}$$

It can be demonstrated that the internal specific energy e of a diatomic molecule can be written as [17]

$$\underbrace{e}_{\text{tot}} = \underbrace{\frac{3}{2}RT}_{\text{translational}} + \underbrace{RT}_{\text{rotational}} + \underbrace{\frac{h\nu/kT}{e^{h\nu/kT}-1}RT}_{\text{vibrational}} + \underbrace{e_{\text{el}}}_{\text{electronic}}, \tag{11.3}$$

where R is the gas constant per unit mass, k is the Boltzmann constant and ν is the fundamental vibration frequency of the molecule.

For an atom, the energy jumps occur only between distinct energy levels of an electronic type and this produces an emission (or absorption) spectrum made up of isolated lines each corresponding to a specific jump (figure 11.13, left). In the case of the molecule, as there are other energetic forms, each electronic level is no longer identified by an isolated value, but by a collection of quantized vibrational levels, organized in quantized rotational levels. This produces a spectrum made of spectral lines given by the combination of electronic, vibrational and rotational levels (figure 11.13, right).

The increase in temperature causes a different distribution at energy levels and this implies a change in the characteristics of the emission spectrum. This concept is of fundamental importance for the technique of measuring the temperature of the hypersonic flow. Note that the experimental spectra in figure 11.13 present a broadening due to the physical sizes of the elements of spectrometers (entrance

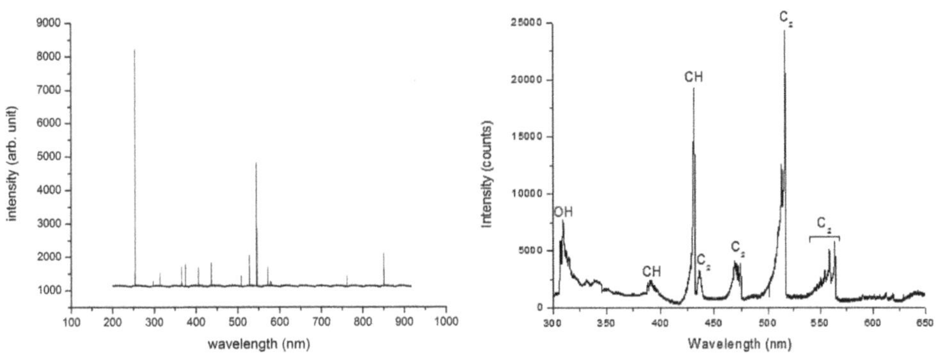

Figure 11.13. Left: typical experimental spectrum of an atomic species (obtained using an Hg vapor lamp as source). Right: typical vibrational bands of combustion products of a Bunsen burner (diffusion flame).

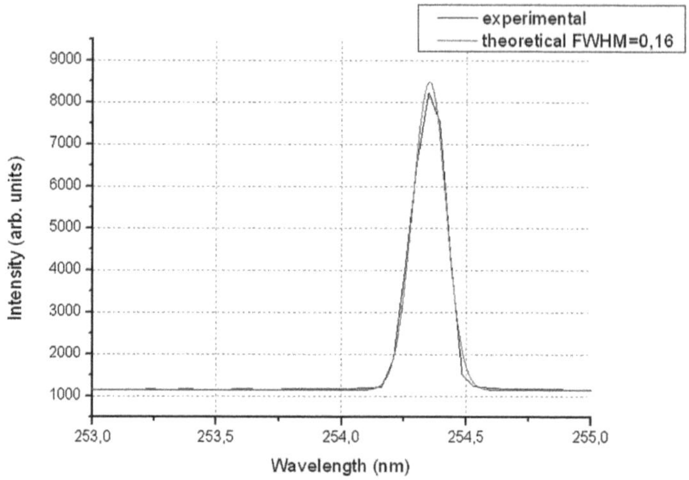

Figure 11.14. Fitting of the experimental mercury emission line with the Gaussian function. The FWHM is related to the spectral resolution of the apparatus used for acquisition (in this specific case 0.075 mm as the entrance slit and a grating of 1200 gr mm^{-1}).

slit, grooves of the diffraction grating, number of pixels of the CCD). For this reason the experimental emission line of an atomic species can be approximated by a Gaussian whose full width at half maximum (FWHM) represents the spectral resolution of the system as observed in figure 11.14. A comparison between experimental spectra and theoretical spectra can be performed when theoretical spectra are convoluted with a slit function of the system (such as the one represented in figure 11.14). The slit function can be deduced from the broadening of an emission line of an atomic species (obtained, for example, by using a vapor mercury lamp as source, for which the spectrum is represented in figure 11.13 (left) and in more detail in figure 11.14, focusing on a single emission line). It corresponds to the Gaussian function that fits the emission peak of an atomic species.

11.3.2 Typical set-up for OES: description of the main components of a spectrometric system

A typical set-up for high-resolution OES measurements is sketched in figure 11.15. It corresponds to the system used for spectrometric measurements performed in plasma wind tunnels SCIROCCO and GHIBLI. The radiation is collected and directed to the entrance slit of the monochromator by a Nikon UV-Nikkor lens [27] and a broadband single mode optical fiber. The monochromator is a Horiba Jobin Yvon HR 460 [28]. It has a planar holographic 1200 grooves/mm grating with an $a \pm 0.15$ nm wavelength position accuracy and a total scanning range of 200–900 nm with a slew rate of 45 nm s^{-1}. Light is finally collected using a Horiba Jobin Yvon Symphony CCD [29] camera (1024 × 256 pixels, each pixel having a 26 × 26 μm size). It has a four-stage thermo-electric cooling system (based on Peltier cells) to work at about 210 K (to ensure a maximum dark current of 0.008 e-/pixel/s). The CCD camera and the monochromator have a remote control.

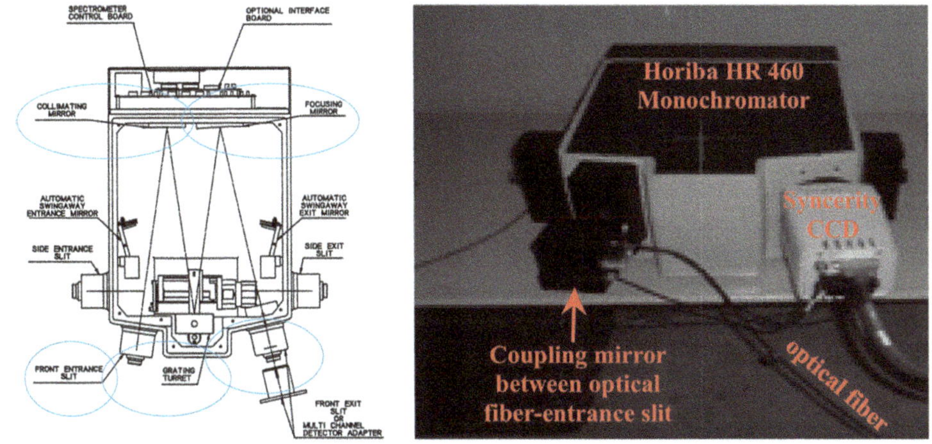

Figure 11.15. Detailed scheme of the Horiba HR 460 monochromator Syncerity CCD (left) and spectrometer assembly composed of an optical fiber (connected to the lens), monochromator and CCD.

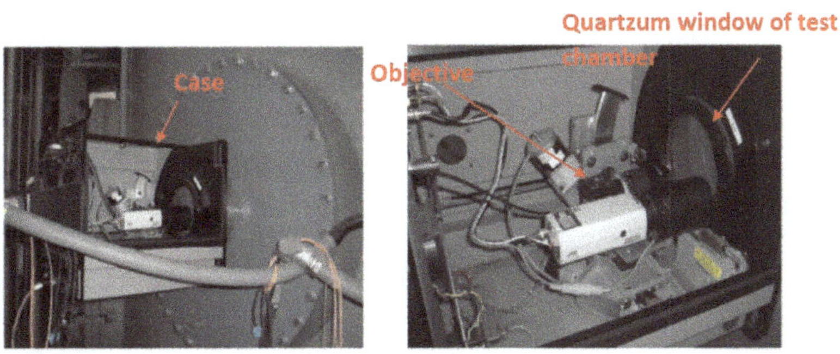

Figure 11.16. Left: the case containing the lens (connected to the monochromator through an optical fiber). Right: a zoom of the case showing the quartz window used as optical access to the PWT test chamber.

In particular, the first element of the measurement chain after the observation window of the test chamber is the lens (figure 11.16). A Nikon UV-Nikkor 105 mm f/4.5 is used (to optimize the UV spectral range acquisition of the spectrum where NO emission of the plasma free jet is predominant), with a focal distance $f = 105$ mm, a focus of 0.48 m–∞ and an aperture of f/4.5–f/32 [29].

The spectroscopic system is composed of a monochromator and a CCD device. The monochromator is a HORIBA JOBIN YVON® HR 460 [28]. The quality of the monochromator is defined, first of all, by the passing band—the range of wavelengths of the beam coming from the exit slit with an energy greater than 50% of the nominal radiation. For the system taken as a reference (Horiba HR 460), there is a bandwidth ranging from 200 nm to 900 nm. Another important element is the spectral resolution—it is not linear and depends on the type of diffracting grating used. The HR 460 is equipped with a grating of 1200 grooves mm^{-1}. Spectral resolution is also heavily affected by the entrance slit (it is inversely proportional to the entrance slit, i.e. a smaller entrance slit gives a better spectral resolution but a worse signal-to-noise ratio (S/N)). The CCD device is instead a Syncerity of HORIBA JOBIN YVON® and is made up of a matrix of 1024 × 256 sensitive elements (pixels). Each pixel (26 μm × 26 μm) measures the radiation intensity it receives and the measured value is associated with the wavelength based on the position of the pixel itself. A Peltier cell brings the CCDs to the operating temperature of −60 °C; only in this case can the thermal noise be considered negligible. An integrated detector controller also has the function of converting the analog signal to digital (ADC), with a 16 bit digitization [29].

11.3.3 Techniques for plasma experimental characterization through OES

The Boltzmann plot is a methodology for the determination of the temperature of the excited electronic levels, the so-called Boltzmann temperature (T_B). In the case of plasma at local thermal equilibrium (LTE) conditions, T_B is equal to the LTE temperature, T [4]. From the general expression of the emission coefficients [4] it is possible to deduce the following relation [30]:

$$\ln\left(\frac{I_{ij}\lambda_{ij}}{A_{ij}g_i}\right) = -\frac{E_i}{kT_B} + C_2. \tag{11.4}$$

Plotting equation (11.4) with E_i (i.e. energy associated with the ith level) on the horizontal axis and $\ln(I_{ij}\lambda_{ij}/A_{ij}g_i)$ on the vertical axis will result in a linear function with slope equal to $-1/kT_B$ where k is the Boltzmann constant 1.3806503 × 10^{-23} m^2 kg s^{-2} K^{-1} (figure 11.17, right). Therefore, taking the relative intensity I_{ij} (the i and j indices referring, respectively, to the upper and lower energy level) of two experimental calibrated atomic lines at wavelength λ_{ij} and using the spectroscopic constants database (deduced from the NIST database [31]) to obtain Einstein coefficient A_{ij} and the degeneracy of the ith level, g_i, it is possible to plot a straight line and deduce T_B from its slope (C_2 is a constant not needed for T_B calculation).

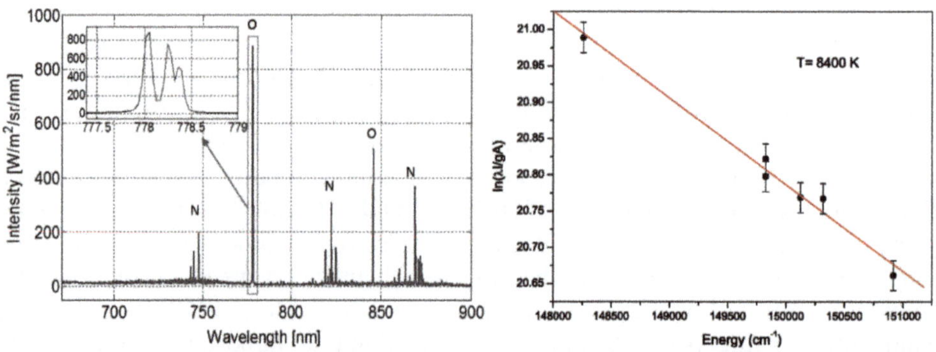

Figure 11.17. Left: nitrogen and oxygen spectral lines emitted by the plasma in the arc-heater of the SCIROCCO facility [25]. Right: the Boltzmann plot obtained from nitrogen lines [25].

The reliability of the method can be enhanced by taking several lines rather than only two, and obtaining the linear function as the least-squares fit of the points in the plot. The advantage of such a method is its simplicity. On the other hand, the major drawback is the huge sensitivity of the result to the spectroscopic database, the uncertainty of the experimental relative intensity and the uncertainty of the Einstein coefficients (A_{ij}) [4]. The methodology described in the current paragraph is valid for optically thin plasmas only. Moreover, in the case of the spectra emitted from the plasma flow in the arc-heater, the Boltzmann plot methodology is applied assuming the plasma jet as a uniform and homogeneous column from the radiation point of view, meaning that the radial profile of the temperature is considered uniform. Calibration of the emitted spectra can be performed in relative intensity using certified deuterium–halogen radiation source as a reference.

OES is the chosen experimental tool to study the free-stream thermodynamic state of the plasma flow, because of its non-intrusivity and the possibility of accessing detailed information about plasma physics and chemistry. At the SCIROCCO Hypersonic facility attention is focused on the NO-γ and NO-δ of vibrational bands, which are the most intense in the whole spectrum. The set-up for high-resolution OES measurements is sketched in figure 11.18.

The hypersonic plasma generated by the PWT at the exit of the conical nozzle is in a non-local thermal equilibrium (NLTE) state. The rotation and translation energy states of the free jet molecules are in equilibrium with each other and are not in equilibrium with the vibration energy state, defining in this way two different

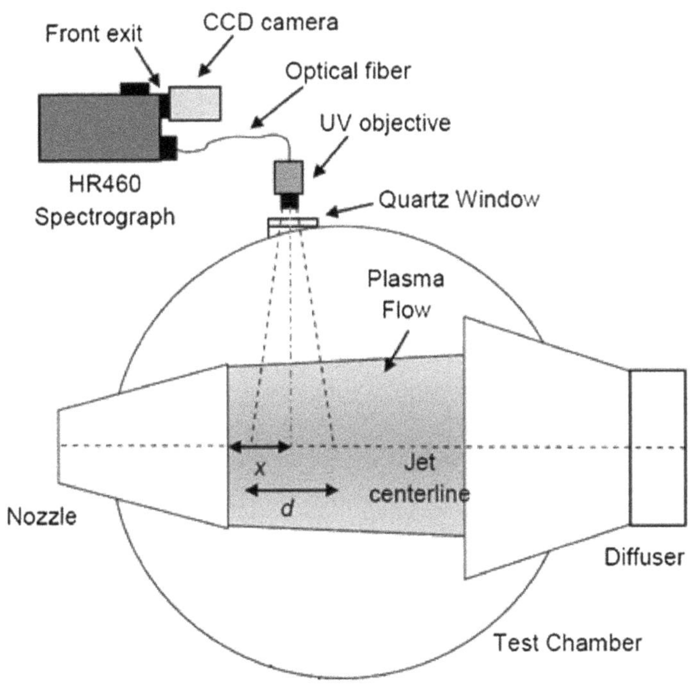

Figure 11.18. Optical system arrangement—top view of the test chamber (elements are not to scale).

(roto-translation and vibration) temperatures associated with the plasma [32], hence a two-temperature approach is usually adopted [33]. This means that rotational energy levels are assumed to be populated according to a Boltzmann distribution ruled by T_{rot}, while electronic and vibrational energy levels are assumed to be populated according to a Boltzmann distribution ruled by T_{vib}. The general expression of the population of the ith level in such a case is

$$n_i(T_{\text{rot}}, T_{\text{vib}}) = \frac{N}{Q(T_{\text{rot}}, T_{\text{vib}})} g_i \exp\left(-\frac{E_i^v}{kT_{\text{vib}}} - \frac{E_i^r}{kT_{\text{rot}}}\right) \tag{11.5}$$

$$E_i = E_i^{ev} + E_i^r, \tag{11.6}$$

where E_i^{ev} and E_i^r are the electronic-vibrational and the rotational contributions to the total energy of the ith level, respectively. N is the population density, $Q(T)$ is the partition function and g_i is the degeneracy of the ith level. An example of an acquired calibrated spectrum is presented in figure 11.19. Emission is dominated by two NO vibrational bands belonging to the $A^2\Sigma \rightarrow X^2\Pi$ (NO-γ) and $C^2\Pi \rightarrow X^2\Pi$ (NO-δ) electronic transitions. In particular, looking at the calibrated spectrum in figure 11.19, vibrational bands from (0,1) to (0,6) belonging to NO-δ and bands from (0,3) to (0,6) belonging to NO-δ are clearly visible [32]. The results of the fitting procedure give rotational and vibrational temperatures (for the case represented in figure 11.20 $T_{\text{rot}} = 614 \pm 23$ K and $T_{\text{vib}} = 1082 \pm 30$ K [33]).

High-resolution predicted spectra compared to the experimental spectra are obtained using SPECAIR software [14, 15]. It is dedicated to predicting the radiative properties of various plasma mixtures made of air H_2O and CO_2. The tool is based on accurate spectroscopic databases and the spectral range is from VUV to IR. SPECAIR allows one to set translational, rotational, electronic (T_{el}) and vibrational temperatures separately [14, 15]. The high-resolution predicted spectra are convolved with a Gaussian slit function (or any arbitrary shape function) in order to be

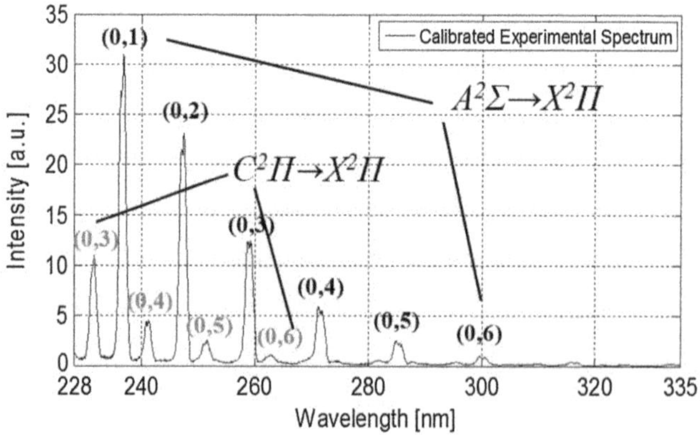

Figure 11.19. Calibrated experimental spectrum collected at 17 cm from the exit nozzle of the SCIROCCO facility.

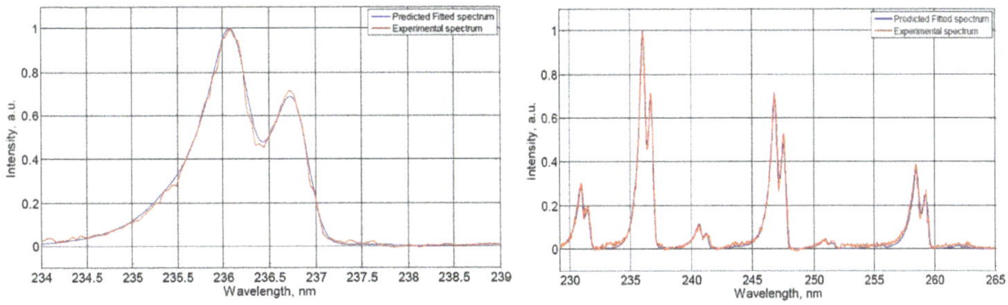

Figure 11.20. Left: results of the fitting procedure on the NO-γ band (0,1) between 234 and 239 nm [14, 15]. Right: results of the fitting procedure between 228–260 nm [14, 15].

able to compare them correctly to the experimental spectra. The Gaussian slit function is characterized by the experimentally determined FWHM using the Hg source that detects the spectral resolution of the used system.

Once both experimental and theoretical spectra have been treated, rotational and vibrational temperatures are determined by fitting the theoretical spectrum on the experimental one, and in particular by minimizing the root mean square error (RMSE) function [32], defined as

$$\mathrm{RMSE}(T_{\mathrm{rot}}, T_{\mathrm{vib}}) = \sqrt{\frac{1}{N}\sum_{l=1}^{N}\left[\bar{S}_{\mathrm{exp}}(l) + \bar{S}_{\mathrm{th}}(l)\right]^2}, \tag{11.7}$$

where N is the total number of spectral points, of both the normalized experimental spectrum $\bar{S}_{\mathrm{exp}}(l)$ and theoretical spectrum $\bar{S}_{\mathrm{th}}(l)$. The final value of the RMSE is an indication of the closeness between the experimental spectrum and the fitted theoretical spectrum, hence it gives an important first indication of the quality of the fitting result.

11.3.4 Meteor characterization through OES

The investigation of meteors offers several topics of considerable interest to a broad scientific community (astronomers, planetologists, geologists, mineralists and spectroscopists).

The knowledge acquired on meteorites has allowed researchers to highlight the extraordinary diversity of the extraterrestrial materials hitherto found and the important role of the impacts on planetary evolution.

Spectroscopy coupled to cameras can provide information on the elements in the meteor.

There is no reference database for the emission spectra produced in the laboratory of the plasmas generated by various types of meteorites crossing the atmosphere. Hence, simulating, through laboratory experiments, meteoric entry into the atmosphere to perform the characterization of their plasma through the use of optical emission spectroscopy can be a way to obtain reference spectra through the characterization of extraterrestrial objects [34].

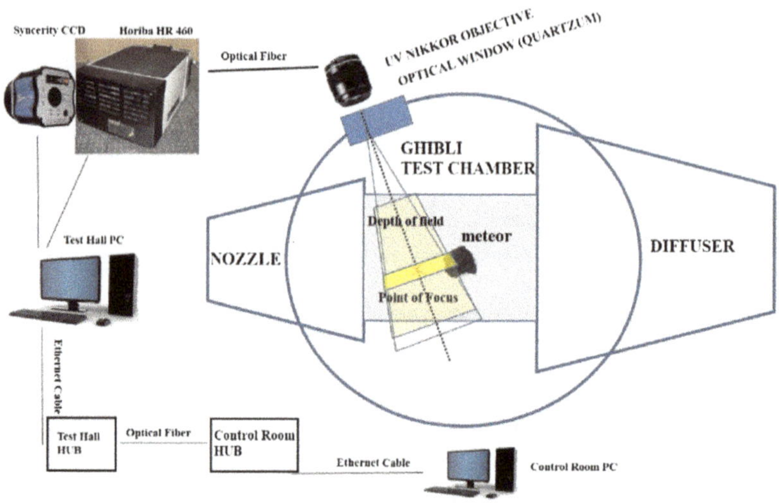

Figure 11.21. Schematic of the acquisition chain used for OES during tests performed at GHIBLI.

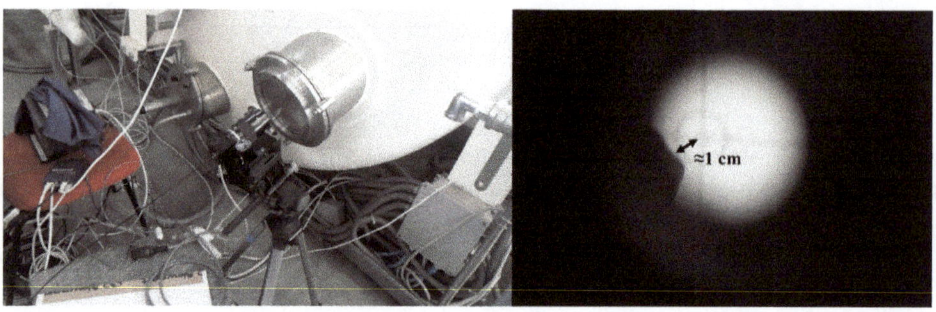

Figure 11.22. Left: objective and window used for OES. Right: point of focus as seen through the ocular of the UV-Nikkor objective.

Preliminary analysis of the feasibility of these kinds of tests has been performed in the GHIBLI plasma wind tunnel.

A schematic of the acquisition chain used for OES during tests performed in GHIBLI is shown in figure 11.21 and figure 11.22 shows photographs of the set-up and point of focus.

The spectrometric system described in section 11.3.2 has been used to capture spectra deriving from the plasma due to the ablation of meteors, specifically, the objective was set in order to focus on a point about 1 cm from the meteor[1] (in order to cover both the meteor and plasma in its proximity) as shown in figure 11.22. Two meteorite fragments over 1 cm in diameter were exposed to a hypersonic jet of plasma composed of air and argon, reaching surface temperatures above 1700 °C, which activated the melting and ablating mechanisms of the material.

[1] The field of view (FOV) of the objective used, described in section 11.3.2.

The two fragments were ordinary chondrites, most likely from the asteroid belt between Mars and Jupiter, collected on Earth in the Sahara desert between 2014 and 2016 and kindly made available by the company Geo Store.

A scan in the 200–900 nm range was performed before and after meteorite insertion (exposure time $t_{exp} = 1$ s, figure 11.23). It highlighted a spectral emission band around 588 nm after meteorite insertion [34]. An acquisition around the emission band detected in the first run (588 nm) was performed, doubling the exposure time ($t_{exp} = 2$ s, figure 11.24) to improve the S/N [34]. The two detected peaks (width ≈ 1 nm) were centered at wavelengths consistent with Na emission lines (table 11.1 shows a shift of about 0.5 nm between the theoretical and experimental results mainly due to the acquisition chain delay and spectral resolution). Emission spectra showing possible Na emission have been successfully acquired. This study has demonstrated the feasibility of such experiments to simulate meteoroid entry into Earth's atmosphere and the chance to investigate the related emission spectra.

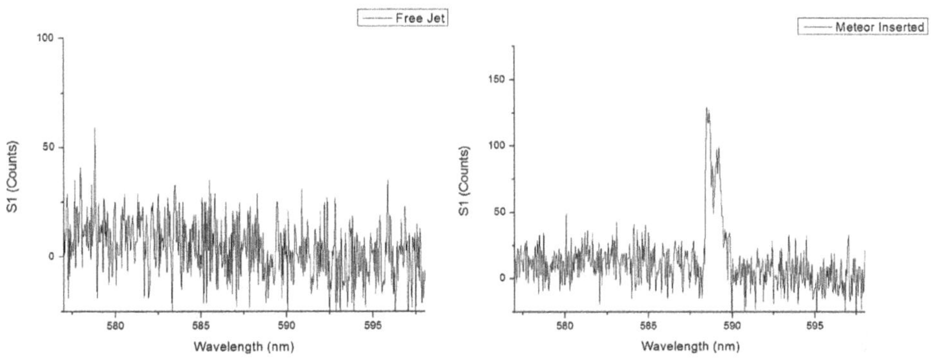

Figure 11.23. Left: emission spectrum of the plasma free jet before meteoroid insertion. Right: emission spectrum of plasma after meteoroid insertion in the first test (exposure time 1 s) [34].

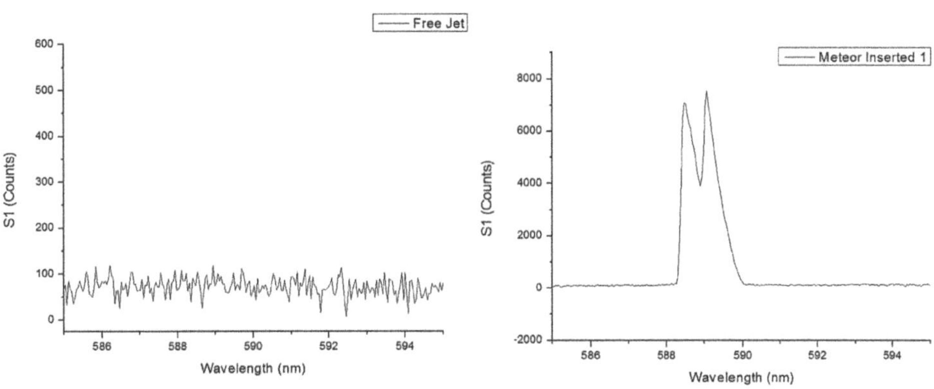

Figure 11.24. Left: emission spectrum of the plasma free jet before meteoroid insertion. Right: emission spectrum of plasma after meteoroid insertion in the second test (exposure time 2 s) [34].

Table 11.1. Comparison between experimental and theoretical lines coming from meteor ablation.

Na-lines expected [nm]	589.0	589.5
Experimental peaks [nm]	588.5	589.0

11.4 Laser induced fluorescence spectroscopy (LIF)

11.4.1 Basic principles of LIF

In the ground facilities where aerospace structures and materials are investigated, it is necessary to obtain a deep knowledge and characterization of the plasma elements interacting with the test samples under investigation. In particular, there are some chemical species such as atomic oxygen or nitric oxide that strongly react with the surfaces of the exposed materials.

Laser induced fluorescence (LIF) has proved to be a powerful experimental technique for the investigation of hypersonic flow fields in high-enthalpy conditions [17]. LIF is the process of light emission from an excited electric state of a species populated upon absorption of laser light radiation. After a certain time (of the order of nano-seconds), this excited species can relax into a lower state by emitting fluorescence radiation [35], as represented in figure 11.25. The main advantages of this technique lie in its high sensitivity and selective analysis of the chemical species, allowing pointwise measurements with a relevant spatial resolution. It also provides access to ground levels not accessible by OES.

With the aim to obtain very important information regarding plasma–material interaction, the characterization and the investigation of the atomic oxygen present in a hypersonic plasma flow is of fundamental importance. For this purpose, LIF spectroscopy has been extensively developed to extract quantitative information on

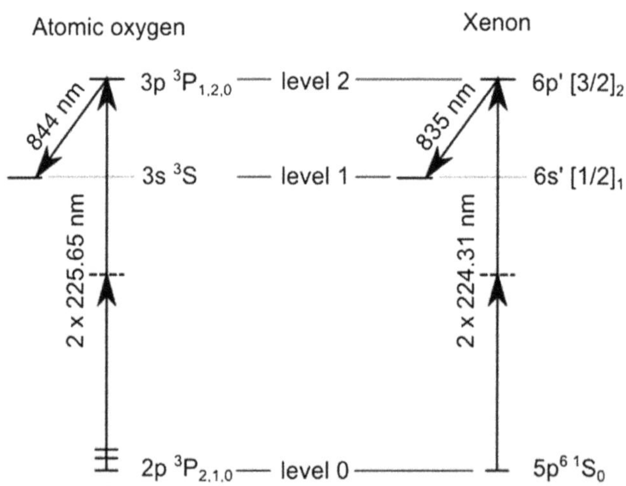

Figure 11.25. Two-photon excitation schemes of atomic oxygen (left) and xenon (right).

the thermodynamic state of atomic oxygen ([36–38] and more recently [39]). Scott [40] surveyed different experimental techniques for measurements of flow properties in arcjets, and the description of the LIF experimental set-up for both nitric oxide and molecular oxygen investigations is illustrated in [41]. Subsequently, Bamford *et al* [36] implemented an LIF-based sensor to characterize the free stream of an arcjet flow generated by the 20 MW NASA Ames aerodynamic heating facility. They investigated the fine structure temperature of atomic oxygen of the level O[$3p^3P$], evaluating the average velocity and measuring the number density for O[$2p^3P$] transition in the free stream. The calibration of the experimental set-up was performed in an earlier work [42] in order to evaluate the two-photon cross section of the atomic oxygen transition of interest. A controlled source via microwave discharge was used to generate the proper amount of atomic oxygen monitored using a NO_2 titration reaction [43]. Del Vecchio *et al* [37] investigated the non-equilibrium effects of the flowfield generated by the high-enthalpy arcjet facility L2K. They used LIF for the estimation of the rotational and translational temperatures of the NO molecules and atomic oxygen in the free stream and behind a shock wave upstream of a blunt body.

11.4.2 Typical set-up for an LIF measurement system

The LIF measurement system set-up needs to follow the specific experimental goal in terms of the specific chemical species to be probed and investigated. In particular, both the excitation scheme and the detection apparatus have to be chosen properly. In the following, the specific test case of LIF investigation on oxygen atoms is reported.

The typical optical set-up for oxygen measurements consists of a tunable laser composed of three units in cascade and a detection system. The first unit is a Nd:YAG laser with an output beam set at 355 nm; the pulse duration is ~10 ns with a repetition rate of 10 Hz. The light from the Nd:YAG laser is used to pump a dye laser in cascade for frequency doubling a beta barium borate (BBO) crystal. An output laser beam with a maximum energy of ~500 μJ per pulse and with a linewidth of 2.3 ± 0.05 pm is produced. The final wavelength ranges in UV radiation between 224.010 ± 0.010 nm for xenon and 225.350 ± 0.010 nm for atomic oxygen with a step width of ~0.5 pm for both studies. The energy of the UV radiation is varied by means of a variable attenuator (VA), which is installed through the optical path at the laser exit. An isometric view of the experimental set-up and the L2K test chamber is shown in figure 11.26. An ultrafast photodiode (PD) monitors the amount of laser energy during its time evolution and it allows for time resolved measurements of the reference signal for the normalization procedure. A periscopic lens system composed of three dichroic mirrors (DMs) directs the laser beam into the L2K test chamber through an optical window. A positive focal lens is installed in the test chamber to focus the laser beam at the pointwise measurement region located ~155 mm downstream from the nozzle exit along the centerline. The final spatial resolution is equal to ~1 mm. The loss of the laser energy through the optical path has been evaluated by relating the energy at the laser exit for the levels 300, 400 and

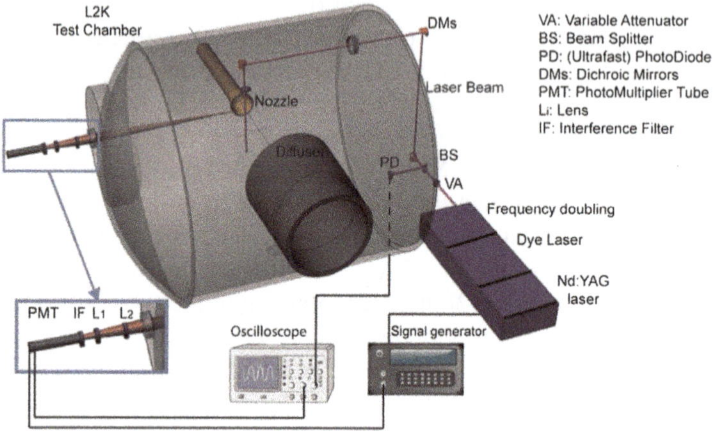

Figure 11.26. L2K test chamber and TALIF experimental set-up. The inset shows a close-up sketch of the detection system.

500 μJ with that measured at the measurement region corresponding to 100, 140 and 180 μJ, respectively. The averaged loss of energy is equal to 65% of that measured at the laser exit.

The fluorescence radiation is observed transversely with respect to the direction of the laser beam and it is collected by means of a detection system (DS), shown in the inset of figure 11.26. The DS is composed of two spherical fused-silica lenses, indicated here as L1 and L2, with diameters of 1 and 2 in and with −50 and 100 mm focal lengths, respectively. By arranging L1 and L2 in series at the relative distance of ~54 mm, it is possible to collect the fluorescence radiation, generated from the measurement volume, to the photocathode of a gated Hamamatsu H7680/-01 photomultiplier tube (PMT). In order to select the light radiation coming from the plasma flow at the wavelengths of interest, an interference filter (IF) with a narrow bandwidth is installed through the optical detection path. As suggested by Niemi *et al* [44], a simplified experimental set-up is arranged by using the same IF for both the atomic oxygen and xenon investigations, without compromising the S/N. The IF is centered at 840 nm with an FWHM equal to 10 nm; in particular, the transmission coefficients are equal to 0.30 and 0.37 for the wavelengths at 844.9 nm and 834.9 nm, i.e. for the atomic oxygen and xenon, respectively.

11.4.3 LIF O atom investigation on hypersonic plasma flow

Figure 11.27 shows the averaged fluorescence signal of xenon at 139 Pa and of atomic oxygen detected in the free stream and behind the shock layer, respectively. An exponential function is superimposed in order to evaluate the measured lifetime τ of the species. The fitting procedure is based on a Levenberg–Marquardt algorithm. In figure 11.27(a), the fluorescence decay of the xenon exhibits a more pronounced scattered signal than that detected for the atomic oxygen. Even though the transmission coefficient of the IF filter in correspondence to the emission

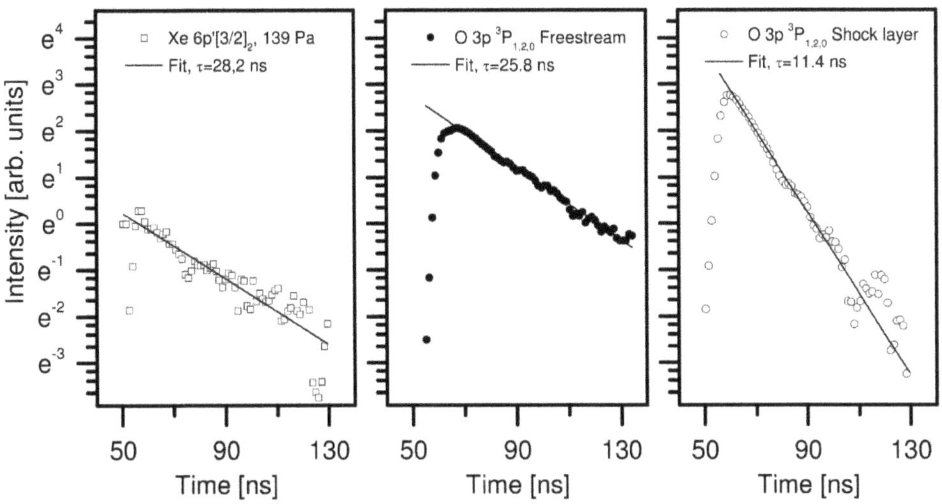

Figure 11.27. Averaged fluorescence signal of the xenon (a) at the pressure of 139 Pa, and atomic oxygen in the free stream (b) and in the shock layer (c); the exponential fitting curve is superimposed.

wavelength of xenon is higher than that of atomic oxygen, the maximum peak of the xenon signal is lower than that detected from atomic oxygen species under the same experimental conditions. The fluorescence signal of the atomic oxygen increases in the shock layer (figure 11.27(c)) with respect to the investigation at the free stream (figure 11.27(b)). Furthermore, it is reasonable to infer, by analyzing the decay rate of the fluorescence signal, that the atomic oxygen concentration increases in the shock layer. In particular, the de-excitation process due to the quenching is inspected by the lifetime τ of the atomic oxygen measured for both cases; for the free-stream and shock layer investigations, the lifetime τ is equal to 25.8 ns and 11.4 ns, respectively.

Following the approach presented in [39], it is possible to characterize the experimental set-up in terms of instrumental line broadening from the calibration with xenon at a constant pressure of 139 Pa and at room temperature, when the spectral profile of the laser is dominated by a Lorentzian shape. Figure 11.28 shows the absorption line profile of the xenon under cold gas conditions; the superimposed fitting curve is based on an approximated Voigt function (pseudo-Voigt), based on the convolution of a Gaussian function with a Lorentzian function [45]. The Doppler (Gaussian) and Lorentzian parts contribute to the Voigt function with the broadening due to the kinetic temperature $(\delta\lambda)_{(D,\mathrm{Xe})}$ and with the instrumental broadening $(\delta\lambda)_{\mathrm{instr}}$, respectively. In the fitting procedure, the Doppler part $(\delta\lambda)_{(D,\mathrm{Xe})}$ is locked at the constant value of 0.12 pm [39]. The resulting instrumental contribution $(\delta\lambda)_{\mathrm{instr}}$ is equal to 3.1 pm, close to that of the laser used for the current experiment. A qualitative inspection of figure 11.28 highlights that the contribution of the Doppler broadening in the absorption profile increases from 1.3 pm to 2.3 pm. As expected, the translational temperature increases in the shock layer with respect

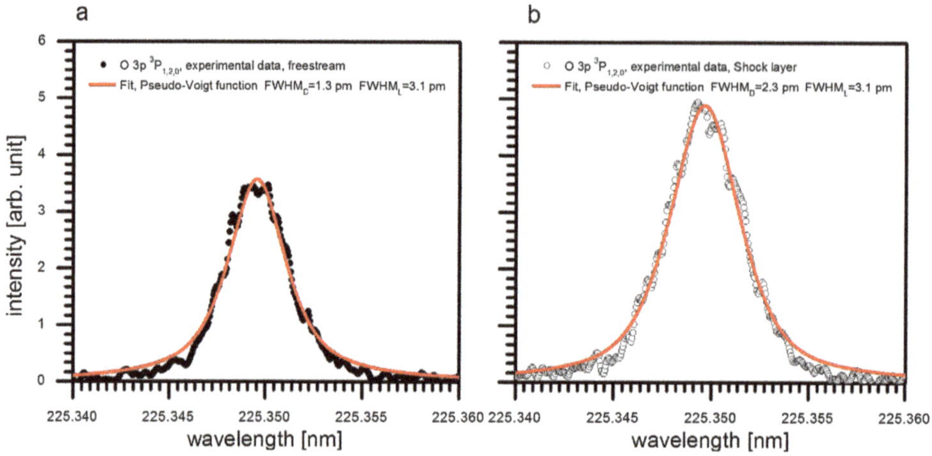

Figure 11.28. Absorption profile of the atomic oxygen $O[3p]^3P_{(1,2,0)}$, the pseudo-Voigt fitting curve (continuous line) is superimposed.

to the free-stream condition, a detalied discussion on the measurement of the translational temperature is presented by Ceglia *et al* [45].

The integral of the absorption line profile, i.e. the fluorescence signal of xenon and atomic oxygen, enables the calculation of the number density of atomic oxygen. For the operating condition of the L2K facility listed in table 11.1, the atomic number density is evaluated in the free stream and in the shock layer, resulting in ~3.33×10^{23} m^{-3} and ~15.6×10^{23} m^{-3}, respectively. The ratio of the standard deviation over the average is equal to ~4%, within the overall uncertainty of ~20% imposed by the estimation of the ratio of the cross sections $\sigma^2_{(\omega,\mathrm{Xe})}/\sigma^2_{(\omega,\mathrm{O})}$ [43]. Thus, the atomic oxygen concentration into the shock layer is about ~4.7 times greater than that measured in the free stream.

11.5 Ion beam analysis (IBA) on meteorites

The interaction of charged particle beams (protons, ν particles or even heavier ions) in the MeV energy range with a material induces a series of phenomena, such as the emission of electromagnetic radiation, charged particles and neutrons of characteristic energies, whose detection is the basis of numerous spectroscopic techniques of elemental analysis called ion beam analysis. In particular, the interaction of the incident particles can occur with electrons or with atomic nuclei.

In the case of interaction with atomic electrons, the energy loss dE of the incident particle can be expressed by the term $S(E)$, the stopping power, given by the Bethe relation, and the interaction of the particle with the atomic electrons can induce the ejection of one or more electrons from atomic shells (ionization). The energy corresponding to the level jump, in order for the atom to return to the initial configuration, is emitted in the form of characteristic energy—X photons. The spectroscopic technique based on the detection of this characteristic radiation is called particle induced x-ray emission (PIXE) [46]. In the case of an elastic interaction of the incident particle with the nuclei of the irradiated material there can be three different phenomena, one of which is the elastic backscattering of the incident particles—Rutherford backscattering spectrometry (RBS) [46]. Moreover, the novel radioactive beam tracer implantation (SLI) by means of a particle accelerator and the novel dual-color thermography technique methodology are used for non-contact, online wear-corrosion and free emissivity temperature measurements, representing extremely powerful tools in aerospace materials science [47–49], which are also applicable to meteorites for on-ground wearing and temperature tests.

PIXE and RBS have been used to determine the average concentration of elements of the so-called 'Paris' meteorite, a fragment from the Musée National d'Histoire Naturelle (MNHN), France, which helped to confirm its classification as a CM (Mighei type) carbonaceous chondrite and provide important information about its carbon content [50]. The primitive meteorites traveled from outer space and they are witnesses to the formation of our solar system about 4.6 billion years ago. These meteorites can be classified into two main families:

- The first group of meteorites comes from differentiated parent bodies which underwent several melting processes.
- The second ground corresponds to undifferentiated meteorites which come from small bodies (planetesimals, asteroids, comets) that have not been differentiated and that have experienced low thermal and aqueous alteration since their formation in the early stages of the solar system.

  Carbonaceous chondrites, such as 'Paris', are members of this family and are the most primitive meteorites.

The elemental composition of the sample was measured using the conventional in-vacuum PIXE and RBS techniques. The 1.7 MV tandem accelerator of the Lebanese Atomic Energy Commission was used to deliver a 3 MeV proton beam (15 mm^2 of beam spot) perpendicular to the sample (a 5×5 mm^2 fragment from the 'Paris' meteorite was provided and sawed by the MNHN) [50]. The PIXE (figure 11.29, left) and RBS (figure 11.29, right) results [50] have been proved to be efficient and applicable to the analysis of meteorites. Global elemental analysis by PIXE and RBS allowed confirmation of the classification of 'Paris' as a carbonaceous chondrite. Furthermore, RBS was used to determine the atomic concentration of carbon and oxygen, and verification of the oxide forms.

11.6 Infrared thermography

The meteorites and space vehicles subjected to hypersonic flow during the entry phase into Earth's atmosphere or during interplanetary trajectory, must withstand huge thermal fluxes due to the high speeds. Plasma wind tunnels (PWTs) allow one to reproduce the thermal conditions and to experimentally simulate the entry phase. In order to monitor the behavior of a tested article, non-intrusive techniques such as thermography and pyrometry are usually used. In particular, thermography allows

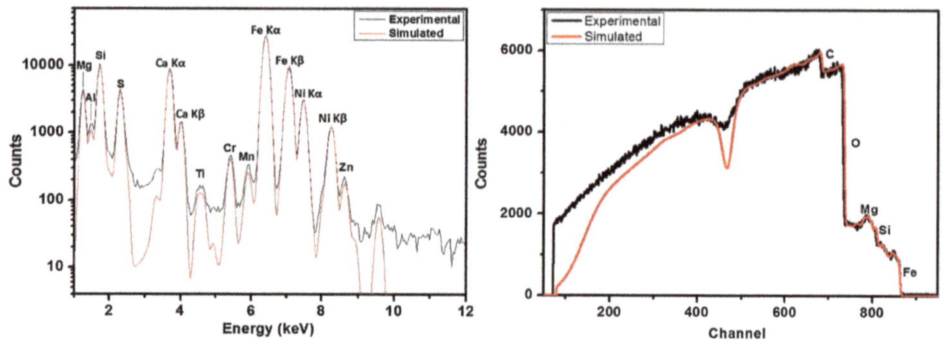

Figure 11.29. Left: PIXE spectrum of an average elemental content of the 'Paris' meteorite at the millimeter scale. PIXE measurements show that the major elemental concentrations, as wt%, are formed by iron (23.6%) and silicon (11.6%), followed by magnesium (9.6%) and calcium (6.4%). Phosphorus, potassium, cobalt and copper were not quantified since their concentration values are just slightly higher than their limit of detection. Right: SIMNRA simulation of the RBS spectrum which is recorded simultaneously with the PIXE spectrum, showing C and O signals. Due to the complex nature of the analyzed fragment (non-homogeneity, structure and matrix), multi-layers are considered in order to obtain a good simulation. From the corresponding results, the average atomic percentages were found to be 7.7% for C and 55% for O, with systematic differences between the simulation and experiment of 1% and 2%, respectively.

one to obtain two-dimensional maps of temperature measurements over surfaces, allowing quantitative and qualitative analyses of temperature distributions over surfaces, detecting the most stressed regions and their temperature [51]. One of the main parameters that affects the quality of the measurement is the emissivity of the material investigated.

Emissivity depends on the temperature, wavelength and surface features of the target investigated. This parameter can be unknown for meteorites or innovative materials such as those employed for vehicles during atmosphere re-entry tests in PWTs. In addition, during tests performed in PWTs, the surface features of the materials investigated can change due to the interaction of plasma species with the material of the tested articles.

Dual-color thermography is a non-intrusive temperature measurement technique based on local gray body hypotheses and on the application of a suitable pair of narrow band filters. It allows one to obtain the ratio between the two input signals to the camera that only depends on the temperature and not on the emissive properties of the investigated surface, thus it can be a powerful tool to avoid the limitations of emissivity knowledge in the thermography used for temperature measurement. However, the application of this technique is far from simple since the choice of two wavelengths compatible with the working range of the IR camera detector is not obvious [48, 49, 52]. The radiation emitted by the target ((1) in figure 11.30), collected by the detector and passing through two filters ((2), (3) in figure 11.30), can be combined through the ratio principle (signal ratio, SR(T)) when the two working wavelengths of the filters are close enough to allow the gray body hypothesis to be as valid as possible.

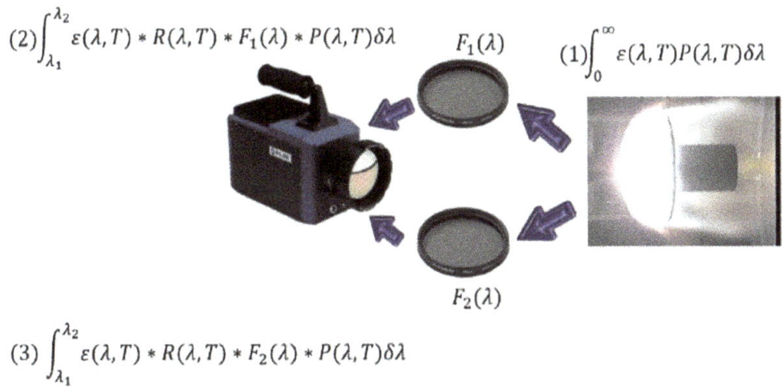

Figure 11.30. Radiation collected by two filters suitable for the dual-color technique.

A numerical tool has been developed, elaborated through the Mathcad code in such a way as to provide an estimation of the radiation captured by a camera detector (sensitive in medium wavelength). The model is based on the integration of the Plank law in the operating spectral range of the IR camera, considering the spectral emissivity trend of the target and the transmissivity functions, respectively, for the sensor $R(\lambda)$ and for the applied filter $F(\lambda)$ (its subscripts 1 and 2 in figure 11.30 are indicative of two close wavelengths where filters work). This tool can be used to define the best two filters in terms of the sensitivity of the SR to temperature and provide an estimate of the accuracy achieved by the dual-color technique. If $G_{n,1}(T)$ and $G_{n,2}(T)$ are the infrared radiations that reach the sensor (assuming a black body as the source) when filters F_1 and F_2 are applied in front of the detector[2], the calibration curves can be obtained by calculating the ratio $SR(T) = G_{n,1}(T)/G_{n,2}(T)$ as a function of only temperature.

Of course, since the gray body condition is ideal, the ratio between emissivities at the two working wavelengths is not 1 and, as consequence, the real temperature (T_{real} on the blue curve corresponding to the signal ratio SR_ε in figure 11.31) can be different from the one measured on the calibration curve (T_{dual} on the red curve corresponding to the signal ratio SR_ε in figure 11.31).

In order to make a estimate of the accuracy of the dual-color technique, different spectral trends of emissivity of materials found in the literature [53], with features similar to those employed for thermal protection systems (TPS), have been used.

Numerical simulation showed that the accuracy in dual-color temperature measurements is governed by the emissivity trend. Components tested in PWTs are generally made of materials that belong to a medium–high emissivity class. For this class of materials the local gray body hypothesis is sufficiently verified, even in the thermal conditions experienced during re-entry simulation. Thus, the dual-color technique can be a valid tool for monitoring the thermal fields that characterize the specimens tested in PWTs.

[2] This means that $G_{n,1}(T)$ and $G_{n,2}(T)$ correspond to the terms (2) and (3) in figure 11.30 when $\varepsilon(\lambda, T) \approx 1$.

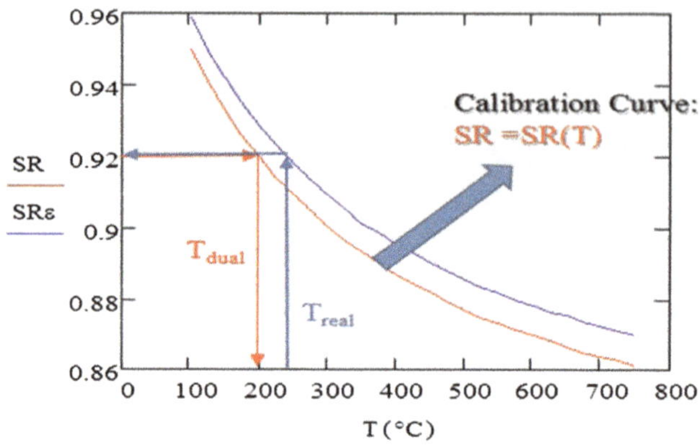

Figure 11.31. Schematic representation of the dual-color technique through comparison between the real signal ratio (SR_ε) and the calibration signal ratio (SR).

11.7 The HEAT facility at SITAEL

The high-enthalpy arc-heated tunnel (HEAT) was originally developed in 1996 with the aim of investigating shock wave/laminar boundary layer interactions in air at different free-stream specific enthalpies and elevated wall surface temperatures [54, 55]. However, its design was conceived with the primary objective of implementing a small-scale, versatile, inexpensive facility, suitable for rapid and relatively low-cost testing and research on a wide number of phenomena occurring in high-enthalpy hypersonic flows. Since its initial set-up, HEAT has been continuously improved and used for numerous experimental campaigns on hypersonics (for example, magnetohydrodynamics interactions in hypersonic flow [56] and investigations of scramjet operation and performance [57]). In 2012 HEAT was used in an explorative activity to assess the feasibility of laboratory simulations of micrometeoroid entry into the atmosphere.

11.7.1 Facility description

In the current configuration, HEAT consists of a vacuum system, a gas feeding system, an arc-heater, a hypersonic nozzle, a power supply system, and a control and data acquisition system, as shown in figure 11.32. The vacuum system is composed of two vacuum vessels. The main chamber, called NDV1, is a 8 m^3 stainless steel cylindrical chamber. The second chamber, called NDV2 or the test chamber, is a 0.5 m^3 stainless steel cylindrical chamber, within which both the hypersonic nozzle and the test item are located. The vacuum system is able to reach an ultimate low pressure of about 1×10^{-2} mbar. The gas feeding system consists of a compressed gas bottle, a manual pressure regulator, two plenum tanks, and two or four inlet solenoid valves. The core of the hypersonic facility consists of the arc-heater and the hypersonic nozzle (figure 11.33). During operation, the system performs a gas-dynamic acceleration of a hot fluid by expansion through the nozzle. The fluid, usually air, argon or nitrogen, is heated by an electrical arc established in the arc-heater. The system is designed to perform fast tests, with a typically 40–500 ms run

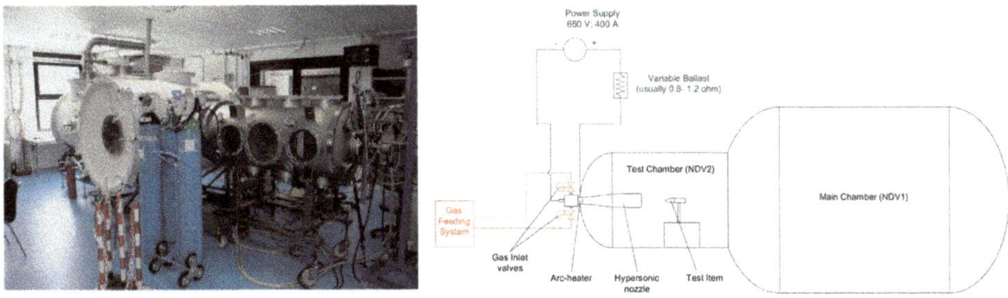

Figure 11.32. The hypersonic wind tunnel HEAT.

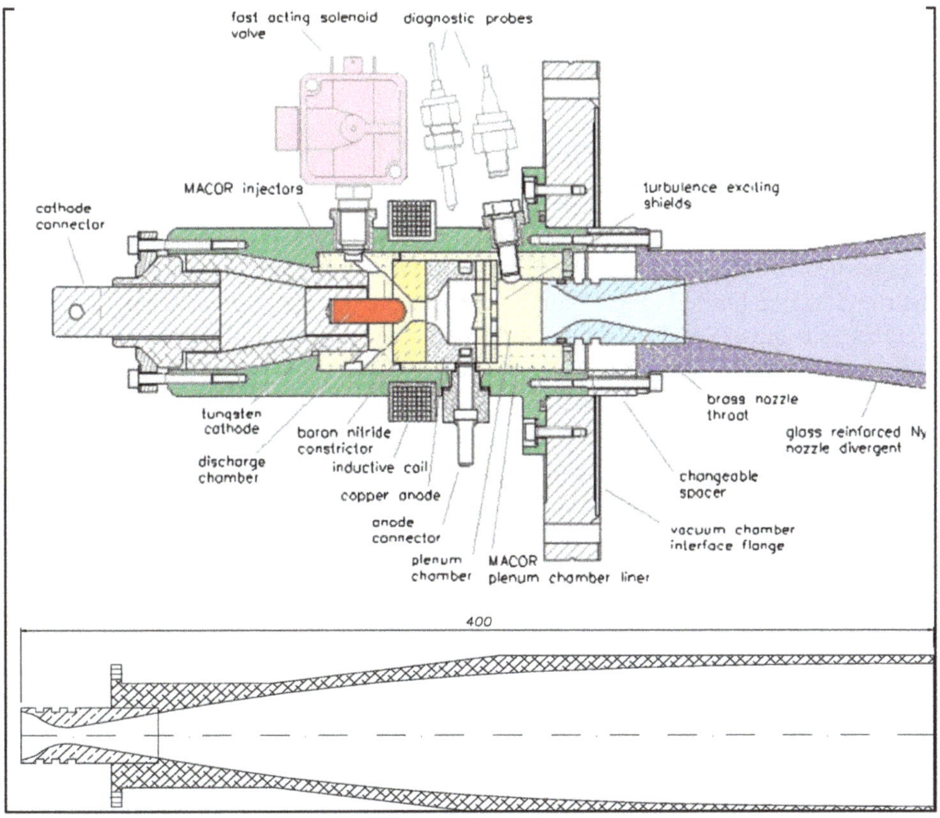

Figure 11.33. The HEAT gas generator and a hypersonic nozzle (nominal Mach number at the exit: 6 in air).

duration. NDV1 operates as buffer, since its volume ensures optimal conditions for the free expansion of the gas flowing in the hypersonic nozzle during each run. The time between two consecutive runs is typically 2–3 min; this allows the back pressure level for the next run to be reached. During each run, after the establishment of a voltage bias between the two electrodes (usually 600–650 V), the gas inlet valves are opened, allowing the gas to flow into the arc-heater chamber and then a gas breakdown occurs. The non-ionized gas is heated by the arc and moves towards the hypersonic nozzle where it expands. During the time between two consecutive runs the gas breakdown cannot take place due to the low pressure inside the chamber ($>5 \times 10^{-2}$ mbar). In order to avoid spontaneous re-ignition of the arc during the pump down phase (just after the run), a voltage bias disengage is commanded. The power supply system is composed of the main power supply and the remote control. The main power supply is a commercial 650 V, 400 A power supply produced by Equipaggiamenti Elettronici Industriali (EEI). The impedance matching of the main power supply is achieved by using one modular resistor bank (typically in the range 0.8–1.2 ohm), positioned in series with the arc-heater. The facility is monitored and controlled by a dedicated laboratory rack in which the main computer and all electronic devices are located, including auxiliary power supplies, diagnostic electronics, chamber pumping system electronics and all data acquisition boards. The control and data acquisition system is based on the National Instruments (NI) architecture and the software is developed using the NI Labview® platform.

11.7.2 HEAT features and performance

Since its first operation, HEAT has been demonstrated to be a relatively inexpensive, flexible, safe and reliable facility. HEAT allows the stagnation pressure (up to 17 bar) and the specific enthalpy (up to 6 MJ kg^{-1}) to be set separately, with different gases. The run duration limited to less than 1 s reduces the pumping rate (the vacuum vessel operates as a buffer) and no active cooling of the components is requested. On the other hand, spotty, unstable arc operation may generate electromagnetic noise, stagnation conditions are inherently less stable than in shock-driven tunnels, and flow contamination from electrode erosion or evaporation is unavoidable (estimated: tungsten 65–250 ppm, copper 1 ppm). The main features of HEAT are summarized in table 11.2. Jet diagnostics at the nozzle exit have been carried out using different techniques, and extensively with temperature and pressure probes [58]. Measurements have shown a core flow of uniform properties extending a few tens of millimeters at the nozzle exit is obtained (from 45 to 90 mm, depending on the nozzle) during the quasi-steady phase of the run, as shown in figure 11.34. Since HEAT can be equipped with different nozzles, quite a wide performance envelope can be covered, as illustrated in figure 11.35. HEAT performance in terms of Mach, Reynolds and Knudsen numbers for test articles with a characteristic length ranging from 10^{-4} to 10^{-2} m is shown in figure 11.36. HEAT is capable of covering a relatively large range in regimes from continuum ($Kn < 0.001$) to transition ($Kn < 10$). In the current configuration, the molecular regime ($Kn > 10$) is outside the facility range.

11.7.3 HEAT for micrometeoroid entry simulation

The use of HEAT to simulate the entry conditions of micrometeoroids (300 μm–1 mm in diameter) was theoretically and experimentally assessed in the framework of a collaboration among the former Alta SpA (now Sitael SpA), the Department of Physics of the University of Florence, the Department of Earth Science and the Department of Physics of the University of Pisa. No report has been published so far on this study and all the documentation available is represented by internal reports of

Table 11.2. Main features of HEAT.

Operating gas	air, N_2, CO_2, Ar, Xe
Test time	between 40 and 500 ms
Total pressure	up to 1.7 MPa
Total enthalpy	up to 6 MJ kg^{-1} for air/N_2
	up to 2 MJ kg^{-1} for Ar with possibility of partial ionization
Electrical power	260 kW DC
	maximum arc voltage: 650 V
	maximum arc current: 630 A
Number of runs per day	up to 50
Nozzle 1	conical nozzle for Mach 6 in Ar, Mach 4 in air
	nozzle exit diameter: 60 mm
	core flow diameter: 50 mm
Nozzle 2	contoured nozzle for Mach 6 in air
	nozzle exit diameter: 70 mm
	core flow diameter: 50 mm
Nozzle 3	conical nozzle for Mach 9 in air
	nozzle exit diameter: 180 mm
	core flow diameter: 90 mm
Nozzle 4	contoured nozzle for Mach 3 in air
	nozzle exit diameter: 60 mm
	core flow diameter: 45 mm

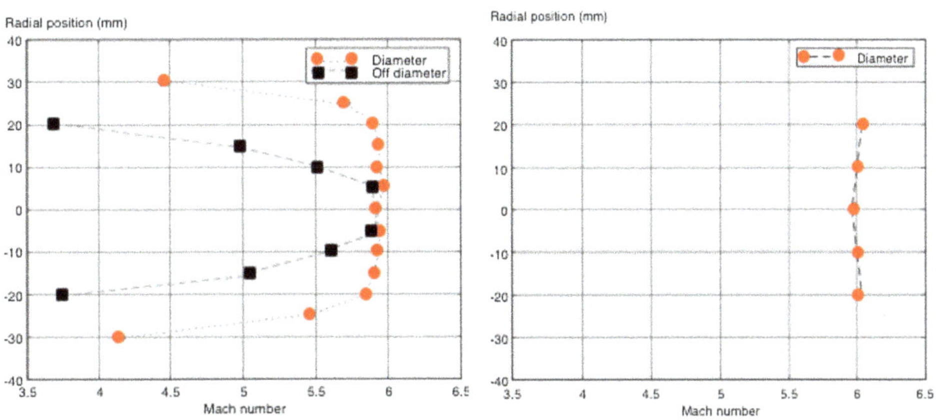

Figure 11.34. Flow characterization at the nozzle exit—air (nominal Mach number at the exit: 6 in air, gas generator total temperature: 293 K (left) 1800 K (right)).

Sitael or other unpublished materials belonging to the departments involved. A theoretical model for the study of the interaction of micrometeoroids with the atmosphere has also been developed. The model and the relevant results are illustrated in [59] and represent support for the definition of the test conditions and data interpretation. Tests were mainly aimed at reproducing the conditions of

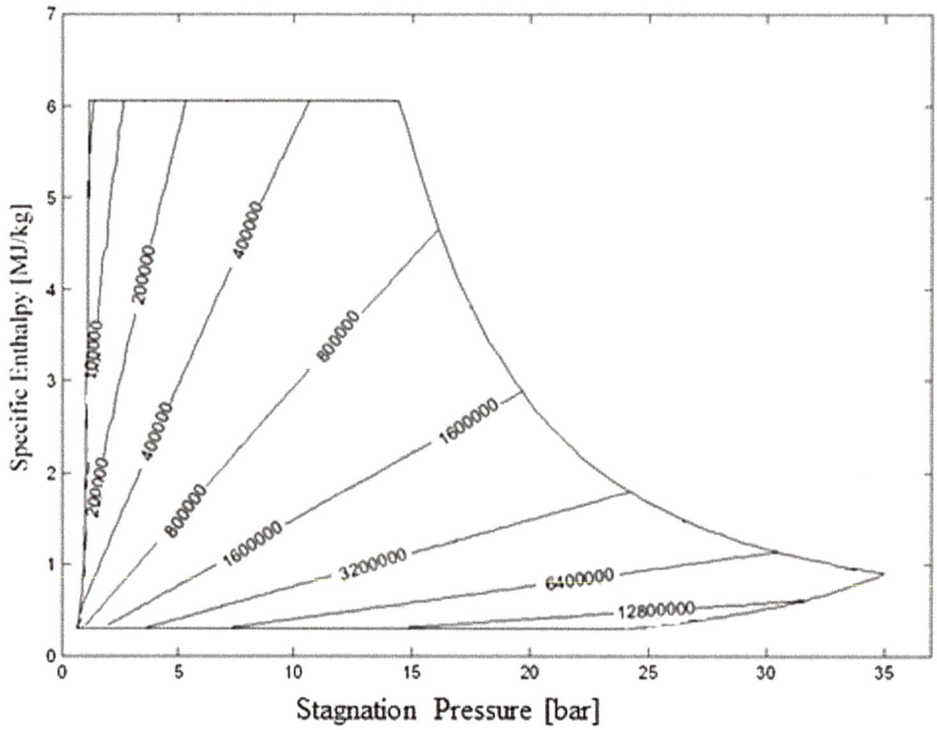

Figure 11.35. HEAT performance envelope. Constant unit Reynolds number lines in the pressure–enthalpy plane (test section Mach number: 6).

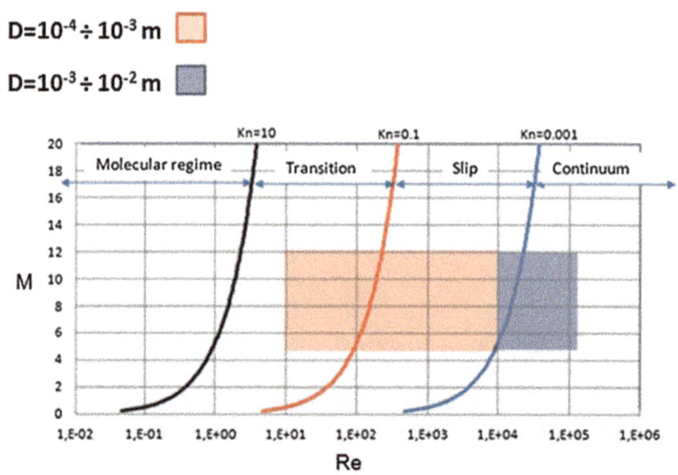

Figure 11.36. HEAT Reynolds–Mach–Knudsen envelope for test article diameter $D = 10^{-4} - 10^{-3}$ m and $10^{-3} - 10^{-2}$ m.

micrometeoroids which survive the interaction with the atmosphere, reach melting temperature and for which melting is the main mass loss mechanism. The model described in [59] indicates chondritic spherical micrometeoroids with a diameter ranging from 200 to 500 μm reach melting conditions for moderate entry velocities (11 km s^{-1}) and grazing incidence angles (45°). Maximum micrometeoroid/atmosphere interaction occurs for altitudes ranging between 90 and 70 km for a time of a few seconds (about 5). For this entry condition, a mean thermal flux in the order of 10^6 W m^{-2} has been estimated, corresponding to an overall deposited thermal energy in the order of 5×10^6 J m^{-2}. Figure 11.37 shows the range of fluid dynamic conditions in terms of Mach, Reynolds and Knudsen numbers (red area) during maximum micrometeoroid/atmosphere interaction.

Tests were carried out in several sessions in 2011 and 2012, both on real stony cosmic spherules (barred olivine type) and on wedge-shaped thin slices obtained from macroscopic ordinary chondrite meteorites, as shown in table 11.3. The thickness at the tip of the slices was about 100 μm. Two typical HEAT run set-ups and results are shown in table 11.4.

Fluid dynamic conditions at the nozzle exit are shown by the blue dot in figure 11.38, considerably far from the real entry conditions. However, as the mean thermal flux produced is in the order of 10^7 W m^{-2} for a run duration of about 350 ms, the overall thermal energy deposited on the test articles in a run as well as on micrometeoroids in real entry conditions are largely of the same magnitude. Hence the thermal conditions for melting can be considered to be reproduced during tests.

11.7.4 Experimental set-up and runs

The test microspherules were glued on stainless steel needles with a zirconia based adhesive (figure 11.39). The needles were placed at the nozzle exit by means of a rack, as illustrated in figure 11.38. The wedge-shaped slices were directly inserted and held in the same rack. During 2011–12 several runs were performed on the test materials. In many runs performed on the microspherules the latter detached from the needle and were lost.

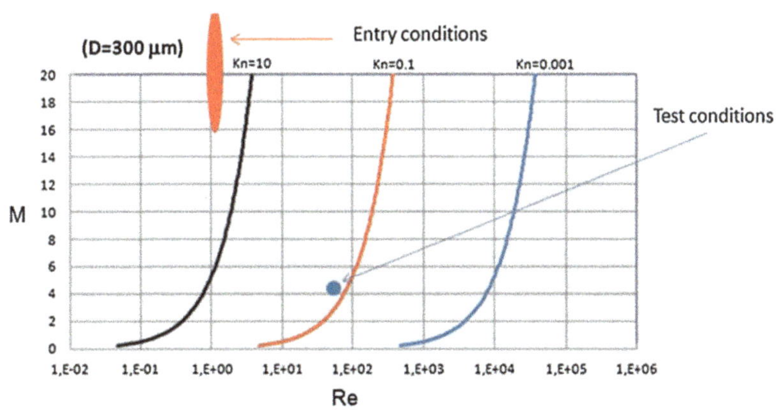

Figure 11.37. Fluid dynamic entry conditions of real micrometeoroids versus HEAT test conditions.

Table 11.3. Ordinary chondrites (OC) tested in the experiments.

Meteorite	Country	Type	Melting (Y/N)
Ghubara	Oman	OC L5	Y
Etter	Texas, USA	OC L5	Y
Juancheng	China	OC H5	Y
Gebel Kamil	Egypt	Iron ungrouped	N
NWA 100	NW Africa	OC L6	Y

Table 11.4. Two typical HEAT run results (tests on microspherules, January 2012).

Run number	Nominal hot test time [ms]	Gas	Mach number	Mean plenum pressure [kPa]	Mean current [A]	Mean voltage [V]	Mean mass flow rate [g s^{-1}]	Mean total enthalpy [MJ kg^{-1}]	Mean plenum temperature [K]
03	350	Air	4.5	147.4	419.9	200.2	14.8	2.43	2309
05	350	Air	4.5	146.9	415.1	208.2	14.0	2.75	2524

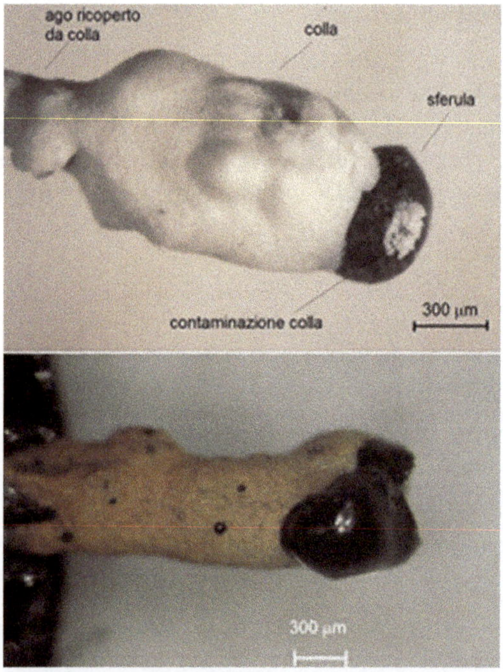

Figure 11.38. Optical observation of a spherule before and after a run.

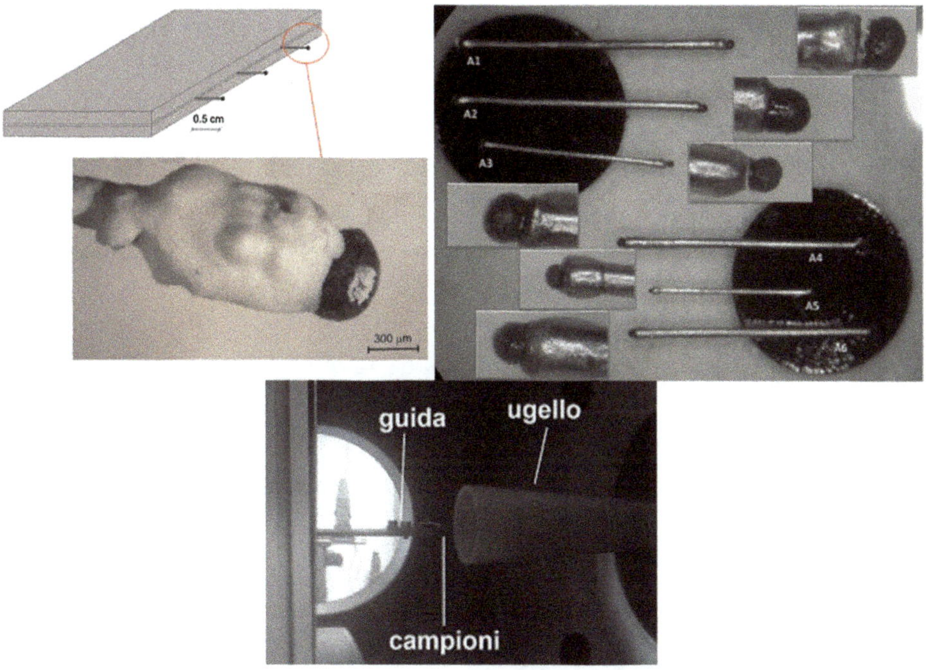

Figure 11.39. Experimental set-up.

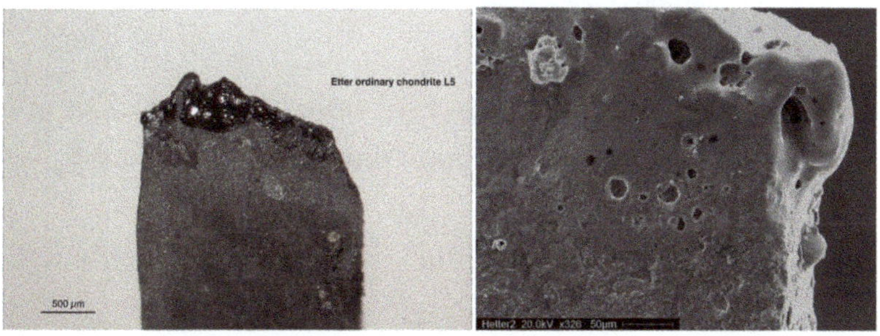

Figure 11.40. Optical and SEM-SE (SE: secondary electrons) observation of a wedge-shaped macrometeorite slice after the run (Etter ordinary chondrite L5).

Visual observations before and after the runs showed that the sample had melted and lost mass (estimated to be about 65% of the initial mass) (see figures 11.38, 11.40 and 11.41). SEM analysis showed dendritic crystals scattered within a glass matrix (figure 11.42); similar dendritic crystals made of magnetite are normally observed on natural micrometeorites. The sample underwent spectroscopic analyses (SEM–XRF–EDS, XRD, μRaman) which confirmed the composition of the dendritic crystals. In particular, SEM–XRF–EDS analyses have shown that the sample surface, in various parts, has a chemical composition similar to natural micrometeorites (figure 11.43); contaminations of zirconium and copper were also detected. The presence of zirconium is probably due to the adhesive used to fix

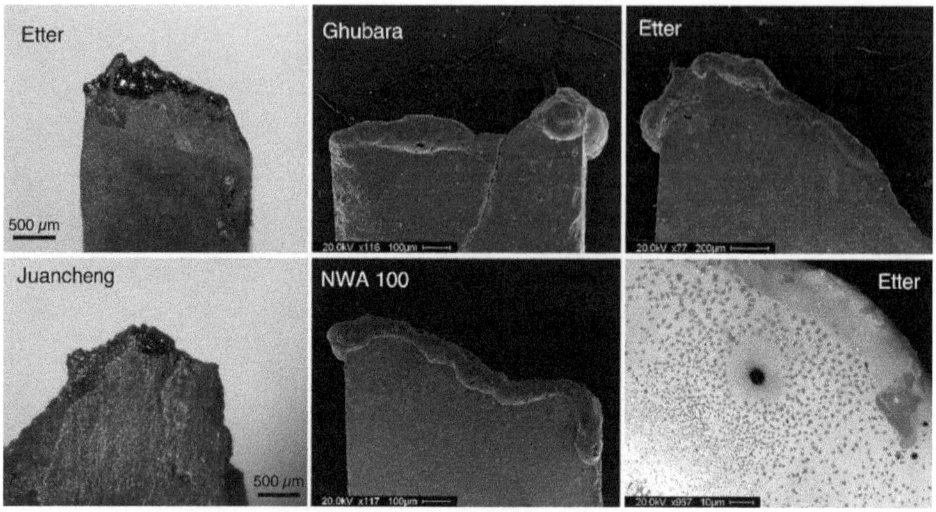

Figure 11.41. Optical and SEM-SE observation of a series of wedge-shaped macrometeorite slices after the run.

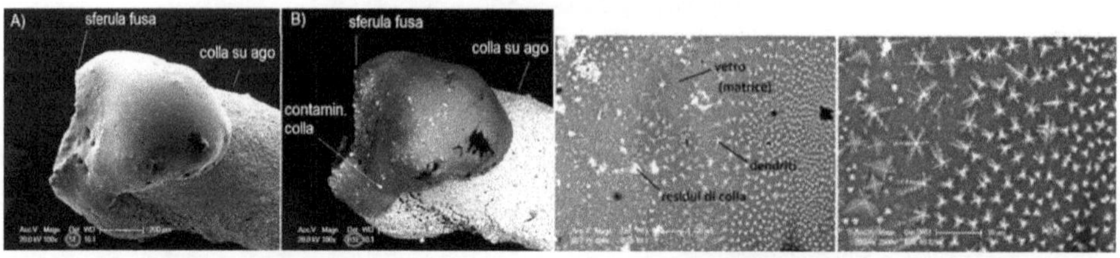

Figure 11.42. SEM analysis of a microspherule. Details of dendritic crystals scattered within a glass matrix.

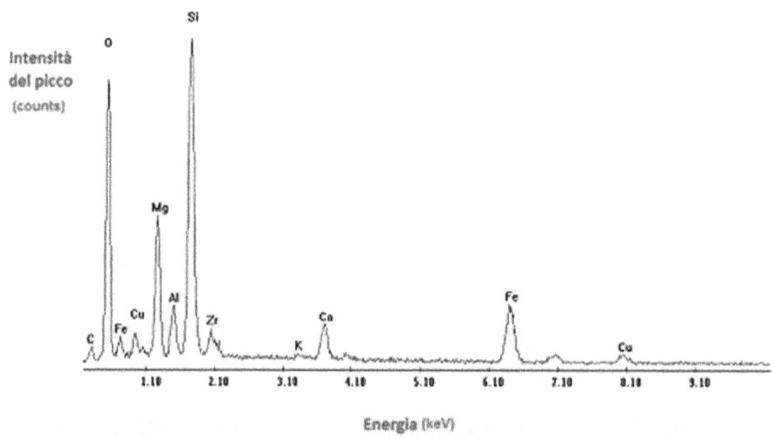

Figure 11.43. SEM–XRF–EDS spectrum of the surface of a cosmic microspherule after ablation.

the sample onto the needle, while the copper probably comes from the electrodes of the arc-heater of the hypersonic facility.

11.7.5 Concluding remarks

Although the fluid dynamic regimes are very different, the overall thermal energy during both the test and entry conditions seems to be of the same order of magnitude. In a first series of tests carried out on natural spherules 300–500 μm in diameter and wedge-shaped thin slices of macroscopic meteorites, the sample reached melting temperature, and a mass loss in line with what occurred in natural micrometeorites was observed. Moreover, the formation of dendritic crystals made of magnetite on the sample surface was observed as in natural micrometeorites. Hence the HEAT facility seems adequate to simulate the entry of 300–500 μm micrometeoroids, while for larger micrometeoroids, a longer (>1 s) running time seems to be necessary.

References

[1] Loehle S *et al* 2017 Experimental simulation of meteorite ablation during Earth entry using a plasma wind tunnel *Astrophys. J.* **837** 112

[2] Agrawal P, Jenniskens P M, Stern E, Arnold J and Chen Y-K 2018 Arcjet ablation of stony and iron meteorites, *2018 Aerodynamic Measurement Technology and Ground Testing Conf.* AIAA paper 2018-4284

[3] Pittarello L, McKibbin S, Goderis S, Soens B, Bariselli F, Barros Dias B R, Zavalan F L, Magin T and Claeys P 2016 Meteorite atmospheric entry reproduced in plasmatron *Meteorit. Planet. Sci.* **51** A518

[4] Laux C O 1993 Optical diagnostics and radiative emission of air plasmas *PhD Thesis* Stanford University, HTGL Report 288, Palo Alto, CA

[5] Zander F, Morgan R G, Sheikh U, Buttsworth D R and Teakle P R 2012 Hot-wall reentry testing in hypersonic impulse facilities *AIAA J.* **51** 476–84

[6] Zander F 2016 Surface temperature measurements in hypersonic testing using digital single-lens reflex cameras *J. Thermophys. Heat Transfer* **30** 919–25

[7] MacDonald M E and Laux C O 2014 Experimental characterization of ablation species in an air plasma ablating boundary layer, *11th AIAA/ASME Joint Thermophysics and Heat Transfer Conf.* AIAA paper 2014-2251

[8] MacDonald M E, Jacobs C M, Laux C O, Zander F and Morgan R G 2014 Measurements of air plasma/ablator interactions in an inductively coupled plasma torch *J. Thermophys. Heat Transfer* **29** 12–23

[9] MacDonald M E, Jacobs C M, Laux C O and Morgan R G 2013 Measurements of air plasma/ablator interactions in a 50 kW inductively coupled plasma torch, *44th AIAA Thermophysics Conf.* AIAA paper 2013-2772

[10] McGuire S D, Tibère-Inglesse A C and Laux C O 2017 Ultraviolet Raman spectroscopy of N_2 in a recombining atmospheric pressure plasma *Plasma Sources Sci. Technol.* **26** 115005

[11] Stancu G D, Kaddouri F, Lacoste D A and Laux C O 2010 Atmospheric pressure plasma diagnostics by OES, CRDS and TALIF *J. Phys. D: Appl. Phys.* **43** 124002

[12] Spence T, Xie J, Zare R, Packan D, Yu L, Laux C O, Owano T and Kruger C 1999 Cavity ring-down spectroscopy measurements of N_2^+ in an atmospheric pressure air plasma, *30th Plasmadynamic and Lasers Conf.* AIAA paper 1999-3433
[13] Laux C O 2007 Spectroscopic challenges in the modelling and diagnostics of high temperature air plasma radiation for aerospace applications *AIP Conf. Proc.* **901** 191–203
[14] Laux C O, Spence T G, Kruger C H and Zare R N 2003 Optical diagnostics of atmospheric pressure air plasmas *Plasma Sources Sci. Technol.* **12** 125
[15] SPECAIR, http://www.spectralfit.com
[16] McGuire S D, Tibère-Inglesse A C and Laux C O 2016 Infrared spectroscopic measurements of carbon monoxide within a high temperature ablative boundary layer *J. Phys. D. Appl. Phys.* **49** 485502
[17] Anderson J D 1989 *Hypersonic and High Temperature Gasdynamics* (New York: McGraw-Hill)
[18] Loehle S *et al* 2017 Experimental simulation of meteorite ablation during Earth entry using a plasma wind tunnel *Astrophys. J.* **837** 112
[19] Prabhu D *et al* 2015 *Workshop on Potentially Hazardous Asteroids Characterization, Atmospheric Entry and Risk Assessment* (Washington, DC: NASA)
[20] Park C 1997 Evaluation of real-gas phenomena in high-enthalpy impulse test facilities: a review *J. Thermophys. Heat Transfer* **11** 10–8
[21] De Filippis F 2004 SCIROCCO arc-jet facility for large scale spacecraft TPS verification, *55th Int. Astronautical Congress 2004* pp 1–8
[22] Russo G, De Filippis F, Borrelli S, Marini M and Caristia S 2002 The SCIROCCO 70-MW plasma wind tunnel: a new hypersonic capability *Advanced Hypersonic Test Facilities* vol 198 (Reston, VA: AIAA) pp 313–51
[23] De Filippis F, Caristia S, Del Vecchio A and Purpura C 2003 The SCIROCCO PWT facility calibration activities, *3rd Int. Symp. Atmospheric Reentry Vehicle and Systems*
[24] De Filippis F, Savino L, Cipullo A and Marenna E 2014 Non-invasive techniques for the diagnosis of aerospace devices *Photonics for Safety and Security* (Singapore: World Scientific) pp 206–21
[25] Cipullo A 2012 Ground testing methodologies improvement in plasma wind tunnels using optical diagnostics *PhD Thesis* Seconda Università degli Studi di Napoli/von Karman Institute for Fluid Dynamics
[26] Purpura C, De Filippis F, Graps E, Trifoni E and Savino R 2007 The GHIBLI plasma wind tunnel: description of the new CIRA-PWT facility *Acta Astronaut.* **61** 331–40
[27] UV nikkor, instruction manual [online]
[28] Yvon H J 1993 HR 460 User Manual—HR460 Rapid Scanning Imaging Spectrograph/ Monochromator
[29] Syncerity CCD Deep Cooled Cameras specification [online]
[30] Ohno N, Razzak M A, Ukai H, Takamura S and Uesugi Y 2006 Validity of electron temperature measurement by using Boltzmann plot method in radio frequency inductive discharge in the atmospheric pressure range *Plasma Fusion Res.* **1** 028
[31] 2009 NIST Database http://www.nist.gov/srd/index.htm
[32] Cipullo A, Savino L, Marenna E and De Filippis F 2012 Thermodynamic state investigation of the hypersonic air plasma flow produced by the arc-jet facility SCIROCCO *Aerospace Sci. Technol.* **23** 358–62

[33] De Filippis F, Marenna E, Savino L and Cipullo A 2011 Experimental characterization of the CIRA plasma wind tunnel flow by optical emission spectroscopy, *17th AIAA Int. Space Planes and Hypersonic Systems and Technologies Conf.* AIAA paper 2011-2213

[34] Pratesi G, Trifoni E, Martucci A, Purpura C, Savino L, Martino M D and Pace E 2018 Investigation of simulated meteoroids entry in the Earth atmosphere by emission spectroscopy meteors, *XIV Congresso Nazionale di Scienze Planetarie*

[35] Candler G, Nompelis I, Druguet M-C, Holden M, Wadhams T, Boyd I and Wang W-L 2002 CFD validation for hypersonic flight—hypersonic double-cone flow simulations, *40th AIAA Aerospace Sciences Meeting & Exhibit* AIAA paper 2002-0581

[36] Bamford D J, O'Keefe A, Babikian D S, Stewart D A and Strawa A W 1995 Characterization of arcjet flows using laser-induced fluorescence *J. Thermophys. Heat Transfer* **9** 26–33

[37] Del Vecchio A, Palumbo G, Koch U and Gülhan A 2000 Temperature measurements by laser-induced fluorescence spectroscopy in nonequilibrium high-enthalpy flow *J. Thermophys. Heat Transfer* **14** 216–24

[38] Takayanagi H, Mizuno M, Fujii K, Suzuki T and Fujita K 2009 Arc heated wind tunnel flow diagnostics using laser-induced fluorescence of atomic species, *47th AIAA Aerospace Sciences Meeting Including The New Horizons Forum and Aerospace Exposition* AIAA paper 2009-1449

[39] Marynowski T, Löhle S and Fasoulas S 2014 Two-photon absorption laser-induced fluorescence investigation of CO_2 plasmas for Mars entry *J. Thermophys. Heat Transfer* **28** 394–400

[40] Scott C D 1993 Survey of measurements of flow properties in arcjets *J. Thermophys. Heat Transfer* **7** 9–24

[41] Arepalli S 1989 Demonstration of the feasibility of laser induced fluorescence for arc jet flow diagnostics *NASA Technical Report* NASA-CR-185595

[42] Bamford D J, Jusinski L E and Bischel W K 1986 Absolute two-photon absorption and three-photon ionization cross sections for atomic oxygen *Phys. Rev.* A **34** 185

[43] Döbele H F, Mosbach T, Niemi K and Schulz-Von Der Gathen V 2005 Laser-induced fluorescence measurements of absolute atomic densities: concepts and limitations *Plasma Sources Sci. Technol.* **14** S31

[44] Niemi K, Schulz-Von Der Gathen V and Döbele H F 2005 Absolute atomic oxygen density measurements by two-photon absorption laser-induced fluorescence spectroscopy in an RF-excited atmospheric pressure plasma jet *Plasma Sources. Sci. Technol.* **14** 375

[45] Ceglia G, Del Vecchio A, Koch U, Esser B and Gülhan A 2018 Two-photon laser-induced fluorescence measurements of atomic oxygen density in hypersonic plasma flow *J. Thermophys. Heat Transfer* https://doi.org/10.2514/1.T5354

[46] Verma H R 2007 *Atomic and Nuclear Analytical Methods* (Berlin: Springer)

[47] De Cesare M, Di Leva A, Del Vecchio A and Gialanella L 2018 A novel recession rate physics methodology for space applications at CIRA by means of CIRCE radioactive beam tracers *J. Phys. D. Appl. Phys.* **51** 09LT01

[48] Musto M, Rotondo G, De Cesare M, Del Vecchio A, Savino L and De Filippis F 2016 Error analysis on measurement temperature by means dual-color thermography technique *Measurement* **90** 265–77

[49] Savino L, De Cesare M, Musto M, Rotondo G, De Filippis F, Del Vecchio A and Russo F 2017 Free emissivity temperature investigations by dual color applied physics methodology in the mid-and long-infrared ranges *Int. J. Therm. Sci.* **117** 328–41

[50] Noun M, Roumie M, Calligaro T, Nsouli B, Brunetto R, Baklouti D, d'Hendecourt L and Della-Negra S 2013 On the characterization of the 'Paris' meteorite using PIXE, RBS and micro-PIXE *Nucl. Instrum. Methods Phys. Res.* B **306** 261–4

[51] De Filippis F, Toscano C, Gallo D, Caruso P and Savino L 2012 Influence of mirrors utilization on the radiation emitted by models subjected to hypersonic flow for surface temperature determination, *11th Int. Conf. on Quantitative InfraRed Thermography*

[52] Möllmann K-P, Pinno F and Vollmer M 2010 Two-color or ratio thermal imaging: potentials and limits *InfraMation* **11** 41–56

[53] Neuer G 1995 Spectral and total emissivity measurements of highly emitting materials *Int. J. Thermophys.* **16** 257–65

[54] Scortecci F, Paganucci F and d'Agostino L 1998 Experimental investigation of shock-wave/boundary-layer interactions over an artificially heated model in hypersonic flow, *8th Int. Space Planes and Hypersonic Systems and Technologies Conf. (Norfolk, VA)* AIAA paper 98-1571

[55] Biagioni L, Scortecci F and Paganucci F 1999 Development of pulsed arc heater for small hypersonic high-enthalpy wind tunnel *J. Spacecr. Rockets* **5** 704–10

[56] Cristofolini A, Borghi C A, Neretti G, Passaro A, Fantoni G and Paganucci F 2008 Magnetohydrodynamics interaction over an axisymmetric body in a hypersonic flow *J. Spacecr. Rockets* **45** 438–44

[57] Liu Q, Passaro A, Baccarella D and Do H 2014 Ethylene flame dynamics and inlet unstart in a model scramjet *J. Propul. Power* **30** 1577–85

[58] Biagioni L, Scortecci F, Paganucci F and Nill L D 1998 Experimental characterization for hypersonic testing, *34th Joint Propulsion Conf. (Cleveland, OH)* AIAA paper 1998-3131

[59] Briani G, Pace E, Shore S N, Pupillo G, Passaro A and Aiello S 2013 Simulations of micrometeoroid interactions with the Earth atmosphere *Astron. Astrophys.* **552** A53

IOP Publishing

Hypersonic Meteoroid Entry Physics

Gianpiero Colonna, Mario Capitelli and Annarita Laricchiuta

Chapter 12

Advanced state-to-state and multi-temperature models for flow regimes

Elena Kustova and Ekaterina Nagnibeda

A currently important problem in high-temperature gas dynamics is the prediction of flow parameters on the trajectories of space vehicles in planet atmospheres. When a space body moving with hypersonic speed enters a planet's atmosphere, a rapid gas compression within the thin shock front causes a jump in temperature and then, behind the shock front, excitation of the internal degrees of freedom of gas molecules and chemical reactions may influence gas flow parameters and transport properties near the body.

Many studies are devoted to modeling non-equilibrium kinetic processes in shock heated gases in Earth or Mars atmospheres in the framework of different approaches (see the references in [1–3]). However, estimations of the accuracy of the proposed models and the limits of their validity under various practically important conditions have not been sufficiently discussed in the literature to date. Kinetic theory methods [2] provide a possibility to derive models from the kinetic equations which describe non-equilibrium flows of multi-component reacting mixtures with different accuracies. In this chapter, we consider a detailed state-to-state description proposed in kinetic theory for reacting air flows [2] and for mixtures containing CO_2 molecules [4], and more simple but sufficiently accurate multi-temperature kinetic models for air and CO_2 containing flows [2, 5, 6]. The state-resolved reaction rate coefficients are considered [7] as well as such effects as vibrational–rotational coupling [8] and the influence of molecular size [9, 10] on the transport coefficients.

A comparison of the results obtained for shock heated air flows in the state-to-state and multi-temperature kinetic theory approaches is shown [11, 12], and the influence of vibrational distributions on flow parameters and chemical reaction rates is discussed [13]. The models for vibrational energy transitions are also discussed in this chapter; for vibrational relaxation time the model [14, 15] which generalizes the classical Landau–Teller formula is used in calculations.

doi:10.1088/2053-2563/aae894ch12

12.1 General kinetic theory method for non-equilibrium flow modeling

For non-equilibrium flows, the generalized Chapman–Enskog formalism is commonly used to develop a closed self-consistent description of fluid dynamics and transport processes [2, 16]. In the case of strong deviation from equilibrium, the main idea is to separate collisional processes into 'rapid' and 'slow', according to the relation between the characteristic times of various internal energy transitions and chemical reactions. The number of slow processes in a system specifies the degree of non-equilibrium and the number of fluid-dynamics variables (and corresponding fluid-dynamics equations) required for the closed flow description. In modern non-equilibrium fluid dynamics, different levels of flow description are used [2]: the conventional one-temperature approach implying weak deviations from thermal equilibrium and highly non-equilibrium chemical reactions; the state-to-state approach considering the internal (vibrational or ro-vibrational) energy exchanges as slow processes and thus suitable for extremely non-equilibrium situations (the price for this is the great number of fluid-dynamics equations which have to be solved for the populations of all internal energy states); and intermediate multi-temperature models suitable for thermal non-equilibrium which are not as detailed as the state-to-state model.

Consider a mixture of gases with translational and internal (rotational, vibrational, electronic) degrees of freedom. Translational degrees of freedom are described classically, whereas internal modes are treated quantum-mechanically. In the mixture, along with elastic collisions, chemical reactions are allowed as well as various inelastic processes occurring due to the presence of internal degrees of freedom: the exchange of rotational and vibrational quanta, transitions of rotational, vibrational and electronic energy into translational energy, etc. Under non-equilibrium conditions, the system tends to restore equilibrium through different thermal and chemical relaxation processes, such as energy redistribution between translational and internal degrees of freedom as a result of inelastic collisions and achieving an equilibrium mixture composition by means of chemical reactions.

The general Chapman–Enskog procedure to derive a closed model of a non-equilibrium flow consists of the following steps [2]:

- Any macroscopic description is based on the fundamental Boltzmann equation for the distribution function of gas particles over velocity and internal energy $f_{ci}(\mathbf{r}, \mathbf{u}_c, t)$ ($\mathbf{r}$ is the space coordinate, $\mathbf{u}_c$ is the particle velocity, t is the time, c is the chemical species and i is its internal state):

$$\frac{\partial f_{ci}}{\partial t} + \mathbf{u}_c \cdot \nabla_r f_{ci} + \mathbf{F}_c \cdot \nabla_{u_c} f_{ci} = J_{ci}, \tag{12.1}$$

$\mathbf{F}_c = \mathbf{F}_c(\mathbf{r}, t)$ is the external force per unit mass, J_{ci} is the collisional operator and ∇_r, ∇_{u_c} are correspondingly gradients taken with respect to the space coordinate and velocity. Under non-equilibrium conditions, the collision operator in the non-dimensional Boltzmann equation can be split into two parts corresponding to the 'rapid' and 'slow' microscopic processes. Rapid processes proceed much faster compared to the time of variation for macroscopic fluid-dynamics parameters (such as gas velocity and density); their characteristic times τ_{rap} are comparable with the mean free time and considerably less than the flow characteristic time θ specified by the gas velocity and length scale, $\tau_{rap}/\theta \ll 1$. Slow processes proceed at the gas-dynamic time scale, $\tau_{sl}/\theta \sim 1$. Depending on flow conditions, rapid and slow processes can be different. For instance, at ambient temperatures, translational and rotational relaxation are fast whereas vibrational relaxation and chemical reactions are slow; under space vehicle re-entry conditions, all the above processes are rapid.

- Collision invariants ψ_{ci} for rapid processes are defined; then the macroscopic variables Ψ corresponding to these invariants are introduced:

$$\Psi(\mathbf{r}, t) = \sum_{ci} \int \psi_{ci} f_{ci}(\mathbf{r}, \mathbf{u}_c, t) d\mathbf{u}_c. \tag{12.2}$$

  Such a choice of macroscopic variables provides a closed flow description under specific non-equilibrium conditions.
- Fluid-dynamics equations are obtained from the Boltzmann equation (12.1) multiplying it by the collision invariants, integrating over velocities, and summing over internal states and chemical species. In the same step, definitions of transport and production terms are introduced. Conservation equations are obtained for invariants of any collision (mass, momentum and energy), whereas relaxation equations are derived for the invariants of the most frequent collisions (rapid processes). Thus a set of governing equations is obtained for chosen non-equilibrium conditions; these equations, however, remain unclosed until the transport fluxes (heat flux, stress tensor, diffusion velocity) and production rates in the equations are defined in terms of the macroscopic flow variables and their derivatives.
- The distribution function is expanded into series in the small parameter, being the ratio of the characteristic times of rapid and slow processes. Transport and production terms are calculated for each approximation. The Euler equations of an inviscid flow correspond to the zeroth-order approximation; the Navier–Stokes equations of a viscous flow are obtained in the first-order approximation. The transport coefficients and reaction rate coefficients can be calculated using the Chapman–Enskog formalism if the molecular interaction potential is specified. Thus, in any approximation of the Chapman–Enskog expansion, a closed set of fluid-dynamics equations can be derived.

This general algorithm is applied in the following sections to specific flow conditions.

12.2 State-to state theoretical model of kinetics and transport properties

The detailed state-to-state flow description is required when the following time relation holds:

$$\tau_{tr} \leqslant \tau_{rot} \ll \tau_{vibr} < \tau_{react} \sim \theta, \tag{12.3}$$

τ_{tr}, τ_{rot}, τ_{vibr}, τ_{react} are the characteristic times of translational, rotational and vibrational relaxation and chemical reactions, respectively. The set of macroscopic variables includes populations of all vibrational states n_{ci} (c stands for the chemical species, i for the vibrational level), the gas velocity $\mathbf{v}$ and temperature T. The governing equations in this case consist of momentum and energy conservation equations coupled to the master equations for vibrational state populations:

$$\begin{aligned} &\frac{dn_{ci}}{dt} + n_{ci}\nabla \cdot \mathbf{v} + \nabla \cdot (n_{ci}\mathbf{V}_{ci}) = R_{ci}^{\text{vibr}} + R_{ci}^{\text{react}} \quad c = 1, \dots, L, \quad i = 0, \dots, L_c \\ &\rho\frac{d\mathbf{v}}{dt} + \nabla \cdot \mathbf{P} = 0, \end{aligned} \tag{12.4}$$

$$\rho\frac{du}{dt} + \nabla \cdot \mathbf{q} + \mathbf{P} : \nabla\mathbf{v} = 0, \tag{12.5}$$

where ρ is the density, u is the total energy per unit mass, $\mathbf{V}_{ci}$ is the diffusion velocity for each vibrational state, $\mathbf{P}$ is the stress tensor, $\mathbf{q}$ is the heat flux, R_{ci}^{vibr}, R_{ci}^{react} are the production terms due to vibrational energy transitions and state-specific chemical reactions, L is the number of chemical species in the mixture and L_c is the number of vibrational states in c species.

The idea of such a detailed approach was proposed in the 1950s by Montrol and Shuler [17]. The first studies of state-to-state non-equilibrium kinetics were carried out much later, first for inviscid flows such as shock waves [18–20] and nozzle expansions [21, 22], and 1D stagnation line flows with simplified transport properties [23, 24]. State-resolved models for transport properties were developed in [25, 26] and then applied to study heat transfer in different flows [27–33]. In this chapter, we discuss several new directions in the development of state-to-state flow modeling.

12.2.1 State-resolved reaction rate coefficients

State-resolved transition and reaction rate coefficients are crucial for correct state-to-state flow simulations. Many models have been developed for the rate coefficients based on analytical theoretical approaches [7–34, 41] and molecular dynamics [42–52]. Quasi-classical trajectory calculations (QCT) are efficient and accurate but highly computationally consuming, which prevents their direct application in computational fluid dynamics (CFD). Existing theoretical models have rather narrow limits of applicability, in particular if they are based on fitting experimental data. Therefore, there is a need for simple but accurate theoretical models suitable for implementation in the CFD codes.

The idea is to generalize the simple Treanor–Marrone model [36] by adjusting its parameters on the basis of the most reliable QCT data. According to the original model, the state-specific dissociation rate coefficient $k_{i,\text{diss}}^{M}$ from the vibrational state i after a collision with a partner M is given by the expression:

$$k_{i,\text{diss}}^{M} = Z_i^{M} k_{\text{diss,eq}}^{M}(T),$$

$k_{\text{diss,eq}}^{M}(T)$ is the thermal equilibrium dissociation rate coefficient and Z_i^M is the non-equilibrium factor:

$$Z_i^M = Z_i(T,\, U) = \frac{Z_{\text{vibr}}(T)}{Z_{\text{vibr}}(-U)} \exp\left(\frac{\varepsilon_i}{k}\left(\frac{1}{T} + \frac{1}{U}\right)\right),$$

ε_i is the vibrational energy of the ith state, k is the Boltzmann constant and $Z_{\text{vibr}}(T)$ is the equilibrium vibrational partition function

$$Z_{\text{vibr}}(T) = \sum_i \exp\left(-\frac{\varepsilon_i}{kT}\right).$$

The model parameter U characterizes the decrease in the dissociation rate with decreasing vibrational level. The commonly used parameters are $U = \infty$ (non-preferential dissociation); $U = D/6k = \text{const}$; the linear function of T, $U = 3T$. None of these parameter values nor its recent modification [50] yield satisfactory agreement with the QCT data for the rate coefficients in the wide range of temperature and vibrational states. In order to overcome this problem, in [7] it was proposed to use the state-dependent parameter in the form

$$U(i,\, T) = \sum_{n=0}^{N} a_n \tilde{\varepsilon}_i^{\,n} \exp\left(T \sum_{k=0}^{K} b_k \tilde{\varepsilon}_i^{\,k}\right), \tag{12.6}$$

where the coefficients a_n, b_k are obtained by fitting QCT data from the Phys4Entry database [53] and are given in [7]. The advantage of this approach is that it does not depend on the vibrational ladder and thus can be used with any vibrational spectrum model.

In figure 12.1, the state-specific rate coefficients of O_2 and N_2 dissociation in a collision with O and N atoms, respectively, calculated using the parameter (12.6) are compared with the QCT data. One can see an excellent agreement of the theoretical and QCT results in the whole range of T and i. It is worth mentioning that the

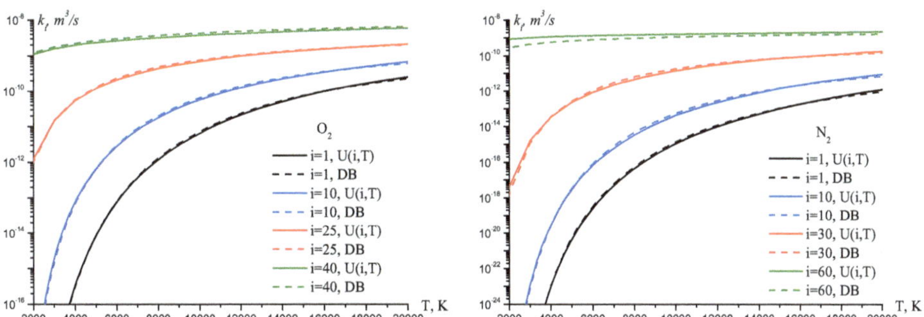

Figure 12.1. Dissociation rate coefficients for fixed vibrational states in O_2 (left) and N_2 (right) as functions of temperature. Comparison with QCT data [53] (curves DB).

proposed model can be quite easily implemented in CFD solvers. Using a similar approach, the rate coefficients for state-resolved dissociation reactions $O_2(i) + O_2$, $N_2(i) + N_2$, $O_2(i) + N_2$ are also obtained.

Let us discuss now the state-resolved rate coefficients of exchange reactions, in particular, those of NO formation:

$$N_2(i) + O = NO(k) + N, \qquad (12.7)$$

$$O_2(i) + N = NO(k) + O. \qquad (12.8)$$

Molecular dynamics calculations for these reactions were carried out in [52, 55, 56]. Theoretical models were proposed by Rusanov and Fridman [38], Polak [37], Warnatz [39], Knab [57] and Aliat [41] (in the latter model, electronic excitation is taken into account). The main limitation of the theoretical models is that they do not take into account vibrational excitation of the reaction product NO.

In figure 12.2, the rate coefficients of reaction (12.7) calculated for the NO ground vibrational state using different models are compared with the QCT rate coefficients from [56]. One can see that the best agreement is obtained when using the model proposed by Aliat [41]. It should be emphasized that in the original work, the expression for the rate coefficient contains an error (or misprint). The formula was corrected by Savelev in [54]; the revised formula has the form

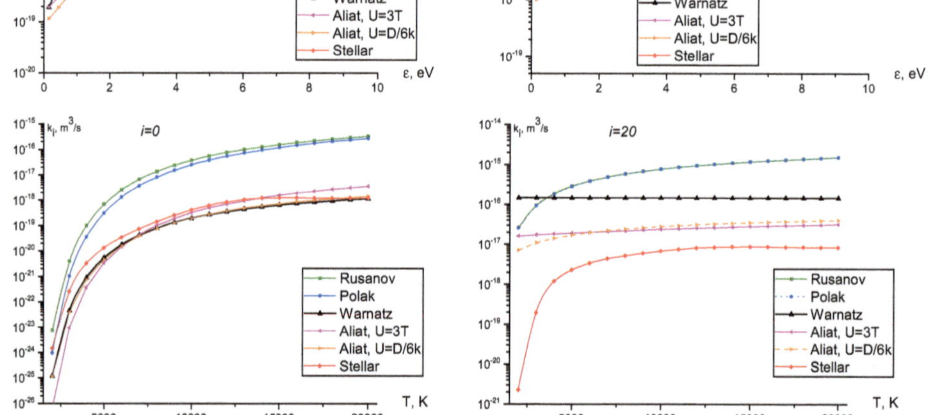

Figure 12.2. Exchange reaction rate coefficients for the $N_2 + O$ reaction for different models. Reproduced with permission from [54]. Copyright 2018 AIP.

$$k_{ci,d}^{\text{exch}}(T,\,U) = \begin{cases} C(T,\,U)\,k_{\text{eq}}^{\text{exch}} \exp\left(-\dfrac{E_{\text{a}}}{kU}\right) \exp\left[\dfrac{\varepsilon_i^c}{k}\left(\dfrac{1}{T} + \dfrac{1}{U}\right)\right], & \varepsilon_i^c < E_{\text{a}} \\ C(T,\,U)\,k_{\text{eq}}^{\text{exch}} \exp\left(\dfrac{E_{\text{a}}}{kT}\right), & \varepsilon_i^c > E_{\text{a}}, \end{cases} \tag{12.9}$$

where $k_{\text{eq}}^{\text{exch}}$ is the Arrhenius thermal equilibrium rate coefficient, E_{a} is the reaction activation energy and the normalizing coefficient $C(T,U)$ is given by

$$C(T,\,U) = Z_c^{\text{vibr}}(T)\left[\sum_{i=0}^{i^*} \exp\left(-\frac{E_{\text{a}} - \varepsilon_i^c}{kU}\right) + \sum_{i=i^*+1}^{L_c} \exp\left(\frac{E_{\text{a}} - \varepsilon_i^c}{kT}\right)\right]^{-1}. \tag{12.10}$$

Here i^* corresponds to the last vibrational level below the activation energy, $\varepsilon_{i^*}^c \leqslant E_{\text{a}}$, $\varepsilon_{i^*+1}^c > E_{\text{a}}$. We strongly recommend using this formula instead of the original one, which yields an error of several orders of magnitude compared to the QCT data.

Generalization of the Aliat model taking into account vibrational excitation was proposed in [54]:

$$k_{M_2(i),NO(k)}^{\text{exch}}(T,\,U) = \begin{cases} C_k(T,\,U)k_{\text{eq},k}^{\text{exch}} \exp\left(-\dfrac{E_{\text{a},k}}{kU}\right) \\ \quad \exp\left[\dfrac{\varepsilon_i^{M_2}}{k}\left(\dfrac{1}{T} + \dfrac{1}{U}\right)\right], & \varepsilon_i^{M_2} < E_{\text{a},k}, \\ C_k(T,\,U)k_{\text{eq},k}^{\text{exch}} \exp\left(\dfrac{E_{\text{a},k}}{kT}\right), & \varepsilon_i^{M_2} > E_{\text{a},k}, \end{cases} \tag{12.11}$$

$$C_k(T,\,U) = Z_{M_2}^{\text{vibr}}(T)\left[\sum_{i=0}^{i^*} \exp\left(-\frac{E_{\text{a},k} - \varepsilon_i^{M_2}}{kU}\right) + \sum_{i^*+1}^{L_c} \exp\left(\frac{E_{\text{a},k} - \varepsilon_i^{M_2}}{kT}\right)\right]^{-1}, \tag{12.12}$$

$$k_{\text{eq},k}^{\text{exch}} = AP_kT^b \exp\left(-\frac{E_{\text{a},k}}{kT}\right), \tag{12.13}$$

where M_2 stands for N_2 or O_2 in equations (12.7) and (12.8), respectively, $E_{\text{a},k}$ is the shifted activation energy, P_k is a function of the NO vibrational state, and $\varepsilon_i^{M_2}$ is the vibrational energy of molecule M_2 at the corresponding level. It should be noted that for the generalized model, the equilibrium rate coefficient, the activation energy and the normalizing constant are introduced for each vibrational state of the NO molecule. The recommended values of parameters in these formulas are given in [54]. Comparison of the results with the QCT data [56] shows very good agreement for the wide range of temperatures and vibrational energy (see figure 12.3). Preliminary calculations carried out in [54] for strong shock waves reveal the significant effect of NO excited states on the mixture composition behind the shock front, which puts in question the usual CFD assumption that NO can be considered only in the ground vibrational state.

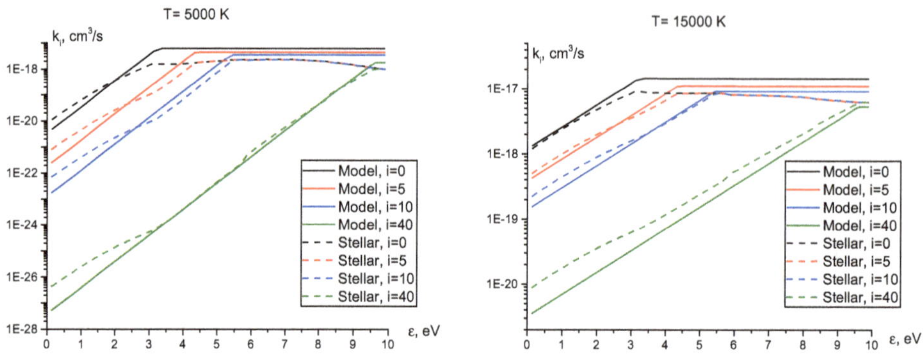

Figure 12.3. State-specific exchange reaction rate coefficients in $N_2 + O$ collision as functions of N_2 vibrational energy for different NO levels. Comparison with the QCT results [56] (Stellar).

Thus, simple and reliable theoretical models for state-resolved rate coefficients of dissociation and exchange reactions are proposed. The advantages of the models are their accuracy, the independence of the vibrational energy spectrum and accounting for the vibrational state of the reaction product. The models can be easily implemented in existing CFD codes without noticeable code modifications.

12.2.2 State-resolved transport coefficients

Simulations of viscous non-equilibrium flows require calculation of the transport coefficients. The algorithm for the state-resolved transport property evaluation developed in [25] is quite computationally expensive. Some simplifications reducing the cost of calculations are proposed in [26]. The reduced algorithm is based on the following assumptions: (1) molecular diameters are independent of the vibrational level and (2) molecules are rigid rotators, that is, their rotational energy is independent of the vibrational state. The validity of these assumptions has been studied recently [8, 58, 59].

First, the effect of varying molecular size on the transport coefficients is estimated. The diameters of vibrationally excited states are calculated using Morse or Tietz–Hua potentials [60]. Collision integrals are calculated using the hard sphere model. It is shown that under thermal equilibrium conditions, the effect of increasing molecular size (if considered for the rigid rotator model) on the viscosity and thermal conductivity of air species is small, but can be noticeable for H_2, Cl_2 and I_2 at high temperatures. Under non-equilibrium conditions, the effect is negligible for both shock heated gases (with low populations of high states) and expanding flows (low-temperature). In figure 12.4, the ratio of the shear viscosity coefficient η to that calculated neglecting varying molecular diameters η_0 is given for shock heated nitrogen and oxygen. It is seen that under non-equilibrium conditions, the effect is negligible.

In order to study the coupled effect of increasing molecular diameters and non-rigidity, the calculations are carried out for various models. The results of thermal conductivity and bulk viscosity coefficient calculation are presented in figure 12.5. First, we consider the rigid rotator model with the constant rotational specific heat $c_{rot} = k/m$ (the black line in figure 12.5; for hydrogen this model is not applicable).

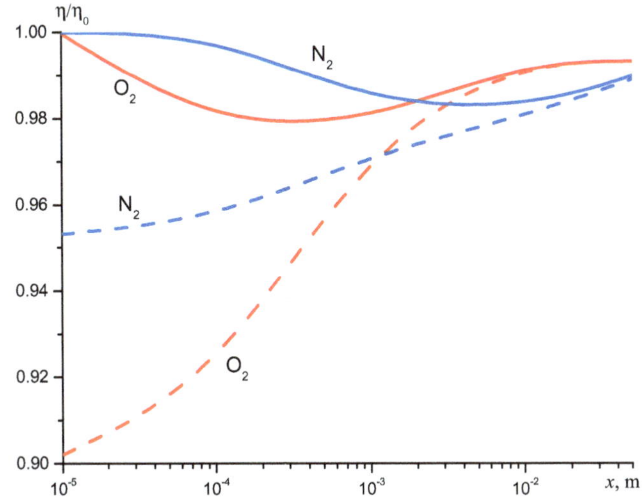

Figure 12.4. Ratios of the shear viscosity coefficients for shock heated N_2 and O_2 as functions of the distance x from the shock front. Solid lines: state-to-state distributions. Dashed lines: thermal equilibrium Boltzmann distributions at the same temperature.

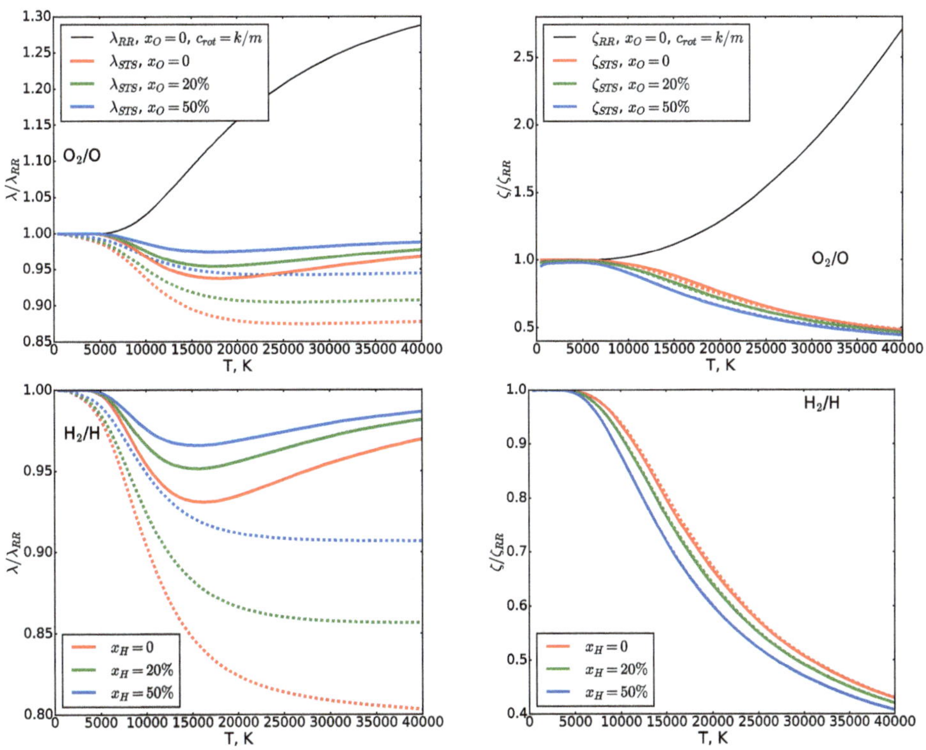

Figure 12.5. Ratio of thermal conductivity λ/λ_{RR} (left) and bulk viscosity coefficients ζ/ζ_{RR} (right); N_2/N mixture (top), H_2/H (bottom). Black line: rigid rotator model with constant c_{rot}; colored lines: full state-specific model; solid and dashed lines correspond to models with constant and variable molecular diameters, respectively. Reproduced with permission from [8]. Copyright 2017 Elsevier.

Then we consider the non-rigid rotator model with constant molecular diameter (solid lines). Finally, a full model of a non-rigid rotator with varying size for vibrationally excited states is applied (dashed lines).

One can see that the simplified algorithm yields an overestimation of the thermal conductivity coefficient. This effect is particularly prominent in hydrogen and to a lesser extent in oxygen. Accounting for increasing diameters of vibrationally excited molecules leads to a further decrease in thermal conductivity. Using the rigid rotator model causes a two-fold increase in the bulk viscosity coefficient; accounting for variable molecular diameters does not have any significant effect on the bulk viscosity coefficient. Assumption of a constant rotational specific heat breaks down at temperatures higher than 10 000 K and leads to an overestimation of transport coefficients, particularly in oxygen flows.

12.2.3 New challenges

For further development of the state-to-state high-temperature flow description, new models taking into account electronic excitation of molecules and atoms are needed. The problem is quite challenging since atomic species have a great number of excited states and data on the electronic energy transitions are scarce. Moreover, the effect of strong variation of atomic size for highly excited electronic states has to be taken into account [9]. Another new challenge for state-resolved models is their application to polyatomic gases, in particular CO_2. The main problems are strong coupling of different vibrational modes, multiple channels of vibrational relaxation, the huge number of vibrational states due to the modes coupling and lack of reliable data on the vibrational energy transitions [4].

Thus, a lot of work has been done in developing state-to-state models of kinetics and transport properties and there are still a lot of unsolved problems. Due to increasing computer power, the state-specific models now represent an efficient tool for in-depth studies of various non-equilibrium flows. However, the implementation of state-to-state models in engineering solvers for 3D viscous flow simulations is still questionable, and developing reduced-order models while maintaining the main advantages of the state-resolved ones is a challenging problem in modern non-equilibrium fluid dynamics.

12.3 Multi-temperature models for reacting air flows

Multi-temperature kinetic models are considerably less computationally expensive compared to the state-to-state approach and therefore are attractive for practical applications. Within these models, vibrational level populations are expressed in terms of vibrational temperatures of different species or various vibrational modes for polyatomic molecules. The expressions for level populations are derived from kinetic theory and are important for the definition of multi-temperature chemical reaction rates with satisfactory accuracy. In the following sections, we consider multi-temperature models for air flows and mixtures containing CO_2 and discuss applications of these models for simulations of non-equilibrium flows behind shock waves.

12.3.1 Vibrational distributions and governing equations

Let us consider multi-temperature models for the flows of a five-component air mixture N_2, O_2, NO, N, O, taking into account vibrational energy transitions, dissociation, recombination and Zeldovich exchange reactions. As explained in section 12.1, the kinetic models are derived from the kinetic equations on the basis of the relations between the relaxation times of the considered processes. It is known from experiments [61] that in vibrationally excited gases, near-resonant vibrational energy exchanges between molecules of the same species (c, c) proceed much faster than non-resonant energy exchanges between different species (c, d), vibrational–translation energy transfer and chemical reactions. Therefore the corresponding relaxation times τ_{VV_1}, τ_{VV_2}, τ_{TRV} satisfy the condition:

$$\tau_{tr} \leqslant \tau_{rot} < \tau_{VV_1}^{(c,c)} \ll \tau_{VV_2}^{(c,d)} < \tau_{TRV} < \tau_{react} \sim \theta. \tag{12.14}$$

In this case, the vibrational level populations are described by the Treanor distributions n_{ci}^{Tr} [62] and depend on the gas temperature T and the first level vibrational temperature T_1^c connected with the total number of vibrational quanta $W_c(T, T_1^c)$ in cth molecules per unit mass:

$$\rho_c W_c(T, T_1^c) = \sum_i i\, n_{ci}^{Tr}(T, T_1^c), \qquad N_2,\ O_2,\ NO, \tag{12.15}$$

where ρ_c is the density of species c.

In the case of the harmonic oscillator model for the vibrational energy spectrum, the level populations are described by the Boltzmann distributions depending on the vibrational temperatures T_v^c. If vibrational relaxation proceeds faster than chemical reactions, the expressions for level populations reduce to the thermal equilibrium Boltzmann distributions with $T_v^c = T$.

Using different quasi-stationary vibrational distributions, one can derive from the kinetic equations (12.1) the closed set of governing equations for the macroscopic air flow parameters: species number densities $n_c(\mathbf{r}, t)$, vibrational temperatures $T_v^c(\mathbf{r}, t)$ of air molecules ($c = O_2,\ N_2,\ NO$), the gas temperature and velocity [2]:

$$\frac{dn_c}{dt} + n_c \nabla \cdot \mathbf{v} + \nabla \cdot (n_c \mathbf{V}_c) = R_c^{react} \quad c = 1, \dots, L,$$
$$\rho \frac{d\mathbf{v}}{dt} + \nabla \cdot \mathbf{P} = 0, \tag{12.16}$$

$$\rho \frac{du}{dt} + \nabla \cdot \mathbf{q} + \mathbf{P} : \nabla \mathbf{v} = 0, \tag{12.17}$$

$$\rho_c \frac{dW_c}{dt} + \nabla \cdot \mathbf{q}_w^c = R_c^W - m_c W_c R_c^{react} + W_c \nabla \cdot (\rho_c \mathbf{V}_c), \quad c = N_2, O_2, NO, \tag{12.18}$$

where $\mathbf{q}_w^c$ is the flux of vibrational quanta for cth molecules, and the production term R_c^W characterizes the change in number of vibrational quanta due to slow vibrational energy transitions and chemical reactions [2]. In this part of the chapter, for simplicity the vibrational excitation of NO molecules is not taken into account and the vibrational temperatures are introduced only for N_2 and O_2 molecules. In this case the flow description is given in the three-temperature approximation.

In the thermal equilibrium flow $T_v^c = T$, and the equations of one-temperature chemical kinetics with the rate coefficients depending only on the gas temperature are coupled to the conservation equations of the momentum and total energy.

12.3.2 Vibrational–chemical coupling

The production terms R_c^W, R_c^{react} in equations (12.16), (12.18) contain the rate coefficients of vibrational energy transitions and chemical reactions. The two-temperature rate coefficients for dissociation and exchange reactions are obtained by averaging the state-resolved reaction rate coefficients with some quasi-stationary vibrational distributions. For exchange reactions (12.7), (12.8) we obtain

$$k_{M_2,\text{NO}}^{\text{exch}}(T,\, T_1^{M_2}) = \frac{1}{n_{M_2}} \sum_i n_{M_2 i}(T,\, T_1^{M_2}) k_{M_2(i),\text{NO}}^{\text{exch}}(T), \tag{12.19}$$

where M_2 corresponds to N_2 for reaction (12.7) and to O_2 for reaction (12.8); NO is considered in the ground vibrational state. The state-specific rate coefficients are introduced in the previous sections.

The temperature dependence of averaged reaction rate coefficients obtained with the use of different models for state-resolved rate coefficients is shown in figure 12.6 for fixed values of vibrational temperatures. Along with the state-dependent models [37, 38, 39] considered in the previous section, the models proposed by Macheret [63] and Park [1] were also used. Close values of coefficients obtained using the Macheret and Park models may be noticed for both reactions; the Warnatz model also provides similar results for reaction (12.7) but gives lower values for reaction (12.8). The models [37, 38] yield noticeably higher values of the two-temperature rate coefficients.

For the calculation of the production term R_c^W in equation (12.18) the vibration energy relaxation rate in the multi-temperature approach should be considered.

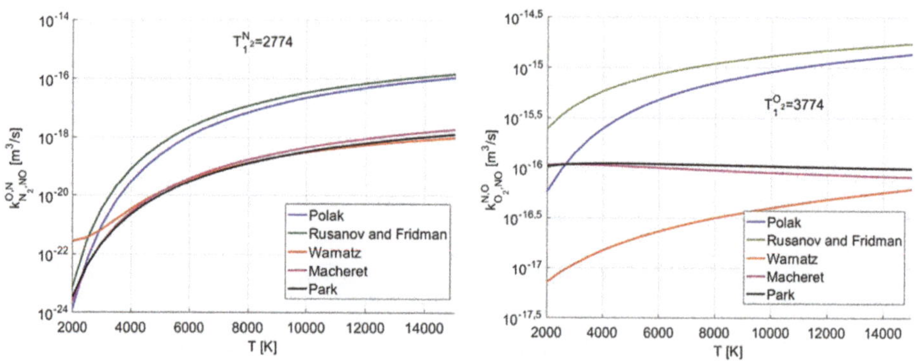

Figure 12.6. Two-temperature exchange reaction rate coefficients for reactions (12.7) (left) and (12.8) (right) as functions of T for fixed vibrational temperatures.

In computational aerodynamics, the classical Landau–Teller model [64], proposed originally for weakly non-equilibrium space homogeneous gas and depending only on the gas temperature, is commonly used for simulations of strongly non-equilibrium gas flows. In [14], the expression for the rate of vibrational energy relaxation in the case of strong deviations from the equilibrium in gas flows was derived from kinetic theory. The obtained expression contains the correction factor T/T_v^c which is important in the flows behind shock waves and in nozzles. This expression is used in the current chapter for the calculation of production terms.

12.3.3 Applications for shock heated air flows

The governing equations in the three-temperature approximation have been applied to study five-component air mixture flows behind shock waves. The results of calculations are presented below for the following conditions in the free stream: $T^{(0)} = 271$ K, $p^{(0)} = 100$ Pa and Mach number M = 16. In this chapter the air mixture in the free stream is considered in the thermal equilibrium state.

The air parameters in the beginning of the relaxation zone were calculated using condition (12.14) for relaxation times of rapid and slow kinetic processes. Thus, within the shock front, slow kinetic processes ($VV_2^{(c,d)}$, TRV energy transitions and chemical reactions) are supposed to be frozen, whereas rapid processes (such as the jump of translational temperature and then translational–rotational energy equilibration and $VV_1^{(c,c)}$ near-resonant exchanges of vibrational quanta between the same species) result in establishing quasi-stationary Treanor vibrational distributions. The flow parameters just behind the shock front are found from the conservation equations for mass, momentum and numbers of vibrational quanta for different species without variation of chemical composition and vibrational temperatures. In the relaxation zone behind the shock front the mixture composition, vibrational temperatures, gas temperature and velocity are found from the differential equations for mixture parameters $n_c(x)$, $T_v^c(x)$, $T(x)$ and $v(x)$.

In figure 12.7, the variation of the gas temperature and vibrational temperatures of N_2 and O_2 molecules found with the use of three kinetic models (two-temperature Treanor distributions of anharmonic oscillators, non-equilibrium Boltzmann distributions of harmonic oscillators depending on the vibrational temperatures and thermal equilibrium Boltzmann distribution depending on the gas temperature) along the relaxation zone is shown. It is seen that in the frame of the one-temperature approximation underestimated values of the gas temperature and vibrational temperatures are obtained.

In figure 12.8, the variation of N_2 and NO number densities in the relaxation zone obtained in the frame of the three kinetic models considered above is presented. It is seen that reactions of NO formation proceed much faster in the one-temperature thermal equilibrium flows. Neglecting anharmonicity leads to overestimated n_{N_2} values close to the shock front and then to the opposite effect with x increasing. Non-monotonic variation of n_{NO} is explained by the competitive effect of dissociation and exchange reactions.

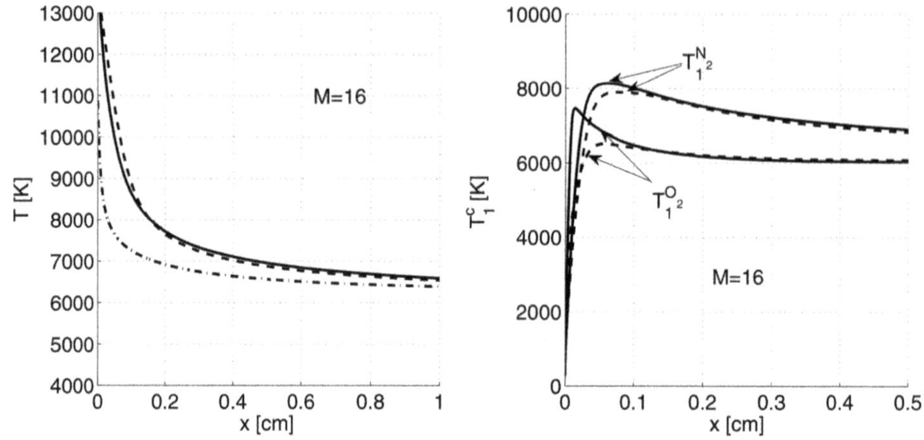

Figure 12.7. Gas temperatures (left) and vibrational temperatures (right) for different kinetic models as functions of x. Solid lines: Treanor distribution; dashed lines: Boltzmann distribution; dash-dotted lines: one-temperature approach.

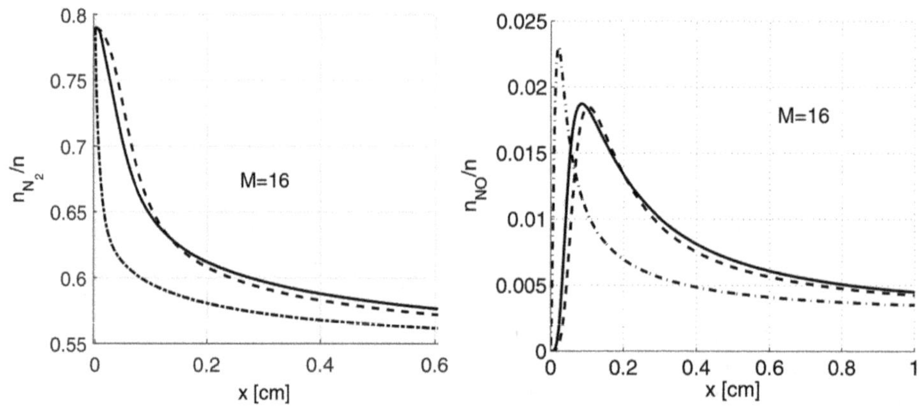

Figure 12.8. Molar fractions of N_2 (left) and NO (right) for different kinetic models as functions of x. Solid lines: Treanor distribution; dashed lines: Boltzmann distributional; dash-dotted lines: one-temperature approach.

The comparison of the gas temperature values and number densities of N_2 molecules found with the use of the most accurate state-to-state approach and in the frame of three-temperature (Treanor) and one-temperature kinetic models is given in figure 12.9. One can see that under the considered conditions the three-temperature model (with accurate reaction rate coefficients and relations for parameters within the shock front) gives results quite close to those obtained in the frame of the state-to-state approximation—the difference in the temperature values does not exceed 3%.

Figure 12.10 illustrates the influence of chemical reaction models on the gas temperature and number density of NO molecules in the relaxation zone. It may be noticed that for both parameters $T(x)$ and $n_{NO}(x)$, the use of Rusanov, Fridman [38] and Polak [37] models leads to very close values of parameters. The same holds for

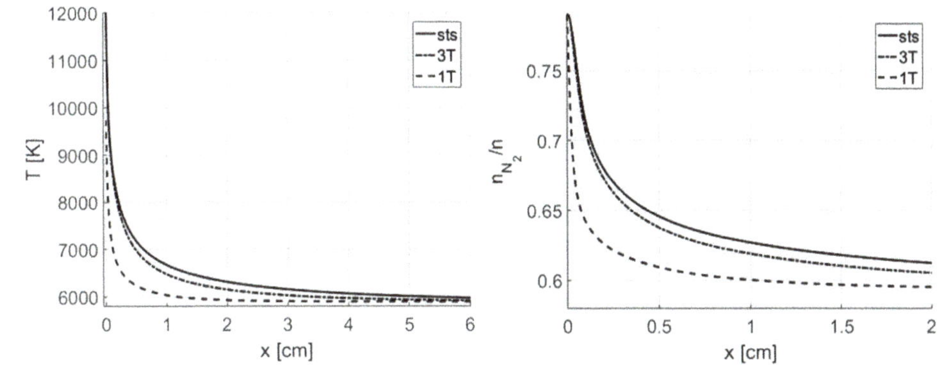

Figure 12.9. The temperature T (left) and N_2 molar fractions (right) behind the shock front as functions of x. Solid lines: state-to-state model; dash-dotted lines: three-temperature model; dashed lines: one-temperature model.

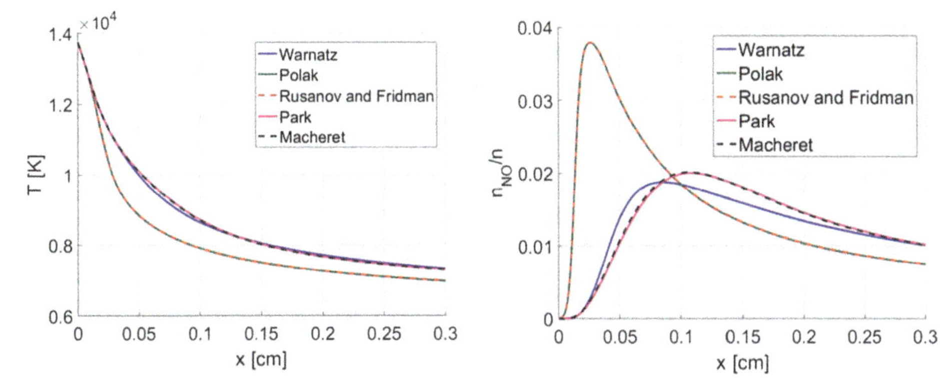

Figure 12.10. Gas temperature (left) and NO molar fraction (right) for different models of chemical reactions as functions of x.

the Macheret [63], Park [1] and Warnatz [39] models which yield similar distributions of fluid-dynamics variables.

One of the important directions for future numerical simulations of five-component air flows is connected with the use of new models for exchange chemical reactions, discussed in section 12.2.1, allowing vibrational excitation of NO molecules. In this case, in the frame of the state-to-state approach, additional equations for NO level populations should be included in governing equations. Using the multi-temperature flow description, the vibrational temperature should be introduced for NO molecules, and the equation for the NO vibrational temperature will be included in the kinetic scheme. In this case the flow is described in the frame of a four-temperature approximation.

12.4 Multi-temperature models for flows containing CO_2

Many papers are devoted to the development of multi-temperature models for vibrational kinetics in CO_2 containing mixtures. These models are very important for applications in gas-dynamic problems because the practical use of the state-to-state approach in polyatomic gases raises serious computational difficulties. The first models for CO_2 kinetics [65–68] were suitable for applications but did not describe interaction between three vibrational CO_2 modes. More accurate modeling of CO_2 flows were proposed on the basis of the kinetic theory [5, 6, 69] and used for calculations of parameters behind shock waves [70] and transport properties in a viscous shock layer near spacecrafts entering the Martian atmosphere [71].

In the present paper, we consider the results obtained on the basis of the model proposed in [5]. The model is based on the relations between the relaxation times of different kinetic processes in the considered mixture. Rapid processes include intra-mode VV_m vibrational energy exchanges within all three (symmetric, bending and asymmetric) modes ($m = 1, 2, 3$) as well as near-resonant VV_{12} energy transitions between the symmetric and bending modes, and make it possible to introduce two vibrational temperatures: T_{12} for the combined symmetric-bending mode and T_3 for the asymmetric mode. Finally, the set of governing equations for the one-dimensional flow of the considered mixture in the relaxation zone behind a plane shock wave includes equations for number densities of species $n_{CO_2}(x)$, $n_{CO}(x)$, $n_O(x)$, the gas temperature $T(x)$ and velocity $v(x)$, as well as for vibrational temperatures $T_{12}(x)$, $T_3(x)$ (x is the distance from the shock front).

The mixture state ahead of the shock front is considered to be in equilibrium. A rapid gas compression within the thin shock front results in a temperature jump which causes fast equilibration of translational and rotational degrees of freedom. Along with these processes, rapid VV_{12}, VV_3 exchanges proceed within the shock front, whereas the mean vibrational energies of combined and asymmetric modes do not change. Slow processes including VT_m vibrational energy transitions and CO_2 dissociation are supposed to be frozen within the shock front and are considered in the relaxation zone behind the shock front, along with changing the macroscopic mixture parameters. In order to solve the governing equations in the relaxation zone, the initial values of mixture parameters just behind a shock should be found. In this chapter, the values of flow parameters in the very beginning of the relaxation zone are connected with those in a free stream by the conservation equations for the momentum, total energy, and two conservation equations of the mean vibrational energies of the combined symmetric-bending and asymmetric mode. Variation of the temperatures T, T_{12}, T_3, flow velocity and mixture composition along the relaxation zone are studied in [70] for the following conditions ahead a shock front: $T_0 = T_{12}^{(0)} = T_3^{(0)} = T_v^{(0)} = 271$ K, $\rho_0 = 3.141 \times 10^{-5}$kg m^{-3}, the Mach number M = 15. The free stream is assumed to consist only of CO_2 molecules.

The obtained results are compared with those found in the frame of the two-temperature and one-temperature CO_2 kinetic models introduced for the cases $T_{12} = T_3 = T_v$ and $T_v = T$. The first case corresponds to the vibrational relaxation of all three CO_2 modes with the same vibrational temperature $T_v(x)$, whereas the second case represents the thermal equilibrium flow of the considered mixture with the gas temperature $T(x)$. In figures 12.11 and 12.12, the influence of vibrational distributions on vibrational temperatures, CO_2 number densities and rate coefficients of CO_2 dissociation is shown.

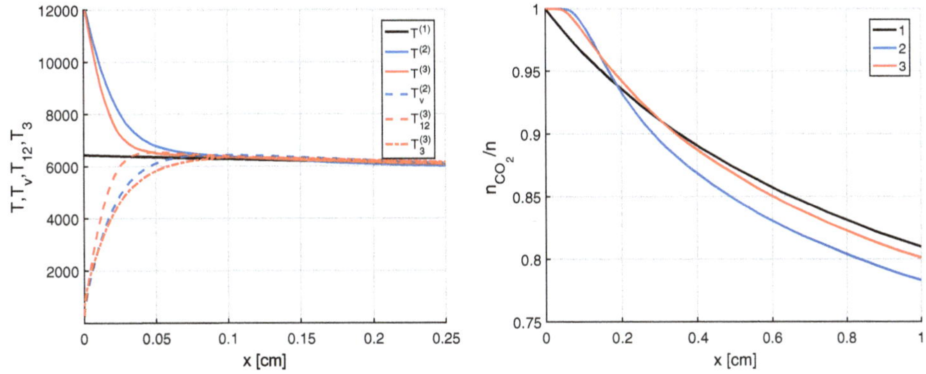

Figure 12.11. Gas temperature T and the vibrational temperatures T_{12}, T_3, T_v (left) and CO_2 molar fraction (right) as functions of x in the one-temperature (1), two-temperature (2) and three-temperature (3) approaches.

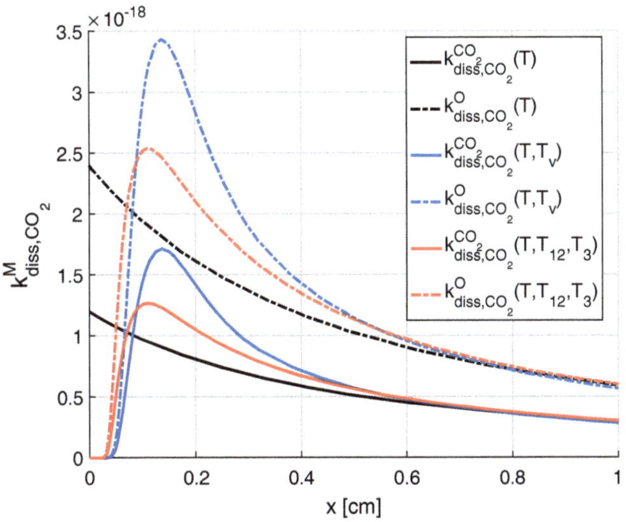

Figure 12.12. Dissociation rate coefficients for different models as functions of x.

In figure 12.11 (left) one can see a decrease in the gas temperature and an increase in vibrational temperatures in the three considered approaches, and in figure 12.11 (right) a delay of dissociation in the three-temperature and two-temperature approaches may be noticed.

The dissociation rate coefficients corresponding to the variation of the parameters behind the shock front are shown in figure 12.12 as functions of x. The dissociation of CO_2 is more efficient in the collision with atoms O, since the coefficient k^{O}_{diss,CO_2} is greater than $k^{CO_2}_{diss,CO_2}$. In the one-temperature approximation, dissociation rate coefficients exceed $k^{M}_{diss,CO_2}(T, T_v)$ and $k^{M}_{diss,CO_2}(T, T_{12}, T_3)$ for small values of x. The non-monotonic behavior of two-temperature and three-temperature dissociation rate coefficients is explained by the increase in vibrational temperatures close to the shock front and then their slow decrease approaching equilibrium.

A three-temperature model of CO_2 vibrational kinetics was also used for the study of transport properties in CO_2 flows [5]. The accurate kinetic theory algorithms were derived on the basis of the Chapman–Enskog method generalized for rapid and slow processes in CO_2 containing mixtures. As a result, the expressions for pressure tensor, total energy flux and diffusion velocities were obtained in a final form suitable for calculations. These theoretical algorithms were implemented directly in numerical codes developed in [72] for calculations of flow parameters in the viscous shock layer near the surface of the spacecraft MSRO entering the Mars atmosphere. Accurate calculations of the bulk viscosity, heat flux and diffusion velocities on the spacecraft trajectory were performed for different regimes [71].

For further studies of non-equilibrium kinetics and transport processes in carbon dioxide, justified models for vibrational transitions and exchange chemical reactions in mixtures containing CO_2 molecules are needed (particularly for mixture CO_2/CO/O_2/C/O). Up to the present time, the lack of such data causes difficulties in numerical modeling of CO_2 containing flows. Overcoming this problem is of crucial importance for CO_2 flow numerical simulations.

12.5 Conclusions

Several challenging problems of non-equilibrium flow modeling are discussed with emphasis on state-to-state and multi-temperature approaches. A general kinetic theory algorithm for deriving a closed set of governing equations is described and applied for specific flow conditions. New models for vibrational state-resolved dissociation and exchange reaction rate coefficients are proposed on the basis of the comparison with recent quasi-classical trajectory calculations. The main advantages of the models are: (1) they are not limited by the temperature range of experimental measurements; (2) they provide a good accuracy; (3) the models do not depend on the vibrational ladder and can be applied for any vibrational energy spectrum; (4) the models for the exchange reaction rate coefficients depend on the vibrational excitation of NO molecules; and (5) all developed models are quite simple and can easily be implemented in CFD solvers.

Some features of state-resolved transport coefficients are discussed. It is shown that the coupled effect of increasing molecular diameters and non-rigid rotations may affect the thermal conductivity coefficients at high temperatures; the bulk viscosity coefficients do not depend on the molecular size but are strongly affected by the rotator model.

New models for multi-temperature description of five-component air and CO_2 containing flows are based on vibrational distributions derived from the kinetic theory and are therefore sufficiently accurate. The multi-temperature approach is much simpler than detailed state-to-state models and may be recommended for numerical flow simulations. It is particularly important for multi-component mixture flows. The parameters of the five-component air and CO_2 containing mixtures in the relaxation zone behind shock waves obtained in the frame of multi-temperature models are shown in this chapter. The comparison of shock heated air flow parameters found with the use of three-temperature and state-to-state kinetic models showed very good agreement.

Some new directions in developing state-to-state and multi-temperature kinetics in dissociation flow regimes are also indicated.

Acknowledgment

This study is supported by Saint Petersburg State University, project 6.37.206.2016, and the Russian Foundation for Basic Research, projects 18-01-00493, 18-08-00707.

References

[1] Park C 1990 *Nonequilibrium Hypersonic Aerothermodynamics* (New York: Wiley)

[2] Nagnibeda E and Kustova E 2009 *Nonequilibrium Reacting Gas Flows. Kinetic Theory of Transport and Relaxation Processes* (Berlin: Springer)

[3] Brun R (ed) 2010 *High Temperature Phenomena in Shock Waves Shock Wave, Science and Technology Reference Library* vol 7 (Berlin: Springer)

[4] Armenise I and Kustova E 2013 State-to-state models for CO_2 molecules: from the theory to an application to hypersonic boundary layers *Chem. Phys.* **415** 269–81

[5] Kustova E and Nagnibeda E 2006 On a correct description of a multi-temperature dissociating CO_2 flow *Chem. Phys.* **321** 293–310

[6] Kustova E and Nagnibeda E 2012 Kinetic model for multi-temperature flows of reacting carbon dioxide mixture *Chem. Phys.* **398** 111–7

[7] Kunova O, Kustova E and Savelev A 2016 Generalized Treanor–Marrone model for state-specific dissociation rate coefficients *Chem. Phys. Lett.* **659** 80–7

[8] Kustova E, Mekhonoshina M and Oblapenko G 2017 On the applicability of simplified state-to-state models of transport coefficients *Chem. Phys. Lett.* **686** 161–6

[9] Istomin V and Kustova E 2017 State-specific transport properties of partially ionized flows of electronically excited atomic gases *Chem. Phys.* **485–86** 125–39

[10] Kremer G M, Kunova O, Kustova E and Oblapenko G 2018 The influence of vibrational state-resolved transport coefficients on the wave propagation in diatomic gases *Physica* A **490** 92–113

[11] Kunova O and Nagnibeda E 2014 State-to-state description of reacting air flows behind shock waves *Chem. Phys.* **441** 66–76

[12] Kunova O, Nagnibeda E and Sharafutdinov I 2016 Non-equilibrium reaction rates in air flows behind shock waves. State-to-state and three-temperature description *AIP Conf. Proc.* **1786** 150005

[13] Kunova O and Nagnibeda E 2015 On the influence of state-to-state distributions on exchange reaction rates in shock heated air flows *Chem. Phys. Lett.* **625** 121–7

[14] Kustova E and Oblapenko G 2015 Reaction and internal energy relaxation rates in viscous thermochemically non-equilibrium gas flows *Phys. Fluids* **27** 016102

[15] Kustova E and Oblapenko G 2016 Mutual effect of vibrational relaxation and chemical reactions in viscous multitemperature flows *Phys. Rev.* E **93** 033127

[16] Giovangigli V 1999 *Multicomponent Flow Modeling* (Boston: Birkhauser)

[17] Montroll E and Shuler K 1957 Studies in nonequilibrium rate processes. I. The relaxation of a system of harmonic oscillators *J. Chem. Phys.* **26** 454–64

[18] Nagnibeda E 1995 The structure of the relaxation zone behind shock waves in the reacting gas flows *Aerothermodynamics for Space Vehicles, ESA SP 367* ed J Hunt (Noordwijk: ESA) pp 299–303

[19] Lordet F, Meolans J, Chauvin A and Brun R 1995 Nonequilibrium vibration–dissociation phenomena behind a propagating shock wave: vibrational population calculation *Shock Waves* **4** 299–312

[20] Adamovich I, Macheret S, Rich J and Treanor C 1995 Vibrational relaxation and dissociation behind shock waves *AIAA J.* **33** 1064–75

[21] Ruffin S and Park C 1992 Vibrational relaxation of anharmonic oscillators in expanding flows *30th Aerospace Sciences Meeting and Exhibit, Aerospace Sciences Meetings* AIAA paper 92-0806

[22] Colonna G, Capitelli M, Tuttafesta M and Giordano D 1999 Non-Arrhenius NO formation rate in one-dimensional nozzle airflow *J. Thermophys. Heat Transfer* **13** 372–5

[23] Armenise I, Capitelli M, Colonna G and Gorse C 1996 Nonequilibrium vibrational kinetics in the boundary layer of re-entering bodies *J. Thermophys. Heat Transfer* **10** 397–405

[24] Capitelli M, Armenise I and Gorse C 1997 State-to-state approach in the kinetics of air components under re-entry conditions *J. Thermophys. Heat Transfer* **11** 570–8

[25] Kustova E and Nagnibeda E 1998 Transport properties of a reacting gas mixture with strong vibrational and chemical nonequilibrium *Chem. Phys.* **233** 57–75

[26] Kustova E 2001 On the simplified state-to-state transport coefficients *Chem. Phys.* **270** 177–95

[27] Armenise I, Capitelli M, Kustova E and Nagnibeda E 1999 The influence of nonequilibrium kinetics on the heat transfer and diffusion near re-entering body *J. Thermophys. Heat Transfer* **13** 210–8

[28] Kustova E, Nagnibeda E, Alexandrova T and Chikhaoui A 2002 On the non-equilibrium kinetics and heat transfer in nozzle flows *Chem. Phys.* **276** 139–54

[29] Kustova E, Nagnibeda E, Armenise I and Capitelli M 2002 Non-equilibrium kinetics and heat transfer in O_2/O mixtures near catalytic surfaces *J. Thermophys. Heat Transfer* **16** 238–44

[30] Armenise I, Barbato M, Capitelli M and Kustova E 2006 State-to-state catalytic models, kinetics and transport in hypersonic boundary layers *J. Thermophys. Heat Transfer* **20** 465–76

[31] Armenise I and Kustova E 2014 On different contributions to the heat flux and diffusion in non-equilibrium flows *Chem. Phys.* **428** 90–104

[32] Josyula E, Burt J, Kustova E, Vedula P and Mekhonoshina M 2015 State-to-state kinetic modeling of dissociating and radiating hypersonic flows *53rd AIAA Aerospace Sciences Meeting* AIAA paper 2015-0475

[33] Kunova O, Kustova E, Mekhonoshina M and Nagnibeda E 2015 Non-equilibrium kinetics, diffusion and heat transfer in shock heated flows of N_2/N and O_2/O mixtures *Chem. Phys.* **463** 70–81

[34] Schwartz R, Slawsky Z and Herzfeld K 1952 Calculation of vibrational relaxation times in gases *J. Chem. Phys.* **20** 1591

[35] Adamovich I, Macheret S, Rich J and Treanor C 1998 Vibrational energy transfer rates using a forced harmonic oscillator model *J. Thermophys. Heat Transfer* **12** 57–65

[36] Marrone P and Treanor C 1963 Chemical relaxation with preferential dissociation from excited vibrational levels *Phys. Fluids* **6** 1215–21

[37] Polak L, Goldenberg M and Levitskii A 1984 *Numerical Methods in Chemical Kinetics* (Moscow: Nauka) (in Russian)

[38] Rusanov V and Fridman A 1984 *Physics of Chemically Active Plasma* (Moscow: Nauka) (in Russian)

[39] Warnatz J, Riedel U and Schmidt R 1992 Different levels of air dissociation chemistry and its coupling with flow models *Advances in Hypersonics: Modeling Hypersonic Flows* (Boston: Birkhäuser)

[40] Aliat A, Kustova E and Chikhaoui A 2005 State-to-state reaction rates in gases with vibration–electronic-dissociation coupling: the influence on a radiative shock heated co flow *Chem. Phys.* **314** 37–47

[41] Aliat A 2008 State-to-state dissociation–recombination and chemical exchange rate coefficients in excited diatomic gas flows *Physica* A **387** 4163–82

[42] Billing G and Fisher E 1979 VV and VT rate coefficients in N_2 by a quantum–classical model *Chem. Phys.* **43** 395–401

[43] Billing G D and Kolesnick R E 1992 Vibrational relaxation of oxygen state-to-state rate constants *Chem. Phys. Lett.* **200** 382–86

[44] Laganà A and Garcia E 1994 Temperature dependence of $N + N_2$ rate coefficients *J. Chem. Phys.* **98** 502–07

[45] Esposito F, Capitelli M and Gorse C 2000 Quasi-classical dynamics and vibrational kinetics in $N_2(v) - N$ system *Chem. Phys.* **257** 193–202

[46] Esposito F, Armenise I and Capitelli M 2006 N–N_2 state to state vibrational-relaxation and dissociation rates based on quasiclassical calculations *Chem. Phys.* **331** 1–8

[47] Esposito F, Armenise I, Capitta G and Capitelli M 2008 O–O_2 state-to-state vibrational relaxation and dissociation rates based on quasiclassical calculations *Chem. Phys.* **351** 91–8

[48] Pogosbekian M and Sergievskaya A 2014 Simulation of molecular reaction dynamics and comparative analysis with theoretical models applied to thermal nonequilibrium conditions *Phys.-Chem. Kinet. Gas Dyn.* **15**

[49] Lombardi A, Faginas-Lago N, Pacifici L and Grossi G 2015 Energy transfer upon collision of selectively excited CO_2 molecules: state-to-state cross and probabilities for modeling of atmospheres and gaseous flows *J. Chem. Phys.* **143** 034307

[50] Andrienko D A and Boyd I D 2015 High fidelity modeling of thermal relaxation and dissociation of oxygen *Phys. Fluids* **27** 116101

[51] Jaffe R, Schwenke D and Chaban G 2009 Theoretical analysis of N_2 dissociation and rotation–vibration energy transfer *47th AIAA Aerospace Sciences Meeting and Exhibit* AIAA paper 2009-1569

[52] Esposito F and Armenise I 2017 Reactive, inelastic and dissociation processes in collisions of atomic oxygen with molecular nitrogen *J. Phys. Chem.* A **121** 6211–19

[53] Planetary entry integrated models, http://phys4entrydb.ba.imip.cnr.it/Phys4EntryDB/

[54] Kustova E V, Savelev A S and Kunova O V 2018 Rate coefficients of exchange reactions accounting for vibrational excitation of reagents and products *AIP Conf. Pro.* **1959** 060010

[55] Bose D and Candler G 1996 Thermal rate constants of the $N_2 + O \rightarrow NO + N$ reaction using *ab initio* $^3A''$ and $^3A'$ potential energy surfaces *J. Chem. Phys.* **104** 2825

[56] Stellar database http://esther.ist.utl.pt/pages/stellar.html

[57] Knab O, Frühauf H and Messerschmid E 1995 Theory and validation of the physically consistent coupled vibration–chemistry–vibration model *J. Thermophys. Heat Transfer* **9** 219–26

[58] Kustova E and Kremer G M 2015 Effect of molecular diameters on state-to-state transport properties: the shear viscosity coefficient *Chem. Phys. Lett.* **636** 84–9

[59] Kornienko O V and Kustova E V 2016 Influence of variable molecular diameter on the viscosity coefficient in the state-to-state approach *Vestnik St Petersburg Univ. Math.* **3** 457–67

[60] Gorbachev Y, Gordillo-Vazquez F and Kunc J 1997 Diameters of rotationally and vibrationally excited diatomic molecules *Physica* A **247** 108–20

[61] Stupochenko Y, Losev S and Osipov A 1967 *Relaxation in Shock Waves* (Berlin: Springer)

[62] Treanor C, Rich I and Rehm R 1968 Vibrational relaxation of anharmonic oscillators with exchange dominated collisions *J. Chem. Phys.* **48** 1798

[63] Macheret S O, Fridman A A, Adamovich I V, Rich J W and Treanor C E 1994 Mechanisms of nonequilibrium dissociation of diatomic molecules, *6th AIAA/ASME Joint Thermophysics and Heat Transfer Conf., Colorado Springs, CO* AIAA paper 94-1984

[64] Landau L and Teller E 1936 Theory of sound dispersion *Phys. Z. Sowjetunion* **10** 34–43

[65] Anderson J 1976 *Gasdynamic Lasers: An Introduction* (New York: Academic)

[66] Cenian A 1989 Study of nonequilibrium vibrational relaxation of CO_2 molecules during adiabatic expansion in a supersonic nozzle. The Treanor distribution—existence and generation *Chem. Phys.* **132** 41–8

[67] Thomson R 1978 The thermal conductivity of gases with vibrational internal energy *J. Phys. D: Appl. Phys.* **11** 2509

[68] Brun R 1988 Transport properties in reactive gas flows *AIAA Thermophysics, Plasmadynamics and Lasers Conf. June 27-29, San Antonio, TX* AIAA paper 88-2655

[69] Kustova E V and Nagnibeda E A 2001 State-to-state theory of vibrational kinetics and dissociation in three-atomic gases *AIP Conf. Proc.* **585** 620–27

[70] Kosareva E V and Nagnibeda E A 2017 Multi-temperature models for shock heated flows of CO_2/CO/O mixture, *7th European Conf. for Aeronautics and Aerospace Sciences (EUCASS)*

[71] Kustova E, Nagnibeda E, Shevelev Y and Syzranova N 2011 Different models for CO_2 flows in a shock layer *Shock Waves* **21** 273–87

[72] Shevelev Y and Syzranova N 2007 The influence of different models of chemical kinetics on supersonic CO_2 flows near blunt bodies *Phys.-Chem. Kinet. Gas Dyn.* **5**

Chapter 13

State-to-state kinetics in CFD simulation of hypersonic flows using GPUs

Giuseppe Pascazio, Francesco Bonelli, Luigi Cutrone, Antonio Schettino, Michele Tuttafesta and Gianpiero Colonna

Advanced thermochemical non-equilibrium models are needed to improve the capabilities of computational fluid-dynamics (CFD) codes in order to predict the consequences of meteoroid and debris entry. Accurate simulations of such events are very challenging since many phenomena are involved, i.e. shock waves, turbulence, thermochemical non-equilibrium, ablation, radiation, etc.

A body entering a planetary atmosphere experiences several flow regimes, at different altitudes, depending on the degree of rarefaction. It proceeds from free molecular flow to hydrodynamic regimes by going through transition and slip flows [1]. In free molecular flow, the fluid is so rarefied that gas particles do not collide with each other but directly strike the vehicle surface [2]. On going deeper into the atmosphere, the transition regime is encountered, where collisions between incoming and reflected particles are no longer negligible and a compression zone is formed with a progressive and smooth increase of density, pressure and temperature of the gas. In this regime, the thickness of the compression region is of the same order as the vehicle characteristic length [1]. Descending even more, collisions progressively increase and the thickness of the compression zone reduces. However, rarefied effects are still not negligible close to the surface, resulting in a temperature jump and non-null velocity at the wall, a condition called the *slip* regime [1]. Finally, the vehicle enters the continuum or hydrodynamic regime where collisions are so frequent and energetic that the compression region becomes a detached bow shock whose thickness is of the order of few nanometers.

Depending on the flow regime, the most suitable physical model and numerical approach must be chosen [2]. In the free molecular and transition regimes, collisions are few and not energetic enough to cause molecular excitation or dissociation and therefore the energy transfer to the vehicle is small, despite its relevance in determining the entry path. For such regimes the fluid-dynamics time-scale is of

doi:10.1088/2053-2563/aae894ch13

the same order as the translational relaxation time and a description based on the Boltzmann equation, such as direct simulation Monte Carlo (DSMC) [3] or high order moment expansion [4], must be used.

On the other hand, solving the Boltzmann equation for the slip and continuum regime is inconvenient or extremely difficult [5] and solving the Navier–Stokes equations is a proper approach, assuming suitable slip boundary conditions when needed [1, 2].

In the continuum regime the compression thickness becomes a shock wave through which the flow kinetic energy is converted into molecular energy, namely, into translational, rotational, vibrational and electronic degrees of freedom [6].

Considering the quantum nature of vibrational and electronic degrees of freedom[1] is fundamental for understanding and predicting macroscopic gas properties in high-enthalpy flows, such as deviations from calorically perfect gas approximation. Moreover, by comparing the different time-scales involved it is possible to introduce the concepts of frozen, equilibrium and non-equilibrium flows. Indeed, due to the quantum behavior of internal modes, at relatively low temperature, e.g. smaller than 600 K for air [6], energy is mainly stored as translational and rotational degrees of freedom, vibrational energy being negligible, validating the calorically perfect gas assumption, i.e. considering a constant value for the isentropic coefficient. When increasing the temperature, vibrational and electronic states are progressively excited, leading to dissociation and ionization; therefore, at the very high temperature downstream of the bow shock, chemical reactions also become important.

All these phenomena have a characteristic relaxation time τ, with the following ordering,

$$\tau_{\text{tr}} < \tau_{\text{rot}} < \tau_{\text{vib}} < \tau_{\text{diss}} < \tau_{\text{el}} < \tau_{\text{ion}}, \tag{13.1}$$

where the subscripts tr, rot, vib, diss, el and ion refer respectively to translational, rotational, vibrational, dissociation, electronic and ionization. Processes with a relaxation time much smaller than the flow characteristic time, τ_{flow}, are in equilibrium, whereas processes with $\tau \gg \tau_{\text{flow}}$ do not occur. When all τ in equation (13.1) are much smaller than τ_{flow} the flow is in local equilibrium, and when they are much larger than τ_{flow} the flow is frozen. If some τ are of the same order as τ_{flow} then the flow is in thermochemical non-equilibrium.

Hence, a wide range of phenomena, with different time-scales and strongly coupled with each other, occur in the shock layer. They can be summarized as follows:

- Just downstream of the bow shock a mixture of vibrationally/electronically excited and chemically reacting non-equilibrium flow is formed.
- Sufficiently downstream of the shock, a local equilibrium condition is reached [7] and, for peak temperatures larger than 10000 K [6], de-excitation of the electronic mode causes a significant amount of radiation.
- Temperature decrease causes recombination in the boundary layer.

[1] Rotational degrees of freedom are also described by discrete levels, but the energy gap is very small, so for aerothermodynamic applications the continuum assumption is a good approximation.

- A huge amount of heat is transferred at the surface of the body.

The huge amount of radiative and convective heat flux can increase the surface temperature up to 3000 K, causing ablation [8].

In this chapter we will focus on entry velocity, such that only vibrational degrees of freedom and dissociation are activated, showing important departures from equilibrium. Certainly, the most popular model is Park's multi-temperature approach [9]. It considers translational and rotational degrees of freedom in thermal equilibrium, characterized by a single roto-translational temperature, whereas the vibrational degree follows a Boltzmann distribution at a different temperature, evolving according to the Landau–Teller law and a removal effect to account for dissociation. The mass action law describes chemical species variations with rate coefficients evaluated with the Arrhenius formula at an effective temperature, which is a weighted geometrical mean of the translational and vibrational ones, to take into account the larger probability of dissociating from higher vibrational levels.

However, the hypothesis of a Boltzmann distribution for the vibrational mode is not always verified, which is responsible for large deviations of the rate coefficients from the Arrhenius law [10]. To overcome this problem the vibrationally resolved *state-to-state* (StS) approach considers all the vibrational levels of molecules, each one treated as a separate species evolving according to internal relaxation and chemical processes [11, 12], thus being able to determine internal distributions even if they depart from the Boltzmann one [13]. However, the number of species and processes involved is much higher than for those of multi-temperature models. For instance, a macroscopic model describes a neutral air mixture by considering five species (N_2, O_2, NO, N, O), 17 reactions and three vibrational temperatures, whereas a vibrationally resolved StS approach requires hundreds of species and thousands of processes, a computationally demanding problem. As a consequence, the StS approach is mainly applied to 1D flows [14–16], with very few 2D [17] and 3D [18] exceptions.

To perform 2D/3D simulations two strategies are possible: developing reduced models from the StS one [19–25] or exploiting the latest advances in computer hardware and software, the approach used in the current chapter.

Indeed, in the last few years a new boost to computational performance has come from the general-purpose computing on graphics processing units (GP-GPU). GPUs were born to handle display rendering with the idea of using many small cores, inherently designed to deal with parallel operations, to alter the color of each pixel. With respect to CPUs, GPUs have an impressive number of transistors (thousands of cores) dedicated to data processing and hiding memory latency with calculations rather than with data caching [26]. On the other hand, they have a very simple control logic, with limited flow control possibilities, and were not designed for general-purpose computing [26]. The 1990s saw the birth of dedicated application programming interfaces (APIs), such as OpenGL and DirectX, to write 2D and 3D rendering applications with GPUs [27]. At first, researchers, attracted by the enormous computing power of GPUs, were forced to use these APIs to implement their applications. The mechanism was extremely convoluted: it required knowing

DirectX or OpenGL, translating the problem in terms of colors, performing computations on the GPU and translating colors back into scientific data [27]. Fortunately, in November 2006 [27] NVIDIA presented a new hardware architecture accompanied by a programming language (CUDA C), which essentially introduced extensions to C/C++, thus allowing an easy way to handle GP-GPU. Subsequently, other programming languages such as OpenCL, OpenACC and OpenHMPP have been released for GP-GPU and heterogeneous CPU/GPU computing. GPUs not only far exceed CPU performance in terms of FLOPs and memory bandwidth, but above all in terms of energy efficiency [28]. This is witnessed by the presence of six clusters, using NVIDIA GPUs, in the first ten positions of the June 2018 Green 500 list [28] and by the fact that the most powerful supercomputers [29], Summit and Sierra installed at Oak Ridge National Laboratory and at Lawrence Livermore National Laboratory, respectively [30, 31], are equipped with NVIDIA GPUs.

In the following sections we will show the feasibility of multi-dimensional CFD simulations coupled with a vibrationally resolved StS approach for a neutral air mixture on a small GPU cluster.

13.1 Physical model

In order to describe the physics of a high-enthalpy flow we consider the 3D Navier–Stokes equations for a neutral air mixture. The system of governing equations reads

$$\int_{V_0} \frac{\partial \mathbf{U}}{\partial t} dV + \oint_{S_0} \mathbf{F} \cdot \mathbf{n} dS = \int_{V_0} \mathbf{W} dV, \tag{13.2}$$

where **U**, **F**, and **W** are the vector of conservative variables, the flux vector and the source term vector, respectively. The expanded equations for both the StS and the Park models are [32–34]

$$\begin{aligned}\mathbf{U} = [&\rho_{1,1}, \dots, \rho_{1,V_1}, \dots, \rho_{S,1}, \dots, \rho_{S,V_S}, \\ &\rho u, \rho v, \rho w, \rho e, \rho_1 \varepsilon_{\text{vib},1}, \dots, \rho_M \varepsilon_{\text{vib},M}]^T,\end{aligned} \tag{13.3}$$

$$\mathbf{F} = (\mathbf{F}_E - \mathbf{F}_V, \mathbf{G}_E - \mathbf{G}_V, \mathbf{H}_E - \mathbf{H}_V) \tag{13.4}$$

$$\begin{aligned}\mathbf{F}_E = [&\rho_{1,1}u, \dots, \rho_{1,V_1}u, \dots, \rho_{S,1}u, \dots, \rho_{S,V_S}u, \\ &\rho u^2 + p, \rho uv, \rho uw, (\rho e + p)u, \rho_1 \varepsilon_{\text{vib},1}u, \dots, \rho_M \varepsilon_{\text{vib},M}u]^T,\end{aligned} \tag{13.5}$$

$$\begin{aligned}\mathbf{G}_E = [&\rho_{1,1}v, \dots, \rho_{1,V_1}v, \dots, \rho_{S,1}v, \dots, \rho_{S,V_S}v, \\ &\rho uv, \rho v^2 + p, \rho vw, (\rho e + p)v, \rho_1 \varepsilon_{\text{vib},1}v, \dots, \rho_M \varepsilon_{\text{vib},M}v]^T,\end{aligned} \tag{13.6}$$

$$\mathbf{H}_E = [\rho_{1,\,1}w, \ldots, \rho_{1,\,V_1}w, \ldots, \rho_{S,\,1}w, \ldots, \rho_{S,\,V_S}w, \\ \rho uw, \rho vw, \rho w^2 + p, (\rho e + p)w, \rho_1\varepsilon_{\text{vib},\,1}w, \ldots, \rho_M\varepsilon_{\text{vib},\,M}w]^T, \tag{13.7}$$

$$(\mathbf{F}_V, \mathbf{G}_V, \mathbf{H}_V) = [-\rho_q\mathbf{u}_q, \boldsymbol{\sigma}, \mathbf{u}\cdot\boldsymbol{\sigma} - \mathbf{q}, \mathbf{q}_m]^T, \tag{13.8}$$

$$\mathbf{W} = [\dot{\omega}_{1,\,1}, \ldots, \dot{\omega}_{1,\,V_1}, \ldots, \dot{\omega}_{S,\,1}, \ldots, \dot{\omega}_{S,\,V_S}, \\ 0, 0, 0, 0, \dot{\omega}_{\text{vib},\,1}, \ldots, \dot{\omega}_{\text{vib},\,M}]^T, \tag{13.9}$$

where $\rho_{s,v}$ is the density of the sth species in the vth vibrational level (no vibrational levels are considered in multi-temperature models), p is the thermodynamic pressure, u, v and w are the flow velocity components in the x, y and z directions, e is the specific total energy, $\varepsilon_{\text{vib},m}$ is the specific vibrational energy of molecule m (used only for multi-temperature models) and M is the total number of molecules. The gas density is $\rho = \sum_s\rho_s$ whereas the density of the component s is $\rho_s = \sum_v\rho_{s,v}$. Finally, the chemical source terms and the vibrational energy source terms (the latter considered exclusively in the case of multi-temperature models) are $\{\dot{\omega}_{s,v}\}$, $\{\dot{\omega}_{\text{vib},m}\}$, respectively.

We model the component diffusion velocities, viscous stress tensor, total energy and vibrational energies' heat flux vectors as follows:

$$\rho_q\mathbf{u}_q = -\rho D_q\nabla Y_q, \tag{13.10}$$

$$\boldsymbol{\sigma} = \mu[\nabla\mathbf{u} + (\nabla\mathbf{u})^T] - \frac{2}{3}\mu\nabla\cdot\mathbf{u}\mathbf{I}, \tag{13.11}$$

$$\mathbf{q} = -\lambda_t\nabla T - \sum_{m=1}^{M}\lambda_{\text{vib},m}\nabla T_{v,m} + \sum_{q=1}^{N}h_q\rho_q\mathbf{u}_q, \tag{13.12}$$

$$\mathbf{q}_m = -\lambda_{\text{vib},m}\nabla T_{v,m} + \varepsilon_{\text{vib},m}\rho_m\mathbf{u}_m, \tag{13.13}$$

where the subscripts m and q indicate the generic molecule and the generic component, respectively, Y is the mass fraction, T and T_v are the translational and vibrational temperatures (the latter used only for multi-temperature models), D_q, μ, λ_t and $\lambda_{\text{vib},m}$ are the mixture component diffusion coefficient, the mixture viscosity, the mixture translational conductivity and the vibrational conductivity of molecule m, respectively.

In order to close the system of governing equations we use a relation between p and e under the approximation of perfect gas [35]

$$p = (\bar{\gamma} - 1)\left[\rho e - \rho(\varepsilon_{\text{vib}} + \varepsilon_{\text{chem}}) - \rho\frac{u^2 + v^2}{2}\right], \tag{13.14}$$

where ε_{vib} and $\varepsilon_{\text{chem}}$ are the total specific vibrational energy and the total specific chemical energy, respectively, $\bar{\gamma}$ is the specific heat ratio of the gas mixture. The expressions of ε_{vib} depend on the thermochemical non-equilibrium model employed; thus, they will be given in the sections concerning the StS and the Park models. The total specific chemical term is given by

$$\varepsilon_{\text{chem}} = \frac{1}{\rho}\sum_{s=1}^{S}\rho_s h_s^f, \tag{13.15}$$

where h_s^f is the specific formation enthalpy of the sth component.

We compute the gas mixture specific heat at constant pressure by considering the degrees of freedom in equilibrium at the translational temperature

$$\bar{c}_p = \alpha\ R\,,$$

where R is the specific gas constant and α is given by

$$\alpha = \sum_s \chi_s \alpha_s, \tag{13.16}$$

where χ_s is the molar fraction of the sth component and α_s is equal to 5/2 for monoatomic and 7/2 for diatomic components, respectively. Finally, by considering the Mayer relation, $R = \bar{c}_p - \bar{c}_v$, where $\bar{c}_v$ is the mixture specific heat at constant volume, we can write the specific heat ratio of the gas mixture as

$$\bar{\gamma} = \frac{\bar{c}_p}{\bar{c}_v} = \frac{\alpha}{\alpha - 1}. \tag{13.17}$$

Further model details can be found in [13, 32].

13.1.1 State-to-state air kinetics

The vibrationally resolved StS model for the five-species neutral air mixture considered here takes into account 68 and 47 vibrational levels for the N_2 and O_2 molecules, respectively, whereas only the ground state is considered for NO, N and O. The formation energies in electronvolts are 4.88195, 2.55764 and 0.941 for N, O and NO, respectively, referred to O_2 and N_2. Each level is considered as a separate species that evolves according to vibrational–vibrational (VV) and vibrational–translational (VT) energy exchanges with atoms (a) and molecules (m) and dissociation–recombination (DR) processes, or when data for the direct dissociation model [36, p 182] are missing, the ladder climbing (LC) approach [36, p 183]. Moreover, NO dissociation and Zeldovich (exchange) [37, 38] reactions were considered:

$$N_2(v) + N_2(w) \leftrightarrow N_2(v-1) + N_2(w+1) \quad (VV) \tag{13.18}$$

$$N_2(v) + N_2 \leftrightarrow N_2(v-1) + N_2 \quad (VTm) \tag{13.19}$$

$$N_2(v) + N \leftrightarrow N_2(v-\Delta v) + N \quad (VTa) \tag{13.20}$$

$$O_2(v) + O_2(w) \leftrightarrow O_2(v-1) + O_2(w+1) \quad (VV) \tag{13.21}$$

$$O_2(v) + O_2 \leftrightarrow O_2(v-1) + O_2 \quad (VTm) \tag{13.22}$$

$$O_2(v) + O \leftrightarrow O_2(v-\Delta v) + O \quad (VTa) \tag{13.23}$$

$$N_2(v) + O_2 \leftrightarrow N_2(v-1) + O_2 \quad (VTm) \tag{13.24}$$

$$N_2(v) + O \leftrightarrow N_2(v-1) + O \quad (VTa) \tag{13.25}$$

$$O_2(v) + N_2 \leftrightarrow O_2(v-1) + N_2 \quad (VTm) \tag{13.26}$$

$$O_2(v) + N \leftrightarrow O_2(v-1) + N \quad (VTa) \tag{13.27}$$

$$O_2(v) + N_2(w-1) \leftrightarrow O_2(v-2) + N_2(w) \quad (VV) \tag{13.28}$$

$$N_2(v) + N_2 \leftrightarrow 2N + N_2 \quad (DRm) \tag{13.29}$$

$$N_2(v) + N \leftrightarrow 2N + N \quad (DRa) \tag{13.30}$$

$$O_2(v) + O_2 \leftrightarrow 2O + O_2 \quad (DRm) \tag{13.31}$$

$$O_2(v) + O \leftrightarrow 2O + O \quad (DRa) \tag{13.32}$$

$$N_2(v_{max}) + O_2 \leftrightarrow N_2(v_{max+1}) + O_2 \equiv 2N + O_2 \quad (LCm) \tag{13.33}$$

$$N_2(v_{max}) + O \leftrightarrow N_2(v_{max+1}) + O \equiv 2N + O \quad (LCa) \tag{13.34}$$

$$O_2(v_{max}) + N_2 \leftrightarrow O_2(v_{max+1}) + N_2 \equiv 2O + N_2 \quad (LCm) \tag{13.35}$$

$$O_2(v_{max}) + N \leftrightarrow O_2(v_{max+1}) + N \equiv 2O + N \quad (LCa). \tag{13.36}$$

$$NO + X \leftrightarrow N + O + X \tag{13.37}$$

$$O_2(v) + N \leftrightarrow NO + O \tag{13.38}$$

$$N_2(v) + O \leftrightarrow NO + N, \tag{13.39}$$

where X is a generic component, i.e. N_2, O_2 NO, N or O. The rate constants of equations (13.18)–(13.39) are only functions of the temperature and of the vibrational quantum number of the molecules involved in the process. The total number of elementary processes is about 10 000.

The total specific vibrational energy is given by

$$\varepsilon_{\text{vib}} = \frac{1}{\rho}\sum_{s=1}^{S}\sum_{v=1}^{V_s}\rho_{s,v}\varepsilon_{s,v}, \tag{13.40}$$

where $\varepsilon_{s,v}$ is the specific energy of the vth level calculated using the anharmonic polynomial expansion

$$\varepsilon_{S,v}W_S = \omega_e\left(v + \frac{1}{2}\right) - \omega_e x_e\left(v + \frac{1}{2}\right)^2 + \omega_e y_e\left(v - \frac{1}{2}\right)^3 + \omega_e z_e\left(v - \frac{1}{2}\right)^4 + \cdots \tag{13.41}$$

whose coefficients are spectroscopic constants [39]. Obviously, the common definition of vibrational temperature does not apply to non-Boltzmann distributions. However, considering that low-energy levels approximately follow a Boltzmann trend, the vibrational temperature can be estimated from the first two levels as

$$T_{Vs} = \frac{(\varepsilon_{s,1} - \varepsilon_{s,0})}{R \ln\left(\dfrac{\rho_{s,0}}{\rho_{s,1}}\right)}. \tag{13.42}$$

13.1.2 Multi-temperature Park's model

For a neutral air mixture the Park's model considers 15 dissociations plus two exchange reactions:

$$N_2 + X \leftrightarrow 2N + X \tag{13.43}$$

$$O_2 + X \leftrightarrow 2O + X \tag{13.44}$$

$$NO + X \leftrightarrow N + O + X \tag{13.45}$$

$$NO + O \leftrightarrow N + O_2 \tag{13.46}$$

$$N_2 + O \leftrightarrow NO + N. \tag{13.47}$$

The law of mass action is used to evaluate the chemical source terms [40]

$$\dot{\omega}_s = W_s \sum_{i=1}^{N_r} \nu_{si} RR_i, \tag{13.48}$$

where W_s is the molecular weight of the sth species, N_r is the total number of reactions, ν_{si} is the difference of product (ν''_{is}) and reactant (ν'_{is}) stoichiometric coefficients of the sth species in the ith reaction, and RR_i is the reaction rate [40]

$$RR_i = k_{f_i} \prod_{s=1}^{S} C_s^{\nu'_{is}} - k_{b_i} \prod_{s=1}^{S} C_s^{\nu''_{is}}, \tag{13.49}$$

where k_{f_i} and k_{b_i} are the rate constants (forward and backward) and C_s is the molar concentration of the sth species. The Arrhenius law is used to evaluate forward coefficients

$$k_{f_i} = A_i T_x^{n_i} \exp\left(-\frac{T_{d_i}}{T_x}\right), \tag{13.50}$$

where constants A_i, n_i and T_{d_i} are taken from [9, p 326] [41]. A geometrically averaged temperature is used as the controlling temperature (T_x) for dissociation reactions

$$T_x = T_v^q T^{1-q}, \tag{13.51}$$

where q is a parameter, here assumed equal to 0.5, whereas for Zeldovich reactions T_x is assumed equal to the translational temperature T [9, p 138] [41]. In our implementation we consider a separate T_v for each molecule; thus a transport equation is solved for each corresponding vibrational energy. The equilibrium constants (whose expressions are given in [9, p 35]) connect forward and backward reaction rates [40]

$$K_{\text{eq}_i} = \frac{k_{f_i}}{k_{b_i}}. \tag{13.52}$$

The vibrational energy source term is decomposed as the sum of a collisional $\{\dot{\omega}_{\text{LT},m}\}$ and chemical $\{\dot{\omega}_{\text{chem},m}\}$ part [9, p 125] [42]. The collisional part is modeled by the Landau–Teller equation

$$\dot{\omega}_{\text{LT},m} = \rho_m \frac{\varepsilon_{\text{vib},m}(T) - \varepsilon_{\text{vib},m}(T_v)}{\tau_m}, \tag{13.53}$$

where $\varepsilon_{\text{vib},m}(T)$ is the equilibrium vibrational energy and τ_m is the relaxation time of molecule m. The latter is evaluated as a weighted harmonic average [43]

$$\frac{1}{\tau_m} = \frac{1}{N_t} \sum_x \frac{N_x}{\tau_{m,X}}, \tag{13.54}$$

where $\tau_{m,X}$ is the relaxation time for collisions between the mth molecule and the collision partner X, N_x is the number density of species x and $N_t = \sum_x N_x$. In turn this is calculated as the sum of a term given by the empirical expression of Millikan–White [44] plus a correction for high temperatures

$$\tau_{m,X} = \tau_{m,X}^{MW} + \tau_{m,X}^{c}. \tag{13.55}$$

The Millikan–White law [9, p 58] [41] and the high temperature correction [9, p 60], [41, 43, 45] are

$$\tau_{m,X}^{\text{MW}} = \frac{p_{\text{atm}}}{p} \exp[A_{m,X}(T^{-1/3} - B_{m,X}) - 18.42] \tag{13.56}$$

$$\tau_{m,X}^{c} = \frac{1}{N_m \sigma \sqrt{\dfrac{8\Re T}{\pi \mu_{m,X}}}}, \tag{13.57}$$

where $p_{\text{atm}} = 101\,325$ Pa is the atmospheric pressure, the constants $A_{m,X}$ and $B_{m,X}$ are given in [41, table 1, p 387], σ is the excitation cross section ($3 \times 10^{-17}(50\,000/T)^2$ cm^2

[41]), $\mathfrak{R}$ is the universal gas constant and $\mu_{m,X}$ is the equivalent molecular weight, i.e. $W_m W_X/(W_m + W_X)$.

The chemical $\{\dot{\omega}_{\text{chem},\,m}\}$ contribution takes into account the loss and gain of vibrational energy due to dissociation and recombination processes that occur preferentially from excited molecules (the preferential removal phenomena) [9, 43]. This term is modeled by means of a harmonic oscillator approach, i.e. the energy lost or recovered during dissociation and recombination processes is equally divided between the vibrational and the translational/rotational modes [9, p 107, 126],

$$\dot{\omega}_{\text{chem},m} = \frac{D_m}{2}\dot{\omega}_m, \tag{13.58}$$

where D_m is the specific dissociation energy of the mth molecules. A harmonic oscillator approach is also used to evaluate vibrational temperatures

$$\varepsilon_{\text{vib},m} = \frac{R_m \theta_m}{\exp(\theta_m/T_V) - 1}, \tag{13.59}$$

where the characteristic vibrational temperatures (θ) are equal to 3393 K, 2273 K and 2739 K for the N_2, O_2 and NO molecules, respectively [9, p 123].

More details on the implemented Park's model can be found in [13, 46].

13.1.3 Multi-temperature CAST model

This model, from hereon called the CAST model, was developed in the framework of CAST, an Italian aerothermodynamic project funded by the Italian Aerospace Agency, as an alternative to Park's model. It was constructed considering thermal rates calculated using the StS dataset described in section 13.1.1, in order to have a consistent comparison with StS results. It is a three-temperature model for five-species air.

13.2 Numerical method

The system of governing equations (13.2) is solved by using an operator splitting approach [47–49] in order to take into account the different time-scales (stiffness [50]) of fluid dynamics and chemistry, increasing the difficulty of the numerical solution. The operator-splitting approach separates fluid-dynamic equations from chemical ones (source terms). In such a way, firstly the homogeneous part of equation (13.2)

$$\frac{\partial \mathbf{U}}{\partial t} + \frac{\partial \mathbf{F}(\mathbf{U})}{\partial x} + \frac{\partial \mathbf{G}(\mathbf{U})}{\partial y} = \mathbf{0}, \tag{13.60}$$

representing the Navier–Stokes equations for a non-reacting flow, is solved, using a finite volume approach on a multi-block structured grid, employing the method of lines to separate space and time discretization. As first step, the convective fluxes are solved by using either the Steger and Warming flux vector splitting [51] or the AUSMPW+ scheme of Kim *et al* [52]; then either a two-step second-order or three-step third-order Runge–Kutta scheme is employed for time integration. To obtain higher spatial accuracy a MUSCL (Monotone Upstream-centered Schemes for Conservation Laws) approach is implemented [53].

Then the homogeneous solution of equation (13.60), $\mathbf{U}^{\text{hom}}(t + \Delta t)$, is used as initial value of the chemical ordinary differential equation (ODE)

$$\frac{d\mathbf{U}}{dt} = \mathbf{W}(\mathbf{U}^{\text{hom}}). \tag{13.61}$$

In order to account for the smaller characteristic times of thermochemical processes, a sub-time step $\Delta t^{(\nu)} = \Delta t/n$ is considered, where Δt is determined on the basis of the Courant–Friedrichs–Lewy (CFL) condition.

In order to obtain the best GPU performance, each thread should perform exactly the same computation; therefore, we use a constant number of subiterations n (chosen on the basis of the case study) for each computational cell.

Equation (13.61), decomposed in the production ($\mathbf{P}$) and loss ($\mathbf{L} \cdot \mathbf{y}$) terms as

$$\frac{d\mathbf{y}}{dt} = \mathbf{P} - \mathbf{L} \cdot \mathbf{y}, \tag{13.62}$$

is solved by using an implicit Gauss–Seidel approach.

Given a time step $\Delta t^{(\nu)}$, we start by discretizing equation (13.62) as

$$(\mathbf{I} + \Delta t^{(\nu)}\mathbf{L}) \cdot \mathbf{y}(t + \Delta t^{(\nu)}) = \Delta t^{(\nu)}\mathbf{P} + \mathbf{y}(t), \tag{13.63}$$

then the iterative Gauss–Seidel approach is used to obtain the generic unknowns

$$y_i^k(t + \Delta t^{(\nu)}) = \frac{\Delta t^{(\nu)} P_i(\mathbf{y}^{k-1}) + y_i(t)}{1 + \Delta t^{(\nu)} L_i(\mathbf{y}^{k-1})}; \qquad i = 0, \dots, N-1. \tag{13.64}$$

This algorithm is suitable for GPU parallelization, maximizing the memory occupancy and operation throughput by fixing the number of iterations. Its stability is assured by the positivity of $\mathbf{P}$ and $\mathbf{L}$ diagonals.

13.3 Computational approach and hardware specifications

In order to run simulations on NVIDIA GPUs, the reactive fluid-dynamics model described above was implemented by using the CUDA C programming language [32]. Moreover, to perform fair comparisons between GPUs and CPUs, exactly the same algorithms in a CPU and a GPU version were considered. The code has been extended to multi-node calculation using Message Passing Interface (MPI) directives [13]. Even for this implementation, both an MPI-CUDA and a pure MPI-CPU version have been developed, emphasizing that in the GPU version (MPI-CUDA) all the relevant computations are performed on GPUs and only MPI data transfer, synchronization and printing are handled by CPUs. The interested reader can find more implementation details in [13, 32].

As concerns the hardware, we employed a small GPU cluster available at 'Politecnico di Bari'. Such a cluster has six nodes, each of one hosts two NVIDIA Tesla K40m and two Intel(R) Xeon(R) CPU E5-2630 v2 2.60 GHz. Nodes communication is allowed by an Infiniband device.

13.4 Results

The model described above is applied to supersonic nozzle expansion and hypersonic flow past a sphere. In both cases the simulations are performed for conditions where experimental data are available. Moreover, results obtained with the StS and Park models are compared.

13.4.1 Simulation of nitrogen supersonic expansion

The nitrogen expansion studied in this experiment is generated in the test section of a reflected shock tunnel at the Electric Arc Shock Tube (EAST) facility at NASA Ames Research Center [54].

The set-up implemented for this experiment is described in detail elsewhere, and it will only be described briefly here. The driven section of the tube has an internal diameter of 10 cm and it is initially filled to 150 Torr with Matheson prepurified grade nitrogen. The tube is converted into a reflected shock tunnel by inserting a nozzle plug section 402 cm from the diaphragm section. This nozzle, with a throat height of 0.64 cm, is 8.3 cm long and has a quadratic profile downstream of the throat

$$A/A^* = 1 + (x/2.54)^2, \tag{13.65}$$

where x is the distance from the throat in centimeters and A^* is the throat area. Moreover, the nozzle insert is equipped with a large optical window on each side of the divergent part of the nozzle. A sketch of the nozzle is given in figure 13.1.

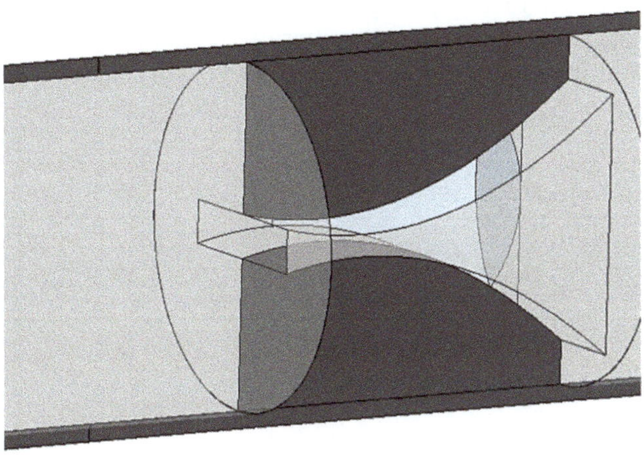

Figure 13.1. Sketch of the 3D nozzle geometry.

Similarly, the driver section is cylindrical, with an internal diameter equal to 10 cm, a length of 76 cm and is initially filled with 325 psia[2] helium.

The test is started by bursting the diaphragm and thus creating a 2.6 km s^{-1} incident shock in the driven section. The shock reflects off the nozzle plug endwall, leaving stagnant nitrogen at approximately 5600 K and 100 atm. The nitrogen then expands through the nozzle.

The stagnation conditions for the test considered are summarized in table 13.1.

Results from reference CFD solver

A standard MPI-CPU Navier–Stokes code [18] was preliminarily set up in order to assess a reference solution and the relative computational time.

The simulation is reduced to a quarter of the physical domain by symmetry, as shown in figure 13.2. The simulation domain includes part of the cylindrical driven section of the shock tube (50 mm upstream of the nozzle insert) and the converging–diverging nozzle with a total length of 108 mm, a throat located 25 mm downstream of the stagnation chamber and a throat section of $6.4 \times 100\,\text{mm}^2$.

The domain was discretized using a 3D multi-block structured grid with 16 blocks and about 270 000 cells on the most refined grid level (L1) and and about 34 000 cells on a coarser grid level (L2). Moreover, a 2D grid was generated by considering only the symmetry plane: such a grid has about 14 000 cells on the L1 grid level and 3500 cells on the L2 grid level, respectively, distributed in three computational blocks. Both grids are depicted in figure 13.3.

A grid convergence study was initially performed on the coarser grids, both 2D and 3D. The results indicated that the L2 grid level is sufficiently refined to satisfactorily capture the solutions.

Comparison of the 3D and 2D solutions (figure 13.4) reveals that the flow inside the convergent–divergent nozzle is substantially 2D, except for local effects close to the wall corners, indicating that the effort of simulating the entire three-dimensional field could have been avoided in this case. It should be noted that the developed methodology is not limited to such simple cases as have been carried out to date. It could also be applied to complex-geometry cases where a 3D discretization is necessary. In fact, the implemented numerical methodology enables calculation times to be kept at reasonable values: a 2D solution with a standard MPI-CPU code

Table 13.1. Stagnation conditions used in the simulation of the EAST nozzle experiments. Reprinted with the permission from [18]. Copyright 2014 AIP Publishing.

	h^0 (J kg^{-1})	p^0 (Pa)	T^0 (K)	ρ (kg m^{-3})	Y_{N_2}
case B	7.30×10^6	1.03352×10^7	5616	6.164	1

[2] Absolute pressure in pounds per square inch.

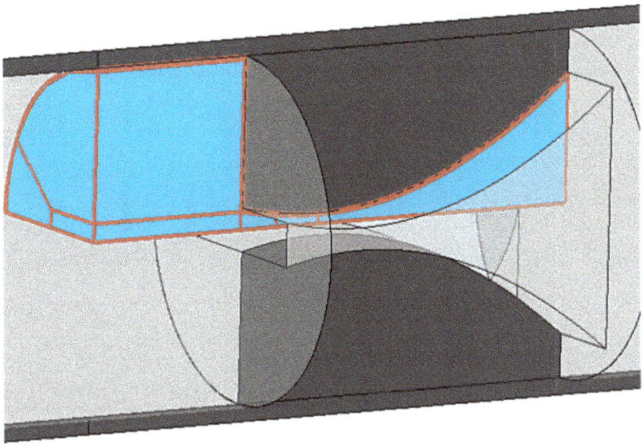

Figure 13.2. Multi-block domain topology used to discretize the 3D nozzle geometry.

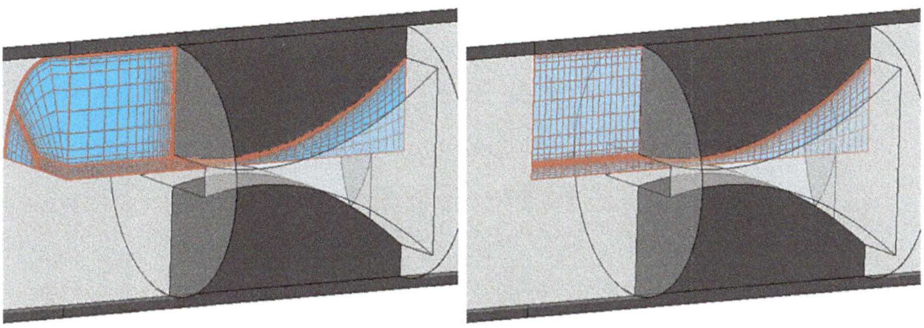

Figure 13.3. Details of the multi-block meshes. Left: 3D computation; right: 2D computation.

on a L2 grid level typically requires 14 days on a single core CPU, when an equivalent solution with macroscopic models is obtainable in a tenth of this time. A complete 3D simulation is also feasible, and comparable to 2D, as long as parallel computational strategies are adopted.

In order to show the enormous advantages of the GPU implementation, the same computation was performed by solving the 2D Euler equations with the MPI-CUDA code. By using a computational grid which includes 65×47 fluid cells, a test case requires about 1.5 h on a single GPU to be completed. Even if the standard MPI-CPU reference code [18] solves the more demanding Navier–Stokes equations, the computational cost reduction obtained with the GPU for the same quality of result, as shown in figure 13.5, is impressive.

The results obtained using StS kinetics were compared to those obtained with the Park (see section 13.1.2) and CAST (see section 13.1.3) multi-temperature models.

Figures 13.6 and 13.7 show N_2 vibrational temperature and mass fraction distributions along the axis calculated with the models considered, where the macroscopic experimental data used for comparison are limited to the vibrational

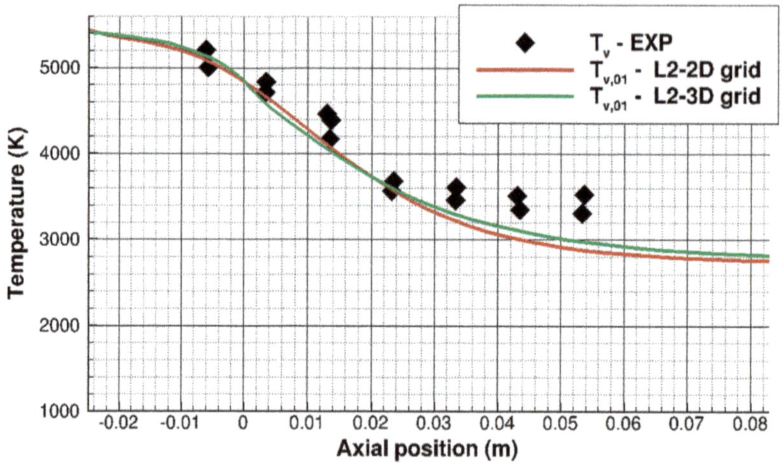

Figure 13.4. Effect of 2D versus 3D discretization of the computational domain (L2 level) on the axial vibrational temperature. Reproduced with permission from [18]. Copyright 2014 AIP Publishing.

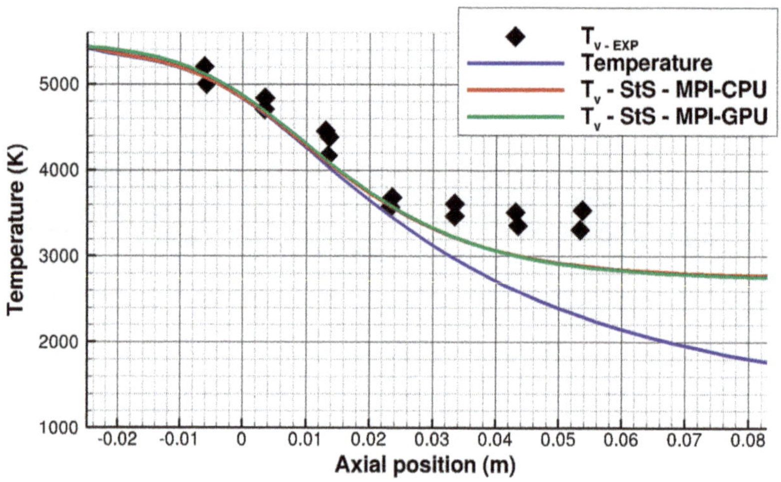

Figure 13.5. Comparison between the axial vibrational temperature profiles obtained with the reference MPI-CPU code and the MPI-CUDA code.

temperature profile along the centerline. In this particular test case, the best result in terms of vibrational temperature (figure 13.6) is obtained with Park's model. It must be pointed out that this model was tuned on these experimental data and this agreement is not a surprise.

Results obtained using the StS and the CAST approaches are comparable, due to the consistency between the data used by the two models, demonstrating that, when constructed on the same dataset, macroscopic and StS models predict the same profile of vibrational temperature.

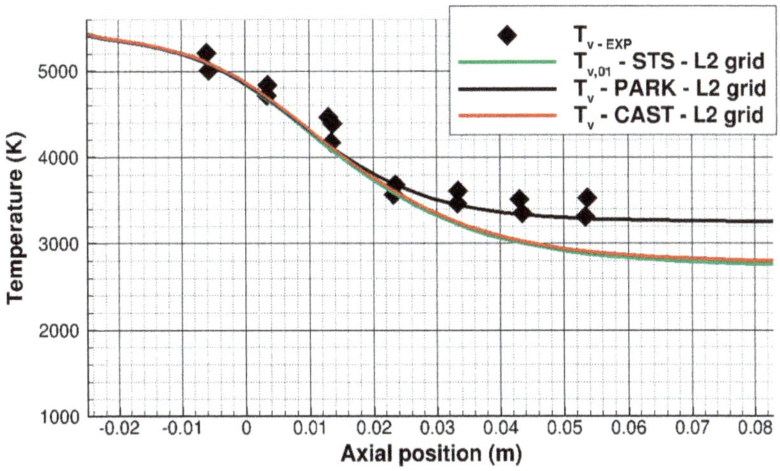

Figure 13.6. Effect of the thermochemical model on vibrational temperature along the centerline. Reproduced with permission from [18]. Copyright 2014 AIP Publishing.

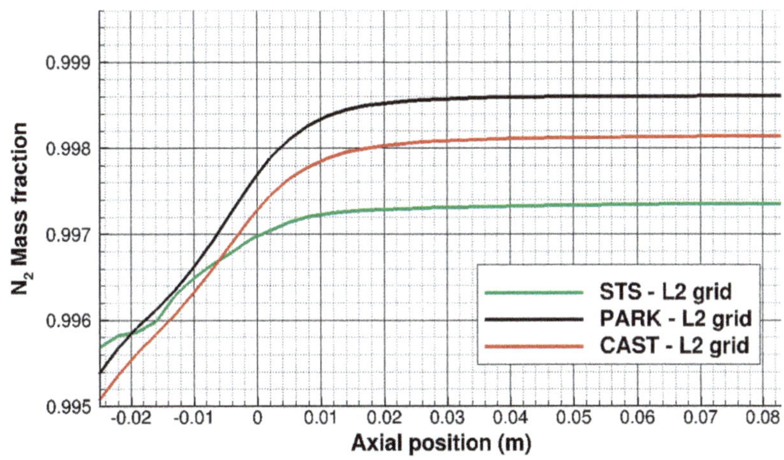

Figure 13.7. Effect of the thermochemical model on the N_2 mass fraction distribution along the centerline. Reproduced with permission from [18]. Copyright 2014 AIP Publishing.

The EAST test case was dedicated to measuring the relaxation of vibrational energy under $N_2 + N_2$ encounters, where large uncertainties are present in the VTm and VV rates at low temperature. These rates are responsible for the relaxation of the low-energy vibrational levels and, as a consequence, the vibrational temperature. The lower temperature predicted by the StS model demonstrate that the VTm considered are larger than the real one. The agreement between macroscopic and StS models is shown in figure 13.8, where the population of the first nine levels of N_2 calculated with the StS model is compared with the experimental results.

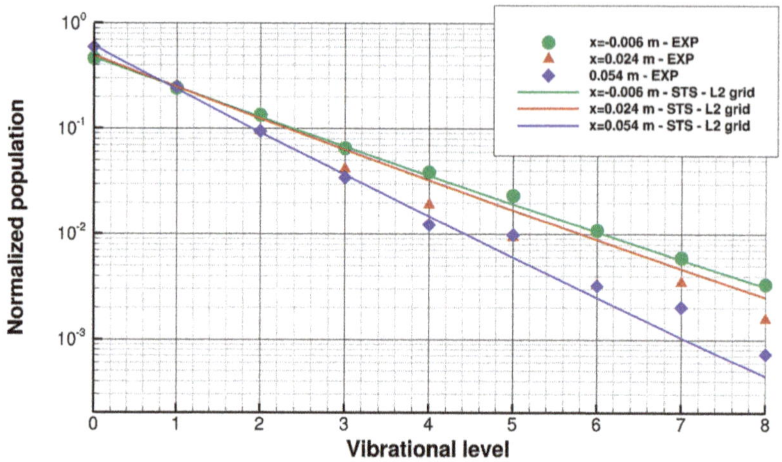

Figure 13.8. Numerical and measured population at $x = -0.6\,\text{cm}$, $2.4\,\text{cm}$ and $5.4\,\text{cm}$. Reproduced with permission from [18]. Copyright 2014 AIP Publishing.

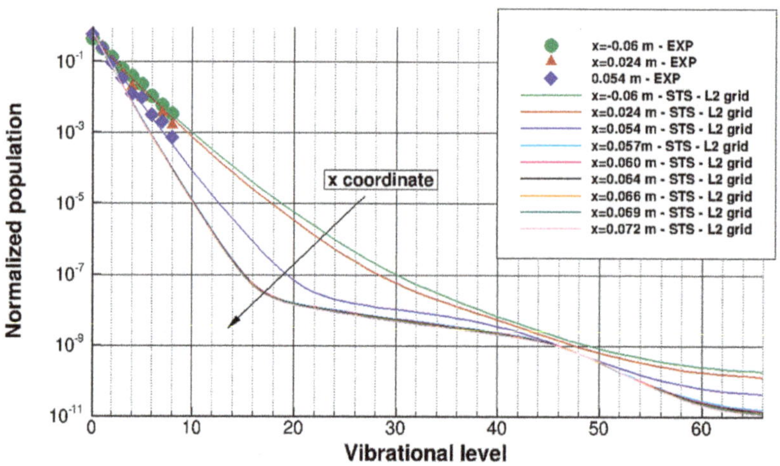

Figure 13.9. Evolution of the vibrational level populations along the centerline. Reproduced with permission from [18]. Copyright 2014 AIP Publishing.

Differences are observed in the compositions (figure 13.7), corresponding to large variations in the atomic molar fractions, N being a minority species (<1%). Nitrogen atoms recombine in the last vibrational level (see equation (13.29)) and then overpopulate the distribution tail as they move toward the nozzle exit (see figure 13.9). Distributions are almost-Boltzmann in the stagnation chamber and close to the wall, but moving from the throat to the exit they strongly deviate towards a non-Boltzmann behavior for high vibrational quantum numbers. As a consequence, in the expansion region, the overpopulated tails enhance the dissociation, resulting in a lower molecular molar fraction predicted by the StS model. Despite the fact that the macroscopic models incorporate a chemistry-to-vibration mechanism that should calibrate the recombination rate in the current experimental conditions, they are based on the hypothesis of a Boltzmann distribution for the vibrational energy levels and cannot account for the overpopulation of the tails. In the presence of higher dissociation this effect becomes more important [19, 55].

GPU calculations

All the results presented in what follows were obtained by solving the Euler equations with the MPI-CUDA code. The GPU implementation makes it possible to also simulate non-equilibrium air mixtures with StS kinetics, requiring excessive computational costs on a CPU, as discussed in the previous example.

As a first test case, air expansion in the EAST nozzle facility was considered for the stagnation conditions summarized in table 13.2 and for the structured computational grid depicted in figure 13.10. In this test case axisymmetric configurations have been simulated.

Figure 13.11 shows the temperature profiles along the axis. The vibrational temperature of nitrogen molecules is much higher than the gas temperature, while almost thermal equilibrium is observed for oxygen. It should be considered that oxygen is mainly present in atomic form, therefore the vibrational distributions are rapidly cooling. But this behavior can lead to the wrong conclusion that oxygen is thermalized at the gas temperature. Looking at the distributions (see figure 13.12(a) and (b)), the tails are much higher for oxygen than for nitrogen due to more relevant atomic recombination in highly excited levels consequence of the higher atomic oxygen concentration. This is the proof that there is no correlation between the low-energy and the distribution tails, as they are affected by different processes.

13.4.2 Hypersonic flows past a sphere

The test case presented in this section concerns a hypersonic flow impinging on a sphere. Thanks to the axisymmetric condition this problem can be solved by considering a 2D configuration see figure 13.13, showing the computational domain along with the boundary conditions and an example of 2 × 4 MPI partitioning.

This problem, based on the experimental set-up of [56, row 18 table 1 and figure 10], was solved for both the StS and Park multi-temperature model, with the following free-stream and geometric conditions: sphere radius R = 7 mm, $u_\infty = 3490\ \mathrm{ms}^{-1}$, $T_\infty = 293$ K and $\rho R = 4 \times 10^{-4}\ \mathrm{kg\ m}^{-2}$.

Table 13.2. Stagnation conditions used in the simulation of the EAST nozzle with an air mixture.

h^0 (J kg^{-1})	p^0 (Pa)	T^0 (K)	ρ (kg m^{-3})	
9.88×10^6	1.016×10^7	5600	5.452	
Y_{N_2}	Y_N	Y_{O_2}	Y_O	Y_{NO}
7.39×10^{-1}	5.27×10^{-3}	4.19×10^{-2}	1.66×10^{-1}	4.78×10^{-2}

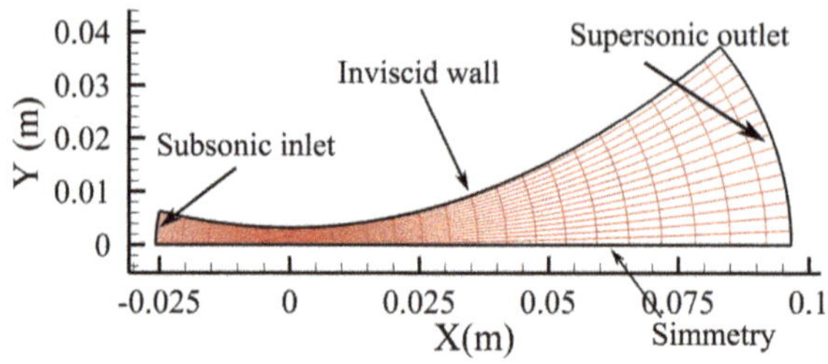

Figure 13.10. The 65 × 47 fluid cell computational grid showing one every three grid lines, and boundary conditions shown.

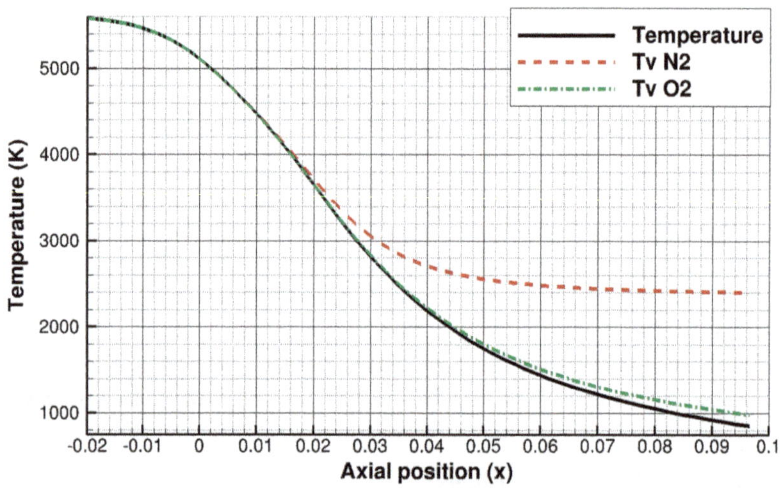

Figure 13.11. Axial temperature profiles for the air flow in the axisymmetric EAST nozzle.

Figure 13.14 compares the temperature contour plots obtained by using both the Park (panel (a)) and the StS (panel (b)) models along with the experimental shock shape [56]. Clearly, the StS model shows better agreement with experiments in terms of stand-off distance.

Figure 13.15(a) reports the temperature profiles along the stagnation line obtained by using the Park and StS models.

The Park model predicts a shorter stand-off distance, resulting from the lower translational temperature, a consequence of the larger dissociation of molecules, in particular for O_2, figure 13.15(b). This difference is the consequence of the non-Boltzmann character of the vibrational distributions obtained by the StS approach, thus pointing out that an accurate description of internal state kinetics also has a relevant impact on macroscopic flow properties.

13.4.3 Code performance analysis

The above results were obtained using the MPI-CUDA implementation [13]. Within this approach 2D fluid-dynamics computations coupled with a vibrationally

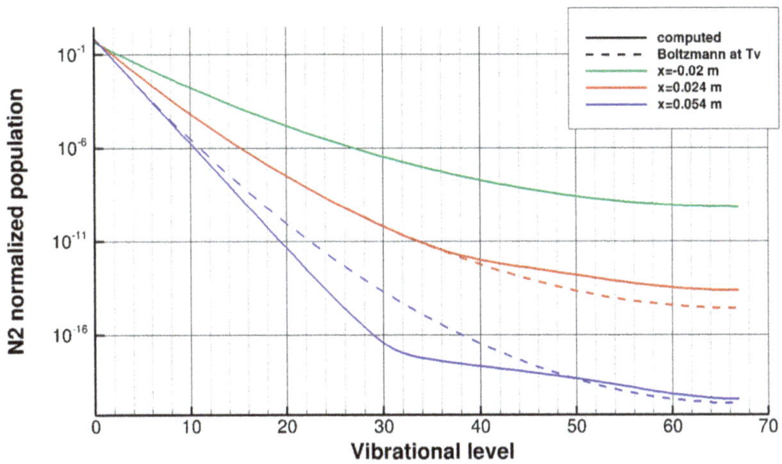

(a) N_2

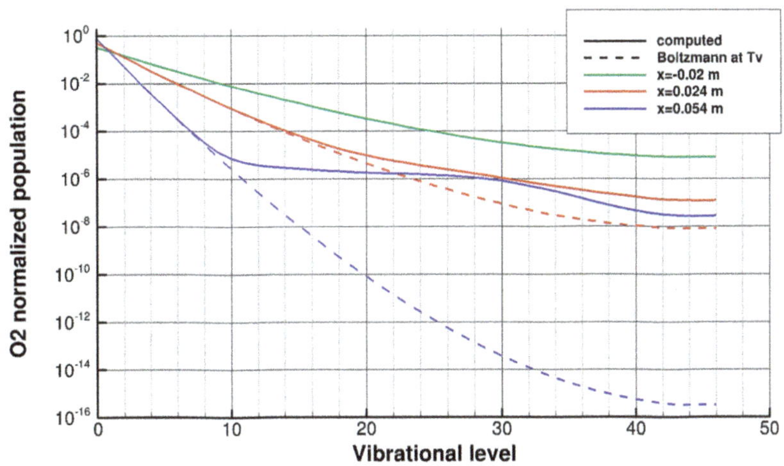

(b) O_2

Figure 13.12. Evolution of the vibrational level populations along the centerline for the axisymmetric nozzle, at $x = -2.0$ cm, 2.4 cm and 5.4 cm: (a) N_2; (b) O_2.

resolved StS air kinetics model can be obtained in a relatively short time on the GPU cluster described in section 13.3. The much better performance of MPI-CUDA compared to MPI-CPU is shown in table 13.3, where the time per iteration (TpI), energy consumption per iteration, speed-up of GPU versus CPU, and energy consumption ratio between CPU and GPU for the case of hypersonic flow past a sphere are reported. Both the StS and Park models were considered by running the parallel simulations on 12 GPUs and on 12 CPUs (72 CPU cores) by varying the grid size.

The use of GPU becomes effective only when the number of fluid cells is large enough. Indeed, looking at the StS simulations, for the first three grids, the time per

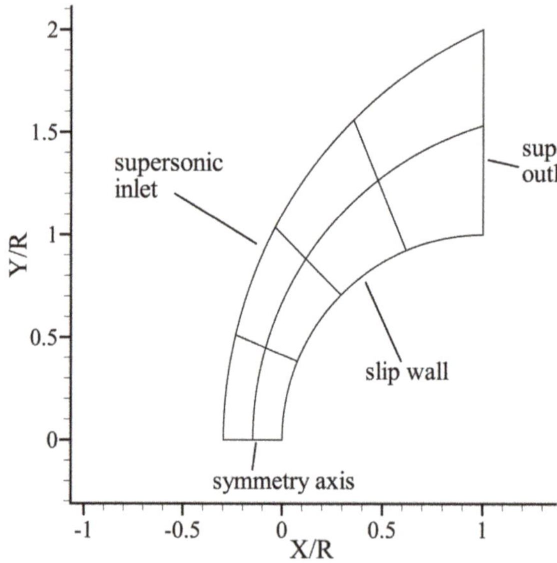

Figure 13.13. Computational domain along with boundary conditions and an example of 2 × 4 MPI partitioning.

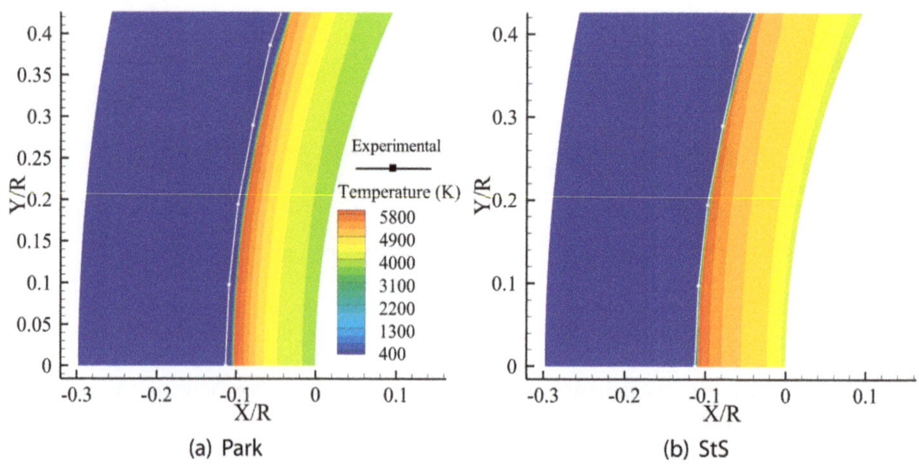

Figure 13.14. CFD temperature contour plots along with experimental shock shape: (a) Park model; (b) StS model.

iteration remains almost constant for GPUs. This is due to the fact that the number of cells is not large enough to saturate the computing power of GPUs, resulting in a small speed-up of GPU versus CPU. However, when all GPU cores are saturated, speed-up values (one GPU versus one core CPU) larger than 100 are reached. Moreover, GPUs show a higher energy efficiency with an energy consumption up to nine times smaller than CPUs when the StS model is employed.

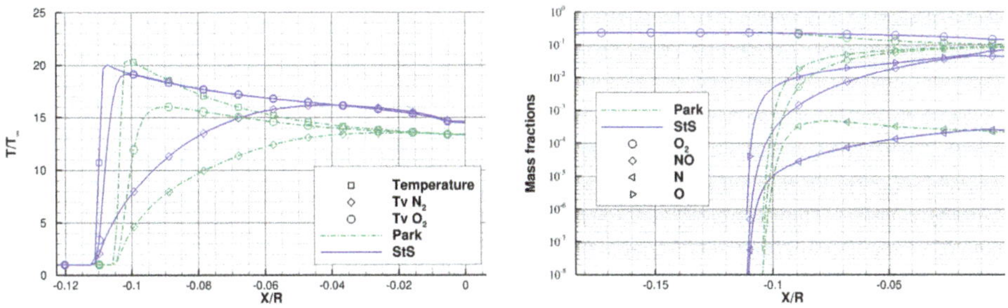

(a) Park (b) StS

Figure 13.15. Stagnation line profiles: (a) temperature profiles; (b) mass fraction profiles.

Table 13.3. Code performance analysis: time per iteration (TpI), energy consumption per iteration, speed-up of GPU versus CPU and energy consumption ratio between CPU and GPU for the case of hypersonic flow past a sphere.

Model	Fluid cells	12 GPUs: TpI (s); energy consumption (J)	12 CPUs: TpI (s); energy consumption (J)	Speed-up (1 GPU versus 1-core-CPU)	CPU/GPU energy ratio
StS	64 × 32	6.33; 1.8×10^4	8.17; 7.8×10^3	1.29 (7.7)	4.33×10^{-1}
	128 × 64	6.36; 1.8×10^4	26.71; 2.56×10^4	4.2 (25.2)	1.42
	256 × 128	6.90; 1.9×10^4	105.9; 10.2×10^4	15.3 (91.8)	5.37
	512 × 256	15.91; 4.5×10^4	419.5; 40.3×10^4	26.4 (158.4)	8.96
	1024 × 512	68.72; 19.4×10^4	1702.1; 163.4×10^4	24.8 (148.8)	8.42
Park	64 × 32	7.50×10^{-3}; 21	1.59×10^{-3}; 1.5	0.21 (1.3)	7.0×10^{-2}
	128 × 64	7.77×10^{-3}; 22	4.55×10^{-3}; 4.3	0.59 (3.5)	1.9×10^{-1}
	256 × 128	7.24×10^{-3}; 20	1.68×10^{-2}; 16	2.32 (13.9)	8×10^{-1}
	512 × 256	1.36×10^{-2}; 38	6.53×10^{-2}; 63	4.8 (28.8)	1.66
	1024 × 512	3.48×10^{-2}; 98	2.46×10^{-1}; 236	7.1 (42.6)	2.41

The same considerations can be made when the Park model is used. However, in this case the speed-up values and energy efficiency of GPUs compared to CPUs are lower than for StS because the Park model is less computationally demanding and it is more difficult to take full advantage of the computing power of GPUs in 2D geometries.

References

[1] Munafò A 2014 Multi-scale models and computational methods for aerothermodynamics *PhD Thesis* CNRS et École Centrale Paris https://tel.archives-ouvertes.fr/tel-00997437/document

[2] Gnoffo P A 1999 Planetary-entry gas dynamics *Annu. Rev. Fluid Mech.* **31** 459–94

[3] Bird G A 1994 *Molecular Gas Dynamics and the Direct Simulation of Gas Flows* (Oxford: Clarendon)

[4] Grad H 1949 On the kinetic theory of rarefied gases *Commun. Pure Appl. Math.* **2** 331–407

[5] Bird G A 1994 *Molecular Gas Dynamics and the Direct Simulation of Gas Flows, Oxford Engineering Science Series* (New York: Oxford University Press)

[6] Anderson J 1988 *Hypersonic and High-Temperature Gas Dynamics* (New York: McGraw-Hill)

[7] Potter D F 2011 Modelling of radiating shock layers for atmospheric entry at Earth and Mars *PhD thesis* The University of Queensland, Brisbane http://cfcfd.mechmining.uq.edu.au/theses/dan-potter-phd-thesis-may-2011.pdf

[8] Schrooyen P 2015 Numerical simulation of aerothermal flows through ablative thermal protection systems *PhD thesis* Université catholique de Louvain, Louvain-la-Neuve https://core.ac.uk/download/pdf/46692835.pdf

[9] Park C 1990 *Nonequilibrium Hypersonic Aerothermodynamics* (New York: Wiley)

[10] Colonna G, Tuttafesta M, Capitelli M and Giordano D 1999 Non-Arrhenius NO formation rate in one-dimensional nozzle airflow *J. Thermophys. Heat Transfer* **13** 372–5

[11] Capitelli M, Celiberto R, Colonna G, Esposito F, Gorse C, Hassouni K, Laricchiuta A and Longo S 2015 *Kinetics Fundamental Aspects of Plasma Chemical Physics* vol 85 (Berlin: Springer)

[12] Colonna G, Pietanza L and D'Ammando G 2016 Self-consistent kinetics *Plasma Modeling: Methods and Applications* ed G Colonna and A D'Angola (Bristol: Institute of Physics Publishing) ch 12

[13] Bonelli F, Tuttafesta M, Colonna G, Cutrone L and Pascazio G 2017 An MPI-CUDA approach for hypersonic flows with detailed state-to-state air kinetics using a GPU cluster *Comput. Phys. Commun.* **219** 178–95

[14] Colonna G, Tuttafesta M, Capitelli M and Giordano D 1999 Non-Arrhenius NO formation rate in one-dimensional nozzle airflow *J. Thermophys. Heat Transfer* **13** 372–5

[15] Colonna G and Capitelli M 2001 The influence of atomic and molecular metastable states in high-enthalpy nozzle expansion nitrogen flows *J. Phys. D: Appl. Phys.* **34** 1812

[16] Colonna G and Capitelli M 2001 Self-consistent model of chemical, vibrational, electron kinetics in nozzle expansion *J. Thermophys. Heat Transfer* **15** 308–16

[17] Giordano D, Bellucci V, Colonna G, Capitelli M, Armenise I and Bruno C 1997 Vibrationally relaxing flow of N past an infinite cylinder *J. Thermophys. Heat Transfer* **11** 27–35

[18] Cutrone L, Tuttafesta M, Capitelli M, Schettino A, Pascazio G and Colonna G 2014 3D nozzle flow simulations including state-to-state kinetics calculation, *AIP Conf. Proc.* **1628** 1154–61

[19] Colonna G, Pietanza L D and Capitelli M 2008 Recombination-assisted nitrogen dissociation rates under nonequilibrium conditions *J. Thermophys. Heat Transfer* **22** 399–406

[20] Guy A, Bourdon A and Perrin M-Y 2013 Consistent multi-internal-temperatures models for nonequilibrium nozzle flows *Chem. Phys.* **420** 15–24

[21] Magin T E, Panesi M, Bourdon A, Jaffe R L and Schwenke D W 2012 Coarse-grain model for internal energy excitation and dissociation of molecular nitrogen *Chem. Phys.* **398** 90–5

[22] Panesi M and Lani A 2013 Collisional radiative coarse-grain model for ionization in air *Phys. Fluids* **25** 057101

[23] Panesi M, Magin T E, Bourdon A, Bultel A and Chazot O 2011 Electronic excitation of atoms and molecules for the FIRE II flight experiment *J. Thermophys. Heat Transfer* **25** 361–74

[24] Liu Y, Panesi M, Sahai A and Vinokur M 2015 General multi-group macroscopic modeling for thermo-chemical non-equilibrium gas mixtures *J. Chem. Phys.* **142** 134109

[25] Munafò A, Liu Y and Panesi M 2015 Modeling of dissociation and energy transfer in shock-heated nitrogen flows *Phys. Fluids* **27** 127101

[26] NVIDIA 2017 CUDA C Programming Guide PG-02829-001-v9.0 https://docs.nvidia.com/cuda/cuda-c-programming-guide/

[27] Sanders J and Kandrot E 2011 *CUDA by Example* (New York: Addison-Wesley)

[28] Green 500 June 2018 https://www.top500.org/green500/lists/2018/06/

[29] TOP 500 June 2018 https://www.top500.org/lists/2018/06/

[30] Oak Ridge National Laboratory (ORNL) Summit https://www.olcf.ornl.gov/summit/

[31] Sierra, Livermore's next advanced technology high performance computing system http://computation.llnl.gov/computers/sierra-advanced-technology-system

[32] Tuttafesta M, Pascazio G and Colonna G 2016 Multi-GPU unsteady 2D flow simulation coupled with a state-to-state chemical kinetics *Comput. Phys. Commun.* **207** 243–57

[33] Colonna G, Tuttafesta M, Capitelli M and Giordano D 2000 NO formation in one dimensional air nozzle flow with state-to-state vibrational kinetics: the influence of $O_2(v)$ + N = NO + O reaction *J. Thermophys. Heat Transfer* **14** 455–6

[34] Colonna G, Tuttafesta M, Capitelli M and Giordano D 1999 Influence on dissociation rates of the state-to-state vibrational kinetics of nitrogen in nozzle expansion, *21th International Symposium on Rarefied Gas Dynamics* **vol 2** Brun R, Campargue R, Gatignol R and Lengrand J-C pp 281–8

[35] Capitelli M, Colonna G and D'Angola A 2011 *Thermodynamics, Atomic, Optical, and Plasma Physics Fundamental Aspects of Plasma Chemical Physics* vol 66 (New York: Springer)

[36] Capitelli M, Celiberto R, Colonna G, Esposito F, Gorse C, Hassouni K, Laricchiuta A and Longo S 2016 *Fundamentals Aspects of Plasma Chemical Physics: Kinetics* (New York: Springer)

[37] Bose D and Candler G V 1996 Thermal rate constants of the $N_2 + O \rightarrow NO + N$ reaction using *ab initio* $^3A''$ and $^3A'$ potential energy surfaces *J. Chem. Phys.* **104** 2825

[38] Bose D and Candler G V 1997 Thermal rate constants of the $O_2 + N \rightarrow NO + O$ reaction based on the $^2A'$ and $^4A'$ potential-energy surfaces *J. Chem. Phys.* **107** 6136

[39] Herzberg G 1963 *Molecular Spectra and Molecular Structure, I. Spectra of Diatomic Molecules* (New York: D. Van Nostrand)

[40] Kenneth Kuan-yun Kuo R A 2012 *Fundamentals of Turbulent and Multiphase Combustion* (Hoboken, NJ: Wiley)

[41] Park C 1993 Review of chemical–kinetic problems of future NASA missions, I: Earth entries *J. Thermophys. Heat Transfer* **7** 385–98

[42] Hao J, Wang J and Lee C 2016 Numerical study of hypersonic flows over reentry configurations with different chemical nonequilibrium models *Acta Astronaut.* **126** 1–10

[43] Park C 1988 Two-temperature interpretation of dissociation rate data for N_2 and O_2, *26th Aerospace Sciences Meeting, Reno, NV, 11–14 January* AIAA-88-0458

[44] Millikan R C and White D R 1963 Systematics of vibrational relaxation *J. Chem. Phys.* **39** 3209

[45] Park C 1985 Problems of rate chemistry in the flight regimes of aeroassisted orbital transfer vehicles *Progress in Astronautics and Aeronautics, Thermal Design of Aeroassisted Orbital Transfer Vehicles* (Reston, VA: American Institute of Aeronautics and Astronautics) pp 511–37

[46] Bonelli F, Tuttafesta M, Colonna G, Cutrone L and Pascazio G 2017 Numerical investigation of high enthalpy flows *Energy Proc.* **126** 99–106
[47] Ran W, Cheng W, Qin F and Luo X 2011 GPU accelerated CESE method for 1D shock tube problems *J. Comput. Phys.* **230** 8797–812
[48] Yee H C 1989 A class of high-resolution explicit and implicit shock-capturing methods *Technical Memorandum* 101088 NASA
[49] Verwer J G 1994 Gauss–Seidel iteration for stiff ODES from chemical kinetics *SIAM J. Sci. Comput.* **15** 1243–50
[50] Bussing T R A and Murman E M 1988 Finite-volume method for the calculation of compressible chemically reacting flows *AIAA J.* **26** 1070–8
[51] Steger J L and Warming R F 1981 Flux vector splitting of the inviscid gasdynamic equations with application to finite-difference methods *J. Comput. Phys.* **40** 263–93
[52] Kim K H, Kim C and Rho O-H 2001 Methods for the accurate computations of hypersonic flows: I. AUSMPW+Scheme. *J. Comput. Phys.* **174** 38–80
[53] van Leer B 1979 Towards the ultimate conservative difference scheme. V. A second-order sequel to Godunov's method *J. Comput. Phys.* **32** 101–36
[54] Sharma S P, Ruffin S M, Gillespie W D and Meyer S A 1993 Vibrational relaxation measurements in an expanding flow using spontaneous Raman scattering *J. Thermophys. Heat Transfer* **7** 697–703
[55] Colonna G, Armenise I, Bruno D and Capitelli M 2006 Reduction of state-to-state kinetics to macroscopic models in hypersonic flows *J. Thermophys. Heat Transfer* **20** 477–86
[56] Nonaka S, Mizuno H, Takayama K and Park C 2000 Measurement of shock standoff distance for sphere in ballistic range *J. Thermophys. Heat Transfer* **14** 225–9

Part III

Elementary processes in hypersonic flows

IOP Publishing

Hypersonic Meteoroid Entry Physics

Gianpiero Colonna, Mario Capitelli and Annarita Laricchiuta

Chapter 14

Thermodynamic and transport properties of reacting air including ablated species

Annarita Laricchiuta, Antonio D'Angola, Fernando Pirani, Lucia Daniela Pietanza, Mario Capitelli and Gianpiero Colonna

The entry of meteoroids into the Earth's atmosphere at high speeds produces a bow shock wave and the high temperatures in the shock layer induce an intense heat flux that melts and vaporizes the body. The shock is then structured in two regions: the 'ablation layer' close to the meteoroid surface and constituted by a vapor in equilibrium with the liquid film at the meteoroid surface, and the air shock layer, separated by an interface, whose thickness depends on the meteoroid's dimensions and its entry conditions (velocity and altitude). Across the layer the temperature changes from around 3000 K at the surface of the body to about 20 000 K at the interface [1], reaching very high temperatures at the shock front.

Any chemical model of meteoric ablation [1–3] should accurately characterize the ablation layer and the interface, deriving the equilibrium composition, the thermodynamic properties and also the transport coefficients for the estimation of the flow characteristics during hypersonic entry, such as friction and surface heat load. The model should describe the transition between the vapor layer, the composition reproducing the elemental fractions characteristic of the meteoroid, and the interface region, where the complexity of the chemistry increases due to the mixing with air components and the properties of the resulting plasma depend on the fraction of the ablated species in the mixture.

The chemical and mineralogical nature of the meteorites (chapter 5) is the basis of their classification and indicates that for chondrites (stony meteorites) the most abundant phases are silicates, producing a differential ablation profile (see figure 4.2 of chapter 4) that shows the dominant ablation of Si, Fe and Mg at an altitude of around 90 km.

In this chapter recent efforts to derive accurate thermodynamic and transport properties of silicon compounds, SiO_2 or SiC, regarded as models for chondritic

doi:10.1088/2053-2563/aae894ch14

meteorites, are reported and the role of ablated silicon species in affecting the properties of air is also investigated, allowing a description of the interface region. The properties are calculated in a wide range of temperatures [3×10^3–5×10^4 K], i.e. using as lower limit the temperature at the melting surface of the meteoroid body. Advanced chemical models are considered, including molecular species such as C_3, O_3, Si_2, Si_3, Si_2N, SiN, NO_2, …, potentially minority species but in some cases important at low temperatures, and also molecular positive and negative ions. The multiply charged atomic ions are included up to the fourth ionization level to ensure the soundness of results for high temperatures, where the plasma is fully ionized.

The calculations are performed with the web-access EquilTheTA tool [4] and core databases, accessed by thermodynamic and transport computational modules, collecting physical–chemical data and transport cross sections for atomic and molecular species. These databases have been extended to include accurate internal partition functions of atomic and molecular silicon-based species and binary collision dynamical information for interactions involving silicon–carbon, silicon–oxygen and silicon–nitrogen compounds.

The thermodynamic and transport properties of plasmas containing silicon-based chemical components represent fundamental information, not only for the simulation of meteoroid thermal ablation during atmosphere entry, but also for the experimental investigation of meteorites. In fact, the composition of the plasma formed in laser-ablation techniques allows, under the assumption of local thermodynamic equilibrium, the reconstruction of synthetic emission spectra that are useful for the elemental analysis of meteorites as well as terrestrial rocks [5–7] through a calibration-free approach [8]. Furthermore, this knowledge offers theoretical support for the design of ablative thermal protection systems for space vehicles [9, 10], as well as arc welding [11] for the production of silica powder.

14.1 The EquilTheTA code

EquilTheTA (EQUILibrium for plasma THErmodynamics and Transport Applications) is a web-based tool [4] which calculates chemical equilibrium product concentrations from any set of reactants and determines the thermodynamic and transport properties for the gas mixture in wide temperature and pressure ranges [12, 13]. The program provides chemical equilibrium, thermodynamic properties and transport coefficients from recent and accurate databases of atomic and molecular energy levels and collision integrals, in the framework of the classical theory of statistical thermodynamics [14] and the Chapman–Enskog theory [15].

In EquilTheTA, the solution of the chemical equilibrium is performed by using a novel, fast, stable and efficient algorithm based on the hierarchical approach [16, 17], which consists in solving one equilibrium equation at a time, after sorting the set of reactions by their decreasing distance from equilibrium, through the *reaction distance*, finding additional reactions to short-cut the path to equilibrium [17]. In the calculations, Debye Hückel corrections to thermodynamic functions and to equilibrium constants [14], and the cutoff for the atomic internal partition functions to avoid divergence, estimated as the largest value between the Fermi and Griem criteria [18], are consistently updated with the composition [19]. The Debye length is calculated considering ions and electrons in the thermodynamic module, while in the transport calculations for the estimation of Coulomb collision integrals only electron density is accounted for. This choice is justified considering that, during the collision, electrons, due to their lower mass, adapt faster than heavy ions to the local potential.

The derivation of transport coefficients relies on the accurate description of the microscopic dynamics of binary collisions in a properly extended temperature range within the Chapman–Enskog method, with a finite Sonine polynomial expansion of the Boltzmann equation, decoupling heavy particles and electrons, and adopting a suitable approximation for different transport coefficients (third-order for the translational thermal conductivity due to free electrons and electrical conductivity, and the first non-vanishing approximation for the translational thermal conductivity of heavy particles and viscosity). The Eucken [20] approximation and the Butler–Brokaw [21] equation have been used for the calculation of, respectively, the internal and reactive thermal conductivities.

14.2 Thermodynamics and equilibrium

The thermodynamic properties in the statistical theory are derived from single-species partition functions [14]:

$$\mathcal{Q}_s = \mathcal{Q}_{\mathrm{tr}}\mathcal{Q}_{\mathrm{int}} = \left[\frac{mk_BT}{2\pi\hbar^2}\right]^{3/2}\frac{Nk_BT}{p} \times \sum_i^{i_{\max}} g_{s,i}e^{-\varepsilon_{s,i}/k_BT}. \tag{14.1}$$

The translation partition function, $\mathcal{Q}_{\mathrm{tr}}$, is derived for a plasma at pressure p in a continuum approximation, while $\mathcal{Q}_{\mathrm{int}}$ is calculated from the internal energy levels $\varepsilon_{s,i}$ for the species, weighted by their statistical weight, $g_{s,i}$. This means that for atomic species the complete spectrum of electronic levels is needed, the number of levels actually included in the summation, $i_{\max}$, depending on the cutoff criterion. For molecules the ro-vibrational levels associated with each bound electronic state also need to be accounted for. For atomic species, the energy levels collected in the NIST database [22] are extended, the completion of series being performed by Ritz–Rydberg formulas [14], avoiding the hydrogenic approximation. For molecular species the spectroscopically detected states are extended with electronic states obtained in *ab initio* electronic structure calculations, including, where available, all the levels correlating with the atomic (neutral or ion) fragments in their ground state and, in a few cases, also states correlating with excited fragments. Moreover, the contribution of the quasi-bound ro-vibrational levels, due to the centrifugal barrier, is accounted for, their effects on the reduced internal specific heat being shown for SiC in [23].

The database of internal levels for O/N/C and for Si/O/C species were already available in EquilTheTA [23–25] and it has been extended to include Si/N chemical systems (Si_2N [22, 26, 27], SiN [22, 28, 29] and SiN^+ [30]) and silicon dimers [22, 31, 32] and trimers [22, 33–35]. A parametric investigation of the equilibrium composition, at constant pressure ($p = 1$ atm), is performed, considering an increasing fraction of silicon dioxide and silicon carbide in air up to pure SiO_2 and SiC plasmas, exploring the range of conditions encountered in the interface region. Table 14.1 presents a complete list of the species included in the chemical model for the two systems, obtained by adding to the air model the chemical species relevant to SiO_2 and SiC and also the chemical species created in the mixing. Five cases, varying the fraction of silicon compound in the initial mixture, are reported in figures 14.1 and 14.2 for SiO_2 and SiC, respectively. The minority species, those with a molar fraction below 10^{-5}, are included in the calculations although they are not displayed in the figures. The case of pure silicon-compound plasmas, already considered in [23] also investigating the pressure dependence, has been updated to include dimers and trimers of carbon and silicon that modify the low-temperature compositions due to the thermodynamic stability of these species. The cases with the addition of ablated species, with different ratios to air components, are particularly tricky in the region below 10^4 K. In fact, depending on the relative fractional ratio of silicon, oxygen, nitrogen and carbon, and on the competitive thermodynamic stability of molecular systems, different oxides are formed.

The electron density is affected by the inclusion of ablated silicon and carbon, characterized by lower ionization potentials with respect to air components ($I_N = 14.53$ eV > $I_O = 13.618$ eV > $I_C = 11.26$ eV > $I_{Si} = 8.15$ eV). Regardless of

Table 14.1. Species included in chemical models for SiO_2–air and SiC–air plasmas.

Air	+ Silicon dioxide	+ Mixed species
N_2 N_2^+	SiO_2 SiO SiO^+	Si_2N SiN SiN^+
O_2 O_2^- O_2^+	Si_3 Si_2 Si_2^+	
NO NO^+	Si Si^{+n}	
N_2O N_2O^+ NO_2		
O_3 O_3^-	**+ Silicon carbide**	**+ Mixed species**
N N^{+n}	SiC SiC^+	SiO_2 SiO SiO^+
O O^{+n} O^-	Si_3 Si_2 Si_2^+	Si_2N SiN SiN^+
Ar Ar^{+n}	C_3 C_2 C_2^+ C_2^-	C_2N C_2O
and electrons	C C^{+n} Si Si^{+n}	CN CN^+ CN^-
		CO_2 CO_2^- CO_2^+
		CO CO^+
		CNO

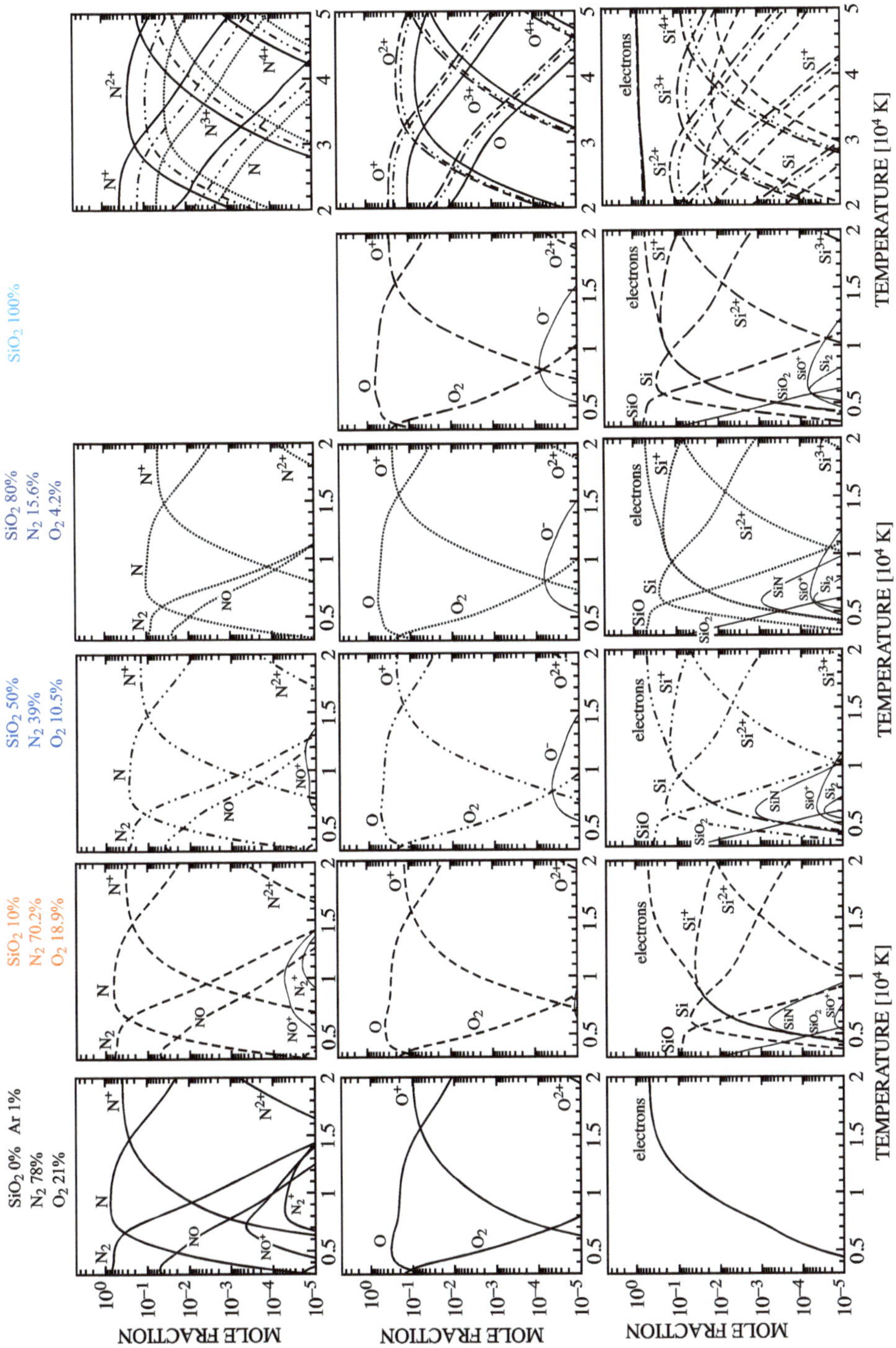

Figure 14.1. Equilibrium composition of SiO_2–air plasmas for varying fractions of SiO_2 in the mixture, at a pressure of 1 atm. The panels from left to right display different fractions of SiO_2, from air to pure SiO_2 plasma. The panels from top to bottom display different chemical species, i.e. nitrogen, oxygen and silicon. The panels on the right-hand side represent the molar fractions of atoms and ion species for temperatures higher than 20 000 K.

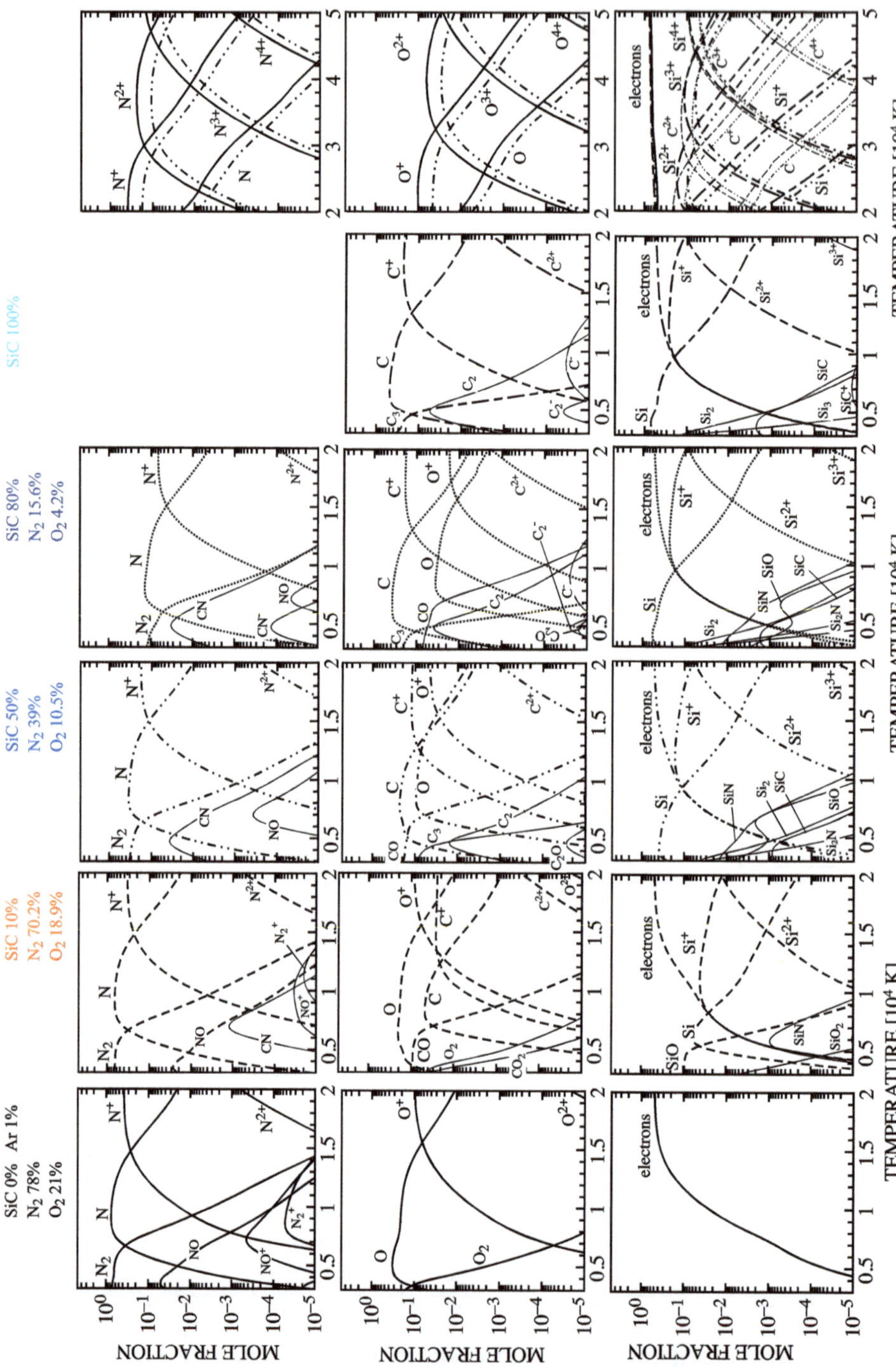

Figure 14.2. Equilibrium composition of SiC–air plasmas for varying fractions of SiC in the mixture, at a pressure of 1 atm. The panels from left to right display different fractions of SiC, from air to pure SiC plasma. The panels from top to bottom display different chemical species, i.e. nitrogen and carbon, oxygen, and silicon. The panels on the right-hand side represent the molar fractions of atoms and ion species for temperatures higher than 20 000 K.

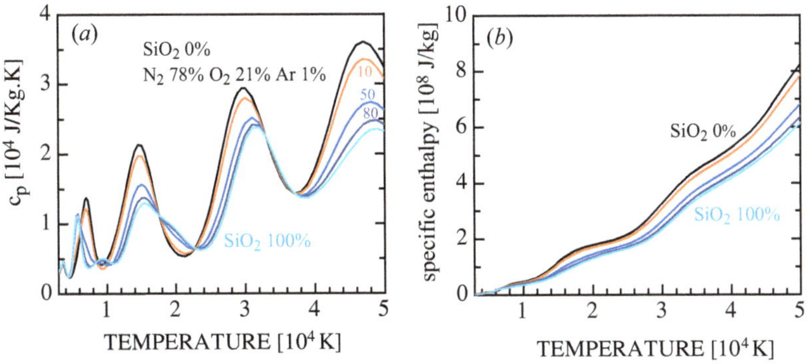

Figure 14.3. (a) Specific heat and (b) specific enthalpy of SiO_2–air plasmas for varying fractions of SiO_2 in the mixture, at a pressure of 1 atm.

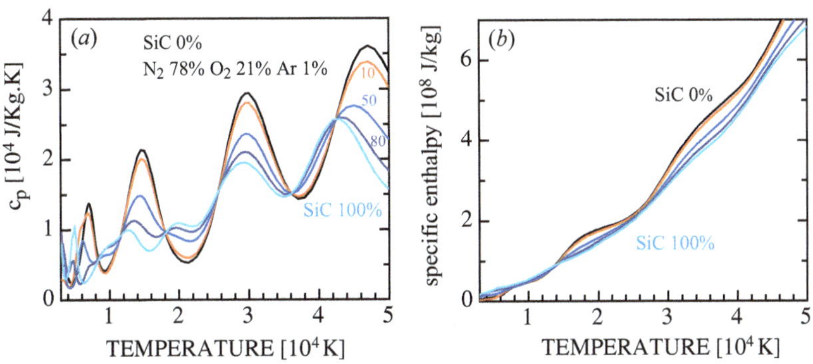

Figure 14.4. (a) Specific heat and (b) specific enthalpy of SiC–air plasmas for varying fractions of SiC in the mixture, at a pressure of 1 atm.

the composition, the molecular ions are minority species. The region above 20 000 K is dominated by atomic ions and the composition is rather simplified and, for that reason, results for the different cases are displayed in separated figures.

The specific heat $c_p = [\partial h / \partial T]_P$ and the specific enthalpy h are reported for the five considered cases in figures 14.3 and 14.4 for the two ablative systems. The peaks in the profile of c_p correspond to inflection points in the composition at the onset of relevant chemical equilibria. In particular, in the case of silicon carbide the temperature profile of the specific heat is significantly different, also affecting the isentropic coefficient γ [19] with respect to the case of air.

14.3 Transport properties

14.3.1 Collision integrals

The fundamental information in the derivation of transport coefficients in the framework of the Chapman–Enskog theory is represented by the collision integrals, describing the elementary elastic collisions governing the transport of mass, energy, momentum and charge [15, 36].

The collision integrals of order (ℓ, s) are defined as

$$\pi\sigma^2\Omega_{i,j}^{(\ell,s)\star} = \frac{\sqrt{2\pi\mu/k_\mathrm{B}T}\,\Omega_{i,j}^{(\ell,s)}}{\frac{1}{2}(s+1)!\left[1 - \frac{1}{2}\frac{1+(-1)^\ell}{1+\ell}\right]} = \frac{4(\ell+1)}{(s+1)![2\ell+1-(-1)^\ell]}\int_0^\infty e^{-\gamma^2}\gamma^{2s+3}Q_{i,j}^{(\ell)}d\gamma, \tag{14.2}$$

where k_B is the Boltzmann constant, μ is the reduced mass of the colliding system, $\gamma^2 = \mu g^2/2k_\mathrm{B}T$ with g the relative collision velocity and σ is the rigid sphere cross section. The transport cross section, $Q^{(\ell)}$, is defined as

$$Q_{i,j}^{(\ell)} = 2\pi\int_0^\infty (1 - \cos^\ell(\vartheta))b\,db, \tag{14.3}$$

with ϑ the deflection angle and b the impact parameter.

The calculation of transport properties needs as input data the collision integrals, of orders depending on the degree of approximation, for all the binary interactions occurring in the plasma.

Heavy-particle interactions

The creation of a complete database of transport cross sections for binary heavy-particle interactions in complex mixtures including large numbers of species is successfully tackled by adopting a hybrid approach that combines the traditional *multi-potential* with the *phenomenological* approach [37, 38]. In the multi-potential approach the effective collision integrals for a given interaction result from the averaging procedure of terms corresponding to each allowed interaction between the two colliding partners, i.e. accounting for the contribution of each electronic state of the quasi-molecular system arising in approaching particles [15]. This approach, although desirable for being the most accurate, even now poses difficulties, particularly when complex mixtures are considered, including somewhat exotic chemical species, requiring knowledge of the complete electronic spectrum of all the relevant molecular systems and relying on computationally demanding quantum mechanical electronic structure methods for the potential energy curves or surfaces. In this context the phenomenological approach is very attractive, allowing the derivation of complete and consistent datasets of collision integrals for possibly any interaction, estimating the interaction potential on a physically sound basis. In fact, the average interaction is modeled by an improved Lennard-Jones (ILJ) potential whose features, such as depth and position of the well, are derived by correlation formulas given in terms of fundamental physical properties of interacting partners (dipole polarizability, charge, number of electrons effective in polarization) [39–43]. The validity of this approach was demonstrated for some benchmark systems relevant to the Earth's atmosphere [37], with results obtained using the ILJ model potential comparing well with those calculated with more accurate methods [44–47]. This approach has been extensively used for the creation of a database for the interactions relevant to planetary atmospheres [12, 13, 25, 48]. Another indubitable advantage of the phenomenological approach is the straightforward derivation of reduced collision integrals, which is accurate in a wide range of temperatures, from a bi-dimensional function that is easily implemented in transport modules [38].

The interactions of the silicon atom and atomic ion with O, N and C have been implemented in the EquilTheTA database within the multi-potential scheme. In table 14.2 the electronic terms predicted by the Wigner and Witmer rules of angular momentum coupling and correlating with the atomic fragments in a given quantum

Table 14.2. Interactions treated with the multi-potential approach in EquilTheTA. Electronic terms correlating with the interacting species (in green the states, out of those theoretically predicted, not included in the calculation) and the references for potential energy curves in the literature are reported.

Interaction	Terms	Reference
$Si(^3P)$–$Si(^3P)$	$^5\Delta_g\ ^3\Delta_u\ ^1\Delta_g$ $^5\Pi_u\ ^5\Pi_g\,^3\Pi_{g,u}\,^1\Pi_{g,u}$ $(1,2)^5\Sigma_g^+\ X^3\Sigma_g^-\ (1,2)^1\Sigma_g^+$ $^5\Sigma_u^-\ (1,2)^3\Sigma_u^+\ ^1\Sigma_u^-$	[31]
$Si(^3P)$–$Si^+(^2P)$	$^4\Delta_{gu}\ ^2\Delta_{gu}$ $(1,2)^4\Pi_g\ ^4\Pi_u\ (2)^4\Pi_u\ (1,2)^2\Pi_{gu}$ $X^4\Sigma_g^-\ (2)^4\Sigma_g^-\ (1,2)^4\Sigma_u^-\ ^4\Sigma_{gu}^+$ $(1,2)^2\Sigma_{gu}^-\ ^2\Sigma_{gu}^+$	[49]
$Si(^3P)$–$C(^3P)$	$^5\Delta\ ^3\Delta\ ^1\Delta$ $(1,2)^5\Pi\ X^3\Pi\ b^1\Pi\ (2)^3\Pi\ (2)^1\Pi$ $(1,2)^5\Sigma^+\ ^5\Sigma^-\ (1,2)^3\Sigma^+ A^3\Sigma^-$ $(c,d)^1\Sigma^+\ ^1\Sigma^-$	[50]
$C(^3P)$–$Si^+(^2P)$	$^4\Delta\ ^2\Delta$ $(1,2)^4\Pi\ (1,2)^2\Pi$ $X^4\Sigma^-\ (2)^4\Sigma^-\ ^4\Sigma^+\ (1,2)^2\Sigma^-$ $^1\Sigma^+$	[51]

(Continued)

$Si(^3P)$–$O(^3P)$	$^5\Delta$ $d^3\Delta$ $D^1\Delta$ $(1,2)^5\Pi$ $b^3\Pi$ $(2)^3\Pi$ $A^1\Pi$ $(2)^1\Pi$ $(1,2)^5\Sigma^+$ $^5\Sigma^-$ $a^3\Sigma^+ c^3\Sigma^+ e^3\Sigma^-$ $X^1\Sigma^+ E^1\Sigma^+ C^1\Sigma^-$	[52–54]
$Si(^3P)$–$N(^4P)$	$^6\Pi$ $b^4\Pi$ $A^2\Pi$ $^6\Sigma^+$ $a^4\Sigma^+$ $X^2\Sigma^+$	[28]
$N(^4P)$–$Si^+(^2P)$	$^5\Pi$ $^3\Pi$ $^5\Sigma^- X^3\Sigma^-$	[30]

state are reported, using the spectroscopic notation together with the references in the literature to the *ab initio* electronic structure calculations of the corresponding potential energy curves. *Ab initio* data are fitted with model potentials (Lennard-Jones, Morse, Hulburt–Hirschfelder, repulsive, …) widely used in the literature. In order to accommodate special features in the interaction potentials, such as secondary minima, shoulders in the repulsive short-range branch or long-range barriers, new functions have been used, characterized by a larger number of parameters and higher flexibility, and then integrated for classical elastic trajectories using a novel recently implemented algorithm that can handle any potential function regardless of the number of extrema [55].

All the other interactions have been modeled within the phenomenological approach. The polarizability values for atomic silicon species have been taken from the literature or estimated through an empirical formula [56] while for polyatomic species an 'effective atomic polarizability-in-molecule' approach [57–59] has been adopted (see table 14.3).

To give an example of the results obtained, in figure 14.5 the collision integrals obtained for Si–Si within the multi-potential approach and for SiO–SiO modeling the interaction with the phenomenological potential are shown.

In the case of ion–parent-atom interactions the odd-order collision integrals must also account for the contribution of the inelastic resonant charge-transfer process [15], the effective diffusion-type collision integrals being defined as $\Omega^{(1,1)\star}_{\rm eff} = [(\Omega^{(1,1)\star}_{\rm el})^2 + (\Omega^{(1,1)\star}_{\rm ex})^2]^{1/2}$. For the $Si(^3P)$–$Si^+(^2P)$ interaction the accurate close-coupling charge-exchange cross section [63], also well predicted by the asymptotic approach [64], has been used for the derivation of the inelastic collision integral with a closed formula [23] (figure 14.6). Analogously, the resonant charge-exchange in the Si–Si^{2+} interaction has been estimated in the framework of asymptotic theory [23], under the hypothesis that the exchange interaction potential

Table 14.3. Polarisability values for silicon compounds. The references for data in the literature are provided, while other values have been estimated.

Species	α_{pol} [Å^3]	Species	α_{pol} [Å^3]
Si	5.38 [60]	Si_2	12.58 [61]
Si^+	2.91 [60]	Si_2^+	8.29
Si^{2+}	1.095	Si_3	15.66 [61]
Si^{3+}	0.76	Si^{4+}	0.0435
SiC	6.63	SiN	3.861 [62]
SiC^+	4.31	SiN^+	3.763 [62]
SiO	4.49	SiO^+	4.095
SiO_2	6.09	Si_2N	9.241

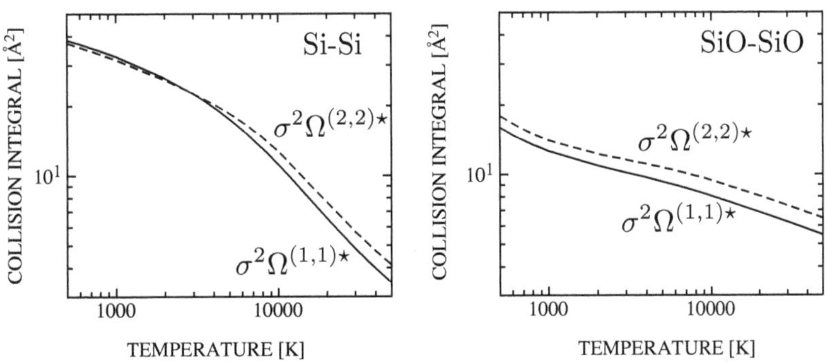

Figure 14.5. Diffusion-type (solid line), $(\ell, s) = (1, 1)$, and viscosity-type (dashed line), $(\ell, s) = (2, 2)$, collision integrals for (left) Si(^3P)–Si(^3P) (multi-potential approach) and (right) SiO–SiO (phenomenological approach) interactions.

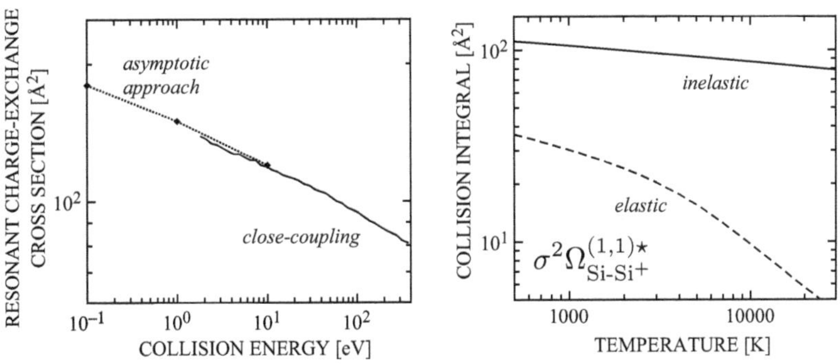

Figure 14.6. Left: Resonant charge-transfer cross section in Si–Si$^+$ collisions by the close-coupling (solid line) [63] and asymptotic approach (dotted line) [64]. Right: Diffusion-type collision integral, $\sigma^2\Omega^{(1,1)\star}$, for the Si–Si$^+$ interaction. Solid line: contribution due to resonant charge transfer; dashed line: elastic contribution.

could be calculated by setting in place of the ionization potential the corresponding value for double ionization of the atom [15, 25]. In principle, also the interactions involving molecular ions and their parent molecules, i.e. SiC–SiC$^+$, SiO–SiO$^+$ or SiN–SiN$^+$, are affected by the resonant charge-exchange process. However, the information on the resonant cross sections for these systems is not available and, moreover, as is evident in figures 14.3 and 14.4, the molecular ions remain minority species in the whole temperature range, the dissociation occurring at lower temperatures than the ionization. Therefore it is expected that neglecting these resonant contributions should not significantly affect the results for the mixture.

Electron–neutral and charged-species interactions

In considering the interaction of electrons with neutrals, quantum effects such as resonances, due to the capture of the electron with the formation of temporary-living negative ions, or the low-energy minimum in the electron scattering due to the Ramsauer effect, must be accounted for. This is the case for electron–silicon scattering, exhibiting a Ramsauer minimum located at ~0.7 eV. Therefore the transport cross section, $Q_{e,j}^{(\ell)}$, has been calculated from theoretical differential cross sections for elastic electron scattering, $\sigma_{e,j}(E, \vartheta)$, obtained in [65] in the optical potential model and extended in the low-energy region for the accurate estimation of collision integrals at low temperatures [23]

$$Q_{e,j}^{(\ell)} = 2\pi \int_0^{\pi} (1 - \cos^{\ell}(\vartheta))\sigma_{e,j}(E, \vartheta)\sin\vartheta \, d\vartheta. \tag{14.4}$$

The temperature profile of collision integrals for the e–Si(^3P) interaction is displayed in figure 14.7 [23], showing a satisfactory agreement with the $\sigma^2\Omega^{(1,1)\star}$ values resulting from the fitting relation given in [66].

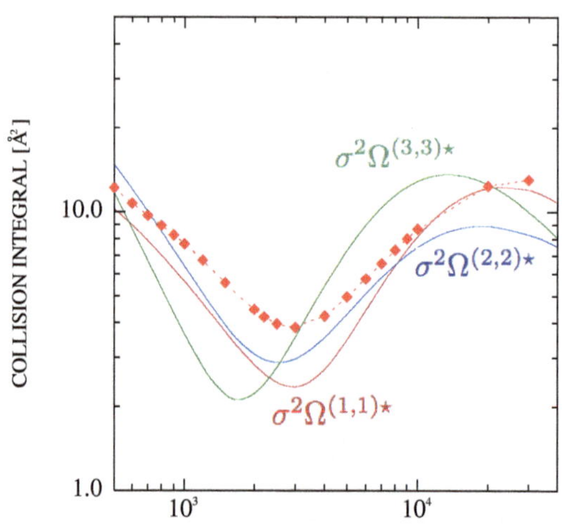

Figure 14.7. Collision integrals up to $(\ell, s) = (3, 3)$ for e–Si(^3P) interaction, compared with theoretical results for $\sigma^2\Omega^{(1,1)\star}$ from [66] (solid diamonds). Reproduced with permission from [23]. Copyright 2018 Institute of Physics Publishing.

Information on the elastic scattering of electrons from silicon molecules is scarce in the literature. However, results for silicon compounds are close to those for atomic silicon and this is also confirmed by accurate calculations of transport cross sections for SiN_2 molecules [67], for this reason the collision integrals for the interaction of electrons with silicon molecules have been approximated by the electron–Si values.

Charged-particle interactions have been modeled assuming a screened Coulomb potential [15, 18].

14.3.2 Transport coefficients

Transport coefficients for the cases of silicon-compound–air plasmas discussed in section 14.2 are reported in figure 14.8.

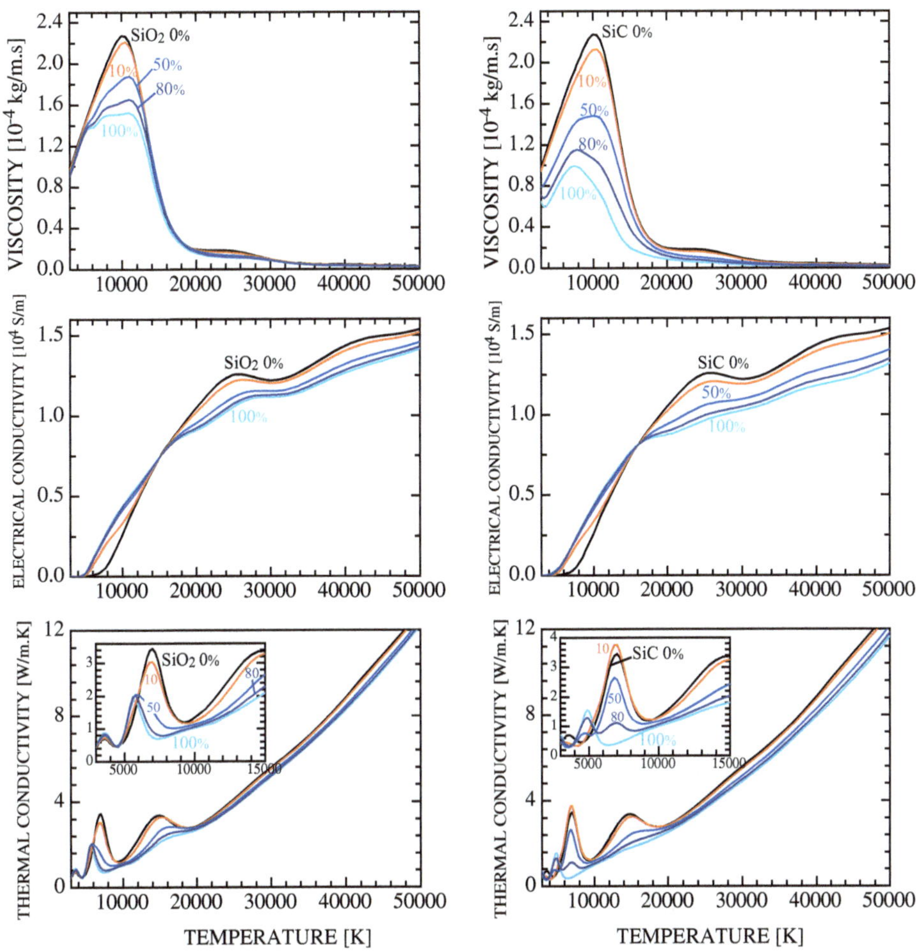

Figure 14.8. Viscosity, electrical conductivity and total thermal conductivity for (left) SiO_2 and (right) SiC–air plasmas for varying fractions of SiO_2 and SiC in the mixture, at a pressure of 1 atm. In the bottom panels the region of $T \leqslant 1.5 \times 10^4$ is magnified.

The analysis starts from the properties of pure SiO_2 and SiC plasmas, the results being updated with respect to [23]. The chemical model for silicon carbide plasma has been extended to include minority species and silicon and carbon trimers. These molecular aggregates dominate the equilibrium composition in the low-temperature region and modify the dissociation equilibria leading to atomic silicon and carbon, reflected in the reactive thermal conductivity and in turn in the total thermal conductivity profile exhibiting two peaks located around 3000 K and 5000 K, associated with the formation of Si and C from Si_3 and C_3. The effect of the inclusion of minority species in the SiO_2 plasma was negligible, the equilibrium composition being dominated by silicon oxides. However, differences in the viscosity profile of pure SiO_2 with respect to that in [23] arise due to the updating of the collision integrals for the Si–O interaction, now including the quintet molecular terms predicted by the momentum coupling (see table 14.2). The interesting aspect is that the newly calculated collision integrals are closer to the values estimated using the phenomenological approach than was expected on the basis of the previous multi-potential results, obtained including only singlets and triplets.

In the case of increasing fractions of silicon compounds in air, the properties change significantly, giving in general larger deviations for silicon carbide.

The viscosity peak, located for air plasma at 10 000 K, is lowered by a factor two in the SiC case and this is largely determined by the anticipated ionization in the presence of silicon atoms, leading to different dominant interactions and to the formation of electrons that do not contribute to viscosity.

The electrical conductivity is consistently higher in the low-temperature region, due to the more efficient ionization in silicon-contaminated plasmas. However, for temperatures above 15 000 K the trend is inverted. In this region the plasma is almost fully ionized and the collisional term is governed by Coulomb interactions. When silicon is present in the mixture the second ionization becomes relevant at these temperatures, thus leading to higher values of collision integrals that hinder the electron diffusion, reducing the electrical conductivity.

The total thermal conductivity presents the most interesting features. In fact, the varying peak structure in the temperature profile below 15 000 K, determined by the reactive component, shows a significant modification of the chemical species relevant to the mixture in the different cases considered, depending on the thermodynamic stability and on the favoring of stoichiometry in the ratio between elements. In going from pure air to pure SiO_2, one observes a gradual transition from a profile dominated by the dissociation of molecular nitrogen, with the peak at 7000 K, to a less pronounced peak located at 5800 K, due to the dissociation equilibrium of SiO. The cases with silicon carbide are more complex. In fact, the addition of a small fraction of silicon (10%) affects the height of the $N_2 \rightarrow 2N$ dissociation peak due to the additional reactive contribution of carbon monoxide dissociation, relevant in the same temperature region. A decrease of the small peak around 3700 K, attributed to O_2 dissociation, is observed and consistent with the formation of stable carbon and silicon oxides at low temperatures. A further increase of the silicon carbide fraction leads to the complete disappearance of the oxygen peak and the appearance of a new peak at a slightly higher temperature, that could be attributed to the dissociation of carbon trimers. In fact, when carbon and silicon become the relevant species in the plasma, the favored low-temperature species is the carbon trimer. The peak associated with the dissociation of CO and N_2 is correspondingly reduced, up to the case of pure SiC plasma, where the only significant feature is the $C_3 \rightarrow 3C$ peak at 5000 K.

14.4 Conclusions

The modeling of the hypersonic entry of meteoroids into Earth's atmosphere in ablative regimes requires the characterization of the thermodynamics and transport properties of the plasma formed in the shock layer, including the ablated species. In this chapter a parametric investigation is performed, deriving the equilibrium composition and properties using the EquilTheTA tool, in the case of different ratios of silicon and air components from pure SiO_2 and SiC to air plasmas.

The core databases of EquilTheTA have been extended to include the internal partition functions of atomic and molecular silicon species, and for collision integrals describing the interactions involving silicon species, with a hybrid multi-potential/phenomenological approach and accounting for the resonant charge-exchange processes.

It should be stressed that in high pressure regimes the excited states of atoms could also be relevant in determining the transport coefficients, as demonstrated for hydrogen and air plasmas [15, 68, 69]. The inclusion of excited states requires their dynamical characterization, i.e. deriving the dependence of collision integrals on the quantum state of colliders, in particular the resonant charge-transfer contribution [68, 70, 71].

The silicon compounds are considered here as a model of ablated species of chondrites, however, a further step will entail the inclusion of iron and magnesium oxides, which are expected to have effects, in particular in the ionization regime, being the metal atoms characterized by lower ionization potentials.

References

[1] Park C 2013 Rosseland mean opacities of air and H-chondrite vapor in meteor entry problems *J. Quant. Spectrosc. Radiat. Transfer* **127** 158–64

[2] Dias B, Bariselli F, Turchi A, Frezzotti A, Chatelain P and Magin T 2016 Development of a melting model for meteors *AIP Conf. Proc.* **1786** 160004

[3] Vondrak T, Plane J M C, Broadley S and Janches D 2008 A chemical model of meteoric ablation *Atmos. Chem. Phys.* **8** 7015–31

[4] EquilTheTA 2018 EquilTheTA http://phys4entrydb.ba.imip.cnr.it/EquilTHETA/

[5] Li W, Gao L, Ma Z and Wang F 2017 Ablation behavior of graphite/SiO_2 composite irradiated by high-intensity continuous laser *J. Eur. Ceram. Soc.* **37** 1331–38

[6] Senesi G S, Tempesta G, Manzari P and Agrosì G 2016 An innovative approach to meteorite analysis by laser-induced breakdown spectroscopy *Geostand. Geoanal. Res.* **40** 533–41

[7] Dell'Aglio M, De Giacomo A, Gaudiuso R, De Pascale O and Longo S 2014 Laser induced breakdown spectroscopy of meteorites as a probe of the early solar system *Spectrochim. Acta* B **101** 68–75

[8] Tognoni E, Cristoforetti G, Legnaioli S and Palleschi V 2010 Calibration-free laser-induced breakdown spectroscopy: state of the art *Spectrochim. Acta* B **65** 1–14

[9] Scoggins J B, Rabinovitch J, Barros-Fernandez B, Martin A, Lachaud J, Jaffe R L, Mansour N N, Blanquart G and Magin T E 2017 Thermodynamic properties of carbon–phenolic gas mixtures *Aerospace Sci. Technol.* **66** 177–92

[10] Monteverde F, Cecere A and Savino R 2017 Thermo-chemical surface instabilities of SiC–ZrB_2 ceramics in high enthalpy dissociated supersonic airflows *J. Eur. Ceram. Soc.* **37** 2325–41

[11] Murphy A B 2010 The effects of metal vapour in arc welding *J. Phys. D: Appl. Phys.* **43** 434001

[12] D'Angola A, Colonna G, Bonomo A, Bruno D, Laricchiuta A and Capitelli M 2012 A phenomenological approach for the transport properties of air plasmas *Eur. Phys. J.* D **66** 1–6

[13] Colonna G, D'Angola A, Laricchiuta A, Bruno D and Capitelli M 2013 Analytical expressions of thermodynamic and transport properties of the Martian atmosphere in a wide temperature and pressure range *Plasma Chem. Plasma Process.* **33** 401–31

[14] Capitelli M, Colonna G and D'Angola A 2011 *Thermodynamics Fundamental Aspects of Plasma Chemical Physics* vol 66 (Berlin: Springer)

[15] Capitelli M, Bruno D and Laricchiuta A 2013 *Transport Fundamental Aspects of Plasma Chemical Physics* vol 74 (Berlin: Springer)

[16] Colonna G 2007 Improvements of hierarchical algorithm for equilibrium calculation *Comput. Phys. Commun.* **177** 493–9

[17] Colonna G and D'Angola A 2004 A hierarchical approach for fast and accurate equilibrium calculation *Comput. Phys. Commun.* **163** 177–90

[18] D'Angola A, Colonna G, Gorse C and Capitelli M 2008 Thermodynamic and transport properties in equilibrium air plasmas in a wide pressure and temperature range *Eur. Phys. J.* D **46** 129–50

[19] Colonna G, D'Angola A and Capitelli M 2012 Electronic excitation and isentropic coefficients of high temperature planetary atmosphere plasmas *Phys. Plasmas* **19** 072115

[20] Eucken A 1913 Über das Wärmeleitvermögen, die spezifische Wärme und die innere Reibung der Gase *Phys.* Z **14** 324–32

[21] Butler J N and Brokaw R S 1957 Thermal conductivity of gas mixtures in chemical equilibrium *J. Chem. Phys.* **26** 1636–43

[22] NIST 2018 NIST Atomic Spectra Database https://www.nist.gov/pml/atomic-spectra-database.

[23] Colonna G, D'Angola A, Pietanza L D, Capitelli M, Pirani F, Stevanato E and Laricchiuta A 2018 Thermodynamic and transport properties of plasmas including silicon-based compounds *Plasma Sources Sci. Technol.* **27** 015007

[24] Capitelli M, Colonna G, Giordano D, Marraffa L, Casavola A, Minelli P, Pagano D, Pietanza L and Taccogna F 2005 High-temperature thermodynamic properties of Mars-atmosphere components *J. Spacecr. Rockets* **42** 980–89

[25] Laricchiuta A *et al* 2009 High temperature Mars atmosphere. Part I: transport cross sections *Eur. Phys. J.* D **54** 607–12

[26] Brugh D J and Morse M D 1997 Resonant two-photon ionization spectroscopy of the 13-electron triatomic Si_2N *Chem. Phys. Lett.* **267** 370–76

[27] Owusu-Ansah E, Wang Y and Shi Y 2016 A theoretical study of the structures and electronic transitions of small silicon nitride clusters (Si_nN_m, $n + m \leqslant 4$) *J. Mol. Spectrosc.* **330** 200–10

[28] Xing W, Shi D, Sun J and Zhu Z 2013 Theoretical study on thirteen Λ-S states of the SiN radical: potential energy curves, spectroscopic parameters and spin–orbit couplings *Eur. Phys. J.* D **67** 228

[29] Cai Z-L, Martin J and François J-P 1998 *Ab initio* study of the electronic spectrum of the SiN radical *J. Mol. Spectrosc.* **188** 27–36

[30] Liu Y, Zhai H and Liu Y 2015 Extensive *ab initio* calculation on low-lying excited states of SiN^+ cation including spin–orbit coupling *Eur. Phys. J.* D **69** 59

[31] Peyerimhoff S D and Buenker R J 1982 Potential energy curves and transition moments for the low-lying electronic states of the Si_2 molecule *Chem. Phys.* **72** 111–18

[32] Bruna P J, Peyerimhoff S D and Buenker R J 1980 Theoretical prediction of the potential curves for the lowest-lying states of the isovalent diatomics CN^+, Si_2, SiC, CP^+, and SiN^+ using the *ab initio* MRD-CI method *J. Chem. Phys.* **72** 5437–45

[33] McCarthy M and Thaddeus P 2003 Rotational spectrum and structure of Si_3 *Phys. Rev. Lett.* **90** 213003

[34] Xu C, Taylor T R, Burton G R and Neumark D M 1998 Vibrationally resolved photoelectron spectroscopy of silicon cluster anions $Si_n^-(n = 3–7)$ *J. Chem. Phys.* **108** 1395–406

[35] McMichael Rohlfing C and Raghavachari K 1992 Electronic structures and photoelectron spectra of Si_3^- and Si_4^- *J. Chem. Phys.* **96** 2114–17

[36] Hirschfelder J O, Curtiss C F and Bird R B 1964 *Molecular Theory of Gases and Liquids* (New York: Wiley)

[37] Capitelli M, Cappelletti D, Colonna G, Gorse C, Laricchiuta A, Liuti G, Longo S and Pirani F 2007 On the possibility of using model potentials for collision integral calculations of interest for planetary atmospheres *Chem. Phys.* **338** 62–8

[38] Laricchiuta A, Colonna G, Bruno D, Celiberto R, Gorse C, Pirani F and Capitelli M 2007 Classical transport collision integrals for a Lennard-Jones like phenomenological model potential *Chem. Phys. Lett.* **445** 133–39

[39] Pirani F, Maciel G S, Cappelletti D and Aquilanti V 2006 Experimental benchmarks and phenomenology of interatomic forces: open-shell and electronic anisotropy effects *Int. Rev. Phys. Chem.* **25** 165–99

[40] Pirani F, Cappelletti D and Liuti G 2001 Range, strength and anisotropy of intermolecular forces in atom–molecule systems: an atom-bond pairwise additivity approach *Chem. Phys. Lett.* **350** 286–96

[41] Cambi R, Cappelletti D, Liuti G and Pirani F 1991 Generalized correlations in terms of polarizability for van der Waals interaction potential parameter calculations *J. Chem. Phys.* **95** 1852–61

[42] Cappelletti D, Liuti G and Pirani F 1991 Generalization to ion–neutral systems of the polarizability correlations for interaction potential parameters *Chem. Phys. Lett.* **183** 297–303

[43] Liuti G and Pirani F 1985 Regularities in van der Waals forces: correlation between the potential parameters and polarizability *Chem. Phys. Lett.* **122** 245–50

[44] Stallcop J R, Partridge H and Levin E 1991 Resonance charge transfer, transport cross sections, and collision integrals for $N^+(^3P)$–$N(^4S)$ and $O^+(^4S)$–$O(^3P)$ interactions *J. Chem. Phys.* **95** 6429–39

[45] Stallcop J R, Partridge H, Pradhan A and Levin E 2000 Potential energies and collision integrals for interactions of carbon and nitrogen atoms *J. Thermophys. Heat Transfer* **14** 480–88

[46] Stallcop J R, Partridge H and Levin E 2001 Effective potential energies and transport cross sections for atom–molecule interactions of nitrogen and oxygen *Phys. Rev.* A **64** 1–12 042722
[47] Levin E and Wright M J 2004 Collision integrals for ion–neutral interactions of nitrogen and oxygen *J. Thermophys. Heat Transfer* **18** 143–47
[48] Bruno D *et al* 2010 Transport properties of high-temperature Jupiter atmosphere components *Phys. Plasmas* **17** 112315
[49] Bruna P J, Petrongolo C, Buenker R J and Peyerimhoff S D 1981 Theoretical prediction of the potential curves for the lowest-lying states of the CSi^+ and Si_2^+ molecular ions *J. Chem. Phys.* **74** 4611–20
[50] Sefyani F L and Schamps J 1994 Theoretical oscillator strengths for infrared electronic transitions of Si_2 and SiC *Astrophys. J.* **434** 816–23
[51] Pramanik A, Chakrabarti S and Das K K 2007 Theoretical studies of the electronic spectrum of SiC^+ *Chem. Phys. Lett.* **450** 221–27
[52] Bauschlicher C W Jr. 2016 The low-lying electronic states of SiO *Chem. Phys. Lett.* **658** 76–9
[53] Oddershede J and Elander N 1976 Spectroscopic constants and radiative lifetimes for valence-excited bound states in SiO *J. Chem. Phys.* **65** 3495–505
[54] Chattopadhyaya S, Chattopadhyay A and Das K K 2003 Configuration interaction study of the low-lying electronic states of silicon monoxide *J. Phys. Chem.* A **107** 148–58
[55] Colonna G and Laricchiuta A 2008 General numerical algorithm for classical collision integral calculation *Comput. Phys. Commun.* **178** 809–16
[56] Alagia M, Brunetti B, Candori P, Falcinelli S, Teixidor M M, Pirani F, Richter R, Stranges S and Vecchiocattivi F 2004 Low-lying electronic states of HBr^{2+} *J. Chem. Phys.* **120** 6985–91
[57] van Duijnen P T and Swart M 1998 Molecular and atomic polarizabilities: Thole's model revisited *J. Phys. Chem.* A **102** 2399–407
[58] Ewig C S, Waldman M and Maple J R 2002 *Ab initio* atomic polarizability tensors for organic molecules *J. Phys. Chem.* A **106** 326–34
[59] Gavezzotti A 2003 Calculation of intermolecular interaction energies by direct numerical integration over electronic densities. An improved polarization model and the evaluation of dispersion and repulsion energies *J. Phys. Chem.* B **107** 2344–53
[60] Hati S and Datta D 1995 Electronegativity and static electric dipole polarizability of atomic species. A semiempirical relation *J. Phys. Chem.* **99** 10742–46
[61] Jellinek J 1999 *Theory of Atomic and Molecular Clusters: With a Glimpse at Experiments* (Berlin: Springer)
[62] NIST 2018 NIST Computational Chemistry Comparison and Benchmark DataBase https://cccbdb.nist.gov/
[63] Sakabe S and Izawa Y 1991 Cross sections for resonant charge transfer between atoms and their positive ions: collision velocity $\leqslant 1$ au *At. Data Nucl. Data Tables* **49** 257–314
[64] Smirnov B M 2001 Atomic structure and the resonant charge exchange process *Phys. Usp.* **44** 221–53
[65] Srivastava R and Williamson W Jr 1989 Differential and total cross sections for the elastic scattering of 1–1000 eV electrons from silicon using the optical model *J. Appl. Phys.* **65** 908–13
[66] André P, Bussiere W and Rochette D 2007 Transport coefficients of Ag–SiO_2 plasmas *Plasma Chem. Plasma Process.* **27** 381–403
[67] Fujimoto M, Michelin S, Mazon K T, Santos A, Oliveira H and Lee M-T 2007 Comparative study of elastic electron collisions on the isoelectronic SiN_2, SiCO, and CSiO radicals *Phys. Rev.* A **76** 012709

[68] Capitelli M, Celiberto R, Gorse C, Laricchiuta A, Pagano D and Traversa P 2004 Transport properties of local thermodynamic equilibrium hydrogen plasmas including electronically excited states *Phys. Rev.* E **69** 026412

[69] Capitelli M 1977 Transport properties of partially ionized gases *J. Phys. Coll.* **38** C3–227

[70] Kosarim A, Smirnov B, Capitelli M, Celiberto R and Laricchiuta A 2006 Resonant charge exchange involving electronically excited states of nitrogen atoms and ions *Phys. Rev.* A **74** 062707

[71] Eletskii A, Capitelli M, Celiberto R and Laricchiuta A 2004 Resonant charge exchange and relevant transport cross sections for excited states of oxygen and nitrogen atoms *Phys. Rev.* A **69** 042718

IOP Publishing

Hypersonic Meteoroid Entry Physics

Gianpiero Colonna, Mario Capitelli and Annarita Laricchiuta

Chapter 15

Electron–molecule processes

Roberto Celiberto, Ratko K Janev, Vincenzo Laporta, Annarita Laricchiuta Zsolt J Mezei, Ioan F Schneider, Jonathan Tennyson and Jogindra M Wadehra

This chapter is intended to give a compendious and comprehensive account of the theoretical description of electron–molecule interactions in molecular plasmas. These processes play a major role, particularly in low-temperature non-equilibrium plasmas, where the population of higher internal quantum states of the molecules may exceed the Boltzmann distribution. In particular, in many plasma conditions of practical interest, electron impact with vibrationally excited molecules can play a critical role in determining the plasma features. *Ab initio* models of such systems require large sets of state-to-state cross section data, which can efficiently be produced by resorting to a wide class of theoretical methods. In the following we will briefly review the existing theoretical cross section data characterizing the electron impact with molecules of interest in meteoroid and spacecraft entry physics.

A conventional classification of electron–molecule processes distinguishes between resonant and non-resonant scattering. In the former case, which occurs mainly at low collision energies, the electron–molecule encounter proceeds through the temporary capture by the molecule of the incident electron, resulting in the formation of an unstable complex in a *resonant quantum state*. As explained below, the evolution of this resonance state usually results in dissociation or excitation of the molecular species. A resonant interaction can also take place between a colliding electron with a positively charged molecular ion, with the formation of a metastable neutral molecule which can evolve again in dissociation or *auto-ionization*, by electron emission, leaving behind an excited cation. Resonant collisions play a role of great importance in gaseous systems due to the relatively high values of their cross sections. They are discussed in section 15.2 for molecular hydrogen, and some details of the mechanism of the resonant collisions involving neutral or positively charged molecules are described in section 15.3 for terrestrial atmosphere molecules.

Conversely, no capture occurs in non-resonant collisions. In this case the impinging electron only interacts with the target by exchanging energy so that, again, excitation or dissociation of the molecule can result. Cross section data for non-resonant collisions are reviewed in section 15.1 for hydrogen and section 15.4 for atmospheric molecules. Theoretical cross sections are given as a result of nuclear-motion calculations which require, in general, preliminary knowledge of electron-scattering parameters such as, for example, the *complex potential functions* resulting from resonant collisions. These, and other quantities, can be obtained by using the well-known R-matrix method. This is based on a close-coupling expansion technique, where the scattering wavefunction for the whole electron–molecule system is expressed in terms of a bound-continuum basis set. This method is briefly discussed in section 15.3.

A central role in electron–molecule scattering is played by molecular hydrogen. The simple structure of the H_2 molecule allows accurate calculations to be performed, which makes this species suitable for testing theories or computational algorithms. This means that continuous, significant attention has been dedicated to this species, as can be testified by the extensive literature dedicated to this subject. In the context of this review, electron–hydrogen collisions are also important if we go beyond Earth's atmosphere and extend our interest to the interactions of celestial bodies with other planets, such as Jupiter or Saturn, whose atmospheres are mainly composed of hydrogen and helium [1]. Electron collisions with H_2 are treated in sections 15.1 and 15.2.

15.1 Non-resonant inelastic e–H_2 collision processes

The inelastic non-resonant processes in e–H_2 collisions are comprehensively discussed in [2, 3], where information about their cross sections is also provided. In the present subsection we discuss only electron-impact excitation and non-dissociative and dissociative ionization processes:

$$e + H_2(X^1\Sigma_g^+;\, v) \rightarrow e' + H_2(N^{1,3}\Lambda_{u,g}) \tag{15.1}$$

$$\rightarrow e' + H_2^+(X^2\Sigma_g^+;\, v') + e \tag{15.2a}$$

$$\rightarrow e' + H_2^+(X^2\Sigma_g^+;\, \epsilon') + e \rightarrow e' + H^+ + H(1s) + e \tag{15.2b}$$

$$\rightarrow e' + H_2^+(B^2\Sigma_u^+;\, \epsilon') + e \rightarrow e' + H^+ + H(1s) + e, \tag{15.2c}$$

where N is the usual spectroscopic symbol of the excited state, Λ is the projection of electron angular momentum on the internuclear axis, and v, v' and ϵ' are vibrational state quantum numbers and continuum energy. Most of the experimental and theoretical calculations for the processes (15.1)–(15.2) have been performed for the ground vibrational state $v = 0$ of H_2 and we confine our discussions to this case. The processes with $v > 0$ are discussed in [2, 3].

15.1.1 Excitation of singlet electronic states

Experimental cross section measurements have been performed for the excitation to the singlet states $B^1\Sigma_u^+$ [2, 3], $C^1\Pi_u$ [4–6] and B', $B''\bar{B}^1\Sigma_u^+$, D, $D'^1\Pi_u$ [5], mainly in the energy region below 350 eV. Theoretical calculations within the Born–Ochkur approximation have been carried out for all these states and also for EF, $H\bar{H}$, $GK^1\Sigma_g^+$ and $I^1\Pi_g$ in [7]. Systematic calculations within the impact parameter version of the Born approximation were carried out for the B', $B''\bar{B}^1\Sigma_u^+$, C, D, $D'^1\Pi_u$ states in [8], while distorted-wave calculations for excitation to $C^1\Pi_u$ and $EF^1\Sigma_g^+$ were carried out in [9]. All available experimental and theoretical cross sections for the symmetry allowed $X^1\Sigma_g^+ \rightarrow N^1\Lambda_u$ transitions mutually agree to within 15%–30% in the energy region around the cross section maximum (~40–50 eV) and below, and within 10% for $E > 80$ eV. They can be fitted to the analytic expression [3]

$$\sigma_{\text{exc}}^{\text{singl-all}} = \frac{\sigma_0}{\Delta E \cdot x}\left(1 + \frac{1}{x}\right)^{\alpha}\left(A_1 + \frac{A_2}{x} + \frac{A_3}{x^2} + A_4 \ln x\right), \tag{15.3}$$

where $x = E/\Delta E$, ΔE is the threshold energy, $\sigma_0 = 5.985 \times 10^{-16}$ cm^2, and the coefficients α, A_i are given in table 15.1. The cross section plots, as a function of the collision energy, are shown in figure 15.1(a).

The cross sections for symmetry-forbidden transitions $X^1\Sigma_g^+ \rightarrow N^1\Lambda_g$ with the same accuracy as for the allowed transitions can be represented by the expression [3]

$$\sigma_{\text{exc}}^{\text{singl-forb}} = \frac{\sigma_0}{\Delta E \cdot x} A_0 \left(1 - \frac{1}{x}\right)^{\beta}, \tag{15.4}$$

where $x = E/\Delta E$ and the threshold energies ΔE and fitting constants A_0 and β are given in table 15.2 for the transitions to the EF, $H\bar{H}$, $GK^1\Sigma_g^+$ and $I^1\Pi_g$ states. The cross section plots are shown in figure 15.1(b).

15.1.2 Excitation of triplet electronic states

Except for the repulsive $b^3\Sigma_u^+$ state, all other lower excited triplet states of H_2 have a bound character in the Franck–Condon region of the ground ($v = 0$) vibrational state of its $X^1\Sigma_g^+$ ground electronic state. The excitation cross section to the dissociative repulsive $b^3\Sigma_u^+$ state has been measured in [10, 11], and to the bound states $a^3\Sigma_u^+$ and $c^3\Pi_u$ in [4]. Theoretical cross section calculations for excitation have been performed for $b^3\Sigma_u^+$ in [12–16], $a^3\Sigma_u^+$ in [12, 14, 16], $c^3\Pi_u$ in [9, 12, 14, 16], and $e^3\Sigma_u^+$ and $d^3\Pi_u$ in [12]. In all the above calculations first-order methods have been

Table 15.1. Values of threshold energies ΔE and fitting parameters in equation (15.3); $a\ (-x)$ denotes $a \times 10^{-x}$.

	B	B'	$B''B$	C	D	D'
ΔE (eV)	12.754	14.85	15.47	13.29	14.996	15.555
α	0.550	0.550	0.550	0.552	0.552	0.552
A_1	3.651 (−2)	6.827 (−3)	2.446 (−3)	3.653 (−2)	8.913 (−3)	3.872 (−3)
A_2	−0.8405	−0.1872	−5.631 (−2)	−0.8398	−0.2049	−8.902 (−2)
A_3	1.2365	0.231 22	6.2846 (−2)	1.2368	0.301 78	0.131 10
A_4	2.5236	0.471 91	0.169 08	2.8740	0.701 26	0.304 64

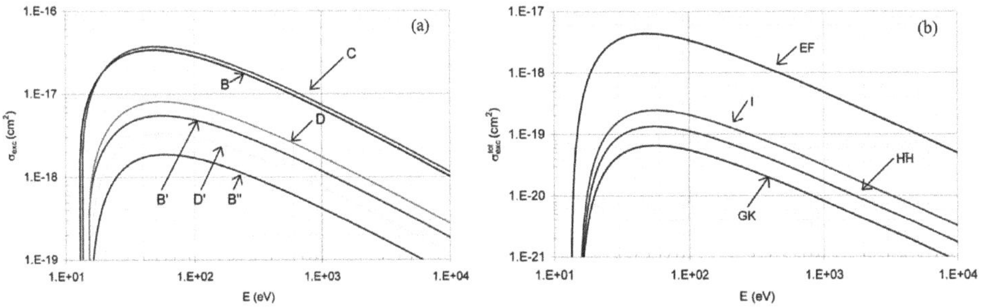

Figure 15.1. Cross sections for allowed (a) and forbidden (b) transitions to the lowest singlet states of H_2.

Table 15.2. Values of threshold energies ΔE and fitting parameters in equation (15.4); $a\ (-x)$ denotes $a \times 10^{-x}$.

	EF	$H\bar{H}$	GK	I
ΔE (eV)	13.13	14.98	14.816	14.824
A_0	0.8322	2.913 (−2)	1.43 (−2)	5.409 (−2)
β	2.71	2.71	2.75	2.80

Table 15.3. Values of threshold energies ΔE and fitting parameters in equation (15.5); $a\ (-x)$ denotes $a \times 10^{-x}$.

	a	b	c	e	d
ΔE (eV)	11.72	7.93	11.72	13.0	13.6
A	0.544	11.16	1.43	0.190	0.145
β	4.5	2.33	5.5	4.5	5.5
γ	1.55	3.78	1.65	1.60	1.75

used, except for [15] and [16] where second-order and R-matrix methods were used, respectively. A critical evaluation of the available cross section data in [3] led to the following analytic fit expression

$$\sigma_{\mathrm{exc}}^{\mathrm{tripl}} = \frac{A}{x^3}\left(1 - \frac{1}{x^\beta}\right)^\gamma (\times 10^{-16}\ \mathrm{cm}^2), \qquad (15.5)$$

where $x = E/\Delta E$, ΔE is the threshold energy, and A, β and γ are fitting constants, given in table 15.3 for the transitions to lower triplet states. The cross section curves are displayed in figure 15.2. We note that the cross sections for the $a \to d$, $c \to h$ and $c \to g$ triplet-to-triplet dipole-allowed transitions have also been calculated in [17] using the impact parameter method.

Finally, we mention that cross section calculations for a dissociative process involving the He_2^+ species have also been performed for the transition from the $X^2\Sigma_u^+$ ground state of the ion to its $A^2\Sigma_g^+$ repulsive state [18].

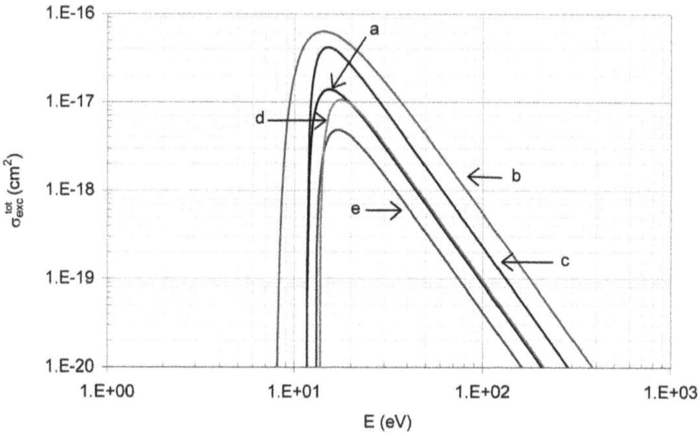

Figure 15.2. Excitation cross sections to the lowest triplet states of H_2.

15.1.3 Non-dissociative and dissociative ionization of H_2

The cross sections for non-dissociative (15.2*a*), dissociative (15.2*b*) and (15.2*c*) ionization of the ground ($v = 0$) state of H_2 have been measured in [19–23] and the contributions of the equation (15.2*b*) and (15.2*c*) channels to the dissociative cross section have been resolved on theoretical grounds [24]. The resulting cross sections for processes (15.2*a*)–(15.2*c*) have been fitted in [3] to the analytic expressions:

$$\sigma^{\text{non-diss}}_{\text{ion, equation (15.2}a)}(x) = \frac{1.628}{x}\left(1 - \frac{1}{x^{0.92}}\right)^{2.19}(C_0 \ln x)(\times 10^{-16}\ \text{cm}^2) \tag{15.6}$$

$$\sigma^{\text{diss}}_{\text{ion, equation (15.2}b)}(x) = \frac{0.02905}{x^{1.25}}\left(1 - \frac{1}{x^{2.78}}\right)^{1.886}(\times 10^{-16}\ \text{cm}^2) \tag{15.7}$$

$$\sigma^{\text{diss}}_{\text{ion, equation (15.2}c)}(x) = \frac{0.5627}{x^{1.20}}\left(1 - \frac{1}{x^{1.22}}\right)^{3.75}(\times 10^{-16}\ \text{cm}^2), \tag{15.8}$$

where $x = E/\Delta E$, $C_0 = 2.05\Delta E$ and the threshold energy ΔE for the reactions (15.2*a*), (15.2*b*) and (15.2*c*) has the values 15.12 eV, 18.15 eV and 30.60 eV, respectively. The cross section plots are shown in figure 15.3.

15.2 Resonant inelastic *e*–H_2 processes

In a system consisting of electrons and molecular hydrogen several processes take place which play a decisive role in understanding the physical properties of the system. Some of these processes are dissociative electron attachment (DEA) to the molecule,

$$e + H_2(v_i, J_i) \rightarrow H + H^-, \tag{15.9}$$

associative detachment (AD) to create the molecule,

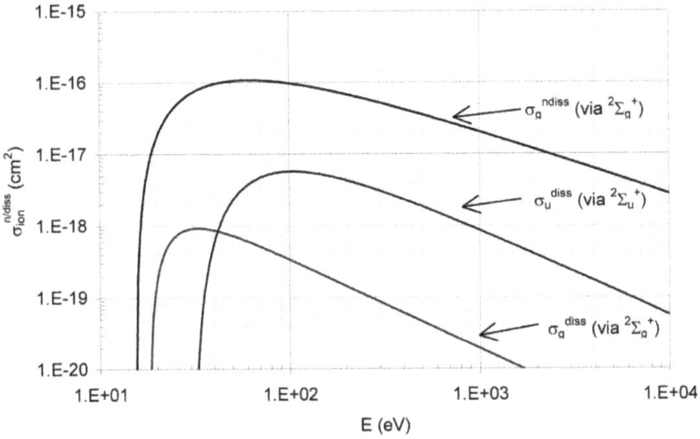

Figure 15.3. Non-dissociative (equation (15.2*a*)) and dissociative (equations (15.2*b*), (15.2*c*)) ionization cross sections of H_2.

$$H + H^- \rightarrow e + H_2(v, J), \tag{15.10}$$

vibrational excitation (VE) of the molecule,

$$e + H_2(v_i, J_i) \rightarrow e + H_2(v_f, J_f), \tag{15.11}$$

and complete dissociative excitation of the molecule,

$$e + H_2(v_i, J_i) \rightarrow e + H + H. \tag{15.12}$$

The collision cross sections for some of these processes are related. For example, the processes of dissociative attachment and of associative detachment of H_2 are the inverse of each other and, therefore, their cross sections are related by the principle of detailed balance. Another relationship exists for cross sections for vibrational excitation and for dissociative excitation of H_2. If a computer program is set up to calculate the cross sections for vibrational excitation to a discrete ro-vibrational level of H_2, then the same program can easily be modified to calculate the cross sections for vibrational excitation to a continuum level of H_2 which amounts to the dissociative excitation of H_2.

Several approaches have been developed for the theoretical description of these four processes. They range from simple phenomenological models to more elaborate calculations taking into account the nonlocality of the nuclear potential arising from the breakdown of the Born–Oppenheimer approximation. In a naïve way one may expect that the above processes involving an electron and a comparatively heavy H_2 molecule would be improbable and have a small cross section, since very little energy could be transferred in the direct collision of an electron with a free H_2 molecule—the electron would bounce off a molecule like a fly would bounce off an elephant. Such a collision would be a non-resonant collision. On the other hand, if due to some mechanism the electron could be entrapped in the H_2 molecule such that it spends more time in the vicinity of the H_2 molecule than the normal transit time, the collision cross section could be enormously large. Such a collision would be a resonant collision. Using again the simile of the fly and the elephant, if the fly were to enter the nose or ear of the elephant where it could spend more time than the normal transit time, the outcome could be enormously turbulent. In this section we review the resonance model, in which the projectile electron is temporarily trapped by the target molecule, and examine its application to the four processes (15.9)–(15.12) in collisions of low-energy electrons with molecular hydrogen.

In the case of electron scattering by molecular hydrogen, the resonance model has been the key in understanding the dissociative attachment and vibrational excitation of the H_2 molecule. Consider a situation in which the potential curves of neutral molecule H_2 and of anion state H_2^- (the resonant state) cross at some internuclear separation R_s. At asymptotically large internuclear separations ($R \to \infty$) the potential curve of H_2 matches the energy state of H + H, while the potential curve of H_2^- matches the energy state of H + H^-. Now, since H^- has a true bound state, the potential curve of the resonant H_2^- state must lie below the potential curve of H_2 for $R > R_s$ and above the potential curve of H_2 for $R < R_s$. Thus, for $R < R_s$ the resonant state can 'shake out' or autodetach the trapped electron, leaving behind a vibrationally excited molecule. In other words, for $R < R_s$ the resonant state will have a finite lifetime which is determined by the width of the resonance. On the other hand, if the nuclei in the resonant state separate out to $R > R_s$, without electron autodetachment having occurred, the electron detachment becomes energetically impossible and the dissociative attachment becomes imminent. Note that R_s, called the stabilization radius, is the internuclear separation beyond which the resonant state becomes stable against electron autodetachment. There are several mechanisms by which the electron could be trapped by the H_2 molecule to form the resonant state of H_2^-. As an example, the incident electron could excite the H_2 molecule upon impact and lose enough energy to prevent its own escape. This is referred to as a closed channel or Feshbach resonance. As another possibility, if the target H_2 molecule is in a configuration of nonzero angular momentum, the incident electron can become trapped in the centrifugal potential barrier from which it tunnels out. This would be an example of an open-channel or shape resonance. A more precise mathematical description of the resonance model can be found in the review article by Wadehra [25]. Stibbe and Tennyson [26] have investigated resonances of the e–H_2 system of various symmetries as a function of bond length. These investigations allow for definitive assignments of the parentage of resonances of this system.

15.2.1 Dissociative electron attachment processes

The process of dissociative electron attachment to molecular hydrogen was investigated experimentally in detail by Schulz and co-workers [27, 28]. In the electron energy range from 3 eV to 18 eV, three prominent peaks are observed in the DEA cross sections around 3.75 eV, 10 eV and 14 eV [29]. Figure 15.4 shows the cross sections for production of a H^- ion from H_2 by electron impact. In this figure, in addition to the three DEA peaks, a rapid increase in the production of H^- is seen for electron energies above 17.2 eV, which is the threshold for the process of polar

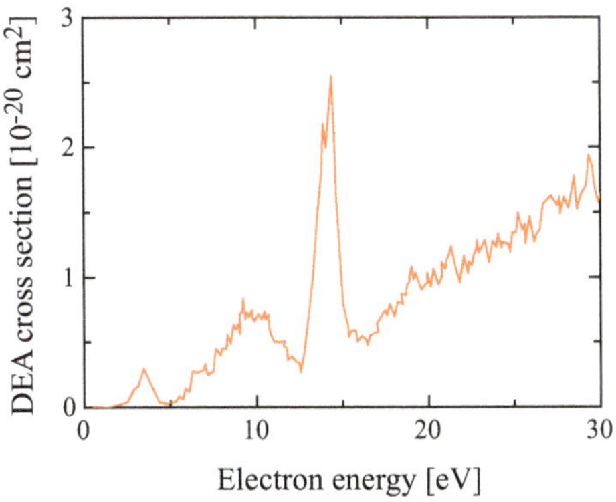

Figure 15.4. Cross section for the production of H^- from H_2 by electron impact [29].

dissociation. On replacing the molecule H_2 by its heavier isotopes HD and D_2 a significant change in the magnitude of cross section for negative ion production was observed, but the location of the three peaks was almost unchanged [31], indicating that each peak corresponds to a distinct resonant state of H_2^-.

The lowest two peaks in the DEA cross sections occur due to the formation of the lowest two resonant states of H_2^-, namely the $^2\Sigma_u^+$ and the $^2\Sigma_g^+$ states, both dissociating into H(1*s*) + H^-. The $^2\Sigma_g^+$ state is repulsive over the complete range of internuclear separations while the $^2\Sigma_u^+$ state is attractive. The third peak in the DEA cross section around 14 eV, which also happens to be the most pronounced peak, is related to the $^2\Sigma_g^+$ resonant state of H_2^- which dissociates into H(n = 2) + H^-. This resonant state, referred to as the excited Rydberg $^2\Sigma_g^+$ state of H_2^-, is a short-lived Feshbach resonance. More accurate measurements indicate that the relative magnitudes of the peak DEA cross sections at low temperatures, when the molecule is at the $v_i = 0$ level, are about 2.8×10^{-21} cm^2 for the peak around 3.75 eV, about 1.2×10^{-20} cm^2 for the peak around 10 eV and about 3.8×10^{-20} cm^2 for the peak around 14 eV. However, at high temperatures, when the H_2 molecule is vibrationally excited, all three peak attachment cross sections are enhanced and their peak values are shifted to lower electron energies, but the enhancement of the lowest peak, arising from the $^2\Sigma_u^+$ resonant state of H_2^-, is most dramatic. Specifically, for a vibrationally excited H_2 molecule, the enhancement of the DEA cross section occurs because the threshold energy of electrons that can dissociatively attach themselves to the molecule is lowered. When the molecule is initially at the $v_i = 10$ level, the threshold energy of the attaching electron becomes zero and the process of dissociative electron attachment occurring through the $^2\Sigma_u^+$ resonant state of H_2^- changes from being an endoergic process to an exoergic process [34, 30]. The most enhanced value of the DEA cross section for this resonant state is 6.2×10^{-16} cm^2, which corresponds to an enhancement factor of 2.2×10^5. The middle peak in the DEA cross section, which is related to the formation of lower $^2\Sigma_g^+$ resonant state, shifts to an electron energy below 4 eV if the H_2 molecule is initially at the $v_i = 5$ level [34, 32]. The peak value of the cross section at this electron energy is 4.5×10^{-20} cm^2 corresponding to an enhancement factor of 3.75. Figure 15.5 shows the contribution of the lower $^2\Sigma_g^+$ resonance of H_2^- to the cross sections for dissociative electron attachment to H_2 in various vibrationally excited levels. Finally, when the DEA process is mediated by the excited Rydberg $^2\Sigma_g^+$ state of H_2^-, the cross section peak increases with increasing vibrational level up to $v_i = 3$ and then begins to decrease as v_i increases [33, 35]. The largest value of the DEA cross section in this case is 4.6×10^{-19} cm^2 for $v_i = 3$ and it corresponds to an enhancement factor of about 12. Figure 15.6 shows the contribution of the excited Rydberg $^2\Sigma_g^+$ resonance of H_2^- to the cross sections for dissociative electron attachment to H_2 with initial vibrational levels of $v_i = 0$, 3 and 6.

15.2.2 Associative detachment processes

The inverse process of associative detachment, namely $H + H^- \rightarrow e + H_2(v, J)$, has not been studied as much as the process of dissociative attachment. The first experimental measurements of the rate constants of associative detachment in the collisions of H^- ions with H atoms were reported by Schmeltekopf *et al* [36] in the context of the application of this reaction to astrophysics. For a gas temperature of 300 K, the rate constant for the AD process, reliable to within a factor of 2, was measured to be 1.3×10^{-9} cm^3 s^{-1}. In all theoretical calculations of the AD process, formation of an intermediate resonant state is assumed. With this assumption, the calculations are simplified to scattering by a complex potential. The two reactants, H^- ions and H(1*s*) atoms, can approach on either of the two potential curves of the resonant state the attractive $^2\Sigma_u^+$ resonant state or the repulsive $^2\Sigma_g^+$ resonant state of H_2^-. If the interaction potential between the ion–atom pair is repulsive, the two particles simply scatter with no reaction taking place. On the other hand, if the interaction between the H^- ions and H(1*s*) atoms is due to the attractive $^2\Sigma_u^+$ resonant state, the electron detachment can take place, resulting in a ro-vibrationally excited H_2 molecule. The calculations of Bieniek and Dalgarno [37] provided cross

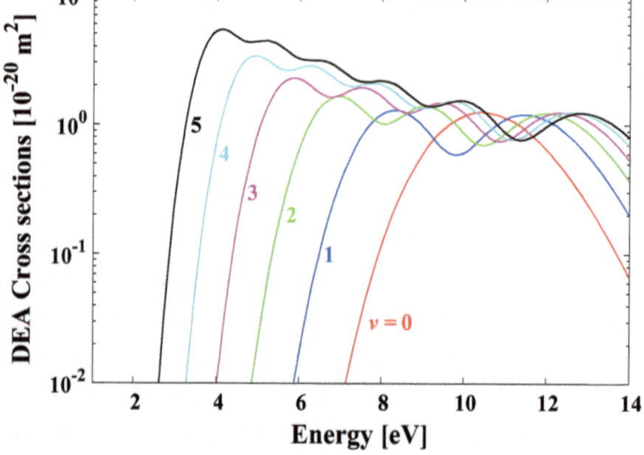

Figure 15.5. Contribution of the $^2\Sigma_g^+$ resonance of H_2^- to DEA cross sections for various vibrationally excited levels of H_2 [31]. Reprinted with permission from IOP Publishing Ltd.

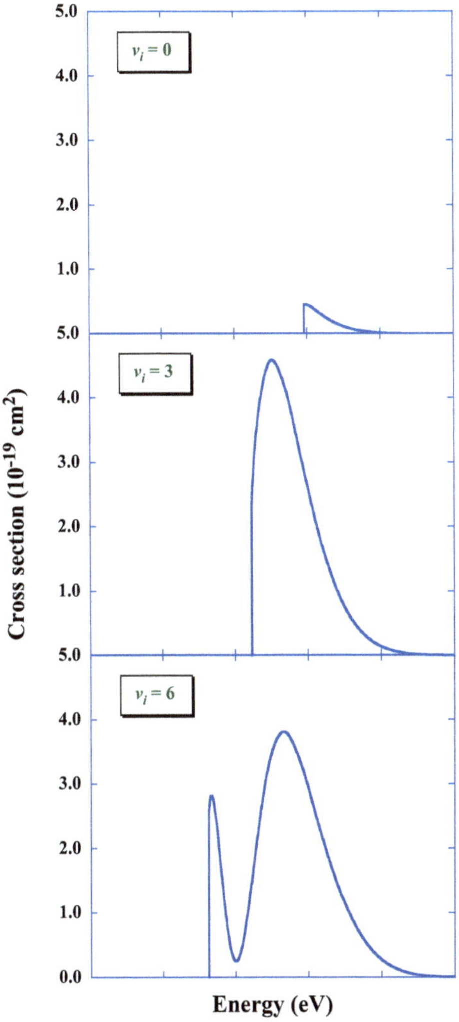

Figure 15.6. Contribution of the Rydberg $^2\Sigma_g^+$ resonance of H_2^- to DEA cross sections for H_2 in $v_i = 0$, 3 and 6 [34].

sections for AD into specific ro-vibrational levels. The process of AD preferentially populates the higher ro-vibrational levels. For the relative ion–atom pair energy of 0.0129 eV and 0.129 eV (or, collision energies corresponding to temperatures of 100 K and 1000 K, respectively), the largest AD cross sections occur for v between 4 and 7 and J between 7 and 17. The calculations of Launay *et al* [38] used the principle of detailed balance to relate the cross sections for dissociative attachment and associative detachment. Their calculated Maxwellian rates for the AD process are consistent with the measurements of Schmeltekopf. Finally, Cizek *et al* [39] used a nonlocal resonance model to calculate the AD cross sections. They noted that both the local as well as the nonlocal complex potential in their resonance model provided essentially the same cross sections and rates. However, their calculated AD rates were significantly larger than the experimental rates of Schmeltekopf. On the other hand, the theoretical results of Cizek *et al* agree quite well with the more recent and more precise measurement of AD rates by Kreckel *et al* [40]. Figure 15.7 shows the experimental rates for associative detachment reaction $H + H^- \rightarrow e + H_2$ as a function of the collision energy of the ion–atom pair measured by Kreckel *et al* [40]. For comparison the solid line in this figure shows rate coefficients derived from theoretical cross sections of Cizek *et al* [39] assuming a Maxwell–Boltzmann energy distribution. The dashed error bands in the figure indicate systematic experimental uncertainty of 1σ standard deviation.

15.2.3 Vibrational excitation processes

The process of vibrational excitation of H_2 by low-energy electrons is also understood by invoking the idea of the formation of an intermediate resonant state of H_2^-. Using a resonance model represented by a local complex potential, Bardsley and Wadehra [41] gave a comprehensive account of vibrational excitation cross sections for electron energies between 2 eV and 11 eV. At low electron energies, between 2 eV and 7 eV, the ground state of the H_2^- ion, namely the $^2\Sigma_u^+$ resonant state, contributes much more to the vibrational excitation cross sections than the $^2\Sigma_g^+$ resonant state for excitation of any level. For electron energies between 7 eV and 11 eV, the $^2\Sigma_u^+$ resonant state still dominates the cross section for excitation of low-lying levels, but the excitation of higher levels ($v_f \geqslant 5$) receives a significant contribution from the $^2\Sigma_g^+$ resonance. Figure 15.8 shows the contributions of the two lowest resonant states of H_2^- to the vibrational excitation cross sections of H_2. The solid lines show the contributions of the $^2\Sigma_u^+$ resonant state, while the dashed lines show the contribution of the next higher resonance, namely, the $^2\Sigma_g^+$ state.

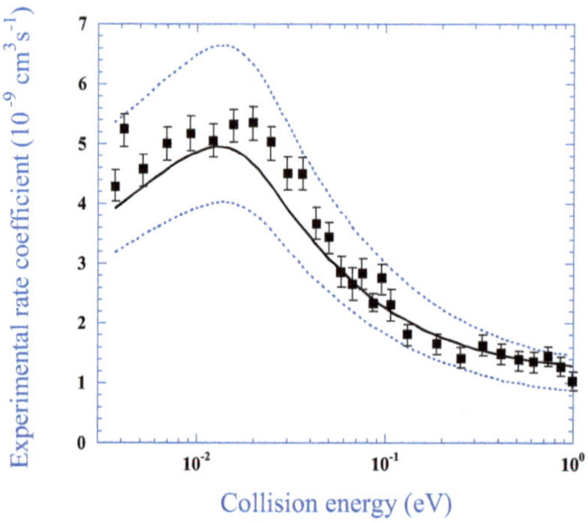

Figure 15.7. Measured rates for associative detachment reaction $H + H^- \rightarrow e + H_2$ as a function of the collision energy of the ion–atom pair due to Kreckel *et al* [40]. The solid line shows the rate coefficients derived from theoretical cross sections of Cizek *et al* [39]. The dashed lines indicate the experimental uncertainty (see the text).

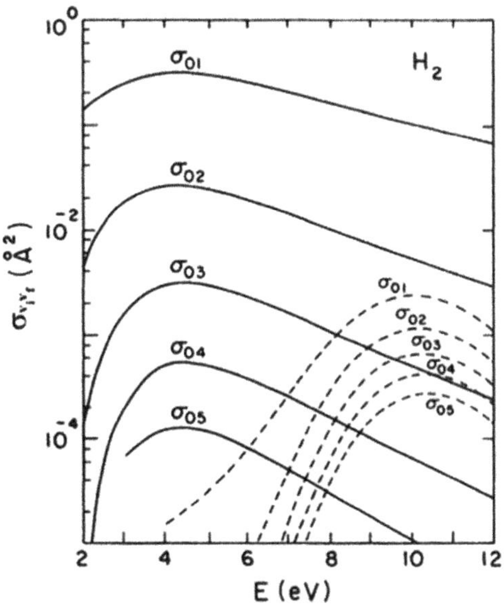

Figure 15.8. Contributions of the two lowest resonant states of H_2^- to the vibrational excitation cross sections of H_2. Solid lines, ${}^2\Sigma_u^+$; dashed lines, ${}^2\Sigma_g^+$ [41].

By recording energy-loss spectra at 10 eV for both H_2 and its heavier isotope D_2, Hall and Andric [10] were able to experimentally demonstrate the interplay between the ${}^2\Sigma_u^+$ and ${}^2\Sigma_g^+$ resonances for vibrational excitation of these two isotopes. From their figure 4, it is possible to obtain the ratios of cross sections for the excitation of two successive vibrational levels of H_2 and D_2. These ratios are in good agreement with those calculated by Bardsley and Wadehra [41]. As an aside, by examining the mass dependence of cross sections for resonant vibrational excitation in H_2 and its heavier isotopes, Atems and Wadehra [42] derived a scaling law. If μ is the ratio of the reduced masses of the heavier isotope to lighter isotope, then the ratio of the cross section for vibrational excitation, from v_i to v_f, for the heavier isotope to lighter isotope is proportional to $\mu^{-|v_i-v_f|/2}$. This scaling law was numerically tested for two molecules, H_2 and HCl as well as their heavier isotopes, for both inelastic and superelastic collisions. This scaling law for the resonant VE process holds very well for small values of $v_f \leqslant 5$ and for small values of μ, even for cross section values ranging over several orders of magnitude.

A convincing proof that the VE process proceeds through the formation of an intermediate resonant state was provided by the experimental work of Allan [43]. Investigations of VE cross sections of H_2 by electron impact in the energy region of the ${}^2\Sigma_u^+$ resonance revealed structures related to the vibrational motion of the resonant anion state of H_2^-. For $v_f \geqslant 3$, the energy dependence of the measured VE cross sections displayed a pronounced vibrational structure which had been predicted earlier in the theoretical work of Mundel *et al* [44]. They postulated this effect as a manifestation of vibration-induced narrowing of resonances. Finally, for electron energies around 11 eV, the excited Rydberg ${}^2\Sigma_g^+$ resonant state of H_2^-

becomes the largest contributor to the vibrational excitation cross sections. Using this resonance Celiberto *et al* [45, 46] calculated the cross sections for vibrationally inelastic and superelastic transitions as a function of the incident electron energy. Interestingly, several sharp narrow peaks appear in the calculated VE cross sections at energies corresponding to the energies of the vibrational levels of the excited $^2\Sigma_g^+$ resonant state of H_2^-. Some scant experimental data about the VE cross sections in this high energy range can be gleaned from the work of Comer and Read [47].

15.2.4 Dissociative excitation processes

Another important process in the system of electrons plus hydrogen molecules is the dissociation of the H_2 molecule by electron impact. This process occurs primarily through excitation of the repulsive $b^3\Sigma_u^+$ state of H_2 that dissociates into $H(1s) + H(1s)$. Schematically, the non-resonant process is

$$e + H_2(X^1\Sigma_g^+) \rightarrow e + H_2(b^3\Sigma_u^+) \rightarrow e + H(1s) + H(1s). \qquad (15.13)$$

This direct process has a threshold of about 8 eV and a cross section of the order of 10^{-17} cm^2 [13]. The possibility of dissociation of H_2 occurring via the formation of the lowest two resonant states of H_2^- (namely the $^2\Sigma_u^+$ and the $^2\Sigma_g^+$ states) was considered by Atems and Wadehra [48]. The $^2\Sigma_u^+$ resonant state decays only to the ground electronic state, $X^1\Sigma_g^+$, of H_2 while the $^2\Sigma_g^+$ resonant state can decay both to the ground electronic state, $X^1\Sigma_g^+$, as well as the first excited state, $b^3\Sigma_u^+$, of molecular hydrogen. Treating dissociation as the process of vibrational excitation into a continuum level, the resonant processes for dissociation of H_2 occur through three channels as follows:

$$e + H_2(X^1\Sigma_g^+, v_i) \rightarrow H_2^-(^2\Sigma_u^+) \rightarrow e + H_2(X^1\Sigma_g^+, \epsilon_f) \rightarrow e + H(1s) + H(1s), \quad (15.14)$$

$$e + H_2(X^1\Sigma_g^+, v_i) \rightarrow H_2^-(^2\Sigma_g^+) \rightarrow e + H_2(b^3\Sigma_u^+, \epsilon_f) \rightarrow e + H(1s) + H(1s), \quad (15.15)$$

$$e + H_2(X^1\Sigma_g^+, v_i) \rightarrow H_2^-(^2\Sigma_g^+) \rightarrow e + H_2(X^1\Sigma_g^+, \epsilon_f) \rightarrow e + H(1s) + H(1s), \quad (15.16)$$

where ϵ_f represents the vibrational continuum energy. Analogous to the process of dissociative attachment, the enhancement of the dissociation cross section occurring through the $^2\Sigma_u^+$ resonant state of H_2^-, namely channel (15.14), is most dramatic when the initial vibrational level v_i is increased. The cross section increases by seven orders of magnitude as v_i is increased from 0 to 12. Figure 15.9 shows the contributions of the $^2\Sigma_u^+$ resonance to the cross section for dissociative excitation of H_2 in various initial vibrational levels for electron energies less than 12 eV. For $v_i = 12$ the dissociation cross section through the $^2\Sigma_u^+$ resonant state of H_2^- becomes comparable in magnitude to the cross section for the non-resonant process (15.13), while its threshold becomes much smaller than the threshold for process (15.13). This suggests that the resonant contribution of channel (15.14) to the rate of

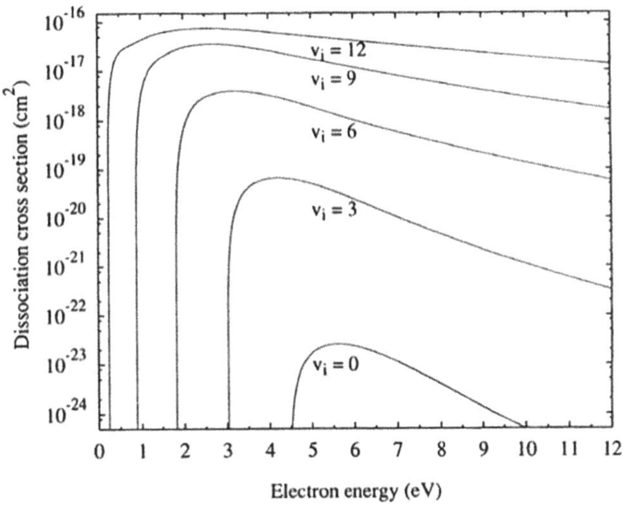

Figure 15.9. Contributions of the $^2\Sigma_u^+$ resonance to the cross section for electron-impact dissociative excitation of H_2 in various initial vibrational levels [48].

dissociation of H_2 can dominate over the non-resonant contribution of channel (15.13). The contributions of the two other resonant channels (15.15) and (15.16) are negligibly small in this range of electron energies.

15.3 Resonant electron-induced reaction cross sections in Earth atmosphere molecules

In this section, the electron–molecular dataset produced for the plasma kinetic modeling application is collected. Full computational details are not reproduced here; rather, the reader is referred to the original papers and to the reviews [32, 49]. The chemical processes considered are summarized in table 15.4. First we briefly describe the methodology we use and then we present a survey of key results for the processes considered in table 15.4.

15.3.1 Methodology

The processes considered in table 15.4 are of three types: (a) electron-impact vibrational excitation which changes the vibrational state, denoted by v/v^+, of the target molecule/molecular cation; (b) electron-impact dissociation which results in the target molecular species dissociating into constituent atoms/atoms and ions, or molecules/molecular ions; and (c) dissociative electron attachment (DEA)/dissociative recombination (DR) where the impacting electron attaches to the target forming a molecular anion/neutral which then dissociates into a neutral and an anionic/neutral fragment. DEA/DR can only occur via the formation of a quasi-bound state of the complex formed by the molecular target and the impacting electron. Such states, as already discussed, are known as resonances; they have both a

Table 15.4. List of the processes considered in this section.

#	Process	Reference
1	$e + N_2(X^1\Sigma_g^+, v) \rightarrow N_2(X^1\Sigma_g^+, v') + e$	[50, 51]
2	$e + N_2(X^1\Sigma_g^+, v) \rightarrow N(^4S) + N(^4S) + e$	[52]
3	$e + N_2^+(X^1\Sigma_g^+, v_i^+) \rightarrow N(^4S) + N(^2D)$	[53]
4	$e + N_2^+(X^1\Sigma_g^+, v_i^+) \rightarrow N(^4S) + N(^2P)$	[53]
5	$e + N_2^+(X^1\Sigma_g^+, v_i^+) \rightarrow N(^2D) + N(^2D)$	[53]
6	$e + N_2^+(X^1\Sigma_g^+, v_i^+) \rightarrow N(^2D) + N(^2P)$	[53]
7	$e + N_2^+(X^1\Sigma_g^+, v_i^+) \rightarrow N_2^+(X^1\Sigma_g^+, v_f^+) + e$	[53]
8	$e + NO(X^2\Pi, v) \rightarrow NO(X^2\Pi, v') + e$	[50]
9	$e + NO^+(X^1\Sigma^+, v_i^+) \rightarrow N(^2D^o) + O(^3P)$	[54–56]
10	$e + NO^+(X^1\Sigma^+, v_i^+) \rightarrow N(^4S^o) + O(^3P)$	[54–56]
11	$e + NO^+(X^1\Sigma^+, v_i^+) \rightarrow NO^+(X^1\Sigma^+, v_f^+) + e$	[54–56]
12	$e + O_2(X\,^3\Sigma_g^-, v) \rightarrow O_2(X\,^3\Sigma_g^-, v') + e$	[57]
13	$e + O_2(X\,^3\Sigma_g^-, v) \rightarrow O(^3P) + O^-(^2P)$	[58, 59]
14	$e + O_2(X\,^3\Sigma_g^-, v) \rightarrow O(^3P) + O(^3P) + e$	[58, 59]
15	$e + CO(X^1\Sigma^+, v) \rightarrow CO(X^1\Sigma^+, v') + e$	[60]
16	$e + CO(X^1\Sigma^+, v) \rightarrow C(^3P) + O^-(^2P)$	[61]
17	$e + CO(X^1\Sigma^+, v) \rightarrow C(^3P) + O(^3P) + e$	[61]
18	$e + CO^+(X^2\Sigma^+, v_i^+) \rightarrow C(^3P) + O(^3P)$	[62, 63]
19	$e + CO^+(X^2\Sigma^+, v_i^+) \rightarrow C(^1D) + O(^1D)$	[62, 63]
20	$e + CO^+(X^2\Sigma^+, v_i^+) \rightarrow C(^1S) + O(^1D)$	[62, 63]
21	$e + CO^+(X^2\Sigma^+, v_i^+) \rightarrow C(^1D) + O(^1S)$	[62, 63]
22	$e + CO^+(X^2\Sigma^+, v_i^+) \rightarrow C(^1D) + O(^3P)$	[62, 63]
23	$e + CO^+(X^2\Sigma^+, v_i^+) \rightarrow C(^1S) + O(^3P)$	[62, 63]
24	$e + CO^+(X^2\Sigma^+, v_i^+) \rightarrow C(^3P) + O(^1S)$	[62, 63]
25	$e + CO^+(X^2\Sigma^+, v_i^+) \rightarrow CO^+(X^2\Sigma^+, v_f^+) + e$	[62, 63]
26	$e + CO_2(X^1\Sigma_g^+, v) \rightarrow CO_2(X^1\Sigma_g^+, v') + e$	[64]

position and width since these states have a finite time to re-emit the temporarily trapped electron.

Electron-impact vibrational excitation can occur through a non-resonant collision, but it is well known that the transition cross section is enhanced by many orders of magnitude in the region of a resonance. Similarly, it is possible for electron-impact dissociation to occur non-resonantly via excitation of an excited electronic state [15], but this process requires much more energy and is therefore highly energetic (generally with more than 5 eV), so below we again consider this process as proceeding via resonances which, as we show, can be efficient at much lower electron temperatures.

Unlike the electron–neutral molecule elementary processes, the electron–molecular cation collisions, i.e. dissociative recombination (DR), elastic collisions (EC) and vibrational excitation/de-excitation (VE/VdE) involve two simultaneous mechanisms: (i) the *direct* mechanism, in which the electron is captured into a doubly excited resonant dissociative state of the neutral system, resulting in neutral fragments or in auto-ionization, and (ii) the *indirect* mechanism, which involves the temporary capture of the electron into a singly excited bound Rydberg state of the neutral, subsequently followed by predissociation or auto-ionization. The quantum interference of the above two mechanisms results in the *total* reactive mechanism.

The calculation of the three processes discussed above requires three elements: (a) accurate representation of the target potential energy curve and associated wavefunction; (b) accurate representation of the resonance curve, which is complex due to the finite lifetime of the resonance; and (c) the ability to perform nuclear-motion calculations on these curves. Our methodology is based on using appropriate, fully quantum-mechanical procedures for each of these steps.

The target curves can be computed using standard molecular electronic structure codes; our code of choice is MOLPRO [65]. It is necessary to use a level of theory which allows the molecule to dissociate correctly. In practice we generally achieve this using multi-reference configuration interaction (MRCI) calculations with large basis sets.

The bound anion state formed as part of the DEA process at large internuclear separations can also be treated in the same fashion.

Resonance parameters, energies and widths are computed using the R-matrix method. This method is designed to treat electron–molecule collisions by dividing space into two regions. An inner region, typically a sphere of radius about $10a_0$, is chosen to contain the entire wavefunction. The inner region problem is independent of the energy of the scattering electron and hence needs to be solved only once for a given problem. The R-matrix links these inner region solutions with the much simpler outer region problem in which the scattering electron is taken to move only in the long-range potential given by the target. The outer region problem is solved as a function of scattering energy which allows the resonance parameters to be characterized [66]. More details of the method can be found in a comprehensive review written by one the authors [67]; details of the individual calculations can be found in the references cited in table 15.4. The UK molecular R-matrix codes used to perform the calculations are well-documented [68] and can be run using the Quantemol-N expert system [69].

In general our studies of resonances in electron collisions with neutral molecules are based on static exchange plus polarization (SEP) calculations. The SEP model is computationally simple and has been shown to give a good, convergent representation of the low-lying resonances which occur in all the systems we consider here [70].

The nuclear-motion method employed for the electron–neutral molecule collisions is based on the projection operator formalism originally developed by Feshbach; a thorough review of the method is provided by Domcke [71]. This procedure allows the nuclei to move on the (complex) resonance curves and was explicitly developed to treat electron-impact vibrational excitation and DEA. In practice we use a program written to work within the so-called local potential approximation, which is designed to use curves computed as described above [45].

For electron collisions with molecular *cations* it is necessary to use coupled-state calculations to capture the many resonance states that characterize this problem. In this case nuclear-motion calculations are generally performed in the framework of multichannel quantum defect theory (MQDT) [54, 72–75], that fully accounts for the two types of fragmentation channels: dissociation channels, for atom–atom scattering, and ionization channels, for electron–cation scattering, including in the latter type the complete series of bound Rydberg states converging to the ground and excited molecular ionic states.

N_2

In this section, the electron-impact collisions involving the ground electronic state of the N_2 molecule are summarized. In particular, the processes of vibrational and dissociative excitation are discussed. They are labeled in table 15.4 as process 1 and 2; at low electron collision energies they both dominantly occur through the resonant $N_2^-(X\,^2\Pi_g)$ state. The nitrogen molecule plays a role of fundamental importance in many scientific and industrial activities. Typical examples are provided by air plasmas studied in a variety of fields such as environmental research, Earth atmospheric phenomena, combustion and aerospace technologies [76–79]. Detailed chemical aspects of the processes involving molecular nitrogen are studied in [80–82]. The first measurements of electron–N_2 vibrational excitation were performed by Schulz [83, 84]. This work was followed by numerous experimental investigations [85–92] and theoretical calculations [86, 93–99].

Figure 15.10 summarizes the state-resolved cross sections and the corresponding rate coefficients for the vibrational excitations and dissociative excitation processes. A comprehensive compilation of electron–N_2 collision cross sections has been provided by Itikawa [100].

NO

The resonant scattering of electrons with nitric oxide involves three electronic states of the NO^- ion of $^3\Sigma^-$, $^1\Delta$ and $^1\Sigma^+$ symmetry. These resonance states have been well studied theoretically [67, 101–103]. Itikawa compiled a comprehensive set of cross sections for electron collisions with NO [104]; they have recently been updated for all nitrogen oxides by Song *et al* [105].

Theoretical results for the vibrational excitation cross sections and rate constants are summarized in figure 15.11.

O_2

Low-energy electron-impact scattering for molecular oxygen involves four electronic states of the O_2^-–molecular ion, identified by the symmetries $^2\Pi_g$, $^2\Pi_u$, $^4\Sigma_u^-$ and $^2\Sigma_u^-$ [106]. In this section we refer to the processes of vibrational excitation, dissociative attachment and dissociative excitation, corresponding, respectively, to the processes 12, 13 and 14 in table 15.4. The first calculations were made by Noble *et al* [107]. More recent papers on vibrational excitation scattering can be found in [108–112]. Although DEA from oxygen molecules has been widely studied [113] there are only

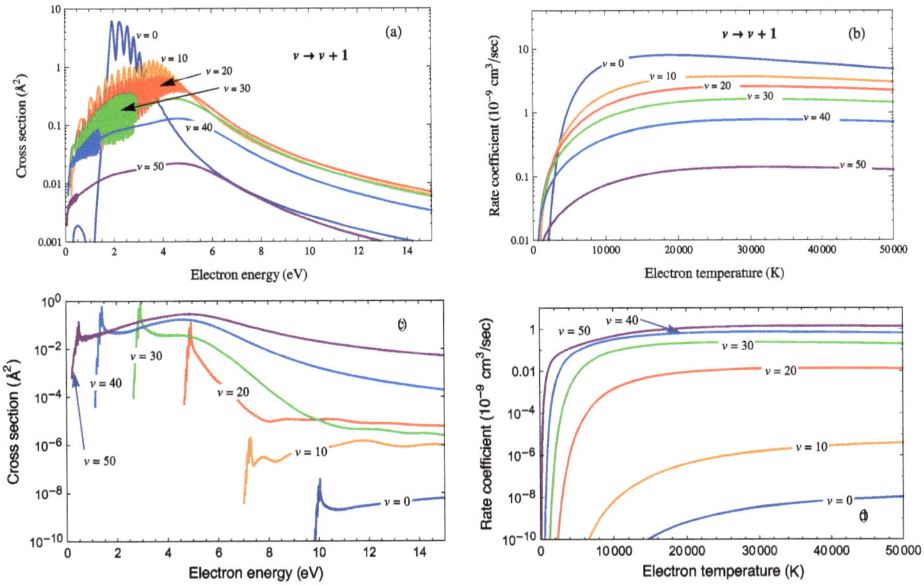

Figure 15.10. Summary of calculated electron–N_2 state-to-state resonant cross sections (left panels) and the corresponding rate coefficients (right panels) for the vibrational excitation and dissociative excitation.

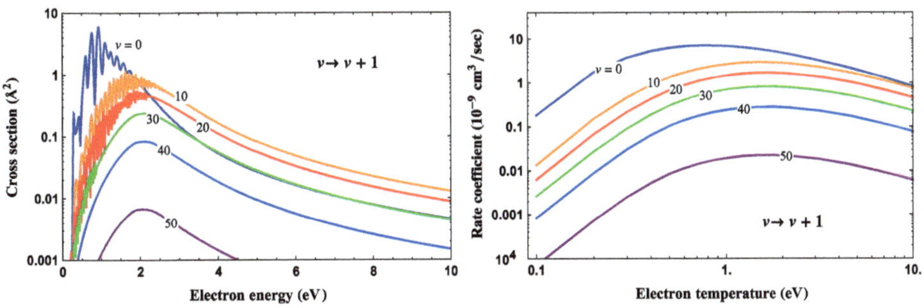

Figure 15.11. Summary of electron–NO vibrational excitation cross sections (left) and on the corresponding rate coefficients (right) for $\Delta v = 1$ transitions.

a few, rather old, papers reporting cross section measurements of the DEA process as a function of the incident electron energy [114–116]. An extensive collection of experimental data can be found in the review by Itikawa [117].

Figure 15.12 summarizes the cross sections and rate coefficients characterizing the electron–oxygen scattering.

CO

In this section the state-to-state results for electron–CO vibrational excitation, dissociative attachment and dissociative excitation (processes 15, 16 and 17 in table 15.4) are presented. Carbon monoxide is a very important molecule playing a fundamental role in many fields. It is the second most abundant species in the interstellar medium after molecular hydrogen and can act as a tracer for

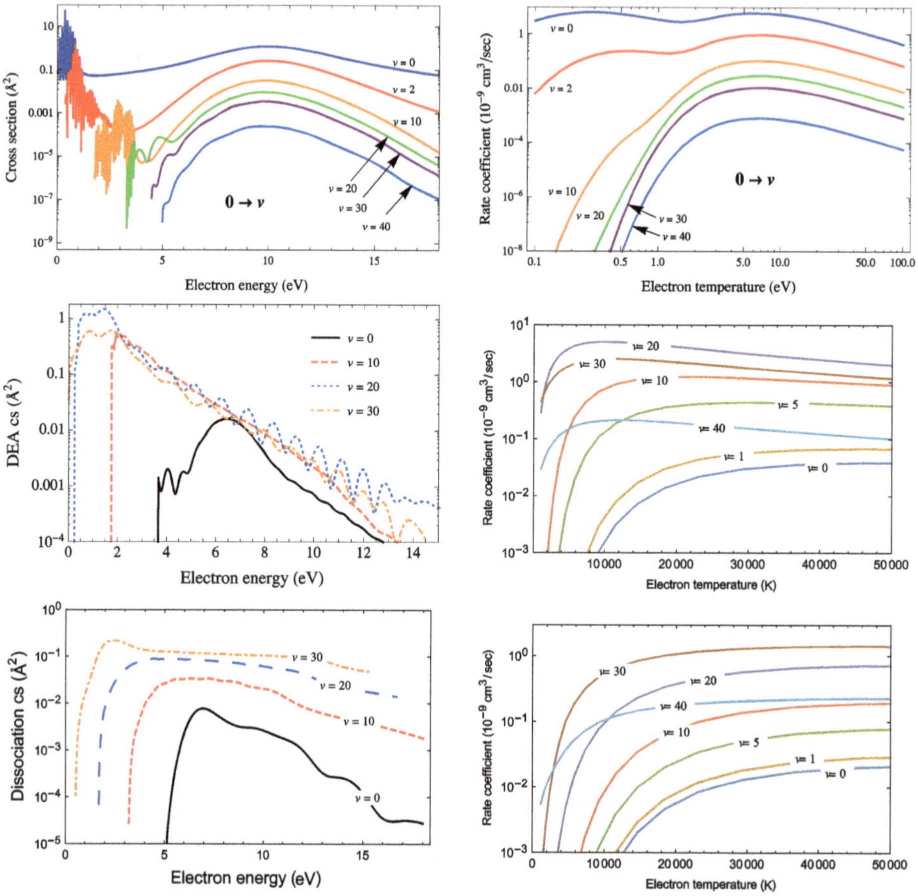

Figure 15.12. Electron–O_2 cross sections (left panels) and the corresponding rate coefficients (right panels) for vibrational excitation, dissociative attachment and dissociative excitation processes.

H_2 molecules, which, due to the lack of ground electronic state transition dipole moment, cannot be observed directly but are detected thorough monitoring CO molecule concentrations [118]. CO is a component of the atmospheres of Mars and Venus and, in the latter case, represents the most abundant chemical species. It therefore also plays a crucial role in space missions in connection to (re-)entry problems [119]. Electron collisions with carbon monoxide are also important for the understanding of processes involved in CO lasers [120] and for studies of CO plasma in the presence of electrical discharge [121, 122].

A complete set of experimental electron–CO cross sections have been compiled by Itikawa [123]. Figure 15.13 summarize the theoretical results for the three processes considered here.

CO_2

One of the technological solutions proposed for the reduction of atmospheric carbon involves the capture at source and storage of CO_2 based on the plasmolysis process

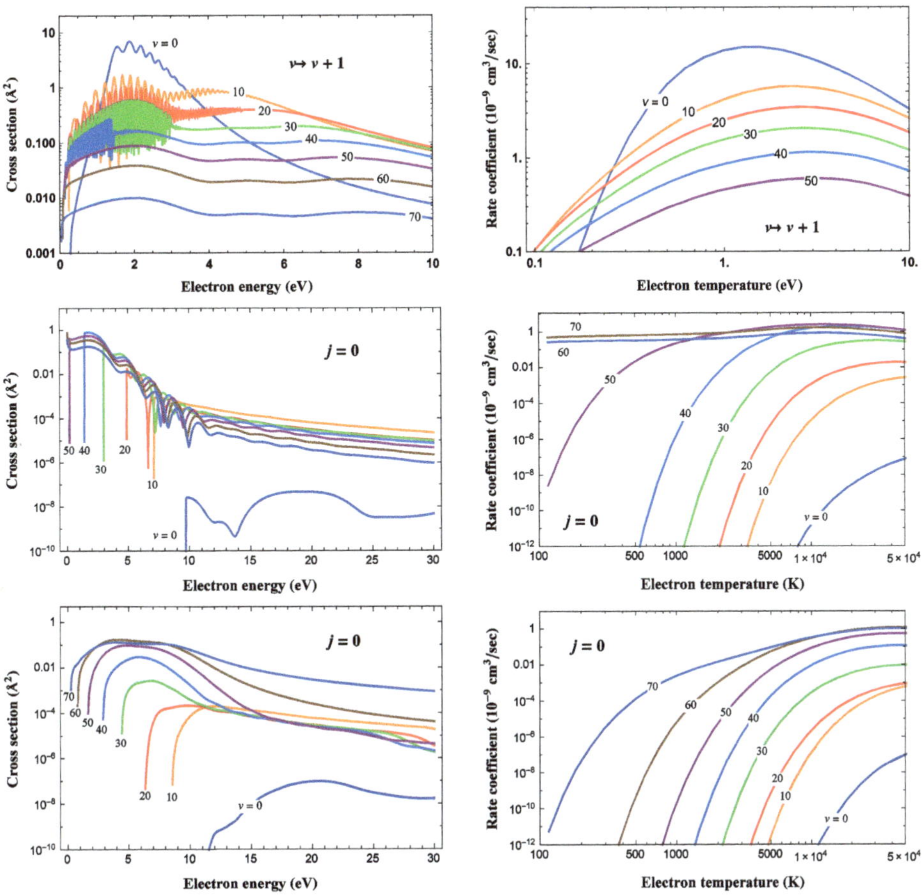

Figure 15.13. Summary of electron–CO cross sections (left panels) and the corresponding rate coefficients (right panels) for vibrational excitation, dissociative attachment and dissociative excitation processes.

which leads to the splitting of CO_2 into CO molecules and atomic or molecular oxygen [124, 125]. The efficiency of the dissociation processes is strongly determined by the vibrational activation of the molecule. Models of CO_2 plasmas, aimed to optimize and clarify this chemical conversion, have recently been constructed [126–131]. The main limitation on these models is the lack of information on electron-impact cross sections or rate coefficients for collisions inducing vibrational transitions in CO_2 molecules; as result modelers usually resort to estimated rates or approximate scaling-laws [130].

Figure 15.14 represents a first attempt at addressing this problem using a series of simplified, one-dimensional treatments of the vibrational motions. The calculated cross sections show two distinctive features observed experimentally: a $^2\Pi_u$ shape resonance around 3.8 eV [132–134] and, at energies below 2 eV, an enhancement due to the presence of the $^2\Sigma_g^+$ symmetry virtual state [135–137]. Both phenomena are explained in terms of a temporary CO_2^- system. A general review of electron–CO_2 cross sections has been provided by Itikawa [138].

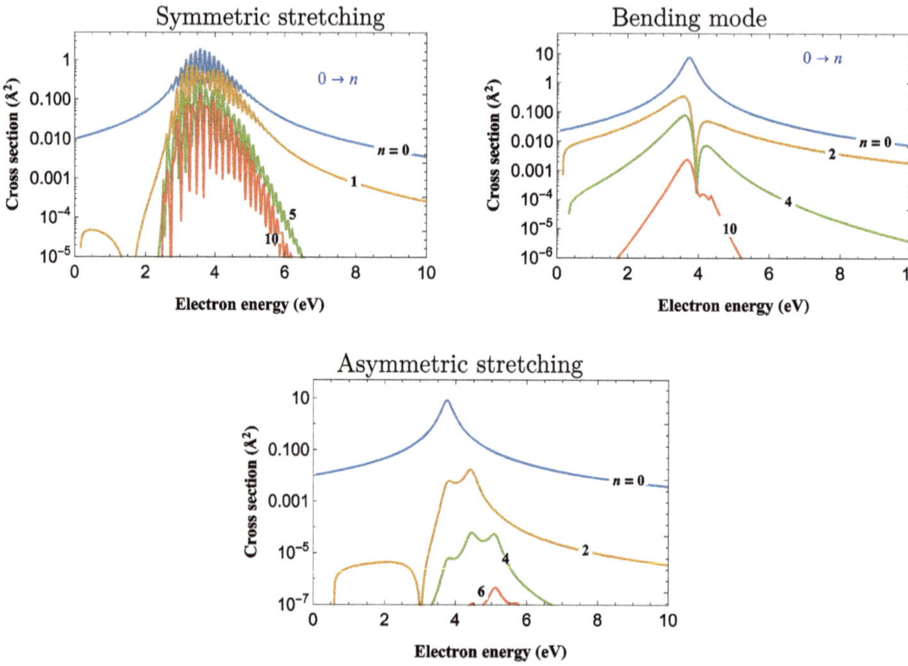

Figure 15.14. Summary of the electron–CO_2 vibrational excitation cross sections.

Figure 15.14 shows selected cross section results, as a function of the incident electron energy, for the three normal modes of CO_2.

NO^+

NO^+ has been proved to be one of the most important molecular ionic species involving air-assisted processes. During the re-entry of a space vehicle in the high shells of the terrestrial atmosphere, a cold plasma is formed close to the wall of the spacecraft. The same elementary processes contribute to the chemistry of other discharges in N_2 and O_2 mixtures such as those involved in the synthesis of nitrogen oxides, cleaning of exhaust gases resulting from combustion, plasma assisted combustion and streamer propagation. Different experimental studies, using various techniques, have been carried out for this molecular system, such as stationary afterglow [139], trapped ion technique [140], merged electron–ion beam technique [141], shock tube [142] and storage-ring measurements [143], complemented by several theoretical approaches [54, 55, 144–146]. Figure 15.15 shows a typical state-to-state, Rydberg resonance dominated, dissociative recombination, resonant elastic collision and vibrational transition cross sections for the NO^+ target in its $v_i^+ = 1$ vibrational level, while figure 15.16 gives the corresponding rate coefficients for the first seven vibrational levels of the molecular cation.

N_2^+

Molecular nitrogen is the most abundant molecule in the terrestrial atmosphere as well as in those of Titan and Triton. Its cation, N_2^+, is therefore prevalent in the

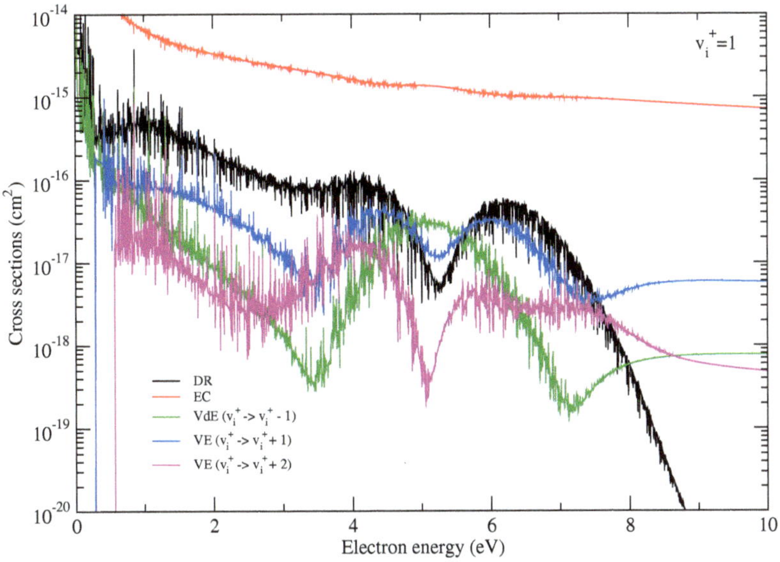

Figure 15.15. Total dissociative recombination (DR), elastic collisions (EC), vibrational excitation (VE) and vibrational de-excitation (VdE) cross sections in electron–NO^+ collisions.

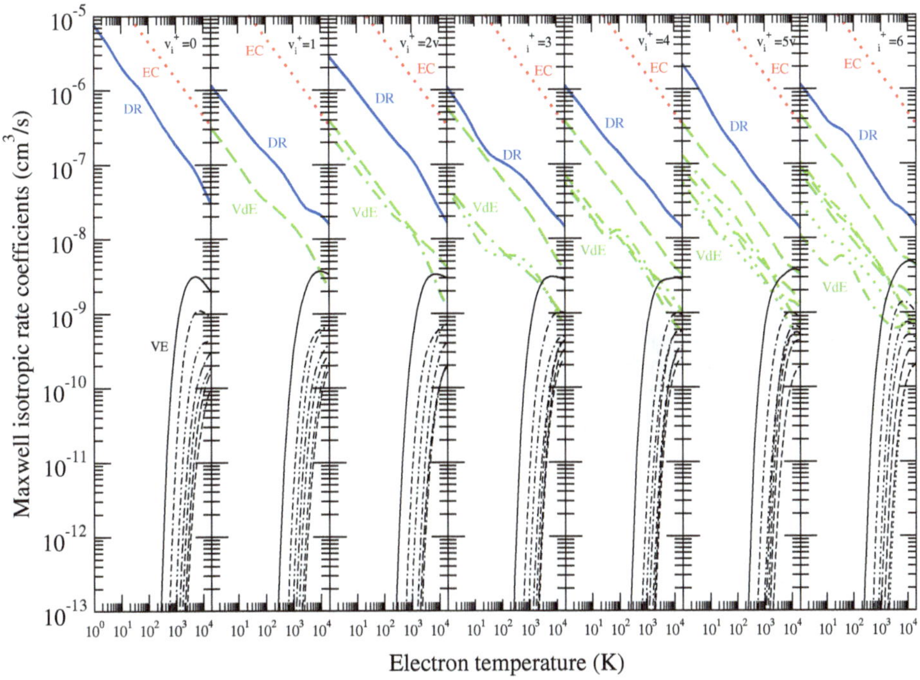

Figure 15.16. Dissociative recombination (DR), elastic collisions (EC), vibrational excitation (VE) and vibrational de-excitation (VdE) rate coefficients in electron–NO^+ collisions.

Earth's ionosphere as well as in nitrogen plasmas produced for reasons varying from lightning strikes to combustion. Therefore there is a strong interest to put together comprehensive nitrogen chemistries for plasmas containing air [80, 82, 147, 148].

The available temperature-dependent electron-induced N_2^+ reaction rate measurements have been performed on vibrationally hot molecules in storage-ring and merged-beam experiments [149–152], completed by flowing afterglow Langmuir probe measurements at 300 K [153–155]. Several MQDT-based calculations have been performed on this cation using different molecular datasets [156, 157].

The results of our most recent calculations using the *ab initio* potential energy curves calculated with R-matrix theory [158] are presented in figures 15.17 and 15.18. Figure 15.17 contains the DR and the monoquantic vibrational transition cross sections for the lowest four vibrational levels of the cation.

CO^+

The carbon monoxide ion CO^+ is one of the most abundant ions detected in the interstellar medium [159], in the coma and tail regions of comets [160], and it is of key relevance for the Martian atmosphere [161]. A pioneering theoretical MQDT-based study [162] was performed for this molecular cation, while the most recent storage-ring experiments [163] complete the earlier afterglow [154, 164] and merged-beam [165] experiments. Our most recent rate coefficients for the DR and its competitive processes calculated at the highest order of complexity are given in figure 15.19. More recent computations of Guberman [166] result in quite similar rate coefficients.

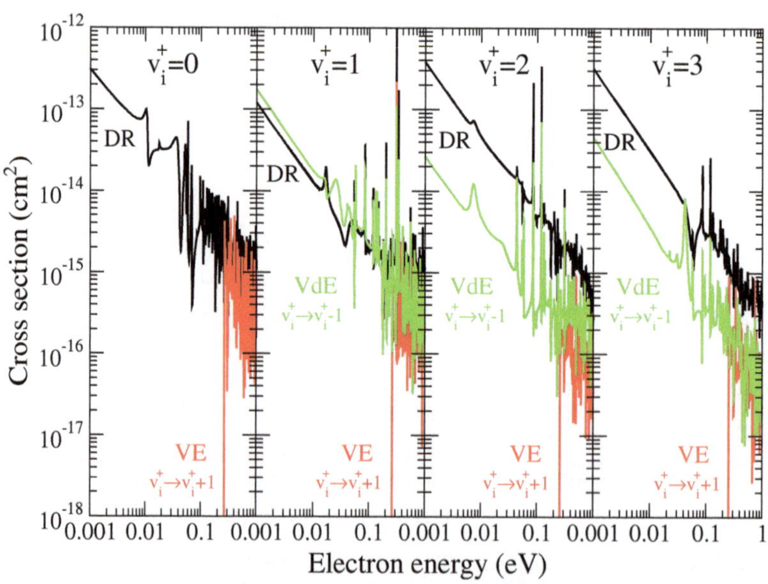

Figure 15.17. Dissociative recombination (DR), vibrational excitation (VE) and vibrational de-excitation (VdE) total cross sections in electron–N_2^+ collisions.

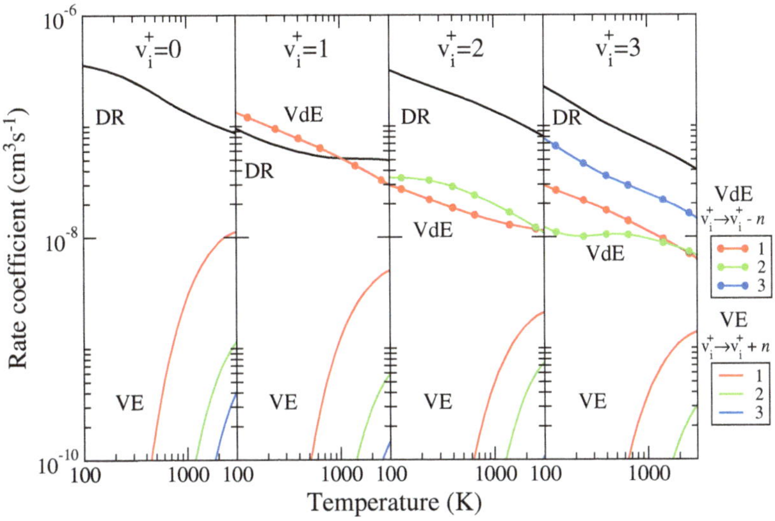

Figure 15.18. Maxwellian dissociative recombination (DR), vibrational excitation (VE) and vibrational de-excitation (VdE) rate coefficients in electron–N_2^+ collisions.

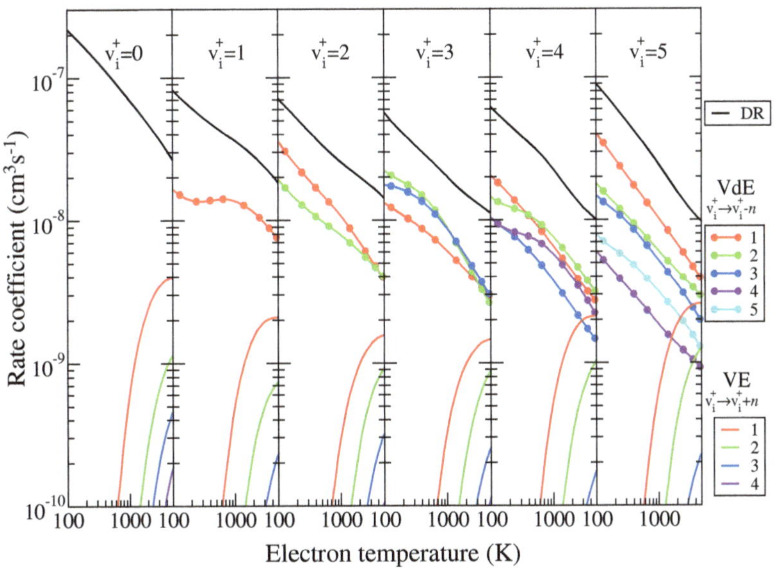

Figure 15.19. Dissociative recombination (DR), vibrational excitation (VE) and vibrational de-excitation (VdE) rate coefficients in electron–CO^+ collisions.

15.4 Non-resonant vibronic excitations in Earth atmosphere molecules

The cross sections for the state-specific vibronic excitations, including dissociative channels and ionization for air molecules, have been investigated in the past in the framework of different theoretical approaches from classical [167] to semi-classical methods [168–171] and have been recently reviewed [32].

The dipole-allowed excitation cross sections in electron–molecule collisions have been shown to be satisfactorily treated in the framework of simplified theoretical approaches which offer the advantage of low-computational cost for providing complete sets of state-to-state cross sections. These cross sections are needed for state-to-state modeling, and the dynamical information provided can be judged to be reasonably accurate when compared with experiments and more accurate scattering theories. This is the case of the BE*f*-scaling method [172], demonstrating that theoretical Born results, overestimating the cross section, can be brought into fairly good agreement with experiments by a simple rescaling procedure with the experimental optical oscillator strengths. However, these approaches lack the description of non-adiabatic effects, i.e. vibronic coupling, strongly affecting the vibrational progression. The inclusion of non-adiabatic coupling is of particular relevance in a state-to-state kinetic scheme as these couplings modify the excitation probability to specific vibrational levels of the excited state, thus affecting the predissociation yields. This is the case for the lowest electronic terms of the $^1\Sigma_u^+$ and $^1\Pi_u$ spectroscopic series of the N_2 molecule. The excitation to these states attracted attention in the past, representing the main dissociative channel, through predissociation mechanisms [173, 174] and also the relevant component of the emission UV spectrum. These states exhibit a strong mixed valence-Rydberg character [175, 176], resulting in significant perturbation of vibronic bands. Also for the O_2 Schumann–Runge continuum [177, 178] the continuum levels of the $B^3\Sigma_u^-$ state are perturbed by the vibronic coupling with the Rydberg $E^3\Sigma_u^-$ state [179], leading to resonances in the profile of the electron energy-loss spectrum, known as *longest band* (LB) and *second band* (SB) transitions. Also, the so-called β- and δ-bands of the NO emission spectrum are known to be characterized by the strong vibronic coupling of mixed valence-Rydberg $B^2\Pi$ and $C^2\Pi$ states.

Recently the electron-impact induced vibronic transitions have been considered in the framework of the *similarity approach* [180], reformulated so as to allow, in quite an easy way, the inclusion of non-adiabatic effects due to vibronic coupling [181, 182]. The similarity cross section has a simplified expression for the state-to-state cross section, i.e.

$$\sigma_{v'v''} = \frac{2\pi e^4}{(\Delta E_{v'v''})^2} f_{v'v''}\,\varphi(x), \tag{15.17}$$

where $\Delta E_{v'v''}$ is the transition energy, $f_{v'v''}$ is the oscillator strength for the vibronic transition, φ is the so-called *similarity function*, a universal function of the reduced incident electron energy $x = E/\Delta E_{v'v''}$, describing the collision dynamics, optimized so as to fit experimental data for electron-induced excitations in atomic systems. The original formulation has been extended to include the vibronic perturbation [181], going through the re-definition of the optical oscillator strength for transitions from the ground to the complex of excited states [183]

$$f_{v'v''} = \frac{2}{3}\frac{g_{ex}}{g_X}\Delta E_{v'v''}|\langle \hat{\chi}^d_{ex}|\hat{M}^d(R)|\chi_X\rangle|^2, \quad (15.18)$$

where g is the statistical weight of electronic terms. $\hat{\chi}^d_{ex}(R) \equiv \begin{pmatrix} \chi^d_a(R) \\ \chi^d_b(R) \\ \chi^d_c(R) \end{pmatrix}$ represents the vector of final-state radial wavefunctions, resulting from the numerical solution of a system of coupled radial Schrödinger equations for states in the diabatic representation, with transition dipole moment vector $\hat{M}^d(R)$, coupling the state-complex with the ground state.

The method has been shown [181] to give state-specific and total cross sections that compare satisfactorily with theoretical and experimental results in the literature [184–187], as can be appreciated in figure 15.20, where the similarity excitation cross sections for the Birge–Hopfield and $X^1\Sigma_g^+ \rightarrow b'^1\Sigma_u^+$ transitions in N_2 are compared with experiments [184, 188, 189]. The comparison is performed accounting properly for two aspects: first the spectral energy window explored by measurements is limited and only collects emission from a finite number of v' levels; second, the

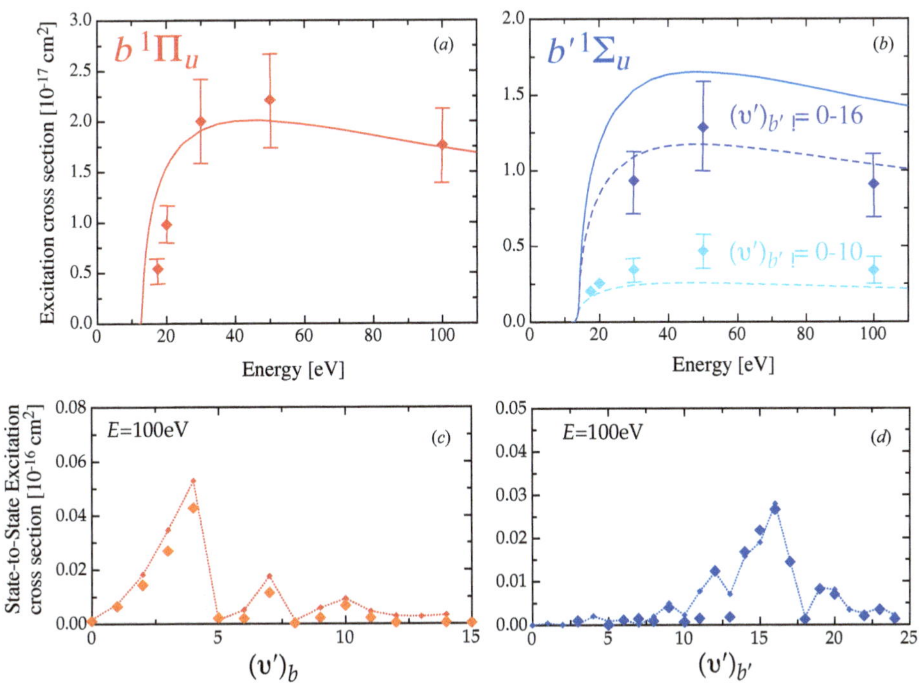

Figure 15.20. (a), (b) Total cross section as a function of collision energy and (c), (d) state-specific cross section as a function of the final vibrational levels, for the excitations to the $b^1\Pi_u$ (Birge–Hopfield system) and $b'^1\Sigma_u^+$ states from the $v'' = 0$ level of the ground state of the N_2 molecule [181] obtained using a decoupled scheme for assignment and compared with experiments (solid diamonds). Total [185, 184] and state-specific [188, 189] cross sections.

experimental vibrational analysis assigned emission intensity in a decoupled frame, while vibronic coupling results in levels of mixed character, therefore the level-character $P_a^{v'} = \langle \chi_a^d | \chi_a^d \rangle$, weighting the state-character of the vibrational wavefunction with respect to the three coupled terms, must switch from 0 to 1.

Figure 15.21(a)–(d) displays the energy profile of the total and state-specific cross sections for the excitation transitions to the mixed valence-Rydberg $B^2\Pi$ and $C^2\Pi$ states from the $v'' = 0$ of the ground state of NO, obtained using the similarity approach [182]. In particular, in figure 15.21(a) the excitation to the Rydberg C state is compared with the experimental integral cross section obtained from electron energy loss in [190] and the BE-scaling cross sections from experimental generalized oscillator strengths in [191]. The similarity cross section does not reproduce accurately the experimental energy profile, however, the agreement still remains remarkable, improved with respect to the results obtained by Adamson *et al* [180] with the original similarity formulation. Also, the comparison with the newest experiments [191] is satisfactory, in particular in the threshold region.

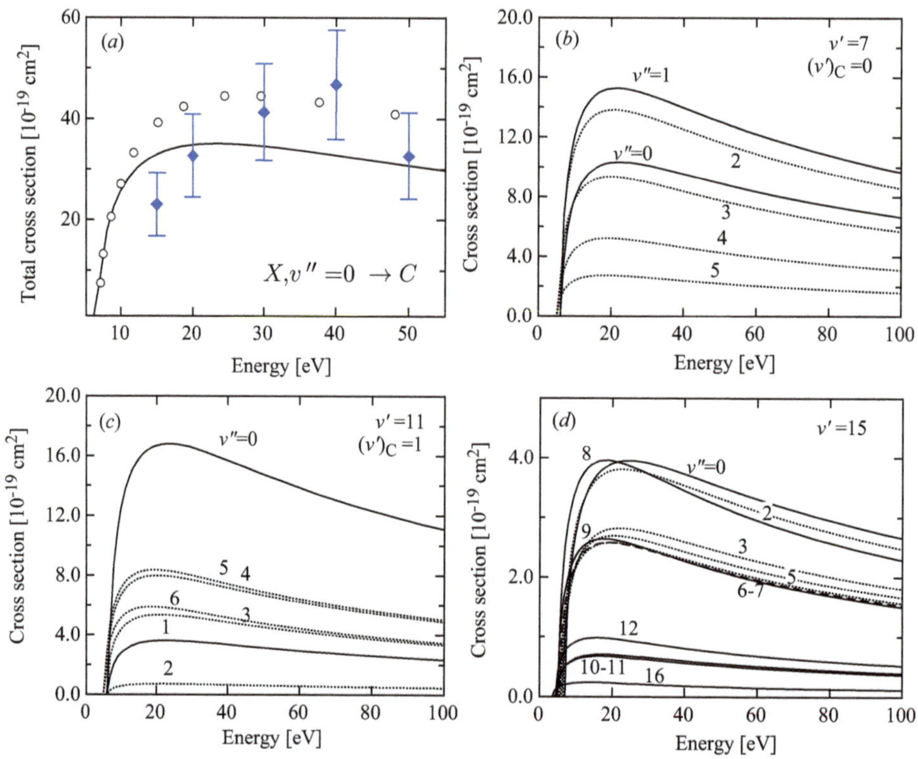

Figure 15.21. (a) Total cross section for excitation to the $C^2\Pi$ states from the $v'' = 0$ level of the ground state of the NO molecule as a function of collision energy (continuous lines) [182], compared to experimental results in the literature. Closed diamonds: [190]; open circles: [191]. (b)–(d) State-specific cross sections for excitation to levels $v' = 7, 11, 15$ from selected v'' levels of the ground state of the NO molecule as a function of collision energy [182]. The corresponding vibrational levels of the C state in the decoupled scheme are indicated, $(v')_C$.

Total cross sections for excitations from a vibrationally excited NO molecule, although obtained in a straightforward manner in the current approach, are affected by the choice to model the vibronic coupling in the excited states by considering only the *B*/*C* interaction and neglecting the coupling with other excited terms of the same symmetry. This limits the level of confidence for transitions to high v' levels that unfortunately give a non-negligible contribution to the excitation cross section already from $v' = 1$. On the other hand, the state-to-state cross sections for excitation from v'' levels to selected v' values lower than 27 can be considered to be unaffected by the limitations of the model. In figure 15.21(b)–(d) selected final levels are assigned to the $C\,^2\Pi$ state, although the state-character analysis, reported in [182], clearly shows that $v' = 15$ has a markedly mixed character. The interest in these transitions is justified by the predissociation branching ratio characterizing the levels $C\,^2\Pi$. In fact, predissociation and spontaneous emission are level-dependent competing channels, and $(v')_C$ levels are expected to undergo predissociation on a nanosecond time-scale, due to the spin–orbit coupling with a quartet electronic state, thus mainly contributing to dissociation channels. The vibrational dependence obviously mimics that of optical oscillator strengths, showing the role of vibrationally excited molecules in contributing to the mechanism. The initial vibrational excitation acts also to reduce the excitation threshold, governing the enhancement with v'' of the corresponding rate coefficients at low average values of the electron energy.

The similarity approach has been shown to be predictive despite the simplicity of its formulation and could be considered as a valuable theoretical tool for the derivation of cross sections to be included in state-to-state kinetics models and also very promising for the treatment of electron-impact induced excitation processes in triatomic molecules, accounting for the dependence of the cross section on the different vibrational modes, which even now poses problems for accurate quantum approaches.

15.5 Conclusion

In this chapter we have reviewed electron–molecule collision processes that are important in planetary atmospheres. Our methodologies have proved to be useful for providing key electron collision cross section data for resonant and non-resonant processes. For the diatomic species considered, H_2, N_2, N_2^+, O_2, NO, NO^+, CO and CO^+, the results presented can be considered to be fairly reliable. For the triatomic CO_2 molecule they must be considered to be less certain. This is because of the approximations thus-far employed for treating the nuclear-motion problem which includes neglect of all couplings between vibrational modes, even that involved in the famous CO_2 Fermi resonance. Work on improving the treatment of electron collisions with CO_2, with a particular focus on the electron-impact dissociation problem, is currently underway.

There are a number of other key triatomic species, such as water, methane, nitric oxides, etc, which would also benefit from having their electron-impact cross sections computed using similar treatments to those considered here.

Acknowledgements

The results on electron–molecular cation collisions reported here have been obtained with the decisive contribution of O Motapon (Douala), K Chakrabarti (Calcutta), D Little (London), A Bultel (Rouen) and Y Moulane (Marrakesh).

IFS, JZsM and VL are grateful for generous financial support from La Région Haute-Normandie, via the GRR Electronique, Energie et Matériaux and the project BIOENGINE, from the Fédération de Recherche 'Energie, Propulsion, Environnement', and from LabEx EMC^3 and FEDER via the projects PicoLIBS (ANR-10-LABEX-09-01), EMoPlaF and CO_2-VIRIDIS.

They acknowledge support from the CNRS via GdR THEMS and the Programme National 'Physique et Chimie du Milieu Interstellaire' (PCMI) of CNRS/INSU with INC/INP co-funded by CEA and CNES. They are grateful to IAEA (Vienna) via the Coordinated Research Project 'Light Element Atom, Molecule and Radical Behaviour in the Divertor and Edge Plasma Regions', and to the French Fédération de Recherche 'Fusion par Confinement Magnétique' (CNRS and CEA).

JZsM acknowledges support from USPC via ENUMPP and Labex SEAM.

References

[1] Hueso R *et al* 2010 First Earth-based detection of a superbolide on Jupiter *Astrophys. J. Lett.* **721** L129

[2] Celiberto R, Janev R, Laricchiuta A, Capitelli M, Wadehra J and Atems D 2001 Cross section data for electron-impact inelastic processes of vibrationally excited molecules of hydrogen and its isotopes *At. Data Nucl. Data Tables* **77** 161–213

[3] Janev R K, Reiter D and Samm U 2003 Collision processes in low-temperature hydrogen plasmas *Forschungszentrum Jülich report* JUEL-4105.

[4] Khakoo M A and Trajmar S 1986 Electron-impact excitation of the $a^3\Sigma_g^+$, $B^1\Sigma_u^+$, $c^3\Pi_u$, and $C^1\Pi_u$ states of H_2 *Phys. Rev.* A **34** 146–56

[5] Ajello J M, Shemansky D, Kwok T L and Yung Y L 1984 Studies of extreme-ultraviolet emission from Rydberg series of H_2 by electron impact *Phys. Rev.* A **29** 636–53

[6] De Heer F J and Carrière J D 1971 Emission of the Werner band system and Lyman-α radiation for 0.05–6-keV electrons in H_2 *J. Chem. Phys.* **55** 3829–35

[7] Arrighini G, Biondi F and Guidotti C 1980 A study of the inelastic scattering of fast electrons from molecular hydrogen *Mol. Phys.* **41** 1501–14

[8] Celiberto R, Laricchiuta A, Lamanna U T, Janev R K and Capitelli M 1999 Electron-impact excitation cross sections of vibrationally excited $X^1\Sigma_g^+$ H_2 and D_2 molecules to Rydberg states *Phys. Rev.* A **60** 2091–103

[9] Mu-Tao L, Lucchese R R and McKoy V 1982 Electron-impact excitation and dissociation processes in H_2 *Phys. Rev.* A **26** 3240–48

[10] Hall R I and Andric L 1984 Electron impact excitation of H_2 (D_2). Resonance phenomena associated with the $X^2\Sigma_u^+$ and $B^2\Sigma_g^+$ states of H_2^- in the 10 eV region *J. Phys. B: At. Mol. Phys.* **17** 3815

[11] Nishimura H and Danjo A 1986 Differential cross section of electron scattering from molecular hydrogen. II. $b^3\Sigma_u^+$ excitation *J. Phys. Soc. Japan* **55** 3031–36

[12] Chung S, Lin C C and Lee E T P 1975 Dissociation of the hydrogen molecule by electron impact *Phys. Rev.* A **12** 1340–49
[13] Rescigno T N and Schneider B I 1988 Electron-impact excitation of the $b^3\Sigma_u^+$ state of H_2 using the complex Kohn method: *R* dependence of the cross section *J. Phys. B: At. Mol. Opt. Phys.* **21** L691
[14] Lima M A P, Brescansin L M, da Silva A J R, Winstead C and McKoy V 1990 Applications of the Schwinger multichannel method to electron–molecule collisions *Phys. Rev.* A **41** 327–32
[15] Stibbe D T and Tennyson J 1998 Near-threshold electron impact dissociation of H_2 within the adiabatic nuclei approximation *New J. Phys.* **1** 2
[16] Chung S and Lin C C 1978 Application of the close-coupling method to excitation of electronic states and dissociation of H_2 by electron impact *Phys. Rev.* A **17** 1874–91
[17] Laricchiuta A, Celiberto R and Janev R K 2004 Electron-impact-induced allowed transitions between triplet states of H_2 *Phys. Rev.* A **69** 022706
[18] Celiberto R, Baluja K L, Janev R K and Laporta V 2016 Electron-impact dissociation cross sections of vibrationally excited He_2^+ molecular ion *Plasma Phys. Controlled Fus.* **58** 014024
[19] Adamczyk B, Boerboom A J H, Schram B L and Kistemaker J 1966 Partial ionization cross sections of He, Ne, H_2, and CH_4 for electrons from 20 to 500 eV *J. Chem. Phys.* **44** 4640–42
[20] Crowe A and McConkey J W 1973 Dissociative ionization by electron impact. I. Protons from H_2 *J. Phys. B: At. Mol. Phys.* **6** 2088
[21] Rapp D and Englander-Golden P 1965 Total cross sections for ionization and attachment in gases by electron impact. I. Positive ionization *J. Chem. Phys.* **43** 1464–79
[22] Corrigan S J B 1965 Dissociation of molecular hydrogen by electron impact *J. Chem. Phys.* **43** 4381–86
[23] Straub H C, Renault P, Lindsay B G, Smith K A and Stebbings R F 1996 Absolute partial cross sections for electron-impact ionization of H_2, N_2, and O_2 from threshold to 1000 eV *Phys. Rev.* A **54** 2146–53
[24] Celiberto R, Cives P, Cacciatore M, Capitelli M and Lamanna U 1990 Direct electron-impact collision cross sections involving vibrationally excited $D_2(v)$ molecules relevant to D^- sources *Chem. Phys. Lett.* **169** 69–73
[25] Wadehra J M 1986 Vibrational excitation and dissociative attachment *Nonequilibrium Vibrational Kinetics—Topics in Current Physics* ed M Capitelli (Berlin: Springer) pp 191–232
[26] Stibbe D T and Tennyson J 1998 Electron–H_2 scattering resonances as a function of bond length *J. Phys. B: At. Mol. Opt. Phys.* **31** 815
[27] Schulz G J 1959 Formation of H^- ions by electron impact on H_2 *Phys. Rev.* **113** 816–19
[28] Schulz G J and Asundi R K 1965 Formation of H^- by electron impact on H_2 at low energy *Phys. Rev. Lett.* **15** 946–49
[29] Srivastava S K and Orient O J 1984 Polar dissociation as a source of negative ions *AIP Conf. Proc.* **111** 56–66
[30] Rapp D, Sharp T E and Briglia D D 1965 Large isotope effect in the formation of H^- or D^- by electron impact on H_2, HD, and D_2 *Phys. Rev. Lett.* **14** 533–35
[31] Fabrikant I I, Wadehra J M and Xu Y 2002 Resonance processes in e–H_2 collisions: dissociative attachment and dissociation from vibrationally and rotationally excited states *Phys. Scr.* **2002** 45

[32] Wadehra J M 1984 Dissociative attachment to rovibrationally excited H_2 *Phys. Rev.* A **29** 106–10

[33] Celiberto R *et al* 2016 Atomic and molecular data for spacecraft re-entry plasmas *Plasma Sources Sci. Technol.* **25** 033004

[34] Celiberto R, Janev R, Wadehra J and Tennyson J 2012 Dissociative electron attachment to vibrationally excited H_2 molecules involving the $^2\Sigma_g^+$ resonant Rydberg electronic state *Chem. Phys.* **398** 206–13

[35] Celiberto R, Janev R K, Wadehra J M and Laricchiuta A 2009 Cross sections for 14-eV $e - H_2$ resonant collisions: dissociative electron attachment *Phys. Rev.* A **80** 012712

[36] Schmeltekopf A L, Fehsenfeld F C and Ferguson E E 1967 Laboratory measurement of the rate constant for $H^- + H \to H_2 + e$ *Astrophys. J.* **148** L155

[37] Bieniek R J and Dalgarno A 1979 Associative detachment in collisions of H and H^- *Astrophys. J.* **228** 635

[38] Launay J M, LeDourneuf M and Zeippen C J 1991 The reversible $H + H^- \rightleftharpoons H_2(v, j) + e^-$ reaction: a consistent description of the associative detachment and dissociative attachment processes using the resonant scattering theory *Astron. Astrophys.* **252** 842–52

[39] Čižek M, Horáček J and Domcke W 1998 Nuclear dynamics of the H_2^- collision complex beyond the local approximation: associative detachment and dissociative attachment to rotationally and vibrationally excited molecules *J. Phys. B: At. Mol. Opt. Phys.* **31** 2571

[40] Kreckel H, Bruhns H, Čižek M, Glover S C O, Miller K A, Urbain X and Savin D W 2010 Experimental results for H_2 formation from H^- and H and implications for first star formation *Science* **329** 69–71

[41] Bardsley J N and Wadehra J M 1979 Dissociative attachment and vibrational excitation in low-energy collisions of electrons with H_2 and D_2 *Phys. Rev.* A **20** 1398–405

[42] Atems D and Wadehra J 1992 Vibrational excitation of H_2 and HCl by low-energy electron impact. An isotope scaling law *Chem. Phys. Lett.* **197** 525–29

[43] Allan M 1985 Experimental observation of structures in the energy dependence of vibrational excitation in H_2 by electron impact in the $^2\Sigma_u^+$ resonance region *J. Phys. B: At. Mol. Phys.* **18** L451

[44] Mündel C, Berman M and Domcke W 1985 Nuclear dynamics in resonant electron–molecule scattering beyond the local approximation: vibrational excitation and dissociative attachment in H_2 and D_2 *Phys. Rev.* A **32** 181–93

[45] Celiberto R, Janev R K, Wadehra J M and Laricchiuta A 2008 Cross sections for 11–14-eV $e - H_2$ resonant collisions: vibrational excitation *Phys. Rev.* A **77** 012714

[46] Celiberto R, Janev R K, Laporta V, Tennyson J and Wadehra J M 2013 Electron-impact vibrational excitation of vibrationally excited H_2 molecules involving the resonant $^2\Sigma_g^+$ Rydberg-excited electronic state *Phys. Rev.* A **88** 062701

[47] Comer J and Read F H 1971 Potential curves and symmetries of some resonant states of H_2^- *J. Phys. B: At. Mol. Phys.* **4** 368

[48] Atems D E and Wadehra J M 1993 Resonant contributions to dissociation of H_2 by low-energy electron impact *J. Phys. B: At. Mol. Opt. Phys.* **26** L759

[49] Laporta V 2017 *Collisions entre électrons et molécules: Mécanismes réactionnels, modèles théoriques et applications aux plasmas hors-équilibre* Habilitation à diriger des recherches Université du Havre (ULH)

[50] Laporta V, Celiberto R and Wadehra J M 2012 Theoretical vibrational-excitation cross sections and rate coefficients for electron-impact resonant collisions involving rovibrationally excited N_2 and NO molecules *Plasma Sources Sci. Technol.* **21** 055018
[51] Laporta V and Celiberto R 2012 Resonant electron–molecule vibrational excitation cross sections and rate coefficients for atmospheric plasmas *J. Phys.: Conf. Ser.* **388** 052036
[52] Laporta V, Little D A, Celiberto R and Tennyson J 2014 Electron-impact resonant vibrational excitation and dissociation processes involving vibrationally excited N_2 molecules *Plasma Sources Sci. Technol.* **23** 065002
[53] Little D A, Chakrabarti K, Mezei J Z, Schneider I F and Tennyson J 2014 Dissociative recombination of N_2^+: an *ab initio* study *Phys. Rev.* A **90** 052705
[54] Schneider I F, Rabadán I, Carata L, Andersen L H, Suzor-Weiner A and Tennyson J 2000 Dissociative recombination of NO^+: calculations and comparison with experiment *J. Phys. B: At. Mol. Opt. Phys.* **33** 4849
[55] Motapon O, Fifirig M, Florescu A, Tamo F O W, Crumeyrolle O, Bultel A, Vervisch P, Tennyson J and Schneider I F 2006 Reactive collisions between electrons and NO^+ ions: rate coefficient computations and relevance for the air plasma kinetics *Plasma Sources Sci. Technol.* **15** 23
[56] Bultel A, Cheron B G, Bourdon A, Motapon O and Schneider I F 2006 Collisional–radiative model in air for Earth re-entry problems *Phys. Plasmas* **13** 043502
[57] Laporta V, Celiberto R and Tennyson J 2013 Resonant vibrational-excitation cross sections and rate constants for low-energy electron scattering by molecular oxygen *Plasma Sources Sci. Technol.* **22** 025001
[58] Laporta V, Celiberto R and Tennyson J 2015 Dissociative electron attachment and electron-impact resonant dissociation of vibrationally excited O_2 molecules *Phys. Rev.* A **91** 012701
[59] Laporta V, Celiberto R and Tennyson J 2014 Rate coefficients for dissociative attachment and resonant electron-impact dissociation involving vibrationally excited O_2 molecules *AIP Conf. Proc.* **1628** 939–42
[60] Laporta V, Cassidy C M, Tennyson J and Celiberto R 2012 Electron-impact resonant vibration excitation cross sections and rate coefficients for carbon monoxide *Plasma Sources Sci. Technol.* **21** 045005
[61] Laporta V, Tennyson J and Celiberto R 2016 Carbon monoxide dissociative attachment and resonant dissociation by electron-impact *Plasma Sources Sci. Technol.* **25** 01LT04
[62] Mezei J Z *et al* 2015 Dissociative recombination and vibrational excitation of CO: model calculations and comparison with experiment *Plasma Sources Sci. Technol.* **24** 035005
[63] Moulane Y, Mezei J Z, Laporta V, Jehin E, Benkhaldoun Z and Schneider I F 2018 Reactive collision of electrons with CO^+ in cometary coma *Astron. Astrophys.* **615** A53
[64] Laporta V, Tennyson J and Celiberto R 2016 Calculated low-energy electron-impact vibrational excitation cross sections for CO_2 molecule *Plasma Sources. Sci. Technol.* **25** 06LT02
[65] Werner H-J, Knowles P J, Knizia G, Manby F R and Schütz M 2012 Molpro: a general-purpose quantum chemistry program package *Wiley Interdiscip. Rev.: Comput. Mol. Sci.* **2** 242–53
[66] Tennyson J and Noble C J 1984 RESON: for the automatic detection and fitting of Breit–Wigner resonances *Comput. Phys. Commun.* **33** 421–24

[67] Tennyson J 2010 Electron–molecule collision calculations using the R-matrix method *Phys. Rep.* **491** 29–76
[68] Carr J M, Galiatsatos P G, Gorfinkiel J D, Harvey A G, Lysaght M A, Madden D, Masin Z, Plummer M, Tennyson J and Varambhia H N 2012 UKRmol: a low-energy electron– and positron–molecule scattering suite *Eur. Phys. J.* D **66** 58
[69] Tennyson J, Brown D B, Munro J J, Rozum I, ambhia H N V and Vinci N 2007 Quantemol-N: an expert system for performing electron molecule collision calculations using the R-matrix method *J. Phys.: Conf. Ser.* **86** 012001
[70] Fujimoto M M, Brigg W J and Tennyson J 2012 R-matrix calculations of differential and integral cross sections for low-energy electron collisions with ethanol *Euro. Phys. J.* D **66** 204
[71] Domcke W 1991 Theory of resonance and threshold effects in electron–molecule collisions: the projection-operator approach *Phys. Rep.* **208** 97–188
[72] Giusti A 1980 A multichannel quantum defect approach to dissociative recombination *J. Phys. B: At. Mol. Phys.* **13** 3867
[73] Jungen C 2011 Elements of quantum defect theory *Handbook of High Resolution Spectroscopy* ed M Quack and F Merkt (Chichester: Wiley)
[74] Motapon O, Pop N, Argoubi F, Mezei J Z, Epee Epee M D, Faure A, Telmini M, Tennyson J and Schneider I F 2014 Rotational transitions induced by collisions of HD^+ ions with low-energy electrons *Phys. Rev.* A **90** 012706
[75] Chakrabarti K, Mezei J Z, Motapon O, Faure A, Dulieu O, Hassouni K and Schneider I F 2018 Dissociative recombination of the CH^+ molecular ion at low energy *J. Phys. B: At. Mol. Opt. Phys.* **51** 104002
[76] Shang J and Surzhikov S 2012 Nonequilibrium radiative hypersonic flow simulation *Prog. Aerosp. Sci.* **53** 46–65
[77] Gordillo-Vázquez F J 2008 Air plasma kinetics under the influence of sprites *J. Phys. D. Appl. Phys.* **41** 234016
[78] Laporta V and Bruno D 2013 Electron-vibration energy exchange models in nitrogen-containing plasma flows *J. Chem. Phys.* **138** 104319
[79] Heritier K L, Jaffe R L, Laporta V and Panesi M 2014 Energy transfer models in nitrogen plasmas: analysis of $N_2(X^1\Sigma_g^+) - N(^4S_u) - e$ interaction *J. Chem. Phys.* **141** 184302
[80] Bultel A and Annaloro J 2013 Elaboration of collisional–radiative models for flows related to planetary entries into the Earth and Mars atmospheres *Plasma Sources Sci. Technol.* **22** 025008
[81] Annaloro J and Bultel A 2014 Vibrational and electronic collisional–radiative model in air for Earth entry problems *Phys. Plasmas* **21** 123512
[82] Dutuit O *et al* 2013 Critical review of N, N^+, N_2^+, N^{++}, and N_2^{++} main production processes and reactions of relevance to Titan's atmosphere *Astrophys. J. Suppl. Ser.* **204** 20
[83] Schulz G J 1962 Vibrational excitation of nitrogen by electron impact *Phys. Rev.* **125** 229–32
[84] Schulz G J 1964 Vibrational excitation of N_2, CO, and H_2 by electron impact *Phys. Rev.* A **135** A988–94
[85] Vicic M, Poparic G and Belic D S 1996 Large vibrational excitation of N_2 by low-energy electrons *J. Phys. B: At. Mol. Opt. Phys.* **29** 1273
[86] Ristić M, Poparić G and Belić D 2007 Rate coefficients for resonant vibrational excitation of N_2 *Chem. Phys.* **331** 410–16

[87] Comer J and Read F H 1971 Electron impact studies of a resonant state of N_2^- *J. Phys. B: At. Mol. Phys.* **4** 1055

[88] Wong C F and Light J C 1984 Application of R-matrix theory to resonant reactive electron–molecule scattering: vibrational excitation and dissociative attachment of N_2 and F_2 *Phys. Rev.* A **30** 2264–73

[89] Allan M 1985 Excitation of vibrational levels up to $v = 17$ in N_2 by electron impact in the 0–5 eV region *J. Phys. B: At. Mol. Opt. Phys.* **18** 4511

[90] Allan M 2005 Electron collisions with NO: elastic scattering, vibrational excitation and $^2\Pi_{1/2}$ $^2\Pi_{3/2}$ transitions *J. Phys. B: At. Mol. Opt. Phys.* **38** 603

[91] Sweeney C J and Shyn T W 1997 Measurement of absolute differential cross sections for the vibrational excitation of molecular nitrogen by electron impact in the $^2\Pi_g$ shape resonance region *Phys. Rev.* A **56** 1384–92

[92] Sun W, Morrison M A, Isaacs W A, Trail W K, Alle D T, Gulley R J, Brennan M J and Buckman S J 1995 Detailed theoretical and experimental analysis of low-energy electron–N_2 scattering *Phys. Rev.* A **52** 1229–56

[93] Dubé L and Herzenberg A 1979 Absolute cross sections from the 'boomerang model' for resonant electron–molecule scattering *Phys. Rev.* A **20** 194–213

[94] Schneider B I, Le Dourneuf M and Lan V K 1979 Resonant vibrational excitation of N_2 by low-energy electrons: an *ab initio* R-matrix calculation *Phys. Rev. Lett.* **43** 1926–29

[95] Berman M, Estrada H, Cederbaum L S and Domcke W 1983 Nuclear dynamics in resonant electron–molecule scattering beyond the local approximation: the 2.3-eV shape resonance in N_2 *Phys. Rev.* A **28** 1363–81

[96] Huo W M, McCoy V, Lima M A P and Gibson T L 1986 Electron–nitrogen molecule collisions in high temperature nonequilibrium air *Thermochemical Aspects of Re-entry Flows, Progress in Astronautics and Aeronautics* vol 103 ed J N Moss and C D Scott (New York: AIAA) pp 152–96

[97] Mihajlov A A, Stojanovic V D and Petrovic Z L 1999 Resonant vibrational excitation/de-excitation of $N_2(v)$ by electrons *J. Phys. D: Appl. Phys.* **32** 2620

[98] Houfek K, Rescigno T N and McCurdy C W 2008 Probing the nonlocal approximation to resonant collisions of electrons with diatomic molecules *Phys. Rev.* A **77** 012710

[99] Houfek K, Čížek M and Horáček J 2008 On irregular oscillatory structures in resonant vibrational excitation cross-sections in diatomic molecules *Chem. Phys.* **347** 250–56

[100] Itikawa Y 2006 Cross sections for electron collisions with nitrogen molecules *J. Phys. Chem. Ref. Data* **35** 31

[101] da Paixao F J, Lima M A P and McKoy V 1996 Elastic *e*–NO collisions *Phys. Rev.* A **53** 1400–06

[102] Zhang Z Y, Vanroose W, McCurdy C W, Orel A E and Rescigno T 2004 Low-energy electron scattering of NO: *ab initio* analysis of the $^3\Sigma^-$, $^1\Delta$, and $^1\Sigma^+$ shape resonances in the local complex potential model *Phys. Rev.* A **69** 062711

[103] Trevisan C S, Houfek K, Zhang Z, Orel A, McCurdy C W and Rescigno T N 2005 Nonlocal model of dissociative electron attachment and vibrational excitation of NO *Phys. Rev.* A **71** 052714

[104] Itikawa Y 2016 Cross sections for electron collisions with nitric oxide *J. Phys. Chem. Ref. Data* **45** 033106

[105] Song M-Y, Yoon J-S, Cho H, Karwasz G P, Kokoouline V, Nakamura Y and Tennyson J Cross sections for electron collisions with oxides of nitrogen *J. Phys. Chem. Ref. Data*

[106] Schulz G J 1973 Resonances in electron impact on diatomic molecules *Rev. Mod. Phys.* **45** 423–86

[107] Noble C J, Higgins K, Wöste G, Duddy P, Burke P G and Teubner P J O 1996 Resonant mechanisms in the vibrational excitation of ground state O_2 *Phys. Rev. Lett.* **76** 3534–37

[108] Krauss M, Neumann D, Wahl A C, Das G and Zemke W 1973 Excited electronic states of O_2^- *Phys. Rev.* A **7** 69–77

[109] Ewig C S and Tellinghuisen J 1991 *Ab initio* study of the electronic states of O_2^- in vacuo and in simulated ionic solids *J. Chem. Phys.* **95** 1097–106

[110] Noble C J and Burke P G 1992 R-matrix calculations of low-energy electron scattering by oxygen molecules *Phys. Rev. Lett.* **68** 2011–14

[111] Higgins K, Noble C J and Burke P G 1994 Low energy electron scattering by oxygen molecules *J. Phys. B: At. Mol. Opt. Phys.* **27** 3203

[112] Higgins K, Gillan C J, Burke P G and Noble C J 1995 Low-energy electron scattering by oxygen molecules. II. Vibrational excitation *J. Phys. B: At. Mol. Opt. Phys.* **28** 3391

[113] McConkey J, Malone C, Johnson P, Winstead C, McKoy V and Kanik I 2008 Electron impact dissociation of oxygen-containing molecules—a critical review *Phys. Rep.* **466** 1–103

[114] Rapp D and Briglia D D 1965 Total cross sections for ionization and attachment in gases by electron impact. II. Negative-ion formation *J. Chem. Phys.* **43** 1480–89

[115] Schulz G J 1962 Cross sections and electron affinity for O^- ions from O_2, CO, and CO_2 by electron impact *Phys. Rev.* **128** 178–86

[116] Christophorou L G, Compton R N, Hurst G S and Reinhardt P W 1965 Determination of electron-capture cross sections with swarm-beam techniques *J. Chem. Phys.* **43** 4273–81

[117] Itikawa Y 2009 Cross sections for electron collisions with oxygen molecules *J. Phys. Chem. Ref. Data* **38** 1–20

[118] Burgh E B, France K and McCandliss S R 2007 Direct measurement of the ratio of carbon monoxide to molecular hydrogen in the diffuse interstellar medium *Astrophys. J.* **658** 446

[119] Campbell L, Allan M and Brunger M J 2011 Electron impact vibrational excitation of carbon monoxide in the upper atmospheres of Mars and Venus *J. Geophys. Res.* **116** A09321

[120] Haddad G and Milloy H 1983 Cross sections for electron carbon monoxide collisions in the range 1–4 eV *Aust. J. Phys.* **36** 473

[121] Gorse C, Cacciatore M and Capitelli M 1984 Kinetic processes in non-equilibrium carbon monoxide discharges. I. Vibrational kinetics and dissociation rates *Chem. Phys.* **85** 165–76

[122] Gorse C and Capitelli M 1984 Kinetic processes in non-equilibrium carbon monoxide discharges. II. Self-consistent electron energy distribution functions *Chem. Phys.* **85** 177–87

[123] Itikawa Y 2015 Cross sections for electron collisions with carbon monoxide *J. Phys. Chem. Ref. Data* **44**

[124] Taylan O and Berberoglu H 2015 Dissociation of carbon dioxide using a microhollow cathode discharge plasma reactor: effects of applied voltage, flow rate and concentration *Plasma Sources Sci. Technol.* **24** 015006

[125] Goede A P, Bongers W A, Graswinckel M F, van de Sanden R M, Leins M, Kopecki J, Schulz A and Walker M 2014 Production of solar fuels by CO_2 plasmolysis *EPJ Web Conf.* **79** 01005

[126] Bogaerts A, Bie C, Snoeckx R and Kozák T 2016 Plasma based CO_2 and CH_4 conversion: a modeling perspective *Plasma Process. Polym.* **14** 1600070

[127] Pietanza L, Colonna G, Laporta V, Celiberto R, D'Ammando G, Laricchiuta A and Capitelli M 2016 Influence of electron molecule resonant vibrational collisions over the symmetric mode and direct excitation–dissociation cross sections of CO_2 on the electron energy distribution function and dissociation mechanisms in cold pure CO_2 plasmas *J. Phys. Chem.* A **120** 2614–28

[128] Kozák T and Bogaerts A 2015 Evaluation of the energy efficiency of CO_2 conversion in microwave discharges using a reaction kinetics model *Plasma Sources Sci. Technol.* **24** 015024

[129] Lombardi A, Faginas-Lago N, Pacifici L and Costantini A 2014 Modeling of energy transfer from vibrationally excited CO_2 molecules: cross sections and probabilities for kinetic modeling of atmospheres, flows, and plasmas *J. Phys. Chem.* A **117** 11430–40

[130] Kozák T and Bogaerts A 2014 Splitting of CO_2 by vibrational excitation in non-equilibrium plasmas: a reaction kinetics model *Plasma Sources Sci. Technol.* **23** 045004

[131] Janeco A, Pinh ao N R and Guerra V 2015 Electron kinetics in $He/CH_4/CO_2$ mixtures used for methane conversion *J. Phys. Chem.* C **119** 109–20

[132] Morrison M A, Lane N F and Collins L A 1977 Low-energy electron–molecule scattering: application of coupled-channel theory to *e*–CO_2 collisions *Phys. Rev.* A **15** 2186–201

[133] Cadez I, Gresteau F, Tronc M and Hall R I 1977 Resonant electron impact excitation of CO_2 in the 4 eV region *J. Phys. B: At. Mol. Phys.* **10** 3821

[134] Allan M 2001 Selectivity in the excitation of Fermi-coupled vibrations in CO_2 by impact of slow electrons *Phys. Rev. Lett.* **87** 033201

[135] Morgan L A 1998 Virtual states and resonances in electron scattering by CO_2 *Phys. Rev. Lett.* **80** 1873–75

[136] Tennyson J and Morgan L A 1999 Electron collisions with polyatomic molecules using the R-matrix method *Philos. Trans. R. Soc. Lond.* A **357** 1161–73

[137] Mazevet S, Morrison M A, Morgan L A and Nesbet R K 2001 Virtual-state effects on elastic scattering and vibrational excitation of CO_2 by electron impact *Phys. Rev.* A **64** 040701

[138] Itikawa Y 2002 Cross sections for electron collisions with carbon dioxide *J. Phys. Chem. Ref. Data* **31** 749–67

[139] Weller C S and Biondi M A 1968 Recombination, attachment, and ambipolar diffusion of electrons in photo-ionized NO afterglows *Phys. Rev.* **172** 198

[140] Walls F L and Dunn G H 1974 Measurement of total cross sections for electron recombination with NO^+ and O_2^+ using ion storage techniques *J. Geophys. Res.* **79** 1911

[141] Mul P M and McGowan J W 1979 Merged electron–ion beam experiments. III. Temperature dependence of dissociative recombination for atmospheric ions NO^+, O_2^+ and N_2^+ *J. Phys. B: At. Mol. Opt. Phys.* **12** 1591

[142] Davidson D F and Hobson R M 1987 The shock tube determination of the dissociative recombination rate of NO^+ *J. Phys. B: At. Mol. Phys.* **20** 5753

[143] Vejby-Christensen L, Kella D, Pedersen H B and Andersen L H 1998 Dissociative recombination of NO^+ *Phys. Rev.* A **57** 3627

[144] Bardsley J N 1968 The theory of dissociative recombination *J. Phys. B: At. Mol. Opt. Phys.* **1** 365

[145] Sun H and Nakamura H 1990 Theoretical study of the dissociative recombination of NO^+ with slow electrons *J. Chem. Phys.* **93** 6491

[146] Vâlcu B, Schneider I F, Raoult M, Strömholm C, Larsson M and Suzor-Weiner A 1998 Rotational effects in low energy dissociative recombination of diatomic ions *Eur. Phys. J.* D **1** 71–8

[147] Capitelli M, Colonna G, D'Ammando G, Laporta V and Laricchiuta A 2014 Nonequilibrium dissociation mechanisms in low temperature nitrogen and carbon monoxide plasmas *Chem Phys.* **438** 31–6

[148] Kadochnikov I N, Loukhovitski B I and Starik A M 2013 Thermally nonequilibrium effects in shock-induced nitrogen plasma: modelling study *Plasma Sources Sci. Technol.* **22** 035013

[149] Peterson J R *et al* 1998 Dissociative recombination and excitation of N_2^+: cross sections and product branching ratios *J. Chem. Phys.* **108** 1978

[150] Sheehan C H and St-Maurice J-P 2004 Dissociative recombination of N_2^+, O_2^+, and NO^+: rate coefficients for ground state and vibrationally excited ions *J. Geophys. Res.: Space Phys.* **109** A03302

[151] Cunningham A J and Hobson R M 1972 Dissociative recombination at elevated temperatures. IV. N_2^+ dominated afterglows *J. Phys. B: At. Mol. Opt. Phys.* **5** 2328

[152] Zipf E C 1980 The dissociative recombination of vibrationally excited N_2^+ ions *Geophys. Res. Lett.* **7** 645

[153] Mahdavi M R, Hadted J B and Nakshbandi M M 1971 Electron–ion recombination measurements in the flowing afterglow *J. Phys. B: At. Mol. Opt. Phys.* **4** 1726

[154] Geoghegan M, Adams N G and Smith D 1991 Determination of the electron–ion dissociative recombination coefficients for several molecular ions at 300 K *J. Phys. B: At. Mol. Opt. Phys.* **24** 2589

[155] Canosa A, Gomet J C, Rowe B R and Queffelec J L 1991 Flowing afterglow Langmuir probe measurement of the N_2^+ ($v = 0$) dissociative recombination rate coefficient *J. Chem. Phys.* **94** 7159

[156] Guberman S L 2013 The vibrational dependence of dissociative recombination: cross sections for N_2^+ *J. Chem. Phys.* **139** 124318

[157] Guberman S L 2014 The vibrational dependence of dissociative recombination: rate constants for N_2^+ *J. Chem. Phys.* **141** 204307

[158] Little D A and Tennyson J 2014 An R-matrix study of singlet and triplet continuum states of N_2 *J. Phys. B: At. Mol. Opt. Phys.* **47** 105204

[159] Fuente A and Martín-Pintado J 1997 Detection of CO^+ toward the reflection nebula NGC 7023 *Astrophys. J. Lett.* **477** L107

[160] Huebner W, Boice D, Schmidt H and Wegmann R 1991 Structure of the coma: chemistry and solar wind interaction *Comets in the Post-Halley Era* ed R L Newburn Jr., M Neugebauer and J Rahevol 2 (Berlin: Springer) p 907

[161] Fox J L and Hac A 1999 Velocity distributions of C atoms in CO^+ dissociative recombination: implications for photochemical escape of C from Mars *J. Geophys. Res.* **104** 24729

[162] Guberman S L 2007 Dissociative recombination of N_2^+, CO^+ and OH^+ *Workshop on Planetary Atmospheres* (Houston, TX: Lunar and Planetary Institute) contribution 1376

[163] Rosén S *et al* 1998 Absolute cross sections and final-state distributions for dissociative recombination and excitation of CO^+ ($v = 0$) using an ion storage ring *Phys. Rev.* A **57** 4462

[164] Laubé S, Lehfaoui L, Rowe B R and Mitchell J B A 1998 The dissociative recombination of CO^+ *J. Phys. B: At. Mol. Opt. Phys.* **31** 4181

[165] Mitchell J B A and Hus H 1985 The dissociative recombination and excitation of CO^+ *J. Phys. B: At. Mol. Opt. Phys.* **18** 547
[166] Guberman S L 2013 Potential curves for the dissociative recombination of CO^+ *J. Phys. Chem.* A **117** 9704
[167] Gryziński M 1965 Classical theory of atomic collisions. I. Theory of inelastic collisions *Phys. Rev.* A **138** A336
[168] Chung S and Lin C C 1972 Excitation of the electronic states of the nitrogen molecule by electron impact *Phys. Rev.* A **6** 988
[169] Hazi A 1981 Impact-parameter method for electronic excitation of molecules by electron impact *Phys. Rev.* A **23** 2232
[170] Redmon M J, Garrett B C, Redmon L T and McCurdy C 1985 Improved impact-parameter method for electronic excitation and dissociation of diatomic molecules by electron impact *Phys. Rev.* A **32** 3354
[171] Celiberto R and Rescigno T 1993 Dependence of electron-impact excitation cross sections on the initial vibrational quantum number in H_2 and D_2 molecules: $X^1\Sigma_g^+ \rightarrow B^1\Sigma_u^+$ and $X^1\Sigma_g^+ \rightarrow C^1\Pi_u$ transitions *Phys. Rev.* A **47** 1939
[172] Anzai K *et al* 2012 Cross section data sets for electron collisions with H_2, O_2, CO, CO_2, N_2O and H_2O *Eur. Phys. J.* D **66** 36
[173] Cosby P C 1993 Electron-impact dissociation of nitrogen *J. Chem. Phys.* **98** 9544
[174] Lewis B R, Gibson S T, Zhang W, Lefebvre-Brion H and Robbe J-M 2005 Predissociation mechanism for the lowest $^1\Pi_u$ states of N_2 *J. Chem. Phys.* **122** 144302
[175] Stahel D, Leoni M and Dressler K 1983 Nonadiabatic representations of the $^1\Sigma^+{}_u$ and $^1\Pi_u$ states of the N_2 molecule *J. Chem. Phys.* **79** 2541–58
[176] Spelsberg D and Meyer W 2001 Dipole-allowed excited states of N_2: potential energy curves, vibrational analysis, and absorption intensities *J. Chem. Phys.* **115** 6438–49
[177] Cosby P 1993 Electron-impact dissociation of oxygen *J. Chem. Phys.* **98** 9560–69
[178] Cosby P and Helm H 1989 Photofragment spectroscopy of O_2: excitation of the Schumann–Runge band system *J. Chem. Phys.* **90** 1434–39
[179] Lewis B, England J P, Gibson S, Brunger M J and Allan M 2001 Electron energy-loss spectra of coupled electronic states: effects of Rydberg-valence interactions in O_2 *Phys. Rev.* A **63** 022707
[180] Adamson S, Astapenko V, Deminskii M, Eletskii A, Potapkin B, Sukhanov L and Zaitsevskii A 2007 Electron impact excitation of molecules: calculation of the cross section using the similarity function method and *ab initio* data for electronic structure *Chem. Phys. Lett.* **436** 308–13
[181] Celiberto R, Laporta V, Laricchiuta A, Tennyson J and Wadehra J M 2014 Molecular physics of elementary processes relevant to hypersonics: electron–molecule collisions *Open Plasma Phys. J.* **7** 33–47
[182] Laricchiuta A, Celiberto R, Capitelli M and Colonna G 2017 Calculation of electron-scattering cross sections relevant for hypersonic plasma modeling *Plasma Process. Polym.* **14** 1600131
[183] Gibson S and Lewis B 1996 Understanding diatomic photodissociation with a coupled-channel Schrödinger equation model *J. Electron Spectrosc. Relat. Phenom.* **80** 9–12
[184] Malone C P, Johnson P V, Liu X, Ajdari B, Kanik I and Khakoo M A 2012 Integral cross sections for the electron-impact excitation of the $b^1\Pi_u$, $c_3^1\Pi_u$, $o_3^1\Pi_u$, $b'^1\Sigma_u^+$, $c_4'^1\Sigma_u^+$, $G^3\Pi_u$, and $F^3\Pi_u$ states of N_2 *Phys. Rev.* A **85** 062704

[185] Khakoo M A, Malone C P, Johnson P V, Lewis B R, Laher R, Wang S, Swaminathan V, Nuyujukian D and Kanik I 2008 Electron-impact excitation of $X^1\Sigma_g^+$ ($v'' = 0$) to the $a''\,^1\Sigma_g^+$, $b^1\Pi_u$, $c_3{}^1\Pi_u$, $o_3{}^1\Pi_u$, $b'^1\Sigma_u^+$, $c_4'^1\Sigma_u^+$, $G^3\Pi_u$ and $F^3\Pi_u$ states of molecular nitrogen *Phys. Rev.* A **77** 012704
[186] Zipf E C and Gorman M R 1980 Electron-impact excitation of the singlet states of N_2. I. The Birge–Hopfield system ($b^1\Pi_u - X^1\Sigma_g^+$) *J. Chem. Phys.* **73** 813–19
[187] Capitelli M and Celiberto R 1998 Electron–molecule cross sections for plasma applications: the role of internal energy of the target *Novel Aspects of Electron–Molecule Collisions* ed K H Becker (Singapore: World Scientific) p 283
[188] Ajello J M, James G K, Franklin B O and Shemansky D E 1989 Medium-resolution studies of extreme ultraviolet emission from N_2 by electron impact: vibrational perturbations and cross sections of the $c_4'^1\Sigma_u^+$ and $b'^1\Sigma_u^+$ states *Phys. Rev.* A **40** 3524
[189] James G K, Ajello J M, Franklin B and Shemansky D E 1990 Medium resolution studies of extreme ultraviolet emission from N_2 by electron impact: the effect of predissociation on the emission cross section of the $b^1\Pi_u$ state *J. Phys. B: At. Mol. Opt. Phys.* **23** 2055
[190] Brunger M J, Campbell L, Cartwright D, Middleton A, Mojarrabi B and Teubner P 2000 Electron-impact excitation of Rydberg and valence electronic states of nitric oxide: II. Integral cross sections *J. Phys. B: At. Mol. Opt. Phys.* **33** 809
[191] Xu X, Xu L-Q, Xiong T, Chen T, Liu Y-W and Zhu L-F 2018 Oscillator strengths and integral cross sections for the valence-shell excitations of nitric oxide studied by fast electron impact *J. Chem. Phys.* **148** 044311

Chapter 16

Heavy-particle elementary processes in hypersonic flows

Fabrizio Esposito, Robyn Macdonald, Iain D Boyd, Kevin Neitzel and Daniil A Andrienko

The fundamental role of ro-vibrational kinetics in the detailed study of hypersonic flows is now well recognized [1–4], where a strong non-equilibrium of molecular internal degrees of freedom is associated with exchanges of ro-vibrational energy in collisions of molecules with other molecules, and with atoms generated by collision-induced dissociation. In order to introduce this kind of information in hypersonic models, it is necessary to know the specific rate coefficients of all these processes. In particular, studying the interaction of diatomic nitrogen with diatomic and atomic nitrogen is of critical importance in the study of non-equilibrium air chemistry. As the most abundant species in Earth's atmosphere, the early collisions between diatomic nitrogen serve to produce the initial concentration of atomic nitrogen which in turn facilitates relaxation and dissociation through exchange processes, resulting in a highly coupled excitation–dissociation process of diatomic nitrogen [5]. Understanding this process requires an accurate characterization of the finite-rate chemical processes, in particular energy transfer and dissociation or recombination reactions.

There are several approaches to calculate the rates for various chemical processes, ranging from quantum scattering techniques which make use of the Schrödinger equation either in an accurate way [6–9], or with semi-classical approximations [10], to classical approaches which make use of Hamilton's equations of motion [11–13]. The focus of this chapter is on the application of the quasi-classical trajectory method for calculating reaction rates for species of interest in non-equilibrium air chemistry. The quasi-classical trajectory (QCT) method makes use of Hamilton's equations of motion for the dynamics of the collision, with the initial and final states of the collisions mapped to quantum states. The interaction between the nuclei is modulated by the collective effects of the electrons via the potential energy surface (PES). Recent work in computational quantum chemistry has allowed for the

doi:10.1088/2053-2563/aae894ch16

calculation of detailed PESs for systems of interest in non-equilibrium air chemistry (e.g. N_2–N, N_2–N_2, N_2–O_2, O_2–O, O_2–O_2, …) [14–22]. The *ab initio* PES can be computed by solving the Schrödinger equation for many configurations of the nuclei, as discussed in [14, 15].

For diatomic oxygen, the most important collisions in terms of thermal relaxation and dissociation at hyperthermal energies are with N_2, O_2, and O. Particularly, the interaction of O_2 with N_2 is of great importance due to the abundance of N_2 in the atmosphere and to the large difference in the dissociation energies and vibrational relaxation times of these species. Currently, only the O_2–O molecular system has been studied in detail at room temperature [23] and under conditions relevant to a flow around a hypersonic vehicle [24, 25]. The interaction of molecular oxygen with a parent atom is important due to the pronounced reactivity and fast vibrational quenching.

The data obtained through QCT calculations can be used to construct a kinetic model for chemical non-equilibrium for application to computational fluid dynamics (CFD). This approach to the modeling of non-equilibrium flows has been applied for studying several systems, including ro-vibrational state-to-state (StS) studies for $N_2(X^1\Sigma_g^+)$ – $N(^4S_u)$ [5] and vibrational StS studies of $N_2(X^1\Sigma_g^+)$ – $N(^4S_u)$ [26], $N_2(X^1\Sigma_g^+)$ – $N_2(X^1\Sigma_g^+)$ [27, 28], O_2–O [29] and O_2–N [30].

16.1 The quasi-classical method

The use of the QCT method for producing the input of hypersonic flow models has now become popular, after the first pioneering efforts more than 20 years ago [1]. It consists in performing classical dynamics simulations of the collision (a trajectory) on the potential energy surface (PES) of the system, starting from the molecule prepared in the internal state corresponding to the ro-vibrational actions of the quantized counterpart, with the respective conjugated variables uniformly distributed, and with a final product analysis that consists in a post-pseudoquantization (i.e. some kind of discretization, see below). The presence of a distribution of initial variables (vibrational and rotational phases, molecular orientation) implies that the final result is obtained by a reactivity average over these distributions.

In fact, the cross section for process r, σ_r, can be defined as

$$\sigma_r = \pi b_{\text{max}}^2 \frac{N_r}{N_{\text{tot}}} = \pi b_{\text{max}}^2 P_r, \tag{16.1}$$

where b_{max} refers to the maximum impact parameter for which the process r can occur, N_r refers to the number of trajectories which resulted in outcome r, N_{tot} refers to the total number of trajectories simulated, and P_r is the probability that a given outcome r was achieved. From the cross section, we can determine the reaction rate by integrating over a thermal distribution in the impact velocity:

$$k_r = \int_{g=0}^{\infty} \sigma_r f^M(T) g \, dg, \tag{16.2}$$

where k_r is the reaction rate for process r, $f^M(T)$ denotes a Maxwellian velocity distribution at translational temperature T and g denotes the impact velocity. The integration over the impact velocity can also be incorporated into the sampling of trajectories, directly yielding a reaction rate.

16.1.1 QCT accuracy and issues

Adding more and more trajectories improves the statistics of the result, and a statistical error can be calculated. However, this is just the numerical error associated with the statistics, it does not warrant that the result is accurate. Indeed, one aspect of particular relevance in this context is the evaluation of the accuracy of these calculations. It is generally believed that at 'sufficiently' high collision and internal energy values classical trajectories should reliably reproduce quantum mechanical processes. But what is the lower limit for obtaining a reliable result with QCT? This limit strictly depends on the collision system and the specific process studied (ro-vibrational energy transfer with or without molecular rearrangement, dissociation/recombination, etc), in a complex way, which is extremely useful to study. In fact, a consequence of this study is the possibility, when needed, of adding to QCT results the results from different, more accurate and much more demanding approaches in molecular dynamics, in order to obtain a final detailed result on large total energy ranges, but at the same time limiting the expensive calculations to a minimum. This has been realized for a quite light collisional reactive system, H + HeH^+ [31], and it is briefly cited here to demonstrate a successful application of the mixing of different methods, and to present the reasons for this success. In [31] both quasi-classical and highly accurate time-independent quantum mechanical (TIQM) calculations have been performed, using the same PES [32]. The result of the comparisons, including ro-vibrational initial state-selected as well as state-to-state reaction cross sections, is extremely good starting from a collision energy of only 0.1 eV and until the dissociation limit. It is also better than the comparisons with results using other quantum mechanical methods with some common approximations [31]. It is a clear indication that the popular *belief* that the use of any quantum mechanical method guarantees a better accuracy than quasi-classical trajectories is invalid. The point is that each method has its own optimal range of applicability, and the best strategy for obtaining detailed data for modeling input is to appropriately merge the best of each method. In the cited case, this operation has been performed with extreme computational efficiency *and* accuracy, because the TIQM calculations are quite fast at (very) low collision energy, while quasi-classical trajectories are extremely fast and accurate (at least for this system) from intermediate to high energy (see [31] for details). Unfortunately, it is impossible to generalize this extremely positive case study, in fact many counterexamples can be shown where the accuracy of QCT is not acceptable in some relatively large energy ranges. For example, in He + $H_2(v, j)$ (v/j vibrational/rotational initial quantum numbers), where the ro-vibrational state is inelastically changed by collision, it is well known that QCT at intermediate collision energy dramatically fails to reproduce accurate QM results, when low lying ro-vibrational states are considered. This is presented for example in [33] and in [34], where it is clear that only for collision energy values higher than about 1 eV do QCT and accurate QM agree. However, the interval between 0.1 (as in the $H + HeH^+$ case) and 1 eV (as in $He + H_2$) includes a large fraction of the chemistry of interest, and covering this interval with accurate QM calculations is generally highly demanding. The $He + H_2$ system, indeed a sort of prototype of an inelastic collisional system, is at first sight quite similar to the previous $H + HeH^+$, at least concerning the masses involved. Why are there two opposite QCT behaviors with these two 'similar' systems? The point is that one is searching for vibrational distributions from two very different processes. In the first case a reaction is involved, while for $He + H_2$ an inelastic process is investigated. In the first case, a reaction necessarily implies a strong interaction of the atomic projectile with the initial diatom, with the formation of a new bond and a strong mixing of reactants' translational and ro-vibrational energies. This means that the final ro-vibrational distribution is expected to be wide and generally smooth, with no or little memory of the initial state. This in turn implies that the standard binning of QCT is generally able to correctly detect the final ro-vibrational actions.

16.1.2 The problem of binning

In more detail, after the collisional event is complete, the final molecular ro-vibrational classical actions are computed and unquantized values of j and v are obtained. These classical results are then binned using the quantized quantities of interest as bin central values. When the classical final vibrational distribution is wider than some bins and smoothly varying, then the histogram binning can even be very accurate, as shown in [31]. If, in contrast, it is (sometimes even asymmetrically) concentrated in only one bin, and/or it shows abrupt variation (such as an exponential peak or a step) with the final vibration, it is more unlikely to obtain an accurate result. These conditions are true, for example, for any inelastic collisional system at sufficiently low total energy, such as $He + H_2$ [34], but also $O + N_2$ [35] and many other air species collisions under the reaction threshold. The reason for this lies in a fundamental aspect of a weak molecular interaction. If the internal state of motion of a molecule is only weakly modified by collision with an atom or another molecule, for example because the collision energy is sufficiently low, using classical mechanics one can just expect a slight modification of the initial (continuous) vibrational value, e.g. from $v = 1$ to $v' = 0.9$ or 1.1. This is not true quantum mechanically, of course, where there is instead *a low probability* for the system to jump to another integer value, e.g. $v' = 0$ or $v' = 2$. The classical vibrational distribution necessarily shows an exponential peak centered at the initial vibration of the system. For sufficiently low interactions, this distribution is restricted to less than one bin (the bin width is one quantum in histogram binning), and the collision appears as classically elastic ($v' = v$), even if it is not.

16.1.3 A classification of trajectories

The binning issue is well represented in figure 16.1, where each point represents the result of one trajectory in the collision of N with $N_2(v = 1, j = 0)$. The abscissa is the

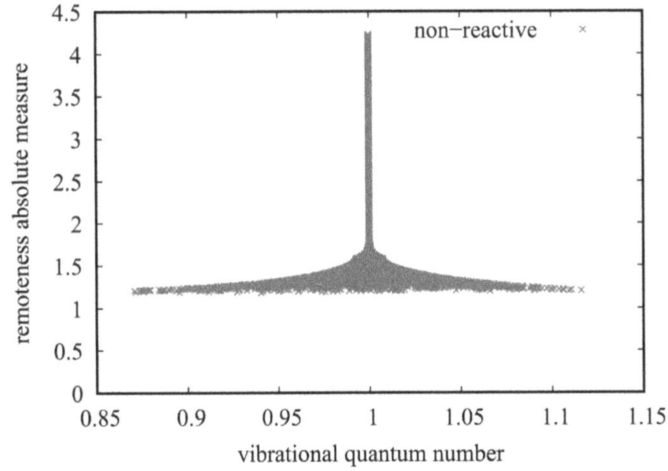

Figure 16.1. N–N_2 results at collision energy $E_{\text{kin}} = 1$ eV.

final vibrational quantum number as obtained classically (i.e. continuous), while on the vertical axis the value z, defined in the following way, is reported:

$$z = \min\left[\frac{\min(R_{AB}/R_{AB}^{\text{eq}},\ R_{AC}/R_{AC}^{\text{eq}})}{(R_{BC}/R_{BC}^{\text{eq}})}\right],$$

where A is the atomic projectile, while B and C are the two atoms in the initial diatom, R_{XY} is the interatomic distance of X from Y, and the 'eq' superscript indicates the equilibrium distance of the corresponding isolated diatom. The minimum is taken along the time evolution of each trajectory. This quantity, inspired by the observations in [36], is a *remoteness absolute measure* (RAM) during each collisional event. The normalization to equilibrium distances is useful in order to consider absolute values of z, in particular when the possible diatoms are different from the initial one. When $z > 1$ the interaction in that particular event (i.e. trajectory) is low, because neither the A–B nor A–C normalized distances become less than the initial diatom BC normalized distance. Practically, it means that the system never goes near the reaction region, there is just a perturbation of the initial diatom, the initial bond is never broken or weakened during that collisional event, which can be indicated as 'purely non-reactive' (PNR). When RAM is about or less than 1, in contrast, there has been a more or less strong interaction, with a reactive or non-reactive effect. A non-reactive event with $z < 1$ is defined as quasi-reactive or recrossing, because the system has entered the strong coupling region (i.e. it has passed the reaction barrier at least a couple of times), even if the final channel is inelastic. In figure 16.1 the results for N + $N_2(v = 1, j = 0)$ collisions are reported for a collision energy E_{kin} of 1 eV and an impact parameter quadratic distribution between 0 and 5 Å. There is only one region to describe, consisting of a sort of bell rapidly expanding for decreasing z values, reducing to a vertical line for RAM > 1.6 centered at the initial $v = 1$ value. For $z < 1.2$ there is no result, that is no trajectory is either quasi-reactive or reactive at the present collision energy for this system.

The reaction barrier is at 1.56 eV for the PES used in these calculations [37], so the conditions of figure 16.1 are inelastic. It is important to note from the figure that the maximum half width of the vibrational distribution is about 0.15 (classical fraction of a quantum number), much lower than the 0.5 value necessary to classify a trajectory as vibrationally inelastic. If the collision energy is increased beyond the reaction threshold, as in figure 16.2 at $E_{\mathrm{kin}} = 4$ eV, the distribution shows three different regions. The higher one is an enlargement of the previous one in figure 16.1, with the vibrational distribution wide by many vibrational quanta, when a RAM value lower than about 1.5 is considered. The second region is approximately delimited by $0.4 \leqslant z < 1$, it includes only inelastic trajectories that have necessarily passed the reaction barrier an even number of times, and showing a smooth, approximately flat vibrational distribution, completely different from the bell distribution of PNR trajectories. The last region in red is the reactive one, characterized by quite a uniform distribution in the form of a thick line in a narrow interval of z values (about $\Delta z = 0.01$). It is clear that all the wide distributions are correctly caught by the histogram binning, so the inelastic QCT results are presumably correct for trajectories with RAM at least lower than 1.5, that is always for reactive and quasi-reactive trajectories, and only partially for purely inelastic trajectories. In figure 16.3 the result for $O + O_2$ in analogous conditions with respect to figure 16.1 are presented. In this case, however, the result is completely different, and similar to $N + N_2$ at much higher collision energy, as in figure 16.2. In particular, in figure 16.3 there are all the three regions of distribution as in figure 16.2, but the quasi-reactive region is much more important in this case, i.e. a lot of inelastic trajectories show a quasi-reactive nature. The important differences between figure 16.1 and figure 16.3 can be easily explained with the absence of a reaction barrier in the $O + O_2$ PES adopted [38]. But the consequences for the QCT vibrational detection accuracy are now clear: whenever the inelastic distribution is dominated by QR trajectories, the quasi-classical inelastic result is in general

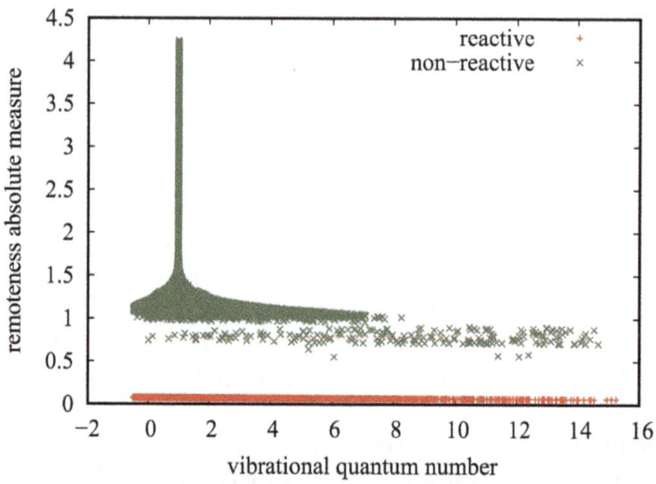

Figure 16.2. $N–N_2$ results at collision energy $E_{\mathrm{kin}} = 4$ eV.

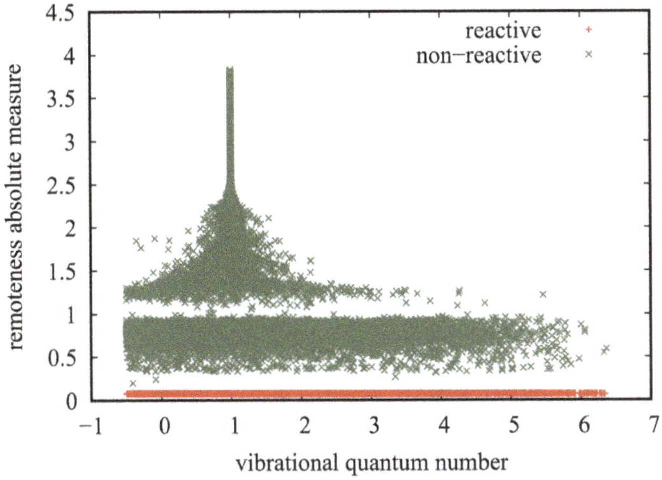

Figure 16.3. O–O_2 results at collision energy $E_{\mathrm{kin}} = 1$ eV.

reliable, because the relative error due to the inaccurate narrow part of the PNR distribution is small. In contrast, if the inelastic distribution is made only of PNR trajectories, the QCT accuracy is much more questionable. Concerning N + N_2 and O + O_2 systems, of great importance in hypersonics, it is clear from the present discussion that in the first case QCT is reliable only for total energy similar or higher than the reaction threshold, while a good approximation is expected also at low energy for O + O_2. For the N + N_2 system the reaction threshold is about 2 eV in more recent and accurate PESs [39–41], therefore in this case the PNR issue in QCT can be expected to be relevant in a range of thousands of kelvins, not just at room temperature. In the O + $N_2(v = 1)$ inelastic rate studied in [35] the minimum temperature for QCT reliable results is estimated to be around 10 000 K. For molecule–molecule collisions this issue can also be more important, because the reactivity is generally lower than in atom–molecule collisions. More studies are needed to assess this point, which are not easy to perform due to the necessity of comparison with accurate QM calculations in four-body collisions. Recently, the vibrational energy transfer in the O_2–Ar molecular system was investigated by means of a fully quantum approach based on the time-dependent Hartree method [42]. The advantage of studying such collisions is in the vast amount of experimental data describing the vibrational relaxation time of oxygen upon interaction with a noble gas. Significant discrepancies, similarly to the ones observed for He + H_2, were obtained at temperatures below 4000 K. At these conditions the QCT method substantially overestimates the O_2–Ar vibrational relaxation time, indicating that this channel of energy transfer is classically closed. It is important to realize that the classical system ‘senses’ the vibrational variation due to collision, and just the final binning is too rough to give a correct final probability of the PNR events. This is one of the cases in which the event is said to be ‘classically forbidden’ [43], and the ‘classical S-matrix’ theory [43] has been proposed for this kind of problem, by using the superposition principle in classical calculations. There are also simple semi-classical methods based on purely classical motion that are able to overcome this QCT deficiency, by exploiting the correspondence between classical and quantum motion of a harmonic oscillator (the DECENT method [44, 45]). Unfortunately, by construction these methods are only useful for purely inelastic systems and only at low vibrational energy. In [46] and [35] a technique is devised to merge DECENT with the QCT method, or more generally any semi-classical method in which the interatomic distances can be evaluated during the collision. Work is in progress to show the feasibility of such a mixed approach. A DECENT-like method would accurately solve the low-energy inelastic process, while QCT would solve the reactive and quasi-reactive parts of the interaction.

16.1.4 The zero point energy issue

QCT accuracy is generally high in the presence of an exothermic reaction without a barrier, as in [31]. This is true for a series of reasons. First of all, the statistics associated with an exothermic process are obviously better than those in the reverse direction, simply because the numerical frequency of a given kind of event follows the probability of occurrence, and this is higher in the exothermic direction. The other reason for choosing (when feasible) the exothermic direction is associated with the zero point energy (ZPE) issue, again in some way linked with the histogram binning. In classical dynamics there is no constraint associated with a minimum energy of a potential well. In a harmonic oscillator, which is the basic model of every molecular oscillator, the oscillation can have any energy, even zero (no motion), if classical motion is considered, but it must have one of the quantized, equidistant energy levels if quantum mechanics is in action. The minimum value is the zero point energy of the (quantum) oscillator. When the classical collision simulation starts, one can always impose any quantized initial condition, but after a classical evolution the ZPE simply loses its meaning. The pseudoquantization is of course done with respect to the quantized values, but if the collisional system has for some reason very low classical vibrational energy, its final distribution can be asymmetric around the lowest vibrational value (i.e. it is all on the left side of the lowest bin, with energy systematically lower than the central bin value). In that case only events not possessing sufficient energy to reach the ZPE are included in the statistics [47]. Various schemes have been proposed to alleviate this problem [48–51]. Of course, if the energy available for products is quite a bit higher than the ZPE, one can expect that the problem will have a limited effect. This explains the good QCT performance in the cited $H + HeH^+$ case concerning ZPE effects, and generally of a collisional system where reactants have internal energy significantly higher than the products' ZPE. In contrast, if the reverse direction is considered for the same reaction, important ZPE effects are found [52], and only for total energy higher than about 2eV ($v = 0$ case) is the QCT agreement with experimental results very good. By 'restricting' in some way the histogram binning by a weight function (Gaussian binning [53]) it is possible in some cases to alleviate this problem [54]. Considering that ZPE decreases as the inverse of the square root of the molecular reduced mass, also for this issue heavier molecules will be easier to treat with QCT.

16.1.5 Other issues

In the case of the H + HeH^+ reaction, there is no barrier to hinder the process, and this is a further reason that explains a good QCT–QM comparison. However, even when a barrier is present, the problem of the lack of tunneling in QCT is rarely an issue in the typical conditions associated with hypersonic flows. In fact, the quasi-classical issue depends on the quantum mechanical penetrability of the barrier. Reaction barriers of light systems are relatively easy to be traversed quantum mechanically. This can be seen in [55], where the effect of using QCT on the H + H_2 reaction can be appreciated with respect to experimental data, but *only at room temperature*, while for heavier isotopic variants it is already difficult to see a relevant discrepancy (see [55], and the detailed discussion in [56]). It is obvious to expect that for heavier species, specifically the air species of interest in aerothermodynamics at temperatures often well above 300 K, tunneling should not be an issue. This is confirmed for example in [39], where the N + N_2 reaction is studied by QCT and compared with accurate time-independent QM on L4 PES [57], with very good agreement in particular at low temperature. In the context of aerothermodynamics modeling, even potential resonances do not constitute a major issue, as in [58], where in addition to the general quite good agreement with wave packet calculations, QCT appears to reproduce at least qualitatively some resonances seen by wave packet dynamics. Another problem is commonly found in chemical kinetics simulations, when assembling data from different molecular processes including some common species, as for example vibrational energy exchange in N + N_2, O + N_2, N_2 + N_2, etc. It consists in the different vibrational energy levels associated with the same molecular species, due to the slightly different asymptotic potentials used in the various calculations. While the lower levels are generally in quite good agreement, significant discrepancies can be found in intermediate to high lying levels (also the total numbers of levels are generally different in different sets of calculations). A rescaling is necessary in this case, taking into account the vibrational energy rather than the vibrational quantum number. A detailed study of this problem is presented in [59].

16.2 Energy transfer and dissociation of N_2

In this section, we present the significant findings of recent work on the nitrogen systems: the $N_2(X^1\Sigma_g^+)$ – $N(^4S_u)$ and $N_2(X^1\Sigma_g^+)$ – $N_2(X^1\Sigma_g^+)$ systems [5, 27, 28, 60]. The results of the $N_2(X^1\Sigma_g^+)$ – $N(^4S_u)$ system are obtained by means of a ro-vibrational StS model using reaction rates determined using the QCT method with the NASA Ames PES for the $N_2(X^1\Sigma_g^+)$ – $N(^4S_u)$ system. The results for the $N_2(X^1\Sigma_g^+)$ – $N_2(X^1\Sigma_g^+)$ system are obtained through the new coarse-grain quasi-classical trajectory (CG-QCT) method which allows for the construction of a reduced order model for systems with unknown StS kinetics. This is necessitated by the prohibitive number of ro-vibrational StS interactions possible for the $N_2(X^1\Sigma_g^+)$ – $N_2(X^1\Sigma_g^+)$ system. The method is based on the multi-group maximum entropy method [61–67].

16.2.1 Nitrogen systems

We are concerned with the chemical non-equilibrium effects due to the interaction between nitrogen atoms and molecules. In particular, we consider atomic nitrogen in the ground electronic state, $N(^4S_u)$, and all the ro-vibrational energy states of molecular nitrogen in the ground electronic state, $N_2(X^1\Sigma_g^+)$.

$N_2(X^1\Sigma_g^+) - N(^4S_u)$ *database*
For the $N_2(X^1\Sigma_g^+) - N(^4S_u)$ system, the reactions we consider are:

$$N_2(i) + N \underset{\kappa_{j-i}^{E}}{\overset{\kappa_{i-j}^{E}}{\rightleftharpoons}} N_2(j) + N, \quad E_i < E_j \tag{16.3}$$

$$N_2(i) + N \underset{\kappa_i^{R}}{\overset{\kappa_i^{D}}{\rightleftharpoons}} 3N, \tag{16.4}$$

where i and j refer to the ro-vibrational state of molecule N_2, κ_{i-j}^{E} and κ_{j-i}^{E} denote the reaction rates for the excitation and de-excitation reactions, respectively, and κ_i^{D} and κ_i^{R} denote the reaction rates for the dissociation and recombination reactions, respectively. The StS data for this system have previously been computed using the QCT method by Jaffe *et al* [14, 15] by running between 1000 and 6000 trajectories from each initial state sampled at nine temperatures between 7500 K and 50 000 K. The error associated with the rate coefficients due to sampling is inversely proportional to the square root of the number of trajectories which resulted in that process. Because exothermic processes (processes where the energy of the final state is lower than that of the initial state) are more likely to occur, these processes have lower statistical error. Therefore, the StS model for $N_2(X^1\Sigma_g^+) - N(^4S_u)$ is constructed using the 13.5 million available reaction rates for the excitation process and the 9380 dissociation rates. The reverse reaction rates are obtained through detailed balance.

$N_2(X^1\Sigma_g^+) - N_2(X^1\Sigma_g^+)$ *database*
For the $N_2(X^1\Sigma_g^+) - N_2(X^1\Sigma_g^+)$ system the reactions we consider are

$$N_2(i) + N_2(j) \underset{\kappa_{kl-ij}^{E}}{\overset{\kappa_{ij-kl}^{E}}{\rightleftharpoons}} N_2(k) + N_2(l), \quad (E_i + E_j) < (E_k + E_l) \tag{16.5}$$

$$N_2(i) + N_2(j) \underset{\kappa_{k-ij}^{ER}}{\overset{\kappa_{ij-k}^{ED}}{\rightleftharpoons}} N_2(k) + 2N, \tag{16.6}$$

where i, j, k and l denote the states of the N_2 molecule, κ_{ij-kl}^{E} and κ_{kl-ij}^{E} denote the reaction rates for the excitation and de-excitation reactions, respectively, and κ_{ij-k}^{ED} and κ_{k-ij}^{ER} denote the combined excitation–dissociation and combined excitation–recombination reaction rates, respectively. In contrast to the $N_2(X^1\Sigma_g^+) - N(^4S_u)$

system which has known StS kinetic data, the magnitude of kinetic data associated with the $N_2(X^1\Sigma_g^+) - N_2(X^1\Sigma_g^+)$ system is too large to compute: there are on the order of 10^{15} possible excitation pathways. Therefore, we must find an alternative to computing the full StS kinetics to study the physics of dissociation and energy transfer. With this objective the coarse-grain quasi-classical trajectory (CG-QCT) method was developed. The CG-QCT method couples the QCT method with the coarse-grain approach for model reduction by sampling initial states for trajectories from within prescribed groups. The grouping is prescribed in this work as either energy-based (considering 40 bins of equal width of bound states and 20 bins of equal width of quasi-bound states) or vibrational-based (considering 61 groups which contain all rotational states of each vibrational state). The temperature within each group is assumed to be in equilibrium with the translational mode. Therefore, the internal energy states within each group are sampled from a Boltzmann distribution at the local translational temperature to compute the group reaction rates. The full CG-QCT method is discussed in [27].

16.2.2 Results

In this section we present the results of the zero-dimensional heat bath studies for the $N_2(X^1\Sigma_g^+) - N(^4S_u)$ and $N_2(X^1\Sigma_g^+) - N_2(X^1\Sigma_g^+)$ systems. For application to CFD, quantifying the macroscopic dissociation rate as well as the energy loss from each mode during the dissociation process is crucial to understanding this process. In addition, understanding the relaxation time associated with each mode allows us to gain insights into the energy transfer process. For both the energy transfer and dissociation processes, the distribution of energy states can provide information about the underlying physics of these processes. For these reasons we will analyze both the macroscopic quantities (e.g. dissociation rates, energy mode coupling terms, relaxation times), as well as analyzing the microscopic distribution of states.

Summary of results for the N_2–N system

For the $N_2(X^1\Sigma_g^+) - N(^4S_u)$ system, let us consider a box composed of 95% $N_2(X^1\Sigma_g^+)$ at 300 K and 5% $N(^4S_u)$ [5]. The pressure is initialized to 10 000 Pa, resulting in a number density of 2.4×10^{24} particles per cubic meter. The box is then instantaneously heated to between 7500 K and 40 000 K and the master equation [68] for each ro-vibrational state of the $N_2(X^1\Sigma_g^+)$ molecule is solved using the reaction rate data obtained from QCT:

$$\frac{dn_N}{dt} = [2\kappa_i^{\mathrm{D}} n_i n_N - 2\kappa_i^{\mathrm{R}} n_N^3] \tag{16.7}$$

$$\frac{dn_i}{dt} = \sum_{j\in\mathcal{I}} \left[-\kappa_{ij}^{\mathrm{E}} n_i n_N + \kappa_{ji}^{\mathrm{E}} n_j n_N\right] + [-\kappa_i^{\mathrm{D}} n_i n_N + \kappa_i^{\mathrm{R}} n_N^3], \quad i \in \mathcal{I}, \tag{16.8}$$

where n_N denotes the number density of nitrogen atoms, n_i denotes the number density of ro-vibrational state i for the $N_2(X^1\Sigma_g^+)$ molecule and the set $\mathcal{I}$ denotes the full set of ro-vibrational states for the $N_2(X^1\Sigma_g^+)$ molecule.

To study the dissociation process at a macroscopic scale, the molar fraction of atomic nitrogen is plotted as a function of time for the 30 000 K test case in figure 16.4. The alternative y-axis in this figure shows the local dissociation rate. This is obtained from the production rate of atomic nitrogen:

$$K_D = \left(\frac{1}{n_{N_2} n_N}\right)\frac{dn_N}{dt}. \tag{16.9}$$

The quasi-steady state (QSS) condition [68] is met when the local dissociation rate (broken red line) reaches a plateau, as seen in figure 16.4 between 2.5×10^{-8}s and 5×10^{-8} s. When the system is in QSS, a global dissociation rate which may be different from the thermal dissociation rate can be defined. However, by the time the QSS condition is met over 40% of the dissociation has occurred. Therefore, the QSS assumption breaks down for the $N_2(X^1\Sigma_g^+) - N(^4S_u)$ system, and using a global QSS dissociation rate will result in inaccurate predictions of the dissociation process.

During the dissociation process energy is lost from both the rotational and vibrational modes. The relative contribution of each of these energy modes to the dissociation energy is plotted in figure 16.5. These chemistry coupling coefficients are computed as

$$\sum_{i \in \mathcal{I}} \dot{\omega}_i^{\mathrm{D}} E_i = C^{\mathrm{DI}} n_{N_2} 2 E_N k^{\mathrm{D}} n_{N_2} n_N \tag{16.10}$$

$$\sum_{v,J} \dot{\omega}_{i(v,J)}^{\mathrm{D}} \Delta\tilde{E}(v, J) = C^{\mathrm{DR}} n_{N_2} 2 E_N k^{\mathrm{D}} n_{N_2} n_N \tag{16.11}$$

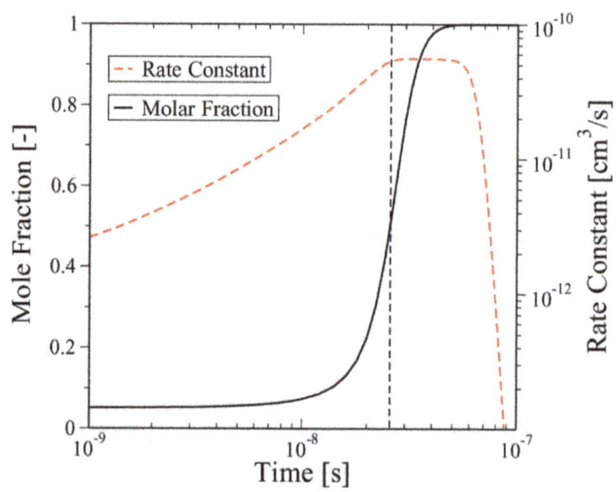

Figure 16.4. Breakdown of the QSS assumption for N_2–N at $T = 30\,000$ K. Reproduced with permission from [5]. Copyright 2013 AIP Publishing.

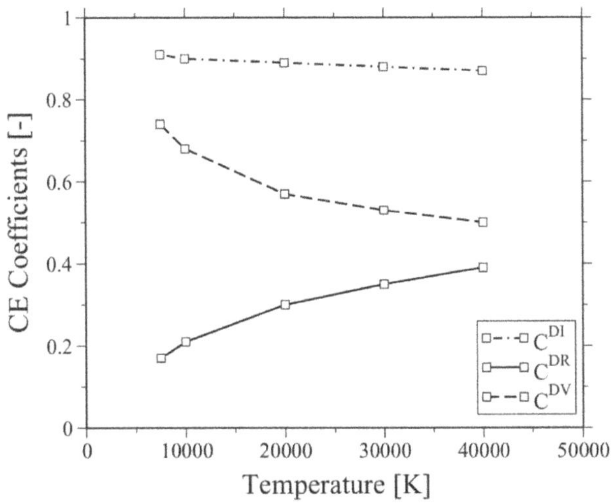

Figure 16.5. Chemistry coupling terms as a function of temperature. Reproduced with permission from [5]. Copyright 2013 AIP Publishing.

$$\sum_{v} \tilde{E}_v \sum_{J} \dot{\omega}^{\mathrm{D}}_{i(v,J)} = C^{\mathrm{DV}} n_{N_2} 2E_N k^{\mathrm{D}} n_{N_2} n_N, \tag{16.12}$$

where $\dot{\omega}_i^{\mathrm{D}} = \sum_{i\in\mathcal{I}}[\kappa_i^{\mathrm{D}} n_i n_N - \kappa_i^{\mathrm{R}} n_N^3]$ encompasses the dissociation and recombination source terms, $E_{i(v,J)} = \Delta\tilde{E}(v, J) + \tilde{E}_v$, where $\tilde{E}_v$ is the energy of the state $(v, J) = (v, 0)$, and C^{DI}, C^{DR} and C^{DV} denote the internal, rotational and vibrational dissociation energy coupling terms. One of the key things to note is the increasing importance of the rotational mode to dissociation as the temperature increases. This contrasts conventional wisdom that capturing the vibrational mode is paramount to capturing the dissociation process [69]. Instead, the relative contributions of rotation and vibration become comparable as the temperature increases. This stems from the fact that at high translational temperatures more dissociation occurs from the quasi-bound states, which are characterized by high rotational energy. This finding demonstrates that the common practice of neglecting quasi-bound states in non-equilibrium dissociation models due to their comparatively low populations can result in inaccurate predictions of the dissociation process.

In contrast to the dissociation process which requires an accurate characterization of the tail of the distribution, the energy transfer process is dictated by the low-energy states. These states hold most of the internal energy of a gas because they are the most populated. When looking at the distribution as a function of time as shown in figure 16.6, there is a distinct separation of vibrational states that occurs for the low-energy states, which is particularly clear at $t = 10^{-7}$ s, denoted by the red circles. This separation of vibrational states is important to characterize and to accurately predict the internal energy of the gas. Moreover, it dictates how the energy transfer process proceeds, as states with the same vibrational quantum number are connected by fast reaction rates and are quick to thermalize, while the different vibrational

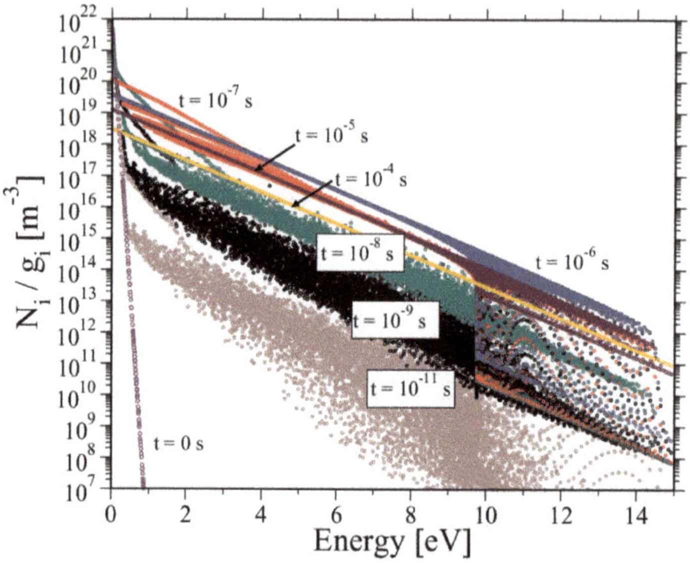

Figure 16.6. Distribution of $N_2(X^1\Sigma_g^+)$ states during the relaxation and dissociation process in a 10 000 K heat bath. Reproduced with permission from [5]. Copyright 2013 AIP Publishing.

strands are connected by much slower transitions, resulting in a significant separation of vibrational strands.

A characteristic relaxation time can be defined during the period during which the internal energy is linearly varying. This characteristic time was extracted from the heat bath simulations for the $N_2(X^1\Sigma_g^+) - N(^4S_u)$ system and compared, in figure 16.7, with the Millikan–White fits developed from empirical experimental correlations in the 1960s. While the behavior of the vibrational relaxation predicted by the StS model at relatively low temperatures is similar to the Millikan–White fit with the high temperature correction applied by Park, the time is an order of magnitude faster than that predicted by Millikan–White. At high temperatures, the rotational, vibrational and internal energy relaxations all asymptote to the same value as the energy available becomes significantly higher than one quanta of rotational or vibrational energy.

Summary of results for the N_2–N_2 system

For the $N_2(X^1\Sigma_g^+) - N_2(X^1\Sigma_g^+)$ system, the CG-QCT method applied to a similar isothermal heat bath. In this case the initial conditions are selected to facilitate a comparison with the direct molecular simulation (DMS) method [27, 28]. The initial pressure and density are set to 760 137 Pa and 1.28 kg m^{-3} with an initial internal temperature of 2000 K. The temperature in the bath is then instantaneously heated to between 10 000 K and 25 000 K. In this case, the conservation equations comprise conservation of mass for atomic nitrogen and conservation group mass for molecular nitrogen:

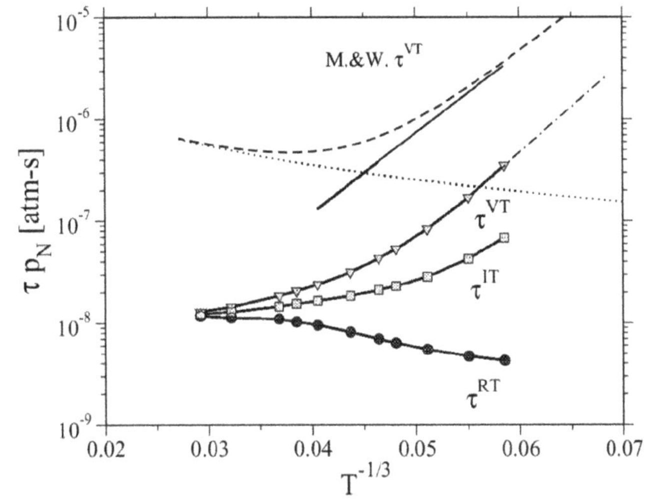

Figure 16.7. Relaxation times for N_2–N as a function of temperature compared to the Millikan–White fit. Reproduced with permission from [5]. Copyright 2013 AIP Publishing.

$$\frac{dn_N}{dt} = 2\sum_{q\in\mathcal{I}}\sum_{r\in\mathcal{I}}\left[K^{\mathrm{ED}}_{pq-r}n_p n_q - K^{\mathrm{ER}}_{r-pq}n_r n_N^2\right] \tag{16.13}$$

$$\begin{aligned}\frac{dn_p}{dt} = &\sum_{q\in\mathcal{I}}\sum_{r\in\mathcal{I}}\sum_{s\in\mathcal{I}}\left[-K^{\mathrm{E}}_{pq-rs}n_p n_q + K^{\mathrm{E}}_{rs-pq}n_r n_s\right]\\ &+\sum_{q\in\mathcal{I}}\sum_{r\in\mathcal{I}}\left[-K^{\mathrm{ED}}_{pq-r}n_p n_q + K^{\mathrm{ER}}_{r-pq}n_r n_N^2\right],\end{aligned} \tag{16.14}$$

where (n_p, n_q, n_r, n_s) denote the number density of group (p, q, r, s) and K^{E}_{pq-rs}, K^{E}_{rs-pq}, K^{ED}_{pq-r} and K^{ER}_{r-pq} denote the grouped reaction rate coefficients for excitation, de-excitation, combined excitation–dissociation and combined excitation–recombination. The results shown are for the 10 000 K case.

In the CG-QCT method, the main assumption is made in the manner in which states are grouped. In this work, two groupings are considered: energy-based and vibrational-based. Figure 16.8 shows a comparison of the composition as a function of time for the two groupings compared with the DMS method. The primary thing to note is the fact that the energy-based grouping matches extremely well with the DMS method, while the vibrational bins, conventionally considered to be very accurate, predict significantly slower dissociation. In addition, it was found that the dissociation process for $N_2(X^1\Sigma_g^+)$ – $N_2(X^1\Sigma_g^+)$ proceeds under the QSS condition.

To understand the differences in the dissociation process between the two grouping methods, the energy lost from the rotational and vibrational mode during the QSS region from the energy- and vibrational-based CG-QCT models are compared in table 16.1. Even at 10 000 K, the CG-QCT method predicts up to 40% of the energy lost in dissociation comes from the rotational mode, higher than

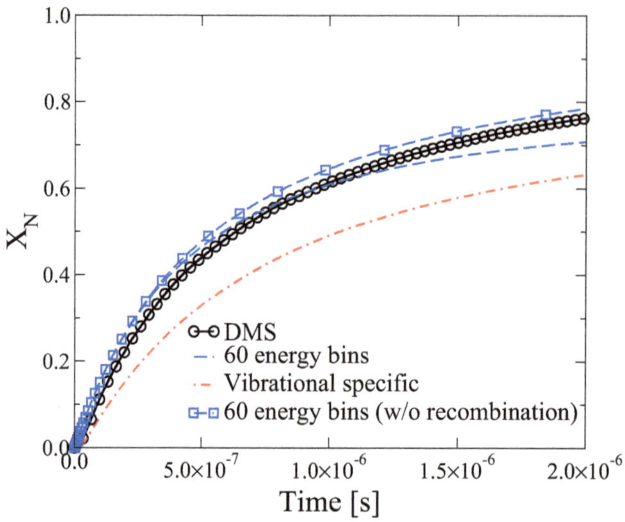

Figure 16.8. Composition as a function of time from energy- and vibrational-based CG-QCT versus DMS at 10 000 K. Reproduced with permission from [28]. Copyright 2018 AIP Publishing.

Table 16.1. Energy mode contribution to dissociation comparison.

	Energy CG-QCT		Vibrational CG-QCT
	Dissociation from QB (%)	$\frac{\Delta e_{rot}^{diss}}{\Delta e_{tot}^{diss}}$ (%)	$\frac{\Delta e_{rot}^{diss}}{\Delta e_{tot}^{diss}}$ (%)
10 000 K	46.9	39.9	10.9
13 000 K	45.0	41.0	14.9
20 000 K	40.8	42.8	24.7
25 000 K	38.6	43.7	31.2

that lost in the $N_2(X^1\Sigma_g^+)$ – $N(^4S_u)$ StS simulation at the same temperature. In contrast, the vibrational specific CG-QCT method predicts that only about 10% of the energy lost in dissociation comes from the rotational mode at 10 000 K. This was found to contradict the findings of the DMS method, meaning that vibrational specific grouping introduces a significant bias towards vibrational energy in dissociation processes. Moreover, a significant amount of energy is lost from high-energy states, particularly quasi-bound states: at 10 000 K up to 47% of the dissociation occurred from quasi-bound states. This supports the theory that accurately predicting the ro-vibrational distribution is crucial to capturing the dissociation process because the tail of the distribution, which is characterized by low vibrational energy and high rotational energy, controls the dissociation process.

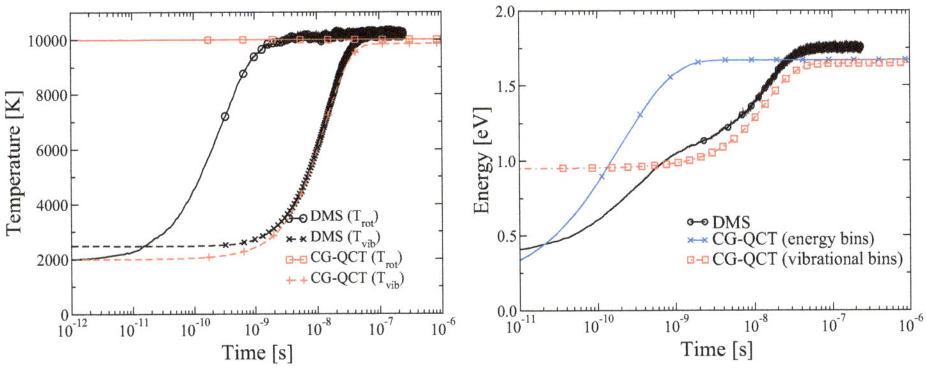

Figure 16.9. Rotational and vibrational temperatures and total internal energy of N_2 predicted by energy- and vibration-based CG-QCT compared with DMS at 10 000 K. Reproduced with permission from [28]. Copyright 2018 AIP Publishing.

The energy transfer process, as seen in the $N_2(X^1\Sigma_g^+) - N(^4S_u)$ results, is controlled by the low-energy states, which are characterized by a strong separation of vibrational states. This is evidenced in figure 16.9 by the agreement in predicting both internal energy and rotational and vibrational temperatures using the vibrational CG-QCT method. The primary deficiency of the vibrational CG-QCT model is in the prediction of the initial energy: the rotational temperature is assumed to be in equilibrium with translation, resulting in significantly higher internal energy initially. Nonetheless, the vibrational relaxation is perfectly matched with the DMS results, while the energy-based CG-QCT method predicts significantly faster relaxation.

16.2.3 Discussion

From the analysis of both the $N_2(X^1\Sigma_g^+) - N(^4S_u)$ and $N_2(X^1\Sigma_g^+) - N_2(X^1\Sigma_g^+)$ systems, we see several similarities. First, in the dissociation process, the rotational mode contributes significantly to the energy loss. In the case of the $N_2(X^1\Sigma_g^+) - N(^4S_u)$ system, the contribution of rotational energy rises to approximately 45% at 40 000 K, while at only 10 000 K the rotational mode contributes up to 40% of the energy lost in dissociation for the $N_2(X^1\Sigma_g^+) - N_2(X^1\Sigma_g^+)$ system. This finding not only indicates the importance of rotation, but also the importance of quasi-bound states in predicting dissociation. Quasi-bound states are often neglected due to their relatively low population, however, for the $N_2(X^1\Sigma_g^+) - N_2(X^1\Sigma_g^+)$ system, they are found to contribute up to 47% of the dissociating molecules. Finally, while the $N_2(X^1\Sigma_g^+) - N_2(X^1\Sigma_g^+)$ system dissociates under the QSS condition, nearly half the dissociation for the $N_2(X^1\Sigma_g^+) - N(^4S_u)$ system occurs under non-QSS conditions.

The energy transfer process for both the $N_2(X^1\Sigma_g^+) - N(^4S_u)$ and $N_2(X^1\Sigma_g^+) - N_2(X^1\Sigma_g^+)$ systems is marked by similar characteristics. In both cases, the separation of vibrational strands governs the internal energy of the molecules during this process. As a result, the vibrational-based CG-QCT method accurately predicted the energy transfer process for the $N_2(X^1\Sigma_g^+) - N_2(X^1\Sigma_g^+)$ system.

16.3 Specifics of O_2–N_2 collisions

Collisions of molecular oxygen with diatomic air species are mostly studied at room temperature. The kinetics of vibrationally hot O_2 in air possibly contributes to the shadow mechanism of ozone extinction [70]. From the computational point of view, a rigorous study of a molecular system requires an accurate potential energy surface (PES). There are several O_4 [71–73] and O_2N_2 [74, 75] PESs, however, they are designed only for low-energy collisions. Typically, these PESs are either limited to a rigid rotor approximation, or utilize empirical data to approximate the intermolecular forces, or do not describe the potential energy in bond-breaking collisions.

One of the first O_2N_2 PESs was proposed by Aquilanti *et al* [74] for studying rigid molecules in their ground vibrational state. This PES was designed specifically for methods of molecular dynamics for the upper atmosphere, and so its use is limited to aerothermodynamics applications. Bartolomei *et al* [76] recently published an improved *ab initio* PES with similar restrictions. The vibration–translation (VT) and vibration–vibration (VV) energy transfer was studied by a semi-classical coupled method by Billing on an empirical PES in the range of temperatures between 250 and 1000 K [75]. Recently, Garcia *et al* [77] reported vibrational–translational (VT) and vibrational–vibrational (VV) rate coefficients using an improved PES for selected O_2 vibrational states at temperatures up to 7000 K. This PES is based on *ab initio* calculations that are interpreted as a combination of a short-range repulsive potential and dispersion, van der Waals and electrostatic interaction terms.

Only recently, the investigation of four-center collisions at higher temperatures received attention when the calculations of an *ab initio* four-body intermolecular potential became computationally tractable. In the context of the present study, the recent six-dimensional O_2N_2 PES by Varga *et al* [78], the so-called the Minnesota PES, is of interest. Full-scale quasi-classical trajectory (QCT) simulations including all ro-vibrational states appear to be computationally intractable using such a high-quality PESs. In addition, it was previously shown that the dissociation rate coefficients have a little sensitivity to the long-range forces between target and projectile species, particularly at high collisional energies. This was demonstrated via studying the kinetics of O_2–O collisions in the temperature range between 3000 and 15 000 K [79]. At the same time, the rate coefficients of bound–bound transitions strongly depend on the intrinsic properties of a particular PES. For example, the vibrational relaxation times of oxygen obtained on the O_3 PES by Varandas and Pais [38] and on an empirical pair-wise PES can differ by a factor of two [79]. Thus, it is necessary to understand the level of fidelity of a particular PES in the temperature range of interest.

To address this question, we investigate the processes of vibrational energy removal and dissociation in the O_2–N_2 molecular system on the accurate Minnesota PES and on the simpler PES developed by Billing [75] at translational temperatures of 10 000 and 20 000 K. To show the difference arising from propagating trajectories on these two different PESs, we present the state-specific cross sections and rate coefficients of the mentioned processes. For the former, we also discuss the dependence of kinetic data on the collisional energy. Finally, we employ the master equation to show the differences in the averaged parameters (such as the vibrational temperature) due to different PESs.

16.3.1 O_2N_2 potential energy surface

An analytical fit to an extensive set of *ab initio* calculated electronic structures of the N_2O_2 triplet was recently published by Varga *et al* [78]. This PES was obtained via the multi-state complete-active space second-order perturbation theory (MS-CASPT2) and was specifically designed to describe internal energy transfer at high collisional energies. Most of the multi-reference calculations of electron correlation energy were conducted in the range of potential energies between 100 and 350 kcal mole^{-1} (76% of the total 54 889 data points). An investigation of the bi-quantum vibrational–vibrational energy exchange involving vibrationally hot O_2 was performed by Andrienko and Boyd [80] and by Garcia *et al* [81]. These studies revealed that the long-range forces of the MS-CASPT2 Minnesota PES were probably obtained with insufficient resolution. In fact, the O_2 quenching by vibrationally cold nitrogen occurs on the Minnesota PES with a rate coefficient that is several orders of magnitude higher than that predicted in experiments [70, 80]. Thus, the Minnesota PES may not be suitable for low-energy collisions, motivating the use in the present study of an older yet more extensively validated PES by Billing [75].

Namely, we propose a study of O_2–N_2 kinetics using a PES that was previously adopted to investigate O_2–N_2 vibrational energy transfer at room temperature. This PES, referred to as the Billing PES, is constructed by the pair-wise summation of short- and long-range forces omitting the electronic terms arising from the many-body interaction. A similar PES was implemented by Garcia and co-workers to study vibrational energy transfer at hyperthermal temperatures [77]. In the current work, we have modified the Billing PES to capture the dissociation process. The total four-body O_2N_2 potential V^{tot} in the Billing PES takes the following form:

$$V^{\text{tot}}\left(R, \mathbf{r}, \theta_A, \theta_B, \phi_{AB}\right) = \sum_{A,B} V^{\text{intra}} + \sum_{i,j} V^{\text{SR}} + \sum_{i,j} V^{\text{disp}} + \sum_{A,B} V^{\text{QQ}},$$

where R is the distance between the two centers of mass (COMs) of molecules A and B, $\mathbf{r} = (r_A, r_B)$ is the vector of A and B intramolecular distances, θ_A, θ_B, ϕ_{AB} are the angular orientations of the two molecules with respect to the line connecting the two COMs and with respect to each other, V^{intra}, V^{SR}, V^{disp} and V^{QQ} are the intramolecular, short-range, dispersion and quadrupole–quadrupole potentials, and i and j are the indices of atoms in the four-body system. The intramolecular potential, V^{intra}, between a pair of atoms forming a molecule is given by the Hulburt–Hirshfelder potential [82]. For the exact expressions of the other terms, we refer to the original work by Billing [75]. Here, we have only modified the dispersion component of the potential as follows:

$$V^{\text{disp}} = \sum_{ij} \frac{C_{6,ij}}{r_{ij}^6}, \tag{16.15}$$

where r_{ij} is the distance between two atoms (i, j) that do not form a molecule. Equation (16.15) takes into account the atomic pair-wise contributions to V^{disp}, while in the original work [75], V^{disp} was computed from the molecule–molecule interaction. The original dispersion term adopted by Billing can potentially lead to a problem of numerical integration in the event of dissociation when the dispersion component does not vanish at infinite separation. Implementation of equation (16.15) eliminates this inconsistency.

16.3.2 Results of O_2N_2 kinetic simulations

The bound–bound and bound–free channels of the O_2–N_2 interaction at translational temperatures of 10 000 K and 20 000 K are studied. Translational–rotational equilibrium is assumed in all calculations: $T = T_{\text{rot}}$. In order to investigate the differences in the Minnesota and Billing PESs and emphasize their influence on the O_2–N_2 kinetic data, we resolve the cross sections and rate coefficients with respect to the initial vibrational state. The former is also presented in the range of collisional energy between 0.1 and 6 eV.

First, the process of oxygen vibrational energy removal by N_2 by means of VT energy transfer is studied:

$$O_2(v) + N_2(T_{\text{vib}}) \rightarrow O_2(v - \Delta v) + N_2, \tag{16.16}$$

with figures 16.10 and 16.11 showing cross sections for $\Delta v = 1$ and 2, respectively. Comparison is performed for $v = 1$, 2, 5 and 10. Solid lines describe the data obtained on the Billing PES and the dashed lines reflect the results for the Minnesota PES. In the process, given by equation (16.16), the N_2 initial vibrational state is

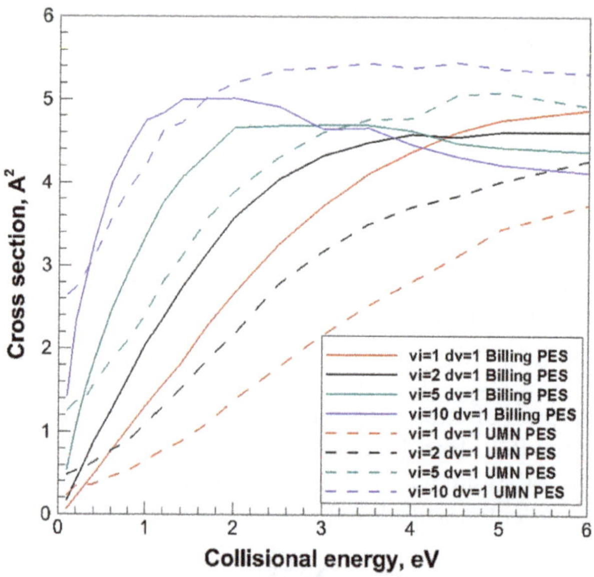

Figure 16.10. Cross section of O_2 mono-quantum vibrational deactivation by N_2 at $T = 10\,000$ K. Solid lines—Billing PES; dashed lines—Minnesota PES.

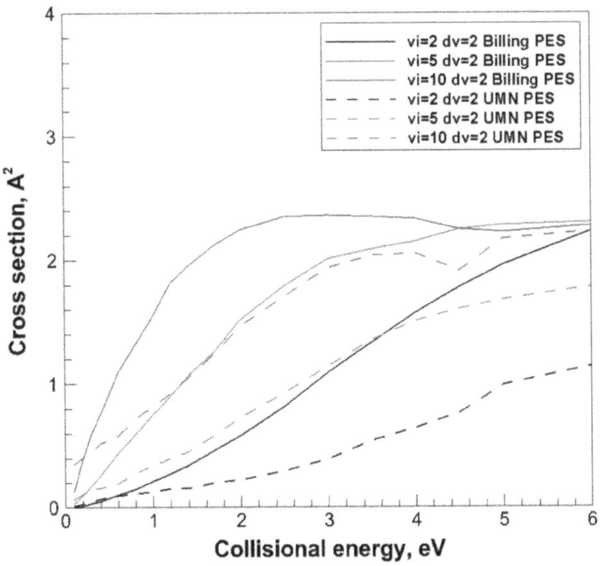

Figure 16.11. Cross section of O_2 bi-quantum vibrational deactivation by N_2 at $T = 10\,000$ K. Solid lines—Billing PES; dashed lines—Minnesota PES.

sampled from the Boltzmann distribution at vibrational temperature $T_{\text{vib}} = T$. The cross sections of the mono-quantum deactivation have a smaller discrepancy between the two studied PESs, compared to the cross sections of the bi-quantum deactivation. Assuming the cross sections generated on the *ab initio* Minnesota PES as the reference data, we expect that the mono-quantum transitions are described more accurately than the multi-quantum transitions when a trajectory is propagated using an atom–atom potential employed in the Billing PES. Such a pair-wise PES may not yield to the correct high order derivatives of the potential with respect to bond distances [75]. Additionally, the Billing PES predicts an earlier decrease of mono-quantum vibrational energy removal for $v = 10$ due to opening of the dissociation channel, compared to the Minnesota PES. Indeed, it will be shown later that the Billing dissociation rate is higher than the Minnesota dissociation rate. Overall, the cross sections of O_2 vibrational energy removal are different between the studied PESs by a factor of two or less. Note the much larger cross sections obtained on the Minnesota PES at small collision energies. Indeed, at collision energies less than 0.2 eV one should expect a pronounced influence of the long-range forces.

The vibrational deactivation of nitrogen is studied by computing the kinetic parameters of the following reaction:

$$O_2(T_{\text{vib}}) + N_2(w) \rightarrow O_2 + N_2(w - \Delta w). \qquad (16.17)$$

The cross sections of mono-quantum deactivation ($\Delta w = 1$) are shown in figure 16.12. The agreement between the Minnesota and Billing PESs in this case is substantially better than for the O_2 vibrational energy removal. As expected, the largest difference is observed at low collisional energies where the long-range forces

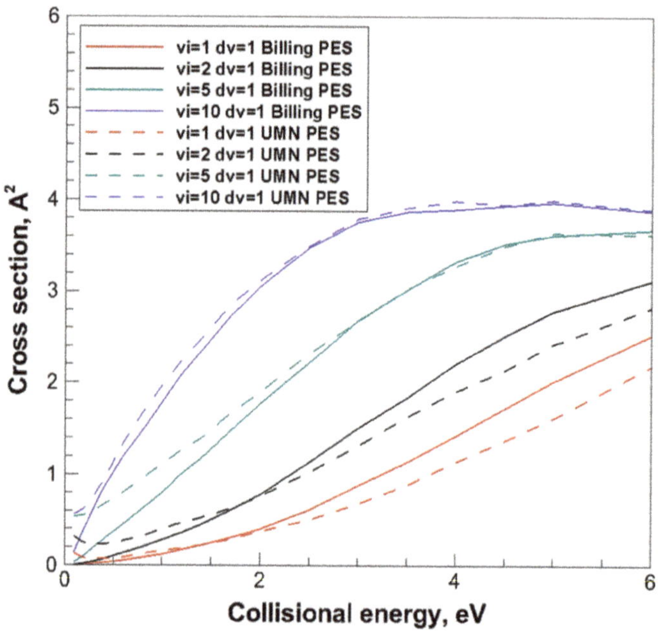

Figure 16.12. Cross section of N_2 mono-quantum vibrational deactivation by O_2 at $T = 10\,000$ K. Solid lines—Billing PES; dashed lines—Minnesota PES.

have a pronounced influence. For collision energies less than 0.4 eV, the cross sections can differ by a factor of two. This discrepancy becomes significantly smaller with increasing collisional energy.

The state-resolved dissociation cross sections are studied in the present work as well. Due to differences in the vibrational ladders between the studied PESs, it seems appropriate to compare the dissociative cross sections for the entire vibrational ladder at a fixed collisional energy. The calculated data are shown in figure 16.13 for collisional energies of 0.1, 1.2, 2.5 and 6 eV. The Billing PES underpredicts the cross sections at low collisional energies, with agreement improving toward higher collisional energies. Note the nonlinear increase of dissociative cross sections for excited vibrational states, which indicates that a simple extrapolation of kinetic data for these states should be done with caution.

The rate coefficients are obtained by integrating the cross sections over the range of collisional energies. The rate of N_2 vibrational energy removal is shown in figures 16.14 and 16.15 for translational temperatures of 10 000 and 20 000 K, respectively. The rate of O_2 vibrational energy removal is shown in figure 16.16 at $T = 10\,000$ K. These bound–bound rate coefficients exhibit a similar dependence on the vibrational energy of the target particle: there is an increase in a rate coefficient due to anharmonicity and a subsequent closing of the bound–bound channel due to dissociation. The comparison between these two PESs of different fidelity shows good agreement for the rates of mono-quantum deactivation. For the multi-quantum jumps, the disagreement can be as high as one order of magnitude. As follows from the previous O_2–N_2 master equation simulations [80] conducted on the

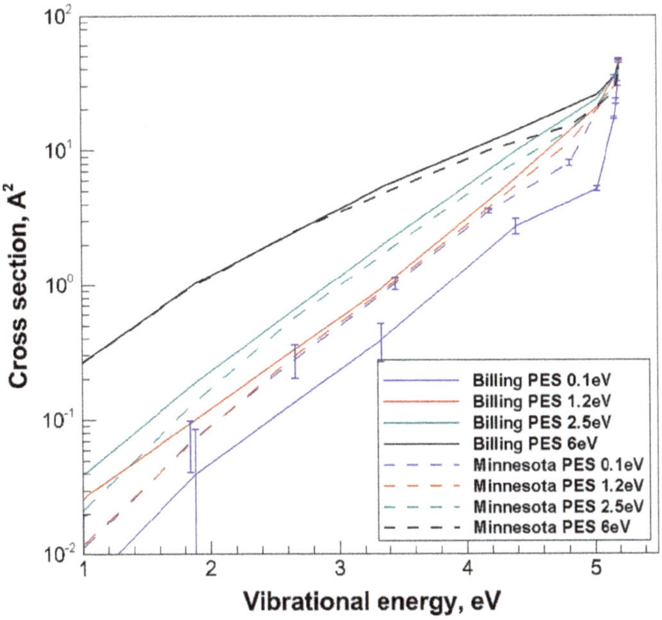

Figure 16.13. Cross section of O_2 dissociation reaction by N_2 at $T = 10\,000$ K. Solid lines correspond to Billing PES, dashed lines correspond to Minnesota PES.

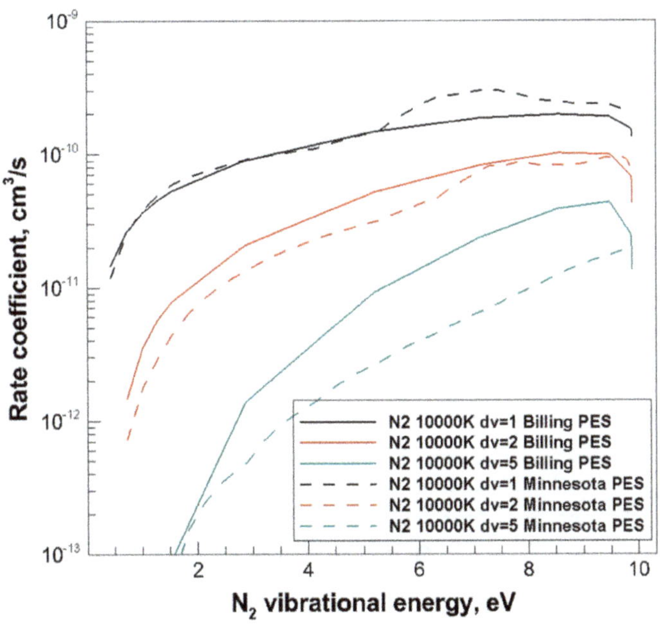

Figure 16.14. Rate coefficient of N_2 vibrational deactivation by O_2 at $T = 10\,000$ K.

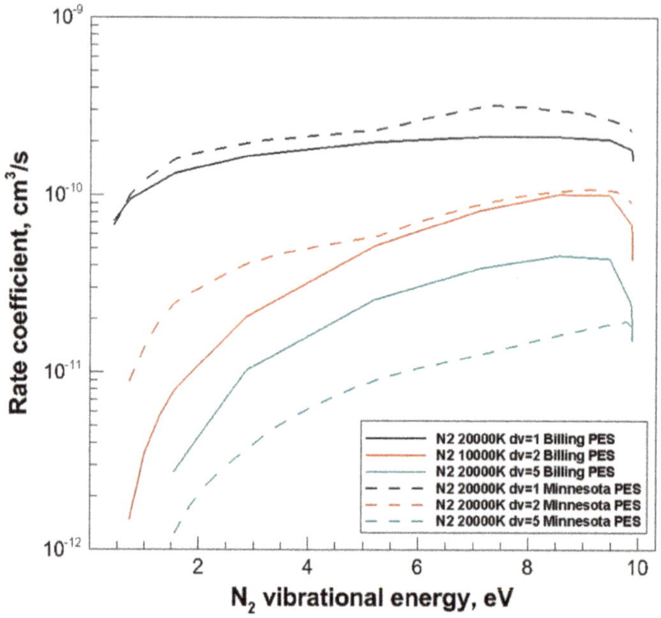

Figure 16.15. Rate coefficient of N_2 vibrational deactivation by O_2 at $T = 20\,000$ K.

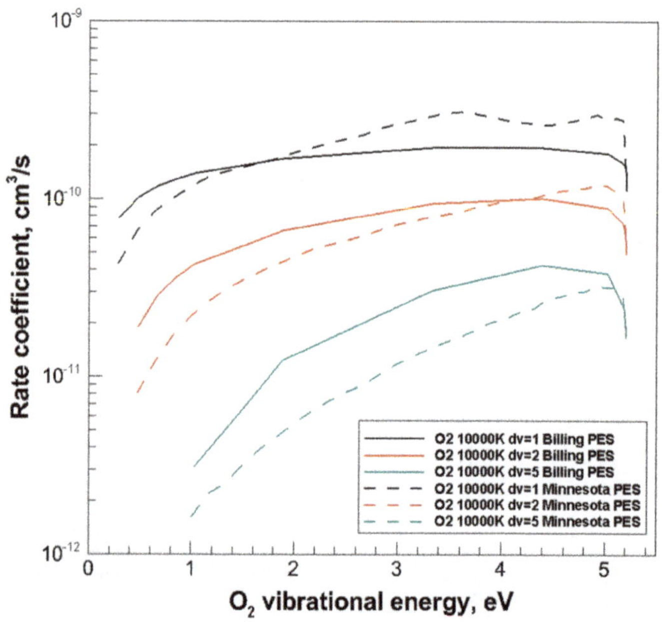

Figure 16.16. Rate coefficient of O_2 vibrational deactivation by N_2 at $T = 10\,000$ K.

Minnesota PES, accounting for the multi-quantum VT processes has a pronounced effect on the O_2 vibrational temperature and is much less important for the N_2 kinetics at given conditions.

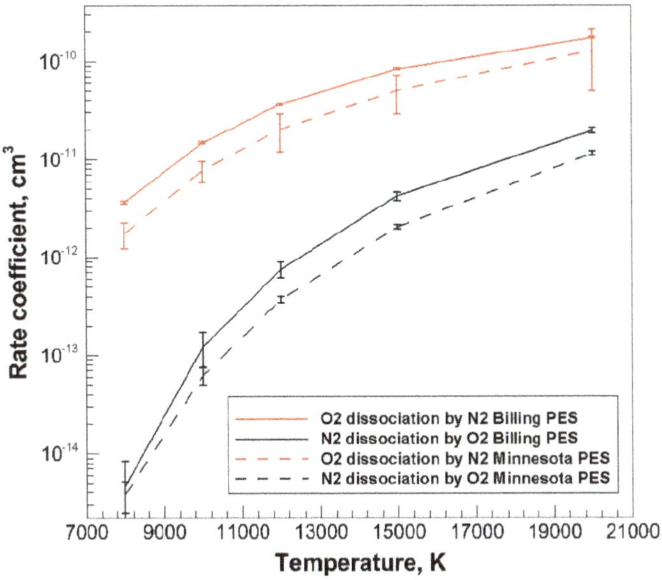

Figure 16.17. O_2 and N_2 thermal equilibrium dissociation rate coefficient obtained on the Billing (solid lines) and Minnesota (dashed lines) PESs.

The vibrationally resolved rate coefficients of O_2 dissociation at $T = 10\,000$ K and N_2 at $T = 20\,000$ K are shown in figure 16.18. The agreement between the Billing and Minnesota PESs is good, however, it is not possible to indicate a consistent difference of one dataset from the other.

Thermal equilibrium dissociation rate coefficient of O_2 and N_2, computed on the Minnesota and Billing PESs by averaging over the entire range of initial vibrational states is presented in figure 16.17. Similarly to the cross sections, the rates are different from each other by no more than a factor of two; this holds for both O_2 and N_2. The largest discrepancy is observed at low temperatures of 8000 and 10 000 K. The Billing PES consistently predicts a higher dissociation rate, which is in agreement with an earlier onset of the dissociation channel, observed in figure 16.10.

The global picture of the differences in reaction rate coefficients, obtained on the two studied PESs, can be drawn by examining the master equation solution. In this approach, each O_2 and N_2 vibrational state is treated as an individual species interacting with all possible vibrational states of a projectile molecule. The formulation of the master equations for the O_2–N_2 heat bath is similar to that of N_2–N. The master equation model could provide the sensitivity of the averaged thermodynamic parameters, such as the vibrational temperature, to a particular dataset of rate coefficients and to the implemented PES.

With dissociation and vibrational relaxation terms introduced in the master equation simulation, it is possible to study the parameters of the quasi-stationary state (QSS) using different datasets of rate coefficients. As follows from figure 16.19, the two studied PESs have a minimal influence on the vibrational temperature of oxygen during the QSS phase. This is because of a fairly good agreement of the

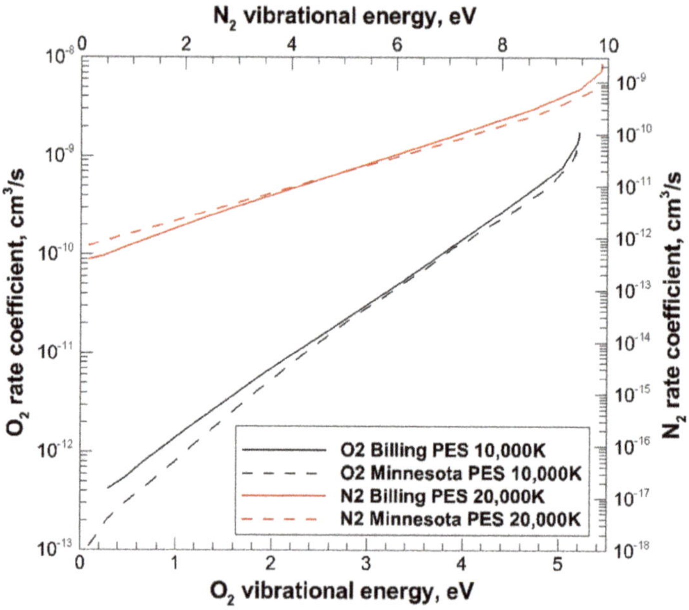

Figure 16.18. Vibrationally resolved rates of O_2 and N_2 dissociation.

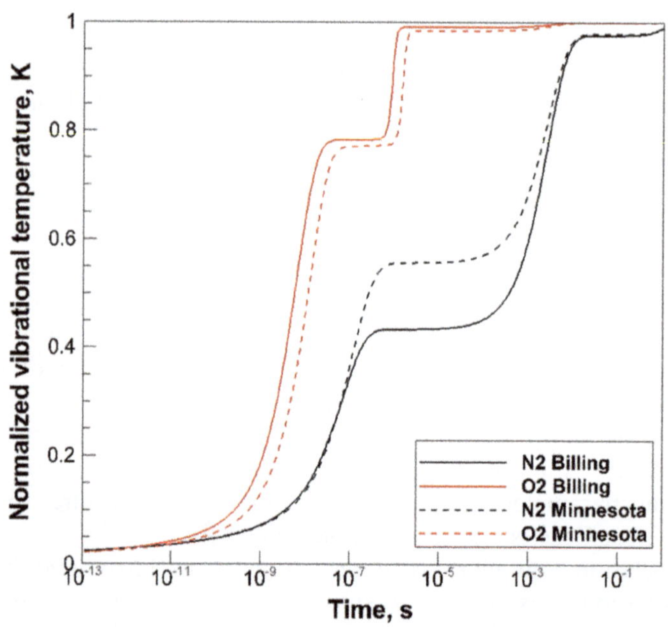

Figure 16.19. Vibrational temperature of O_2 and N_2 obtained via the master equation simulation of a heat bath at $T = 10\,000$ K.

O_2 dissociation rates between the Billing and Minnesota PESs. A more drastic influence is observed for nitrogen: the Billing PES predicts the N_2 vibrational temperature by approximately a thousand degrees lower than that on the Minnesota PES. Indeed, since the N_2 equilibrium dissociation rate is larger on the Billing PES, as can be seen from figure 16.17, the onset of N_2 dissociation in a heat bath should occur at a lower vibrational temperature.

16.3.3 Summary of O_2–N_2 kinetic study

Vibrational energy transfer and dissociation in O_2–N_2 collisions are studied by means of the QCT method at hyperthermal energies. Special attention is paid to the fidelity of the O_2N_2 PES used in the QCT method and to the resulting differences in cross sections and rate coefficients. Toward this end, a high-quality *ab initio* PES generated at the University of Minnesota as well as a much simpler yet more computationally efficient PES proposed by Billing were employed. On average, the QCT method with the Minnesota PES is an order of magnitude more expensive compared to that with the Billing PES. The cost of simulations on the former PES increases for excited vibrational states and at high collision energies.

Vibrational energy removal in oxygen is found to be sensitive to the PES used in the QCT method. This is observed in the entire range of collisional energies between 0.1 and 6 eV and for the initial vibrational state between 1 and 10. Nitrogen is shown to be less sensitive in terms of vibrational energy removal to the details of the PES. This can be attributed to the larger vibrational spacing in N_2 molecules which reduces the importance of the differences in the long-range intermolecular forces. The rate coefficients, generated by the Minnesota and Billing PESs, resemble similarities in the dependence from the initial vibrational energy of a target particle. For the mono-quantum VT rate coefficients, the difference is limited to a factor of two, while the multi-quantum rates can have a discrepancy as high as one order of magnitude.

Dissociation cross sections and reaction rates are also compared on the PESs of different fidelity. It is shown that the bound–free transition rates are less sensitive to the PES adopted for trajectory simulations. Dissociative cross sections show excellent agreement with each other at collisional energies higher than 1 eV. The Billing PES predicts thermal equilibrium rate coefficients to be larger than that on the Minnesota PES by a factor of two. It is desirable to study the departure in macroscopic parameters caused by these two PESs by means of master equation simulations.

The master equation simulation of a heat bath at a translational temperature of 10 000 K has been conducted as well. This step estimates the influence of a particular PES on thermodynamic parameters, such as the vibrational temperature. Again, the Billing set of rates produces a faster relaxation of oxygen; however, the vibrational temperature of O_2 during the QSS phase is captured quite accurately. This is because of a fairly good agreement of the O_2 dissociation rates between the Billing and Minnesota PESs. In contrast, the N_2 vibrational temperature during the QSS phase using the Billing data is lower by a thousand degrees than the vibrational temperature obtained using the Minnesota dataset. The onset of the N_2 QSS phase strongly depends on the specifics of vibrational–translational and vibrational–vibrational energy exchange with oxygen.

References

[1] Capitelli M 1996 *Molecular Physics and Hypersonic FLows, NATO Science Series C* (Amsterdam: Springer)

[2] Capitelli M *et al* 2011 Plasma kinetics in molecular plasmas and modeling of reentry plasmas *Plasma Phys. Controlled Fus.* **53** 124007

[3] Laganá A, Lombardi A, Pirani F, Gamallo P, Sayòs R, Armenise I, Cacciatore M, Esposito F and Rutigliano M 2014 Molecular physics of elementary processes relevant to hypersonics: atom–molecule, molecule–molecule and atoms–surface processes *Open Plasma Phys. J.* **7** 48–59

[4] Celiberto R *et al* 2016 Atomic and molecular data for spacecraft re-entry plasmas *Plasma Sources Sci. Technol.* **25** 033004

[5] Panesi M, Jaffe R L, Schwenke D W and Magin T E 2013 Rovibrational internal energy transfer and dissociation of $N(^4S_u) + N_2(^1\Sigma_g^+)$ system in hypersonic flows *J. Chem. Phys.* **138** 044312

[6] Launay J M 1991 Computation of cross sections for the $F + H_2$ $(v = 0, j = 0)$ $\rightarrow FH(v'j) + H$ reaction by the hyperspherical method *Theor. Chim. Acta* **79** 183–90

[7] Skouteris D, Castillo J F and Manolopoulos D E 2000 ABC: a quantum reactive scattering program *Comput. Phys. Commun.* **133** 128–35

[8] Gray S K and Balint-Kurti G G 1998 Quantum dynamics with real wave packets, including application to three-dimensional $(J = 0)D + H_2 \rightarrow HD + H$ reactive scattering *J. Chem. Phys.* **108** 950

[9] Meyer H-D, Manthe U and Cederbaum L 1990 The multi-configurational time-dependent Hartree approach *Chem. Phys. Lett.* **165** 73–8

[10] Billing G D 2003 *The Quantum Classical Theory* (New York: Oxford University Press)

[11] Karplus M, Porter R N and Sharma R D 1965 Exchange reactions with activation energy. I. Simple barrier potential for (H, H_2) *J. Chem. Phys.* **43** 3259–87

[12] Truhlar D G and Muckerman J T 1979 Reactive scattering cross sections III. Quasiclassical and semiclassical methods *Atom-Molecule Collision Theory: A Guide for the Experimentalist* ed R B Bernstein (New York: Plenum) pp 505–66

[13] Jaffe R L, Schwenke D W and Panesi M 2015 First principles calculation of heavy particle rate coefficients *Hypersonic Nonequilibrium Flows: Fundamentals and Recent Advances* ed E Josyula (Reston, VA: American Institute of Aeronautics and Astronautics) pp 103–58

[14] Jaffe R L, Schwenke D W, Chaban G and Huo W 2008 Vibrational and rotational excitation and relaxation of nitrogen from accurate theoretical calculations, *46th AIAA Aerospace Sciences Meeting and Exhibit , Reno, NV* AIAA Paper 2008–1208

[15] Chaban G, Jaffe R L, Schwenke D W and Huo W 2008 Dissociation cross-sections and rate coefficients for nitrogen from accurate theoretical calculations, *46th AIAA Aerospace Sciences Meeting and Exhibit , Reno, NV* AIAA Paper 2008-1209

[16] Paukku Y, Yang K R, Varga Z and Truhlar D G 2013 Global *ab initio* ground-state potential energy surface of N_4 *J. Chem. Phys.* **139** 044309

[17] Bender J D, Valentini P, Nompelis I, Paukku Y, Varga Z, Truhlar D G, Schwartzentruber T and Candler G V 2015 An improved potential energy surface and multi-temperature quasiclassical trajectory calculations of $N_2 + N_2$ dissociation reactions *J. Chem. Phys.* **143** 054304

[18] Varga Z, neda R M-P, Song G, Paukku Y and Truhlar D G 2016 Potential energy surface of triplet N_2–O_2 *J. Chem. Phys.* **144** 024310

[19] Lin W, Varga Z, Song G, Paukku Y and Truhlar D G 2016 Global triplet potential energy surfaces for the $N_2(X^1\Sigma) + O(3P) \rightarrow NO(X^2\Pi) + N(^4S)$ reaction *J. Chem. Phys.* **144** 024309

[20] Varga Z, Paukku Y and Truhlar D G 2017 Potential energy surfaces for O + O_2 collisions *J. Chem. Phys.* **147** 154312

[21] Paukku Y, Yang K R, Varga Z, Song G, Bender J D and Truhlar D G 2017 Potential energy surfaces of quintet and singlet O_4 *J. Chem. Phys.* **147** 034301

[22] Paukku Y, Varga Z and Truhlar D G 2018 Potential energy surface of triplet O_4 *J. Chem. Phys.* **148** 124314

[23] Schinke R, Grebenshchikov S Y, Ivanov M and Fleurat-Lessard P 2006 Dynamical studies of the ozone isotope effect: a status report *Annu. Rev. Phys. Chem.* **57** 625–61

[24] Esposito F and Capitelli M 2007 The relaxation of vibrationally excited O_2 molecules by atomic oxygen *Chem. Phys. Lett.* **443** 222–6

[25] Andrienko D A and Boyd I D 2016 Rovibrational energy transfer and dissociation in O_2–O collisions *J. Chem. Phys.* **144** 104301

[26] Esposito F, Armenise I and Capitelli M 2006 N–N_2 state-to-state vibrational relaxation and dissociation rate coefficients based on quasi-classical calculations *Chem. Phys* **331** 1–8

[27] Macdonald R L, Jaffe R L, Schwenke D W and Panesi M 2018 Construction of a coarse-grain quasi-classical trajectory method I. Theory and application to N_2–N_2 system *J. Chem. Phys.* **148** 054309

[28] Macdonald R L, Grover M S, Schwartzentruber T E and Panesi M 2018 Construction of a coarse-grain quasi-classical trajectory method. II. Comparison against the direct molecular simulation method *J. Chem. Phys.* **148** 054310

[29] Esposito F, Armenise I, Capitta G and Capitelli M 2008 O–O_2 state-to-state vibrational relaxation and dissociation rates based on quasiclassical calculations *Chem. Phys.* **351** 91–8

[30] Andrienko D A and Boyd I D 2016 Thermal relaxation of molecular oxygen in collisions with nitrogen atoms *J. Chem. Phys.* **145** 014309

[31] Esposito F, Coppola C M and De Fazio D 2015 Complementarity between quantum and classical mechanics in chemical modeling. The $H + HeH^+ \rightarrow H_2^+ + He$ reaction: a rigorous test for reaction dynamics methods *J. Phys. Chem.* A **119** 12615–26

[32] Ramachandran C, De Fazio D, Cavalli S, Tarantelli F and Aquilanti V 2009 Revisiting the potential energy surface for the $He + H_2^+ \rightarrow HeH^+ + H$ reaction at the full configuration interaction level *Chem. Phys. Lett.* **469** 26–30

[33] Balakrishnan N, Vieira M, Babb J, Dalgarno A, Forrey R and Lepp S 1999 Rate coefficients for ro-vibrational transitions in H_2 due to collisions with He *Astrophys. J.* **524** 1122

[34] Celiberto R *et al* 2017 Elementary processes and kinetic modeling for hydrogen and helium plasmas *Atoms* **5** 18

[35] Esposito F and Armenise I 2017 Reactive, inelastic, and dissociation processes in collisions of atomic oxygen with molecular nitrogen *J. Phys. Chem.* A **121** 6211–9

[36] Smith I W 1977 Reaction and relaxation of vibrationally excited molecules: a classical trajectory study of Br + HCl(v') and Br + DCl(v') collisions *Chem. Phys.* **20** 437–43

[37] Laganá A and Garcia E 1994 Temperature dependence of nitrogen atom–molecule rate coefficients *J. Phys. Chem.* **98** 502–7

[38] Varandas A and Pais A 1988 A realistic double many-body expansion (DMBE) potential energy surface for ground-state O_3 from a multiproperty fit to *ab initio* calculations, and to

experimental spectroscopic, inelastic scattering, and kinetic isotope thermal rate data *Mol. Phys.* **65** 843–60
[39] Caridade P, Galvão B and Varandas A 2010 Quasiclassical trajectory study of atom-exchange and vibrational relaxation processes in collisions of atomic and molecular nitrogen *J. Phys. Chem.* A **114** 6063–70
[40] Wang D, Huo W M, Dateo C E, Schwenke D W and Stallcop J R 2004 Quantum study of the $N + N_2$ exchange reaction: state-to-state reaction probabilities, initial state selected probabilities, Feshbach resonances, and product distributions *J. Chem. Phys.* **120** 6041
[41] Rampino S, Skouteris D, Laganà A, Garcia E and Saracibar A 2009 A comparison of the quantum state-specific efficiency of $N + N_2$ reaction computed on different potential energy surfaces *Phys. Chem. Chem. Phys.* **11** 1752
[42] Ulusoy I S, Andrienko D A, Boyd I D and Hernandez R 2016 Quantum and quasi-classical collisional dynamics of O_2–Ar at high temperatures *J. Chem. Phys.* **144** 234311
[43] Miller W H 1970 The classical S-matrix: a more detailed study of classically forbidden transitions in inelastic collisions *Chem. Phys. Lett.* **7** 431–5
[44] Giese C F and Gentry W R 1974 Classical trajectory treatment of inelastic scattering in collisions of H^+ with H_2, HD, and D_2 *Phys. Rev.* A **10** 2156
[45] Blais N C and Truhlar D G 1978 *Ab initio* calculation of the vibrational energy transfer rate of H_2 in Ar using Monte Carlo classical trajectories and the forced quantum oscillator model *J. Chem. Phys.* **69** 846
[46] Capitelli M, Celiberto R, Colonna G, Esposito F, Gorse C, Hassouni K, Laricchiuta A and Longo S 2016 Reactivity and relaxation of vibrationally/rotationally excited molecules with open shell atoms *Fundamental Aspects of Plasma Chemical Physics* (Berlin: Springer) pp 31–56
[47] Mandy M and Martin P 1991 Integral cross sections for atomic $H + H_2(0, 0) \rightarrow H_2(0, j') + H$: comparison of quasiclassical and quantum results *J. Phys. Chem.* **95** 8726–31
[48] Alimi R, García-Vela A and Gerber R 1992 A remedy for zero-point energy problems in classical trajectories: a combined semiclassical/classical molecular dynamics algorithm *J. Chem. Phys.* **96** 2034–8
[49] Xie Z and Bowman J M 2006 Zero-point energy constraint in quasi-classical trajectory calculations *J. Phys. Chem.* A **110** 5446–9
[50] Varandas A 2007 Trajectory binning scheme and non-active treatment of zero-point energy leakage in quasi-classical dynamics *Chem. Phys. Lett.* **439** 386–92
[51] Bonhommeau D and Truhlar D G 2008 Mixed quantum/classical investigation of the photodissociation of NH_3 ($\tilde{A}$) and a practical method for maintaining zero-point energy in classical trajectories *J. Chem. Phys.* **129** 014302
[52] Tang X N, Xu H, Zhang T, Hou Y, Chang C, Ng C, Chiu Y, Dressler R and Levandier D 2005 A pulsed-field ionization photoelectron secondary ion coincidence study of the $(H_2^+(X, v^+ = 0 - 15, N^+ = 1) + He$ proton transfer reaction *J. Chem. Phys.* **122** 164301
[53] Bonnet L 2008 The method of Gaussian weighted trajectories. III. An adiabaticity correction proposal *J. Chem. Phys.* **128** 044109
[54] Pérez de Tudela R, Aoiz F, Suleimanov Y V and Manolopoulos D E 2012 Chemical reaction rates from ring polymer molecular dynamics: zero point energy conservation in $Mu + H_2 \rightarrow MuH + H$ *J. Phys. Chem. Lett.* **3** 493–7

[55] Mayne H R and Toennies J P 1981 Quasiclassical trajectory studies of the $H + H_2$ reaction on an accurate potential-energy surface. III. Comparison of rate constants and cross sections with experiment *J. Chem. Phys.* **75** 1794–803

[56] Aoiz F J, Bañares L, Herrero V J, Sáez Rábanos V and Tanarro I 1997 The $H + D_2 \rightarrow HD + D$ reaction. Quasiclassical trajectory study of cross sections, rate constants, and kinetic isotope effect *J. Phys. Chem.* A **101** 6165–76

[57] Garcia E, Saracibar A, Gómez-Carrasco S and Laganà A 2008 Modeling the global potential energy surface of the $N + N_2$ reaction from *ab initio* data *Phys. Chem. Chem. Phys.* **10** 2552–8

[58] Akpinar S, Armenise I, Defazio P, Esposito F, Gamallo P, Petrongolo C and Sayòs R 2012 Quantum mechanical and quasiclassical Born–Oppenheimer dynamics of the reaction $N_2 + O \rightarrow N + NO$ on the N_2O a^3A'' and b^3A' surfaces *Chem. Phys.* **398** 81–9

[59] Armenise I and Esposito F 2015 N_2, O_2, NO state-to-state vibrational kinetics in hypersonic boundary layers: the problem of rescaling rate coefficients to uniform vibrational ladders *Chem. Phys.* **446** 30–46

[60] Panesi M, Munafò A, Magin T E and Jaffe R L 2014 Nonequilibrium shock-heated nitrogen flows using a rovibrational state-to-state method *Phys. Rev.* E **90** 013009

[61] Panesi M and Lani A 2013 Collisional radiative coarse-grain model for ionization in air *Phys. Fluids* **25** 057101

[62] Munafò A, Liu Y and Panesi M 2015 Modeling of dissociation and energy transfer in shock-heated nitrogen flows *Phys. Fluids* **27** 127101

[63] Munafò A, Mansour N N and Panesi M 2017 A reduced-order NLTE kinetic model for radiating plasmas of outer envelopes of stellar atmospheres *Astrophys. J.* **838** 126

[64] Johnston C O and Panesi M 2018 Impact of state-specific flowfield modeling on atomic nitrogen radiation *Phys. Rev. Fluids* **3** 013402

[65] Liu Y, Panesi M, Sahai A and Vinokur M 2015 General multi-group macroscopic modeling for thermo-chemical non-equilibrium gas mixtures *J. Chem. Phys.* **142** 134109

[66] Sahai A, Lopez B, Johnston C O and Panesi M 2017 Adaptive coarse graining method for energy transfer and dissociation kinetics of polyatomic species *J. Chem. Phys.* **147** 054107

[67] Munafò A, Panesi M and Magin T E 2014 Boltzmann rovibrational collisional coarse-grained model for internal energy excitation and dissociation in hypersonic flows *Phys. Rev.* E **89** 023001

[68] Macdonald R L, Munafò A, Johnston C O and Panesi M 2016 Nonequilibrium radiation and dissociation of CO molecules in shock-heated flows *Phys. Rev. Fluids* **1** 043401

[69] Park C 1993 Review of chemical–kinetic problems of future NASA missions, I: Earth entries *J. Thermophys. Heat Transfer* **7** 385–98

[70] Park H and Slanger T 1994 $O_2(X, v = 8\text{–}22)$ 300 K quenching rate coefficients for O_2 and N_2, and $O_2(X)$ vibrational distribution from 248 nm O_3 photodissociation *J. Chem. Phys.* **100** 287–300

[71] Varandas A and Pais A 1991 Double many-body expansion potential energy surface for $O_4(^3A)$, dynamics of the $O(^3P) + O_3(^1A_1)$ reaction, and second virial coefficients of molecular oxygen *Theoretical and Computational Models for Organic Chemistry* (Berlin: Springer) pp 55–78

[72] Coletti C and Billing G D 2002 Vibrational energy transfer in molecular oxygen collisions *Chem. Phys. Lett.* **356** 14–22

[73] Billing G D and Kolesnick R 1992 Vibrational relaxation of oxygen. State to state rate constants *Chem. Phys. Lett.* **200** 382–6

[74] Aquilanti V, Bartolomei M, Carmona-Novillo E and Pirani F 2003 The asymmetric dimer N_2–O_2: characterization of the potential energy surface and quantum mechanical calculation of rotovibrational levels *J. Chem. Phys.* **118** 2214–22

[75] Billing G D 1994 VV and VT rates in N_2–O_2 collisions *Chem. Phys.* **179** 463–7

[76] Bartolomei M, Carmona-Novillo E, Hernández M I, Campos-Martínez J and Moszynski R 2014 Global *ab initio* potential energy surface for the $O_2(^3\Sigma_g^-) + N_2(^1\Sigma_g^+)$ interaction. Applications to the collisional, spectroscopic, and thermodynamic properties of the complex *J. Phys. Chem.* A **118** 6584–94

[77] Garcia E, Kurnosov A, Laganà A, Pirani F, Bartolomei M and Cacciatore M 2015 Efficiency of collisional $O_2 + N_2$ vibrational energy exchange *J. Phys. Chem.* B **120** 1476–85

[78] Varga Z, Meana-Pañeda R, Song G, Paukku Y and Truhlar D G 2016 Potential energy surface of triplet N_2O_2 *J. Chem. Phys.* **144** 024310

[79] Andrienko D and Boyd I D 2015 Investigation of oxygen vibrational relaxation by quasi-classical trajectory method *Chem. Phys.* **459** 1–13

[80] Andrienko D and Boyd I D 2017 State-resolved O_2–N_2 kinetic model at hypersonic temperatures, *55th AIAA Aerospace Sciences Meeting* , *Grapevine, TX* AIAA Paper 2017-0659

[81] Garcia E, Pirani F, Laganà A and Martí C 2017 The role of the long-range tail of the potential in $O_2 + N_2$ collisional inelastic vibrational energy transfers *Phys. Chem. Chem. Phys.* **19** 11206–11

[82] Steele D, Lippincott E R and Vanderslice J T 1962 Comparative study of empirical internuclear potential functions *Rev. Mod. Phys.* **34** 239

IOP Publishing

Hypersonic Meteoroid Entry Physics

Gianpiero Colonna, Mario Capitelli and Annarita Laricchiuta

Chapter 17

Non-empirical analytical model of non-equilibrium dissociation in high-temperature air

Sergey Macheret, Han Luo and Alina Alexeenko

In the atmospheric entry of bodies such as spacecraft or meteorites flying at orbital or superorbital velocities, the translational temperature T behind the strong shock reaches very high values, up to several tens of thousands of kelvins. At the same time, the vibrational modes of molecules require a number of collisions to reach thermal equilibrium. In a rarefied gas, this results in the existence of a significant zone between the shock and the surface where the molecules are translationally hot but vibrationally cold ($T_v < T$). This thermal non-equilibrium affects chemistry, radiation and heat fluxes.

Moreover, the molecular dissociation itself results in thermal non-equilibrium. Indeed, consider that the vibrational mode is activated primarily in vibrational–translational (VT) collisional relaxation processes, and the vibrational energy is spent primarily on dissociation. Since the Landau–Teller temperature dependence of the VT relaxation time τ_{VT} is approximately $\tau_{VT} \sim \exp(B/T^{1/3})$, where B is a constant, and the Arrhenius temperature dependence of the dissociation reaction time is approximately $\tau_R \sim \exp(D/T)$, where D is the dissociation energy in kelvins, then the dissociation rate increases with T faster than the activation rate does. Additionally, the mean energy transferred in the VT event $\Delta E \ll D$, whereas each dissociation event removes energy $\sim D$. Thus, above a certain temperature T_0, $T > T_0$, the vibrational temperature T_v can never reach T due to the fast vibrational energy removal in dissociation. Simple estimates show that depending on species, $T_0 = 6000 - 12\,000$ K. Below T_0, vibrational relaxation (activation) happens first, and then the dissociation occurs in thermal equilibrium. Above T_0, the dissociation proceeds at $T_v < T$ until so many atoms are accumulated that their recombination equilibrates T_v and T. This high-temperature phenomenon is known as the vibration–dissociation coupling, and modeling this phenomenon has been a subject of research since the dawn of the space age.

In most simulations of non-equilibrium hypersonic flows, the models used for reaction probabilities in DSMC and two-temperature rate coefficients in CFD are empirical. Examples include the Park model for CFD and the total collision energy (TCE) model for DSMC. These models are simple and therefore computationally efficient, but they are not based on the physics of reactive collisions and can be quite inaccurate. On the other hand, with advances in quantum chemistry that can now generate potential energy surfaces (PESs) for simple colliding molecules and the rapidly growing computational power, quasi-classical trajectory (QCT) simulations are becoming almost routine, but incorporation of QCT-generated data into CFD and DSMC greatly increases complexity and requires extensive additional computational resources. Additionally, uncertainties or errors inherent in QCT include the uncertainties of *ab initio* electronic structure calculations, fitting errors of the PESs, statistical errors of QCT calculations, etc. Quantifying all such uncertainties is a difficult task. In addition, there are no *ab initio* PESs for many molecular systems, and since the sensitivity of the results to the details of the PESs has not been studied in detail, the use of model PESs is associated with an unknown error.

However, in addition to the two extremes—empirical models and QCT-based approximations—there is a third approach: theoretical (physical) models based on reasonable and physically justified assumptions and approximations. Algebraic, i.e. closed-form, formulas based on such physical models can be almost as simple as the Park and TCE models and can be conveniently used in both DSMC and CFD, while also ensuring accuracy of the modeling and providing insight into the underlying physics.

In this chapter, we review one such approach: the classical impulsive model of dissociation, known also as the Macheret–Fridman model [1–5].

17.1 Description of the Macheret–Fridman model

The theoretical model for dissociation, the so-called Macheret–Fridman model (MF) [1–5] was originally proposed by Macheret *et al* [1–3] based on the classical impulsive limit. The MF model [1–5] simplifies the mechanism of dissociation at very high temperatures and collision energies with a few important assumptions. Like QCT calculations, it assumes that translational, rotational and vibrational motion can be described by classical mechanics. Additionally, because of the high velocity required for dissociation, in particular for low-vibrational states, the Massey parameter $= \frac{\Delta U \cdot R}{\hbar V} \ll 1$, where ΔU is the energy transferred into the vibrational mode, R is the characteristic radius of interaction, V is the relative velocity of the colliding particles, and $\hbar$ is the reduced Planck's constant. Therefore, the collision can be viewed as instantaneous, i.e. close to the impulsive limit. The details of the PES are not important in such an approximation; additionally, the molecules do not experience three-body interactions before and after the collision. The last approximation is that the reaction rates are only determined by the energy threshold and the behavior of the cross sections just above the threshold. That follows from the exponential dependence of the distribution function on energy.

Recently [4], we made corrections to the earlier derived formulas and compared the theoretical predictions to QCT and to some experimental data. We have also developed a new implementation of the MF model for DSMC [5].

In the classical impulsive limit, the energy transfer from kinetic energy to vibrational energy of dissociating molecules is determined by the masses of colliding particles and by the set of angles and phases in the instant when the chemical rearrangement occurs. Detailed derivations of closed-form reaction probabilities and rates can be found in [2] (see also [4]). The collision geometries for diatom–atom and diatom–diatom collisions are shown in figure 17.1. Molecule AB is the dissociating molecule and CD is the colliding partner. A right-handed coordinate system is defined in the following manner: atoms A, B and C form the xy plane and the y-axis is in the direction of BC. In the center-of-mass frame, the velocities of molecules AB and BC are v and u accordingly. The angles γ_1 and γ_2 are the polar and azimuth angles for velocity u. v_0 and v_1 are the maximum velocities of intramolecular vibration. We assume that both molecules are harmonic oscillators (HO). In such a case, the magnitudes of vibrational velocities are $v_0 \cos \varphi_0$ and $v_1 \cos \varphi_1$ given phase angles φ_0 and φ_1. The angle between the BC and AB vibrational velocity is θ. The polar and azimuth angles of the CD vibrational velocity are β_1 and β_2. v_r is the rotational velocity of molecule AB. Its direction is perpendicular to the vibrational velocity of molecule AB with a polar angle β. To define the direction of rotational velocity v_r' for molecule CD, two reference vectors $\vec{c}_{\text{ref1}}$ and $\vec{c}_{\text{ref2}}$ are introduced. $\vec{c}_{\text{ref1}}$ is in the xy plane and perpendicular to the vibrational velocity v_1. $\vec{c}_{\text{ref2}}$ is perpendicular to both $\vec{c}_{\text{ref2}}$ and v_1. Then, the direction of rotational velocity v_r' is uniquely determined by the angle δ between v_r' and $\vec{c}_{\text{ref2}}$ and it must be in the plane formed by $\vec{c}_{\text{ref1}}$ and $\vec{c}_{\text{ref2}}$.

Denote the masses of atoms A, B, C and D as m_A, m_B, m_C and m_D; the reduced masses of molecules AB and BC become $\mu_{AB} = m_A m_B/(m_A + m_B)$ and $\mu_{CD} = m_C m_D/(m_C + m_D)$. For the case of diatom–atom collision, μ_{CD} is equal to the mass of atom C. We can also calculate the reduced mass of the whole system:

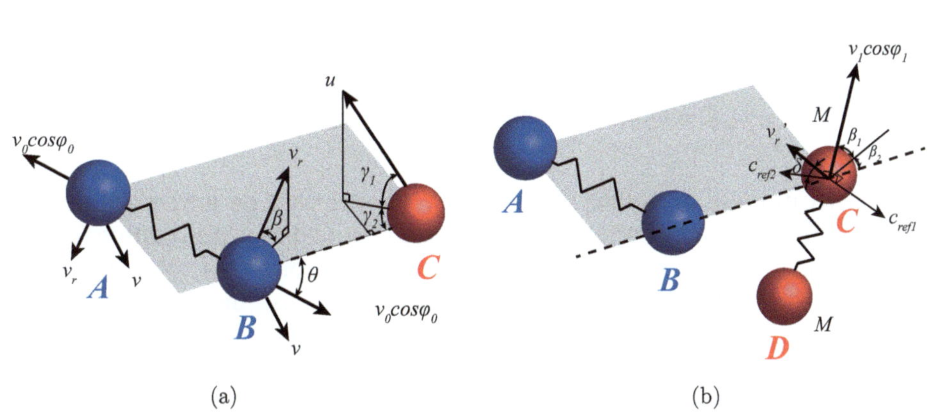

Figure 17.1. Collision geometries: (a) diatom–atom and (b) diatom–diatom.

$$\mu = \begin{cases} (m_A + m_B)m_C/(m_A + m_B + m_C), & \text{for diatom–atom} \\ (m_A + m_B)(m_C + m_D)/(m_A + m_B + m_C + m_D), & \text{for diatom–diatom} \end{cases}. \tag{17.1}$$

The relations between velocities and energies become

$$\begin{gathered} E_v = 2\mu_{AB} v_0^2, \quad E_v' = 2\mu_{CD} v_1^2, \quad E_r = 2\mu_{AB} v_r^2 \\ E_r' = 2\mu_{CD} v_r'^2, \quad E_t = \frac{(m_A + m_B)^2 v^2}{2\mu}, \end{gathered} \tag{17.2}$$

where E_v, E_v' are the vibrational energies of molecule AB and CD, E_r, E_r' are the rotational energy of each molecule and E_t is the collisional energy.

The two-body problem is solved for the elastic collision between atoms B and C. Energy and momentum transfer are only allowed in the direction of the collision. By making the post-collision vibrational energy of molecule AB equal to the dissociation energy D, we can obtain the threshold function $F = E_{t,\min}$, which is the minimum collisional energy for the dissociation reaction $AB + CD \rightarrow A + B + CD$ to occur. For diatom–atom collision, the threshold function is

$$\begin{aligned} F = \frac{\mu/\mu_{AB}}{4\cos^2\gamma_1 \cos^2\gamma_2} \Bigg\{ & \frac{\sqrt{D - E_v \sin^2\varphi_0} + \sqrt{E_v}\cos\varphi_0}{\cos\theta \cdot m_C/(m_B + m_C)} \\ & - \frac{2\mu_{AB}}{m_B}\left[\sqrt{E_v}\cos\varphi_0\cos\theta + \sqrt{E_r}\cos\beta\sin\theta\right]\Bigg\}^2. \end{aligned} \tag{17.3}$$

For diatom–diatom collision, the threshold function is

$$\begin{aligned} F = \frac{\mu/\mu_{AB}}{4\cos^2\gamma_1 \cos^2\gamma_2} \Bigg\{ & \frac{\sqrt{D - E_v \sin^2\varphi_0} + \sqrt{E_v}\cos\varphi_0}{\cos\theta \cdot m_C/(m_B + m_C)} \\ & - \frac{2\mu_{AB}}{m_B}\left[\sqrt{E_v}\cos\varphi_0\cos\theta + \sqrt{E_r}\cos\beta\sin\theta\right] \\ & - \frac{2\sqrt{\mu_{AB}\mu_{CD}}}{m_C}\Big[\sqrt{E_r'}(\cos\delta\cos\beta_2\sin\beta_1 + \sin\delta\sin\beta_2) \\ & - \sqrt{E_v'}\cos\varphi_1\cos\beta_1\cos\beta_2\Big]\Bigg\}^2. \end{aligned} \tag{17.4}$$

The derivations of equations (17.3) and (17.4) are based on the assumption that collisions are close to the impulsive limit. It can be found that the rotational energy of dissociating molecule E_r contributes to the dissociation by its component along the direction of collision. However, rotational energy also increases the effective dissociation energy due to the centrifugal barrier, which arises as the rotational energy decreases when the molecule is stretched during dissociation. We adopt the strategy proposed in [2] by using an effective dissociation energy D_{ef}, that includes both effects. D_{ef} is evaluated with a long-range potential energy curve

$V(r) = -2D(R_{eq}/r)^6$, where R_{eq} is the equilibrium distance of the molecular bond and r is the internuclear distance

$$D_{ef} = D - E_r + \frac{2E_r^{3/2}}{3\sqrt{6D}}. \tag{17.5}$$

The final expressions of threshold functions for diatom–atom and diatom–diatom collisions are

$$F = \frac{\mu/\mu_{AB}}{4\cos^2\gamma_1\cos^2\gamma_2}\left[\frac{\sqrt{D_{ef} - E_v\sin^2\varphi_0} + \sqrt{E_v}\cos\varphi_0}{\cos\theta \cdot m_C/(m_B + m_C)} - \frac{2\mu_{AB}}{m_B}\sqrt{E_v}\cos\varphi_0\cos\theta\right]^2, \tag{17.6}$$

$$\begin{aligned} F = \frac{\mu/\mu_{AB}}{4\cos^2\gamma_1\cos^2\gamma_2}\Bigg\{&\frac{\sqrt{D_{ef} - E_v\sin^2\varphi_0} + \sqrt{E_v}\cos\varphi_0}{\cos\theta \cdot m_C/(m_B + m_C)} \\ &- \frac{2\mu_{AB}}{m_B}\sqrt{E_v}\cos\varphi_0\cos\theta \\ &- \frac{2\sqrt{\mu_{AB}\mu_{CD}}}{m_C}\Big[\sqrt{E_r'}(\cos\delta\cos\beta_2\sin\beta_1 + \sin\delta\sin\beta_2) \\ &- \sqrt{E_v'}\cos\varphi_1\cos\beta_1\cos\beta_2\Big]\Bigg\}^2. \end{aligned} \tag{17.7}$$

It should be noted that, for a diatom–diatom colliding pair such as $O_2 + N_2$, there is more than one possible dissociation reaction channel. With the same definitions of generalized coordinates as in figure 17.1, one can derive the threshold function for the dissociation of CD molecule as well:

$$\begin{aligned} F' = \frac{\mu/\mu_{CD}}{4\cos^2\gamma_1\cos^2\gamma_2}\Bigg\{&\frac{\sqrt{D_{ef}' - E_v'\sin^2\varphi_1} + \sqrt{E_v'}\cos\varphi_1}{\cos\beta_1\cos\beta_2 \cdot m_B/(m_B + m_C)} \\ &- \frac{2\mu_{CD}}{m_C}\sqrt{E_v'}\cos\varphi_1\cos\beta_1\cos\beta_2 \\ &+ \frac{2\sqrt{\mu_{AB}\mu_{CD}}}{m_B}\Big[\sqrt{E_r}\cos\beta\sin\theta + \sqrt{E_v}\cos\varphi_0\cos\theta\Big]\Bigg\}^2. \end{aligned} \tag{17.8}$$

17.2 Macheret–Fridman model for CFD

To develop MF model formulas for CFD, i.e. multi-temperature rate coefficients, the next step is to determine the optimum configuration with the minimum threshold energy by setting all the partial derivatives of the threshold function to 0. The probability of dissociation can then be obtained based on Taylor expansion of the threshold function near the optimum configuration. The probability of the dissociation reaction is then

$$P = \begin{cases} \dfrac{4(E_t - F)^{3/2}}{\left(3\pi^2 F\sqrt{D_{\rm ef}}\big/\sqrt{1-\alpha}\right) \times \sqrt{(1-\sqrt{\alpha E_v/D_{\rm ef}})[1-(2-\sqrt{\alpha})\sqrt{E_v/D_{\rm ef}}]}}, \\ \quad \text{if } E_v \leqslant \alpha D_{\rm ef} \\ \dfrac{(E_t - F)^2}{\left(2\pi^2 F D_{\rm ef}\sqrt{\alpha}\big/\sqrt{(1+\sqrt{\alpha})/(1-\sqrt{\alpha})}\right)\sqrt{(1-E_v/D_{\rm ef})(E_v/\alpha D_{\rm ef}-1)}}, \\ \quad \text{if } E_v > \alpha D_{\rm ef} \end{cases} \tag{17.9}$$

where α is the mass factor that, for the case $m_A = m_B = m$, $m_C = m_D = M$, is

$$\alpha = \left(\frac{m}{m+M}\right)^2. \tag{17.10}$$

Multi-temperature reaction rate coefficients can then be calculated by integrating the probability with distribution functions of translational, rotational and vibrational energy. To remain consistent with the basic assumptions in CFD, Maxwell–Boltzmann distribution functions are used as a first approximation. However, it should be noted that the dissociation reaction favors high-vibrational levels and thus depletes their population, making the vibrational energy distribution (VDF) deviate from the Boltzmann distribution. The prediction of reaction rates with non-Boltzmann distribution requires state-to-state calculations, which so far have not been carried out for the MF model.

It is important to note that the MF model analyzes the energy transformation in reactions and calculates the energy threshold and the probability of dissociation assuming that the collision has occurred. Thus, in order to obtain the rate coefficient or the cross section of dissociation, the dissociation probability calculated by the model should be multiplied by the 'total' rate coefficients or total cross sections of collisions; such total cross sections or collisional rate coefficients should be taken from sources other than the model itself. In [1–4], instead of attempting to calculate the total cross sections and the collisional rates $k_{\rm coll}(T)$, the focus was on the non-equilibrium factor $Z = k(T,\ T_v)/k(T)$, i.e. on the behavior of the dissociation rate in thermally non-equilibrium conditions normalized by its thermally equilibrium (Arrhenius) value,

$$k(T) = AT^n \exp\left(-\frac{D}{T}\right), \tag{17.11}$$

where the dissociation energy D is in kelvins. Accordingly, we use the condition $k(T = T_r = T_v) = k(T)$. In such a context, the multi-temperature rates are calculated as

$$k(T, T_v, T_r) = k_{\mathrm{coll}}(T) \times \int_{E_v} \int_{Et=F}^{+\infty} \int_{E_r} P \cdot f_{T_r}(E_r)\mathrm{d}E_r f_T(E_t)\mathrm{d}E_t f_{T_v}(E_v)\mathrm{d}E_v. \tag{17.12}$$

The extreme complexity of the reaction probability in equation (17.9) precludes the direct analytical integration in equation (17.12), and the following approximations were employed in [2] and [4]. The two-temperature rate coefficient was presented as the sum of the contributions for low, $E_v \leqslant \alpha D_{\mathrm{ef}}$, and high, $E_v > \alpha D_{\mathrm{ef}}$, vibrational levels:

$$k(T, T_v) = k_l + k_h. \tag{17.13}$$

A simple version of the steepest descent method can be applied to calculate the rates for $E_v \leqslant \alpha D_{\mathrm{ef}}$. The method approximates the pre-exponential factor by its value in the vicinity of the exponent's maxima. The two-temperature rate coefficient for dissociation from low-vibrational states is

$$k_l(T, T_v) = AT^n \times L \exp\left(-\frac{D}{T_a} + \Delta D\left(\frac{1}{T_a} - \frac{1}{T}\right)\right), \tag{17.14}$$

and the contribution from the high-vibrational levels is approximately

$$k_h(T, T_v) = AT^n \times (1 - L)\frac{1 - \exp(-hv/T_v)}{1 - \exp(-hv/T)} \exp\left(-\frac{D}{T_v}\right), \tag{17.15}$$

where for diatom–atom collisions

$$L = \frac{\sqrt{1 - \alpha}}{\pi^{3/2}}\sqrt{\frac{D}{D^*}}\left(\frac{T}{D}\right)^{1-n}\left(1 + \frac{5(1 - \alpha)T}{2D^*}\right)\left(12\pi b\alpha(1 - \alpha)\frac{D}{T}\right)^{1/2}, \tag{17.16a}$$

whereas for diatom–diatom collisions

$$L = \frac{2(1 - \alpha)}{\pi^2\alpha^{3/4}}\left(\frac{D}{D^*}\right)\left(\frac{T}{D}\right)^{3/2-n}\left(1 + \frac{7(1 - \alpha)(1 + \sqrt{\alpha})T}{2D^*}\right) \left(12\pi b\alpha(1 - \alpha)\frac{D}{T}\right)^{1/2}. \tag{17.16b}$$

In equations (17.14)–(17.16), T_a is the 'average' temperature

$$T_a = \alpha T_v + (1 - \alpha)T, \tag{17.17}$$

D^* is the approximated value of D_{ef} (assuming a small value of α, which is usually the case),

$$D^* = D - \Delta D = D - 3b\alpha^2 D, \tag{17.18}$$

and v is the vibrational frequency of the dissociating molecule. The average vibrational energy removed by dissociation is

$$E_{v,\mathrm{rem}} = \frac{(\alpha D^* \times (T_v/T_a)^2) \cdot k_l + D \cdot k_h}{k_l + k_h}. \tag{17.19}$$

Figures 17.2 and 17.3 show the comparison of the non-equilibrium factor $k(T, T_v)/k(T)$ in two versions of the MF model with QCT simulations for N_2–O [6] and O_2–O [7] collisional dissociation. The simplified version [3] that ignores the rotational correction to the dissociation energy is denoted as 'Macheret–Fridman S' and the full version, with reduction of the dissociation barrier due to rotational energy, is denoted as 'Macheret–Fridman F'. Figure 17.4 compares the full and simplified MF models with QCT results [8] for N_2–N_2 collisional dissociation. For N_2–N_2 dissociation, Singh and Schwartzentruber's algebraic interpolation formula [9] fitted to the QCT data and an algebraic fit to experimental measurements [10, 11] are also presented. Although it might appear that for N_2–N_2 dissociation (figure 17.4) both QCT and the full MF model diverge from the fit to experimental results at vibrational temperatures below 7000–8000 K, it must be kept in mind that the actual experimental data have been obtained at vibrational temperatures of at least 7000–8000 K, and simple extrapolation of those experimental results to low-vibrational temperature would be incorrect.

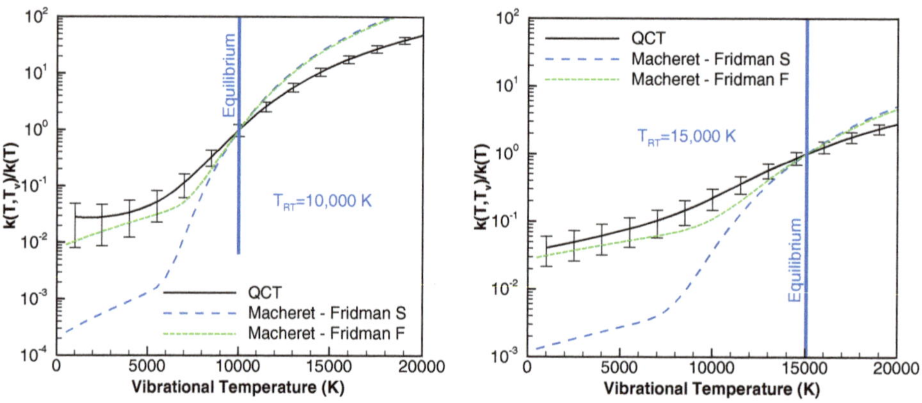

Figure 17.2. Non-equilibrium factor for $N_2 + O \rightarrow N + N + O$.

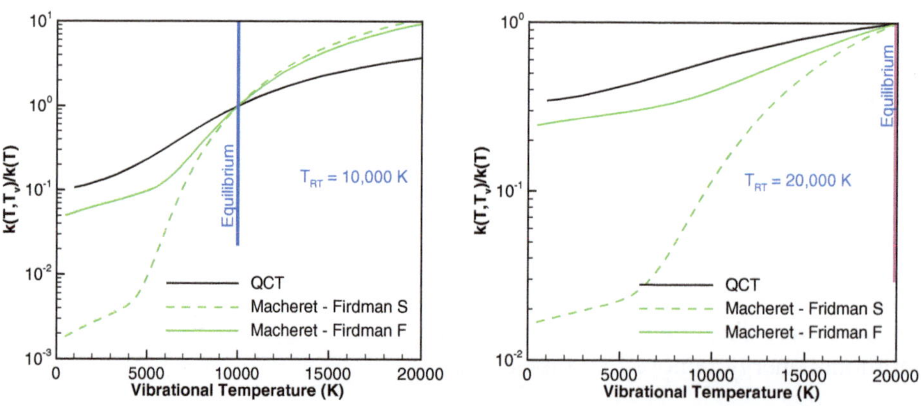

Figure 17.3. Non-equilibrium factor for $O_2 + O \rightarrow O + O + O$.

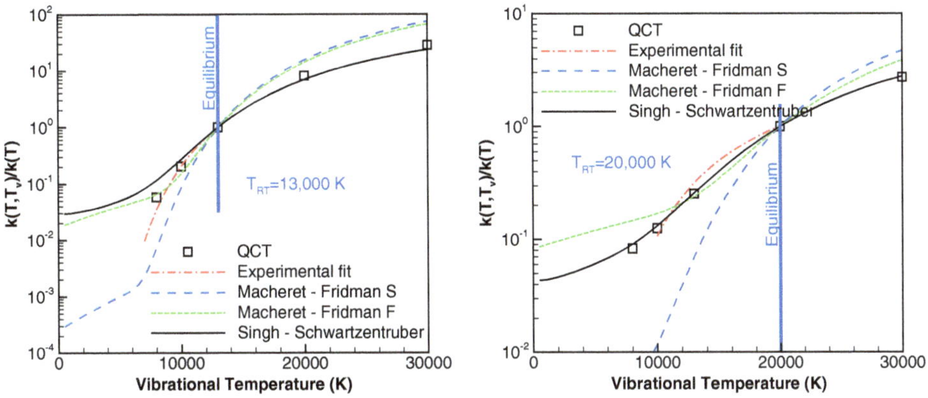

Figure 17.4. Non-equilibrium factor for $N_2 + N_2 \rightarrow N + N + N_2$.

As seen in figures 17.2–17.4, the simplified version of the model which ignores rotational effects underpredicts the dissociation rate in vibrationally cold conditions by orders of magnitude, whereas the full version of the model shows very good agreement with QCT, in particular since the uncertainty of QCT results at low-vibrational temperatures is a factor of 3 at best. Thus, the vibration–rotation coupling is seen to be very important in vibrationally cold conditions, as the rotation lowers the effective dissociation energy (although not by the full rotational energy due to the 'centrifugal barrier' stemming from the conservation of angular momentum). This effect, predicted in [1, 2] and validated in [4], is quite consistent with the observations made in [8] from the QCT analysis.

More information on the comparison between the MF model and QCT for other temperatures and colliding pairs can be found in [4].

17.3 Macheret–Fridman model for DSMC

With the complete descriptions of threshold functions (17.6)–(17.8), the MF-DSMC model can be formulated. As mentioned in section 17.2, the MF model calculates reaction probabilities $P(E_t)$ by expanding equations (17.6) and (17.7) in Taylor series of phase angles and calculating the probability of $E_t \geqslant F$. The mathematical approximation introduced by the Taylor expansion results in singularities of reaction probabilities [2]. Here, instead of using the Taylor expansion, we use a different technique. Because DSMC is a stochastic method and the collision model does not change the distributions of phase angles, we can directly sample collision geometries and use the threshold function to determine whether dissociation occurs. The ensemble average will then produce the correct reaction probabilities. The pseudocodes are listed in algorithm 17.1.

Algorithm 17.1. MF-DSMC model.

1. Generate phase angles γ_1, γ_2, β_1, β_2, θ, φ_0, φ_1, θ, δ, β.
2. Compute effective dissociation energy $D_{\mathrm{ef},i}$ with equation (17.5).

3. Compute threshold energy $F_i(E_v, E_r, E_v', E_r')$ with equations (17.6)–(17.8) for each possible dissociation reaction channel i.
4. Select the reaction pathway i with the minimum threshold energy F_i.
5. If $E_t \geqslant F_i$ then reaction i occurs (i.e. $P_{j=i} = 1$, $P_{j\neq i} = 0$).
6. Otherwise no dissociation occurs (i.e. $P_j = 0$).
7. End if.

It should be noted that D_{ef} and the term $D_{ef} - E_v \sin^2 \varphi_0$ in equations (17.6)–(17.8) may become less than 0 for certain values of energies of the colliding particles. This is due to the particular form of the potential energy curve used to derive D_{ef}. We have used a more realistic potential energy curve, i.e. the Morse potential, to remove this singularity, but found that the change in final results is not significant. Thus, a simple approach is to make the reaction probability P equal to 1 when this condition occurs.

To validate the new MF-DSMC model, we implemented the model in a heavily modified DS1V code [12]. A standard no-time-counter (NTC) scheme was used. Elastic collisions are modeled according to the variable hard sphere (VHS) model [12]. Regarding the modeling of internal energy, a discrete vibrational energy model was used with a continuous model for rotational energy. All the calculations were started at either thermal equilibrium ($T = T_r = T_v$) or non-equilibrium ($T = T_r \neq T_v$) conditions with a Maxwell–Boltzmann distribution of collisional energy and internal energy. In order to isolate the influence of reaction models from the VT and RT energy relaxations, the Larsen–Borgnakke (LB) model with constants $Z_R = 1$ and $Z_V = 10^4$ was used. To maintain the steady state and isothermal condition, dissociations were only tracked and molecules did not actually dissociate.

A very important feature of the MF model is that it calculates the probability of dissociation reaction provided that a collision occurs. Therefore, in order to calculate the total probability and rate of dissociation, one needs to multiply the MF probability by the rate of collisions which has to be taken from outside the MF model. For example, the rate of collisions can be calculated with the VHS model or from matching the thermal equilibrium rate to the experimental or QCT data. In this context, we first validated the MF-DSMC model by comparing the non-equilibrium characteristics such as state-specific rates and non-equilibrium factors to the available data, and then examined the absolute dissociation rates that require knowledge of the collision rates.

In [5], the vibrational state-specific reaction rates predicted by MF-DSMC for different colliding pairs were compared to the rates obtained by QCT and direct molecular simulation (DMS). Good agreement was found for most systems investigated [5]. The MF-DSMC model correctly captures the linear dependence of $\log(k(T, E_v))$ on vibrational energy and also the relationship between the slope of the logarithm of the rates and translational temperature T. For the $N_2 + O$ system, the model predicts rates almost identical to the QCT results multiplied by the electronic degeneracy factor. The agreement improves with increasing temperature, which confirms the expectation that collision dynamics are weakly sensitive to the details of the PES at high temperature and in high-energy collisions.

However, the MF-DSMC model fails to correctly predict the vibrational state-specific rates for very high-vibrational states [5] because it assumes harmonic vibrations. Compared to a harmonic oscillator (HO), the actual particles spend more time at the outer turning point, which is the preferential phase angle for dissociation. Thus, the HO assumption results in underestimation of the dissociation rates and probabilities for molecules with high-vibrational energy. The issue becomes less significant for either high translational temperature or low-vibrational temperature because of the increasing contributions from molecules with low-vibrational energy.

The comparison of non-equilibrium factors computed with the two-temperature MF model intended for CFD (denoted as MF in figures 17.5 and 17.6; this is the 'full' model that includes rotational effects, see section 17.2) and the MF-DSMC model with QCT data is shown in figures 17.5 and 17.6. (More comparison data can be found in [5].) As can be seen in the figures, the results predicted with the MF-DSMC model and QCT calculations match closely. Compared to the MF-CFD model, the agreement is particularly good under vibrationally hot conditions. For vibrationally cold conditions, the non-equilibrium factor calculated from the MF-DSMC model agrees with the QCT data very well, in particular given the uncertainty of QCT calculations (with the error bars illustrated in figure 17.5).

Further validation of the MF-DSMC model in reference [5] was to compare the calculated equilibrium dissociation rates with QCT calculations, experimental measurements and empirical estimations. It was observed that the rates calculated

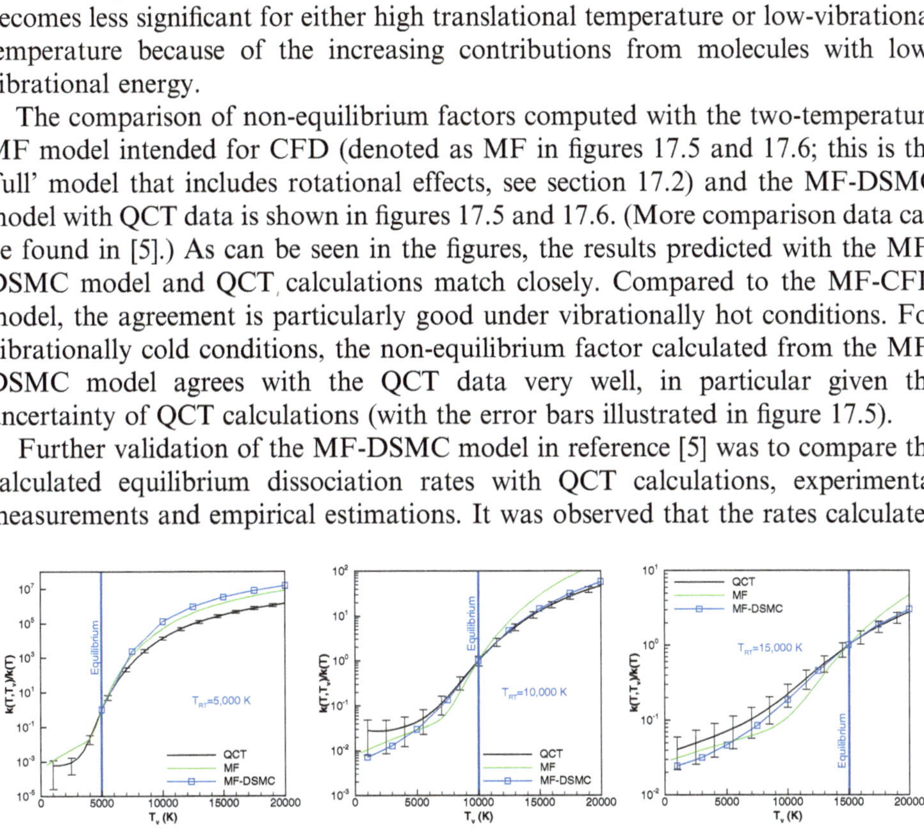

Figure 17.5. Non-equilibrium factor for $N_2 + O \rightarrow N + N + O$ in comparison to QCT data [6].

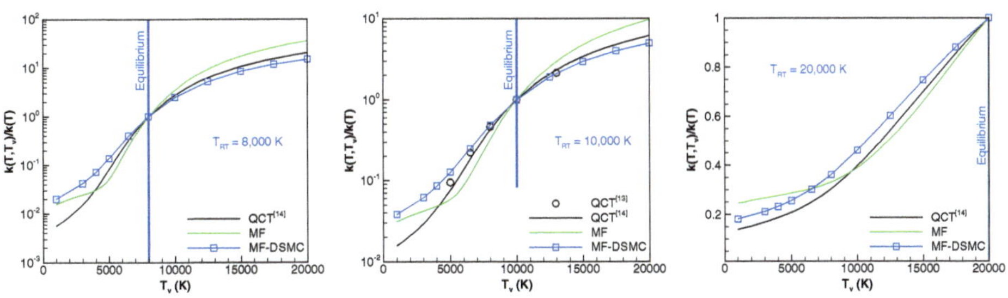

Figure 17.6. Non-equilibrium factor for $O_2 + N_2 \rightarrow O + O + N_2$ in comparison to QCT data [13, 14].

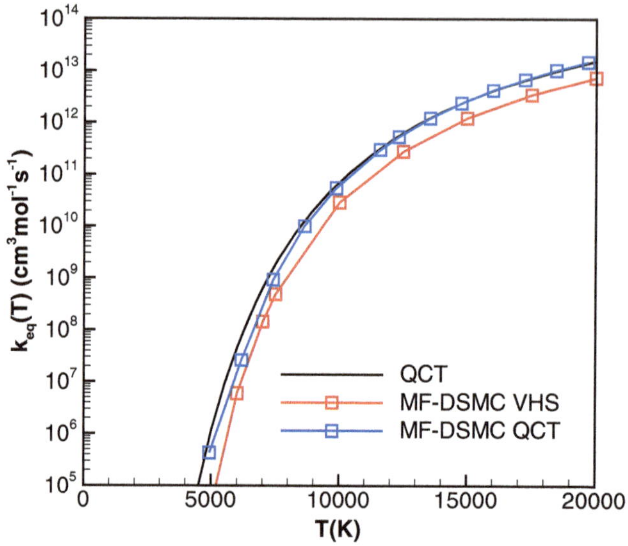

Figure 17.7. Comparison of equilibrium dissociation rates for N_2 dissociation in collisions with atomic oxygen.

with the MF-DSMC model are lower than the QCT data by an approximately constant factor. Since the non-equilibrium factors calculated with the MF-DSMC model were found to agree very well with QCT data, the disagreement between the calculated equilibrium rates and the QCT data is clearly due to the underestimation of total collision cross sections and rates by the VHS model commonly used in DSMC rather than to the MF dissociation model itself. Therefore, to improve the model, MF-DSMC should be paired with a better collision model. In [5], such a collisional model for N_2 dissociation in collisions with atomic oxygen was developed based on calibration against QCT scattering data. Equilibrium dissociation rates given by the MF-DSMC model with the VHS collision model and QCT-based collision model are compared in figure 17.7. The corrected collision model results in a much better agreement between MF-DSMC and QCT. For $T > 10\,000$ K, the difference between MF-DSMC and QCT is negligible, which again justifies the assumption of impulsive limit for high-energy collisional dissociation. For other colliding partners, similar procedures of correcting total collision cross sections and scattering law can be used. It should be noted that, since the collision rates are changed by collision model, vibrational and rotational relaxation numbers should also be calibrated consistently in order to reproduce the correct relaxation times.

17.4 Concluding remarks

The goal of the work described here was to develop theoretical models of non-equilibrium dissociation at high temperatures so that these models would have, without adjustable parameters, accuracy matching that of QCT while avoiding the complexity of QCT and ensuring the simplicity and computational efficiency of the Park model in CFD and the TCE model in DSMC. This goal has been largely achieved. The results indicate the potential of the MF model to be used as the standard CFD and DSMC dissociation reaction model.

To further improve the MF-DSMC model, the probability calculation can be performed for an anharmonic molecular oscillator. This would improve the predictions for high-vibrational levels, albeit at the expense of reduced computational efficiency.

We found that although the MF model describing the dissociation probability once the collision occurred works well, the popular VHS model used in DSMC to calculate collision rates needs improvement. A QCT-calibrated collision model developed for a particular case of N_2 + O collisions ensured very good match of thermally equilibrium dissociation rate calculated by MF-DSMC and QCT. A more physical solution would involve simplified collision dynamics based on model algebraic potential energy functions such as Lennard-Jones or Born–Mayer potentials.

Finally, if the classical impulsive approach could be extended to exchange reactions, such as $O + N_2 \rightarrow NO + N$, and simple physical models of those reactions could be developed, that would greatly increase the ability to accurately and efficiently model high-temperature non-equilibrium flows.

References

[1] Macheret S O and Rich J W 1993 Nonequilibrium dissociation rates behind strong shock waves: classical model *Chem. Phys.* **174** 25–43

[2] Macheret S O, Fridman A A, Adamovich I V, Rich J W and Treanor C E 1994 Mechanisms of nonequilibrium dissociation of diatomic molecules, *6th Joint Thermophysics and Heat Transfer Conf.* AIAA Paper 1994-1984

[3] Chernyi G G, Losev S A, Macheret S O and Potapkin B V 2002 *Physical and Chemical Processes in Gas Dynamics, Vol. I: Cross sections and Rate Constants* (Reston, VA: AIAA)

[4] Luo H, Alexeenko A A and Macheret S O 2018 Assessment of classical impulsive models of dissociation in thermochemical nonequilibrium *J. Thermophys. Heat Transfer* **32** 861–68

[5] Luo H, Sebastião I B, Alexeenko A A and Macheret S O 2018 Classical impulsive model for dissociation of diatomic molecules in DSMC *Phys. Rev. Fluids* **3** 113401

[6] Luo H, Kulakhmetov M and Alexeenko A 2017 *Ab initio* state-specific N_2+O dissociation and exchange modeling for molecular simulations *J. Chem. Phys.* **146** 074303

[7] Kulakhmetov M, Gallis M and Alexeenko A 2016 *Ab initio*-informed maximum entropy modeling of rovibrational relaxation and state-specific dissociation with application to the O_2+O system *J. Chem. Phys.* **144** 174302

[8] Bender J D, Valentini P, Nompelis I, Paukku Y, Varga Z, Truhlar D G, Schwartzentruber T and Candler G V 2015 An improved potential energy surface and multi-temperature quasiclassical trajectory calculations of $N_2 + N_2$ dissociation reactions *J. Chem. Phys.* **143** 054304

[9] Singh N and Schwartzentruber T E 2017 Coupled vibration–rotation dissociation model for nitrogen from direct molecular simulations, *47th AIAA Thermophysics Conf., Denver, CO* AIAA Paper 2017-3490

[10] Losev S A, Makarov V N, Pogosbekyan M J, Shatalov O P and Nikolsky V S 1994 Thermochemical nonequilibrium kinetic models in strong shock waves on air, *6th Joint Thermophysics and Heat Transfer Conf.* AIAA Paper 1994-1990

[11] Yalovik M S and Losev S A 1972 Vibrational kinetics and dissociation of nitrogen molecules at high temperature, *Sci. Proc. Inst. Mech. Mosc. State Univ.* **18** 3–34

[12] Bird G A 1994 *Molecular Gas Dynamics and the Direct Simulation of Gas Flows* (New York: Oxford University Press)

[13] Chaudhry R S, Grover M S, Bender J D, Schwartzentruber T E and Candler G V 2018 Quasiclassical trajectory analysis of oxygen dissociation via O_2, O, and N_2, *2018 AIAA Aerospace Sciences Meeting* AIAA Paper 2018-0237

[14] Andrienko D A and Boyd I D 2018 Dissociation of oxygen and nitrogen in a bimolecular reaction at hypersonic temperatures, *2018 AIAA Aerospace Sciences Meeting* AIAA Paper 2018-0240

IOP Publishing

Hypersonic Meteoroid Entry Physics

Gianpiero Colonna, Mario Capitelli and Annarita Laricchiuta

Chapter 18

The role of vibrational activation and bimolecular reactions in non-equilibrium plasma kinetics

Lucia Daniela Pietanza, Vincenzo Aquilanti, Patricia Barreto, Mario Capitelli, Gianpiero Colonna, Andrea Lombardi, Sergey Macheret and Federico Palazzetti

The hypersonic entry of meteoroids into Earth's atmosphere induces ablation with the production of molecular and atomic species in the environment. The formed species are subjected to the action of several collision processes, including electron–molecule and heavy-particle–heavy-particle collisions. The high level of gas temperature achieved by the medium promotes reactive channels which can be aided by the presence of vibrationally excited states in either the flow or in the boundary layer surrounding the meteoroid. The characterization of the relevant processes requires, in particular, the dependence of cross sections and rate coefficients on the vibrational quantum number of the reactants. This dependence has been extensively studied in the case of electron–molecule processes (see chapter 15), while the corresponding behavior of heavy-particle–heavy-particle reactive collisions is still an unsolved problem so that their characterization still requires the use of simplified methods. The aim of this chapter is to present the application of an old but still widely used method to analyze the reactive channels promoted by gas temperature and vibrational temperature and their role in affecting the properties of the whole medium excited during the meteoroid impact. We select, in this case, a plasma formed in the activation of both CO_2 and CO molecular species and their influence on the whole kinetics. These species are produced during the ablation of selected meteoroids (see chapter 6) [1]. The plasma kinetics code to understand the role of vibrational excitation in promoting bimolecular reactions, developed in the last few years by some of the current authors, has been used to this end. A second case study considers the activation of the different processes by an instantaneous increase of gas temperature followed by the formation of well-structured vibrational distribution functions and the opening of reactive channels aided by this excitation. The chapter

doi:10.1088/2053-2563/aae894ch18

is essentially divided in two parts, the first dedicated to the state-to-state rates in molecular collisions, while the second examines the use of these rates in modeling the global rates of reactive processes involving heavy-particle collisions as compared to the corresponding rates promoted by electron-impact collisions. In the first part, particular emphasis is given to the Boudouard reaction rates and to the dissociation rates, which are strongly assisted by the presence of vibrational excitation of the medium. Different plasma kinetics are discussed in the second part, emphasizing situations where non-equilibrium vibrational distributions and non-equilibrium electron energy distribution functions play an important role in affecting the rate coefficients of reactive processes.

18.1 Reactive channels promoted by heavy-particle collisions

18.1.1 The Boudouard reaction as a prototype

We consider the Boudouard reactions written in the absence of vibrational excitation as

$$CO(X^1\Sigma_g, v = 0) + CO(X^1\Sigma_g, w = 0) \rightarrow CO_2 + C \tag{18.1}$$

$$CO(X^1\Sigma_g, v = 0) + CO(a^3\Pi, w = 0) \rightarrow CO_2 + C \tag{18.2}$$

$$CO(X^1\Sigma_g, v = 0) + CO(X^1\Sigma_g, w = 0) \rightarrow C_2O + O \tag{18.3}$$

$$CO(X^1\Sigma_g, v = 0) + CO(a^3\Pi, w = 0) \rightarrow C_2O + O. \tag{18.4}$$

The energetics of these processes, calculated by a quantum chemistry approach, can be visualized in figure 18.1, as recently reported by Barreto *et al* [2].

The corresponding activation energies are reported in table 18.1.

This table reports the classical barrier, $V^{\neq}$, without the zero-point energy and V_a^G, which includes the zero-point energy for the four considered processes in the absence of ro-vibrational energy in both reactants and products. The transition-state-theory rate constants are then determined using the equation

$$k^{TST}(T) = \left(\frac{k_B T}{h}\right)\left(\frac{Q^{\neq}}{Q_{CO}^{A} Q_{CO}^{B}}\right)\exp\left(-\frac{V_a^{G\neq}}{RT}\right), \tag{18.5}$$

where k_B and h are the Boltzmann and Planck constants and T is the temperature. The terms Q_A^{CO} and Q_B^{CO} are the partition functions of the CO reactants and $Q^{\neq}$ is the partition function of the transition state. In this latter case, the contribution from the translational mode along the reaction coordinate is not considered, corresponding to an imaginary frequency and accounted for the Eyring term $k_B T/h$. Details can be found in Barreto *et al* [2]. The vibrationally adiabatic ground-state potential-energy curve is given by the equation

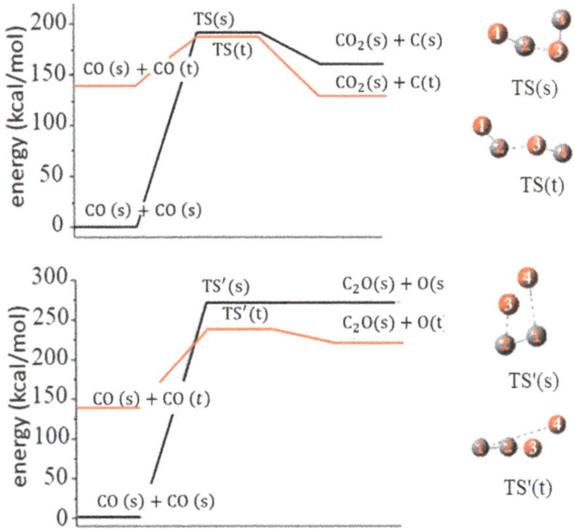

Figure 18.1. Energy (in kcal mol^{-1}) diagrams for the four reactions CO (s) + CO (s) → CO_2+ C, CO(s) + CO (t) → CO_2 + C, CO (s) + CO (s) → C_2O + O and CO (s) + CO (t) → C_2O + O, where s and t correspond to singlet and triplet states, i.e. the structure and the relative energy of the activated complex or transition state is reported for each reaction, indicated, respectively, as TS(s), TS(t), TS′(s) and TS′(t). Reprinted from [2]. With permission of Springer Verlag.

Table 18.1. *Ab initio* quantities (in kcal mol^{-1}) for forward and backward processes.

Reaction	$V^{\neq}$ Forward	V_a^G Forward	$V^{\neq}$ Backward	V_a^G Backward
CO(s) + CO(s) → CO_2 + C	191.99	192.75	31.57	31.28
CO(s) + CO(t) → CO_2 + C	47.25	46.90	58.44	56.53
CO(s) + CO(s) → C_2O + O	271.87	273.46	−0.33	1.69
CO(s) + CO(t) → C_2O + O	100.39	100.98	17.87	18.37

$$V_a^{G\neq} = V^{\neq} + \epsilon_{\text{ZPE}}, \tag{18.6}$$

where $V^{\neq}$ is the classical potential energy of the saddle point measured from the overall zero of energy and ϵ_{ZPE} is the sum of harmonic zero-point energies. In figure 18.2, we report the rate constants of the four processes, calculated by equation (18.5) (see [2]), as a function of $1/T$ and fitted by the Arrhenius equation

$$k = A\exp(-E_a/RT). \tag{18.7}$$

The parameter E_a will be referred to here as the apparent activation energy to be considered in the following discussion. Fitted values A and E_a have been reported in table 18.2 to reproduce the rate constants. Rate constants involving the triplet state,

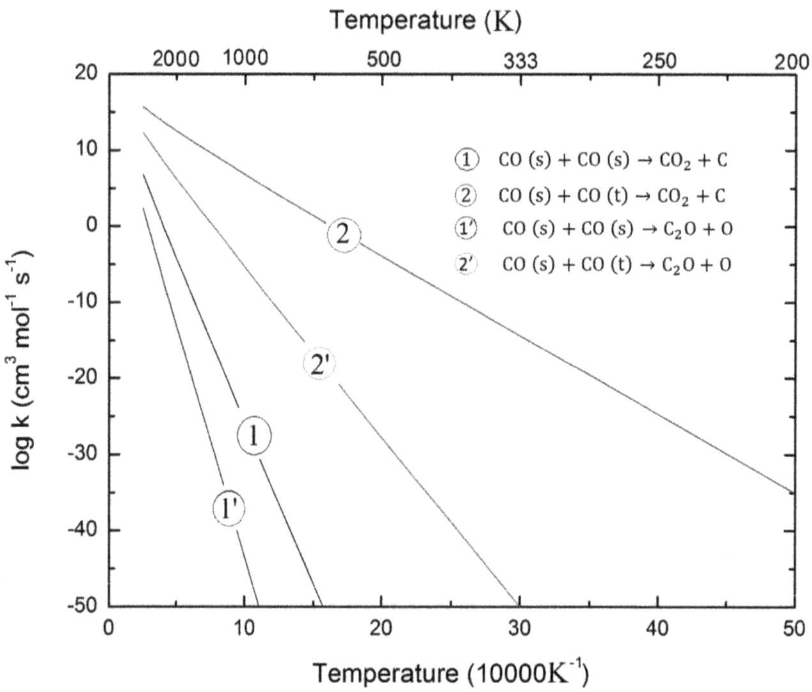

Figure 18.2. CO VDF as a function of the vibrational quantum number v during (a) the discharge and (b) the post-discharge in the MW test case.

Table 18.2. Values of E_a (in kcal mol^{-1}), A (in cm^3 mol^{-1}s^{-1}) and α for the reactions (18.1)–(18.4)

Reaction	E_a kcal mol^{-1}	A (cm^3 mol^{-1} s^{-1})	α
$CO(s) + CO(s) \rightarrow CO_2 + C$	191.99	192.75	31.57
$CO(s) + CO(t) \rightarrow CO_2 + C$	47.25	46.90	58.44
$CO(s) + CO(s) \rightarrow C_2O + O$	271.87	273.46	−0.33
$CO(s) + CO(t) \rightarrow C_2O + O$	100.39	100.98	17.87

i.e. the reactions (18.2) and (18.4), are orders of magnitude higher than the values involving singlet states due to the lower activation energies.

At low temperatures ($T < 1000$ K), the rate constants for the four processes without considering vibrational excitation of reactants present very small values. The dependence of the Arrhenius rates on the vibrational content of reactants will be reported in the next section 18.1.2. At this point, we can compare only the activation energy of process 1 (i.e. 8.3 eV in the current study) with the indirect experimental value recently obtained by Essenigh *et al* [3] (11.6 eV), as well as with a value of about 6 eV previously obtained by Rusanov *et al* [4] (see also Fridman *et al* [5]).

18.1.2 The role of vibrational excitation

An estimation of the role of vibrational excitation in the endothermic reactions can be made by using the calculated (or existing experimental) forward (E_a) and backward (E_a^-) activation energies for $v = w = 0$. As an example, for the reaction

$$A + BC(v) \rightarrow AB + C \tag{18.8}$$

one can write the dependence of the rate constant on the vibrational energy as

$$k_f^v = A \exp\left(-\frac{E_a}{k_B T}\right) \exp\left(\frac{\alpha E_v}{k_B T}\right) \tag{18.9}$$

if

$$\alpha E_v < E_a, \tag{18.10}$$

while the equation reduces to

$$k_f^v = k_T^0 \tag{18.11}$$

for

$$\alpha E_v > E_a, \tag{18.12}$$

where α is the coefficient of utilization of vibrational energy and E_v is the vibrational energy of the vth level. The parameter α [6–8] is calculated by the activation energies of the forward and backward reactions, see [2] according to the approximate relation (see [7])

$$\alpha \approx \frac{E_a}{E_a + E_a^-} \tag{18.13}$$

linking forward and backward activation energies. The calculated α values of the four reactions are reported in table 18.2. The reported α values are different for the four reactions following the rules of the Macheret–Fridman α model. These general rules can be summarized as follows: (i) exothermic reactions with practically no activation energy should have $\alpha = 0$, i.e. the corresponding rates are independent of vibrational energy of reactants; (ii) endothermic reactions should present α values close to 1, i.e. the corresponding rates are strongly dependent on the vibrational energy; (iii) thermo-neutral reactions should have an α value around 0.5. In our case, using equation (18.13) in combination with the forward and backward activation energies from table 18.1 gives for the endothermic reactions (18.1)–(18.4) α values, respectively, of 0.86 for (18.1), 0.45 for the quasi-thermo-neutral reaction 18.2, 1 for (18.3) and 0.85 for (18.4). The α values of the endothermic reactions (18.1)–(18.2) depend on the magnitude of the backward activation energies, while the α value of 1 for reaction (18.3) is typical of reactions with negligible backward activation energy.

Reversing reaction (18.3) from products to reagents, i.e. $O + C_2O \rightarrow CO + CO$, one obtains an exothermic reaction with a value of α close to 0 ($\alpha = 1.69/(273.46 + 1.69) = 0.0006$) in agreement with the rule (i).

Reversing process (18.1), i.e. by considering the reaction $C + CO_2 \rightarrow 2CO$ we obtain $\alpha = 31.28/(192.75 + 31.28) = 0.14$. In the case of two vibrationally excited molecules participating in the reaction, the simplest approximation consists in considering the total vibrational energy $E_v + E_w$, so that the vibrational energies of both reactants are characterized by equal efficiencies in the reaction. In such a way, the state-to-state rate constant becomes $k_f^{v,w}$ and the effect of vibrationally excited states on the total rate is given by

$$k_f^{v,w} = f_v f_w A \exp\left(-\frac{E_a}{kT}\right) \exp\left(\frac{\alpha(E_v + E_w)}{kT}\right), \tag{18.14}$$

where f_v and f_w are the populations (expressed in molar fractions) of the vibrational states of reactants. The parameter A entering in equation (18.14) represents the pre-exponential term in the Arrhenius equation (see table 18.2). An alternative to the previous treatment, first suggested by Fridman *et al* [5], calculates the reaction rates of process (18.1) by a statistical model of chemical reactions [4] that assumes formation of a long-lived intermediate complex where the energy moves freely among the modes. The forward rate constant k_f for process (18.1) can be generally written as [3]

$$k_f = \sum_{v=0}^{v_{\max}} \sum_{w=0}^{w_{\max}} k_f^{v,w}, \tag{18.15}$$

where in the statistical theory approximation the state specific rate constants $k_f^{v,w}$ are expressed as

$$k_f^{v,w} = v(T)\theta_{v,w} S f_v f_w \left[1 - \frac{E_a}{E_v + E_w}\right]^2 \left(\frac{\omega_{CO}}{\omega_{CO_2}}\right)^2. \tag{18.16}$$

In equation (18.16), $v(T) = 3 \times 10^{-10}(T/300)^{1/2}$ is the frequency of gas kinetic collisions of CO molecules, E_a is the activation energy of the process, E_v and E_w are the vibrational energies of the reactants, ω_{CO} and ω_{CO_2} are the vibrational frequencies of CO and CO_2 (in the latter case the one of the asymmetric mode), $\theta_{v,w}$ is a step function (i.e. $\theta_{v,w} = 1$ when $E_v + E_w \geqslant E_a$, $\theta_{v,w} = 0$ when $E_v + E_w < E_a$) and S is a steric factor. The rate in this case depends on the activation energy of the process. The value of the activation energy, initially estimated at 6 eV by Rusanov *et al* [4] and largely used previously [9–11] to reproduce the vibrational distributions of CO under discharge and post-discharge conditions, has been reassessed during the years toward higher values. An interesting experimental study [3] involved detailed measurements of both CO vibrational distributions and CO_2 yield. Equations (18.15) and (18.16) in combination with experimental dissociation yields and experimental vibrational distributions were used [3] to obtain an activation energy of 11.6 eV, approximately amounting to the dissociation energy of a CO molecule [3]. Unfortunately, no calculations exist for reaction (18.2), while experimental data have been presented previously [12–15] without specifying the vibrational states of both reactants. Equations (18.9)–(18.12) can be directly used for estimating the role of vibrational excitation in the dissociation rates of diatomic molecules (see section 18.2).

18.2 The plasma kinetic model

In this section, we want to emphasize the role of the reaction (18.1), i.e. the Boudouard reaction leading to CO_2 and C in a reacting CO plasma alimented by a millisecond microwave discharge followed by the corresponding afterglow. We will show results obtained by applying a millisecond microwave discharge to the CO plasma to focus on conditions in which the electric discharge induces a strong vibrational excitation in the CO molecule. In the first case, we use a model based on the solution of a time-dependent zero-dimensional Boltzmann equation for the calculation of the electron energy distribution function (EEDF), coupled to the non-equilibrium vibrational kinetics of CO levels, to the kinetics of electronically excited states of CO, C and O species, as well as with a simple dissociation–recombination and ionization–recombination kinetics describing the plasma mixture [16, 17]. The plasma mixture considered is composed of the following species: CO, CO_2, C, O, CO^+, CO_2^+, C^+, O^+ and e^-. The energy level diagrams of CO, C and O are reported in [16, 17]. The plasma chemistry model includes the following two reactive channels assisted by vibrational excitation (pure vibrational mechanisms, PVM) [16, 17]:
(1) Direct dissociation (PVM_1)

$$CO(v) + M \rightarrow C + O + M. \tag{18.17}$$

(2) Boudouard or disproportionation reaction (PVM_2)

$$CO(v) + CO(w) \rightarrow CO_2 + C. \tag{18.18}$$

The direct dissociation rate coefficients can be calculated by applying the Fridman–Macheret α-model [18] to the $v = 0$ Arrhenius dissociation rate coefficients ($K_{v=0}$) taken from [19], i.e.

$$K_d^{\text{direct}}(v) = K_{v=0}\exp(\alpha E_v/T) = AT^{\eta}\exp(-\theta/T)\exp(\alpha E_v/T), \tag{18.19}$$

with $A = 1.74 \times 10^{-9}\text{cm}^3\,\text{s}^{-1}$, $\eta = 0.19$, $\theta = 123\,661$ K and α the parameter that determines the efficiency of vibrational energy in lowering the reaction barrier according to the Fridman–Macheret α-model. When $\alpha\ E_v = \theta$, $K_d^{\text{direct}}(v) = K_{v=0}$. According to Macheret and Fridman rules, $\alpha = 1$ for strongly endothermic reactions, as in the case of reaction (18.17). The rate for process (18.18) has been previously discussed (see section 18.1). In addition to the two PVM mechanisms, CO dissociation occurs through the direct electron-impact mechanism (DEM), i.e.

$$e + CO(v) \leftrightarrow e + C + O, \tag{18.20}$$

as well as through a resonant dissociation mechanism (RES) [20, 21], i.e.

$$e + CO(v) \rightarrow CO^-(^2\Pi) \rightarrow e + C(^3P) + O(^3P). \tag{18.21}$$

Electron-impact ionization of CO, C and O and reverse reactions are discussed in [16, 17],

$$e + \mathrm{CO}(v) \leftrightarrow e + \mathrm{CO}^{+} + e \tag{18.22}$$

$$e + \mathrm{C}(^{3}P) \leftrightarrow e + \mathrm{C}^{+} \tag{18.23}$$

$$e + \mathrm{O}(^{3}P) \leftrightarrow e + \mathrm{O}^{+}. \tag{18.24}$$

The formed C and O atoms are allowed to recombine according to the process

$$\mathrm{C} + \mathrm{O} + \mathrm{M} \rightarrow \mathrm{CO} + \mathrm{M}. \tag{18.25}$$

Ion losses occur mainly by the dissociative recombination process

$$\mathrm{CO}^{+} + e \rightarrow \mathrm{C} + \mathrm{O} \tag{18.26}$$

and by the following recombination processes

$$\mathrm{C}^{+} + e \rightarrow \mathrm{C} \tag{18.27}$$

$$\mathrm{O}^{+} + e \rightarrow \mathrm{O}. \tag{18.28}$$

The electronic excited state kinetics of CO, C and O take into account electron-impact excitation and de-excitation processes. For CO, we consider only excitation from the ground state, i.e.

$$e + \mathrm{CO}(X^{1}\Sigma^{+}, v = 0) \leftrightarrow e + \mathrm{CO}(X), \tag{18.29}$$

where $X = a^{3}\Pi$, $a'^{3}\Sigma^{+}$, $b^{3}\Sigma^{+}$, $A^{1}\Pi$, $B^{1}\Sigma^{+}$, $C^{1}\Sigma^{+}$ and $E^{1}\Sigma^{+}$ are the CO electronic excited states included in the model. We assume that the CO electronic excited state emission lines are totally reabsorbed (optically thick plasma), i.e. the escape factor (λ_{ij}) describing such transitions and their corresponding effective Einstein coefficients (A_{ij}^{*}) are assumed equal to 0:

$$A_{ij}^{*} = \lambda_{ij} A_{ij} = 0. \tag{18.30}$$

In this way, we consider conditions in which the role of the CO electronic excited states in affecting the EEDF is maximized. The CO VDF is obtained by solving the vibrational master equations describing the 80 vibrational level kinetics under the action of electron–vibration (e–V), vibration–vibration (V–V) and vibration–translation (V–T) energy exchange processes, dissociation and ionization by electron impact involving each vth vibrational level (see, respectively, equations (18.20), (18.21) and (18.22)), the two PVM processes of equations (18.17) and (18.18) and available radiative transitions for CO vibrational levels. Details and equations can be found in [16, 17]. The rate coefficients of the electron-impact processes entering the chemical and vibrational kinetics are calculated from the corresponding electron-impact cross sections, introduced in the electron Boltzmann equation, by integration over the calculated EEDF. This procedure assures the self-consistent coupling between the electron Boltzmann equation and the kinetics of excited states.

18.2.1 MW test case

The MW test case study is characterized by the following values of pressure, gas temperature, pulse duration and reduced electric field value: $P = 5$ Torr, $T_{gas} = 500$ K, $\tau_{pulse} = 2.5$ ms and $E/N = 60$ Td. At the beginning of time evolution ($t = 0$), all CO molecules are in the $v = 0$ level, while the ionization degree is 10^{-6}. In this test case, the recently calculated value of 8.3 eV [2] is used for the activation energy of the Boudouard process. Macroscopic quantities, such as vibrational and electron temperatures and molar fractions, for this case study are discussed in [16, 17]. In the present contribution, we want to discuss the form of the vibrational distribution function (VDF), of the EEDF and of the dissociation rate coefficients by heavy-particle collisions. In particular, figure 18.3 shows the time-dependent CO VDF in (a) discharge and (b) post-discharge conditions.

During the discharge, the VDF is heated as a result of e–V processes, i.e. by the resonant processes

$$e + CO(v) \rightarrow CO^{-}(^{2}\Pi) \rightarrow e + CO(w) \tag{18.31}$$

acting over all the vibrational ladder, and by V–V (vibration–vibration energy transfer) processes, i.e.

$$CO(v) + CO(w) \leftrightarrow CO(v - 1) + CO(w + 1) + \Delta\epsilon. \tag{18.32}$$

At the end of the pulse ($\tau_{pulse} = 2.5$ ms), a plateau appears extending up to $v = 30$ as a result of V–V processes. The plateau is maintained in the post-discharge up to 5 ms, and starts to be deactivated by V–T processes from 8 ms on. The length of the V–V plateau depends on the competition between the V–V up-pumping mechanism

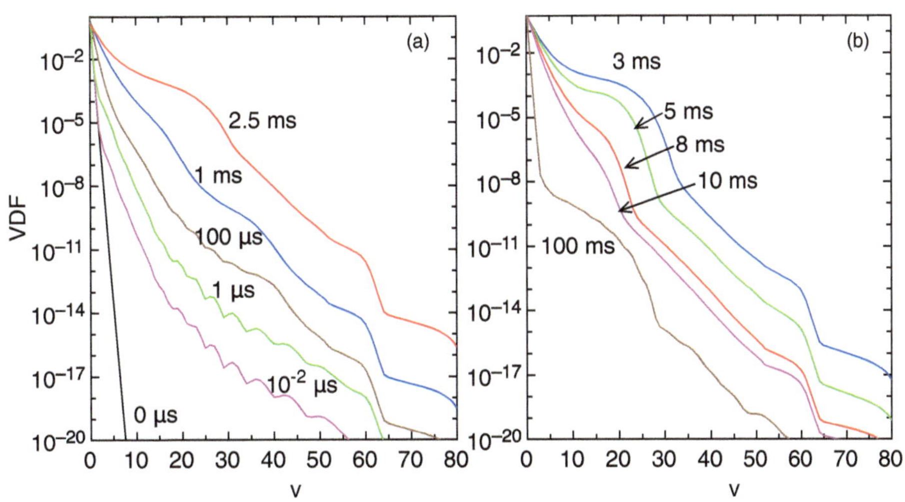

Figure 18.3. CO VDF as a function of the vibrational quantum number v during (a) the discharge and (b) the post-discharge in the MW test case. Reproduced with permission from [16]. Copyright 2017 Institute of Physics Publishing.

and the most relevant deactivation processes in CO plasma, the Boudouard process and the V–T deactivation by C and O atoms, i.e.

$$CO(v) + C \leftrightarrow CO(v-1) + C + \Delta\epsilon_T \tag{18.33}$$

$$CO(v) + O \leftrightarrow CO(v-1) + O + \Delta\epsilon_T. \tag{18.34}$$

By changing the activation energy of the Boudouard dissociation process, the depletion zone will move and the length of the plateau will change: the higher the activation energy, the longer the V–V plateau. To understand this point, we report in figure 18.4 the VDF in the previous MW test case at the end of the discharge (a) $t = 2.5$ ms and at two different times during the post-discharge, (b) $t = 3$ ms and (c) $t = 10$ ms, calculated by using three different values for the Boudouard activation energies: $E_a = 6$ eV, 8.3 eV and 11.6 eV. A lower activation energy ($E_a = 6$ eV) results in a more deactivated VDF plateau, the reverse being true for the highest used value of 11.6 eV. The $E_a = 8.3$ eV case, instead, is intermediate between the previous two case studies. The corresponding Boudouard (or PVM_2) rate coefficients largely depend on the choice of the activation energy, as shown in figure 18.5.

These rate coefficients and the direct dissociation (or PVM_1) rate coefficients are calculated from the VDF through

$$PVM_1 = \frac{n_{CO}}{n_e}\sum_v k_D^{direct}(v)(\alpha, T_{gas})f_v \tag{18.35}$$

$$PVM_2 = \frac{n_{CO}}{n_e}\sum_{v,w} k_D^{Boud}(v, w)(E_a, T_{gas})f_v f_w, \tag{18.36}$$

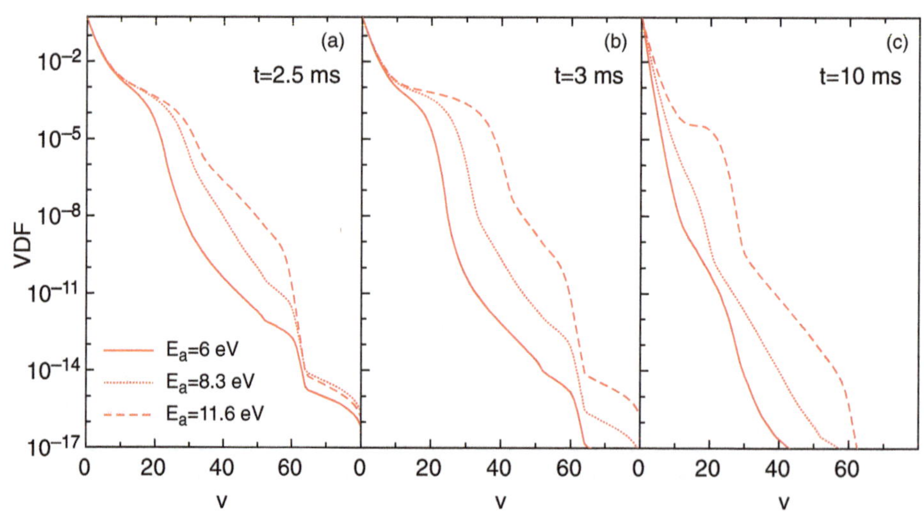

Figure 18.4. VDF in the MW test case at three different times (a) $t = 2.5$ ms, (b) $t = 3$ ms and (c) $t = 10$ ms calculated by assuming three different values of the Boudouard (or PVM_2) activation energy $E_a = 6$ eV, 8.3 eV and 11.6 eV. Reproduced with permission from [16]. Copyright 2017 Institute of Physics Publishing.

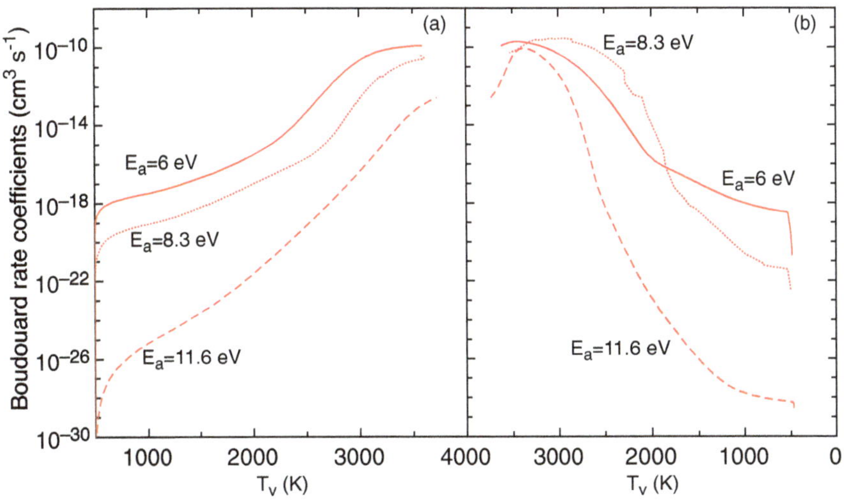

Figure 18.5. Boudouard (or PVM_2) rates as a function of the CO vibrational temperature in (a) discharge and (b) post-discharge conditions at the three different values of the Boudouard (or PVM_2) activation energy E_a = 6, 8.3 and 11.6 eV.

where $k_D^{\text{direct}}(v)$ and $k_D^{\text{Boud}}(v, w)$ are the rate coefficients of equations (18.17) and (18.18), which depend on the gas temperature and on the α coefficient for the direct dissociation and on the activation energy E_a for the Boudouard process.

During the discharge phase, lower activation energies result in higher Boudouard (or PVM_2) rate coefficients, while in the post-discharge regime (figure 18.5), the rate coefficients calculated with $E_a = 6$ eV and 8.3 eV show a similar behavior, both being much higher than the 11.6 eV case. The PVM_1 rate coefficients, instead, are indirectly affected by the E_a choice only through the corresponding change of the VDF, see equation (18.35). Figure 18.6 shows the PVM_1 rate coefficients calculated in (a) discharge and (b) post-discharge conditions at the three E_a values of the Boudouard process. Under discharge conditions (figure 18.6(a)), the PVM_1 rate coefficients depend on the E_a choice only at high vibrational temperature. A similar behavior is observed in the post-discharge conditions (figure 18.6(b)) although in this case the dependence of PVM_1 rate coefficients on E_a is much more important. Figure 18.7 shows the corresponding EEDF time evolution in (a) discharge and (b) post-discharge conditions. The corresponding Boltzmann solver is discussed by [16, 17].

During the discharge, the EEDF increases due to the effect of the electric field and very soon reaches ($10^{-2}\mu$s) the quasi-stationary distribution. At longer times ($t > 1$ ms), the EEDF still increases due to the effect of superelastic (vibrational) collisions, i.e.

$$e(\epsilon) + \mathrm{CO}(v) \rightarrow e(\epsilon + \Delta_{vw}) + \mathrm{CO}(w), \tag{18.37}$$

which transfer vibrational excitation to the electrons. When the pulse is turned off, the EEDF cools down quickly showing, as in the case of CO_2 [22–24], the characteristic peaks due to superelastic processes involving the metastable electronic excited states $a^3\Pi$, $a'^3\Sigma^+$, $A^1\Pi$, $B^1\Sigma^+$:

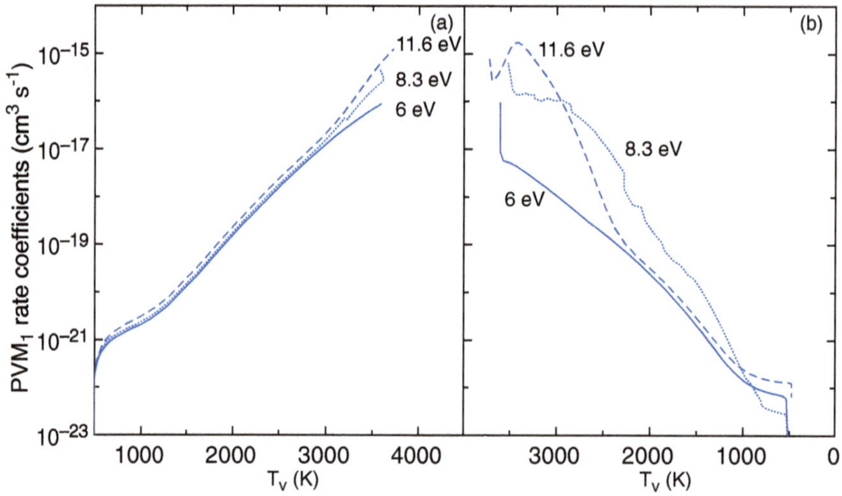

Figure 18.6. PVM$_1$ rates as a function of the CO vibrational temperature calculated in (a) discharge and (b) post-discharge conditions at the three values of the Boudouard activation energy E_a.

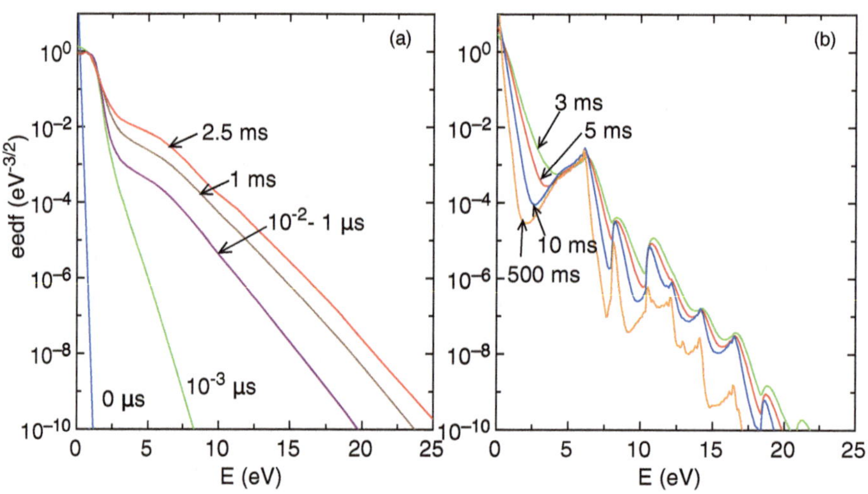

Figure 18.7. EEDF time evolution in discharge (a) and post-discharge (b) conditions in the MW test case. Reproduced with permission from [16]. Copyright 2017 Institute of Physics Publishing.

$$e(\epsilon) + \mathrm{CO}(a^3\Pi) \rightarrow e(\epsilon + \Delta\epsilon_1 = 6\ \mathrm{eV}) + \mathrm{CO} \tag{18.38}$$

$$e(\epsilon) + \mathrm{CO}(a'^3\Pi) \rightarrow e(\epsilon + \Delta\epsilon_2 = 6.8\ \mathrm{eV}) + \mathrm{CO} \tag{18.39}$$

$$e(\epsilon) + \mathrm{CO}(A^1\Pi) \rightarrow e(\epsilon + \Delta\epsilon_3 = 8.03\ \mathrm{eV}) + \mathrm{CO} \tag{18.40}$$

$$e(\epsilon) + \mathrm{CO}(B^1\Sigma^+) \rightarrow e(\epsilon + \Delta\epsilon_4 = 10.78\ \mathrm{eV}) + \mathrm{CO}, \tag{18.41}$$

creating several peaks in the EEDF at the corresponding electronic excited state energies, i.e. 6, 6.8, 8.03 and 10.78 eV. For higher electron energies ($\epsilon > 12$ eV), the peaks' structure is characterized by the superposition of more than one superelastic electronic process (see also [25, 26]), which can involve the same or different electronic states, i.e.

$$e(\epsilon + \Delta E_{X_i}) + \mathrm{CO}(X_j) \rightarrow e(\epsilon + \Delta E_{X_i} + \Delta E_{X_j} + \mathrm{CO}(X^1\Sigma^+)). \tag{18.42}$$

The magnitudes of the peaks depend on the concentration of the electronic excited states, which in our model is calculated only by considering electron-impact excitation and de-excitation processes and certainly overestimated, due to the neglect of the deactivation mechanisms, such as quenching processes and radiative processes. The results reported in figure 18.7 characterize a situation where the optical allowed transitions are optically thick and therefore completely reabsorbed. A comparison of the different rate coefficients for this case study is reported in figure 18.8 (see also [16]) for discharge and post-discharge conditions. In the figure, we report DEM(0) and DEM which are the DEM dissociation rate coefficients taking into account the contribution of, respectively, only the ground state and all the CO vibrational levels. In an equivalent way, RES(0) and RES are the resonant dissociation rate coefficients and, finally, PVM_1 and PVM_2 are the PVM dissociation rate coefficients for the direct and Boudouard processes. Definitions of previous dissociation rate coefficients can be found in [16]. In both cases, DEM and PVM_2 mechanisms assume the same importance, in particular when the system presents high vibrational concentrations, i.e. high vibrational temperatures in the relevant plots.

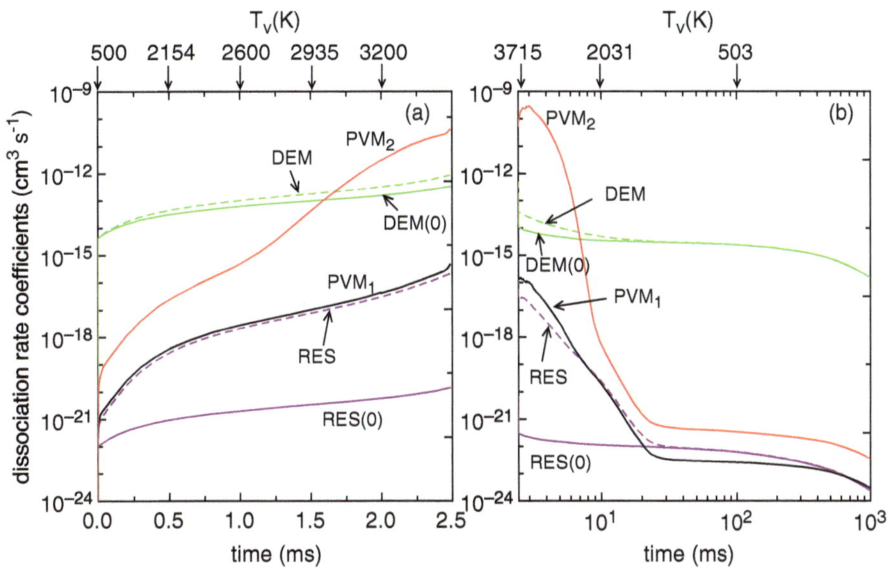

Figure 18.8. PVM and DEM rate coefficients as a function of time in discharge (a) and post-discharge (b) in the MW test case. The second x-axis reports the corresponding vibrational temperature values.

18.2.2 Case study 2: sudden increase of the translational temperature

The previous case study is not directly linked to the meteoroid entry physics, even though we can expect that similar non-equilibrium distributions for both the VDF and EEDF can result in some points of the flow. This last consideration can be justified by inspection of the following second numerical example, i.e. the instantaneous heating of a mixture of CO_2/CO followed by the population of vibrational levels by V–T and V–V processes, which in turn can promote the dissociation processes by the PVM_1 and PVM_2 mechanisms. The test conditions are the following: $P = 1$ atm, $T_{gas} = 20\,000$ K, $T_v(0) = T_e(0) = 500$ K, initial molar fractions $\chi_{CO_2} = 1$ and $\chi_e = \chi_{CO_2^+} = 10^{-4}$. The considered large initial electron molar fraction reflects to a given extent the easy ionization of metallic vapors during the meteoroid–atmosphere interaction. In this case, we use a similar approach applied to the CO_2/CO mixture adding to the model of CO_2 developed by Pietanza *et al* [22–24], the kinetics of CO reported in the previous test case. Figure 18.9 shows the time evolution of the molar fractions and of the electron and vibrational temperatures of CO_2 and CO. At the beginning, CO_2 is the prevalent species of the mixture but very soon, due to the very high gas temperature, dissociation occurs leading to CO and O through the reaction

$$CO_2(000) + CO_2 \rightarrow CO + O. \tag{18.43}$$

Then CO dissociation starts through the different mechanisms described in section 18.2. The ionization degree of the mixture decays from 10^{-4} up to 10^{-6} at the end of the time evolution (see figure 18.9(a)). The vibrational temperatures of CO_2 and CO are reported in figure 18.9(b). Both the quantities present an increase as a function of the time at a maximum value of $T_v(CO_2) = 20\,000$ K and $T_v(CO) = 8000$ K. A slight decrease is observed after the corresponding maxima followed by a quasi-plateau in both cases. These plateaux result from the interplay of forward and backward V–T processes involving, in particular at longer times, C and O atoms, which in this

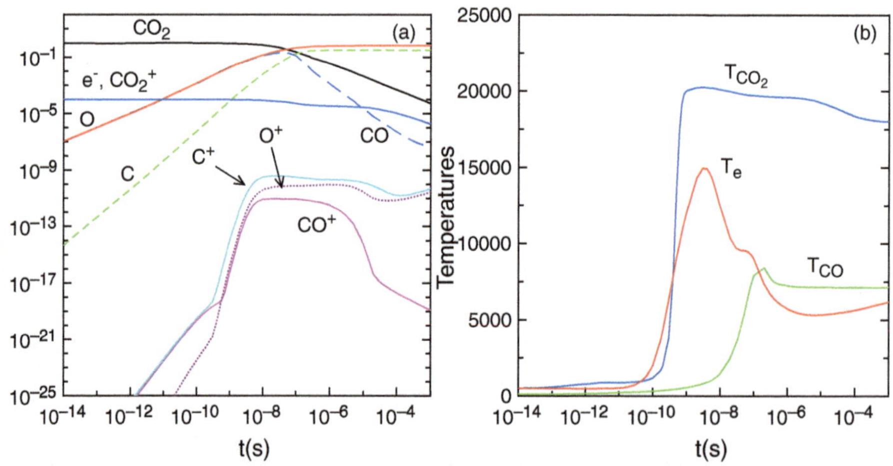

Figure 18.9. (a) Molar fraction and (b) temperature time evolution in case study 2.

time-scale are the most important species. Also, the electron temperature first increases due to the superelastic e-V (SeV) processes with CO_2 and CO molecules. In particular, the first peak at approximately 15 000 K is due essentially to SeV processes with CO_2 molecules, while the second peak at nearly 10 000 K to SeV processes with CO. However, once both CO_2 and CO start dissociating, forming an atomic (C and O) mixture, SeV processes become less efficient and the electron temperature starts decreasing, reaching, at the end of the time evolution, a temperature value of 6200 K. Figure 18.10 shows the time evolution of the EEDF. It should be stressed that the EEDF time evolution is controlled by SeV up to 10^{-8}s, while the

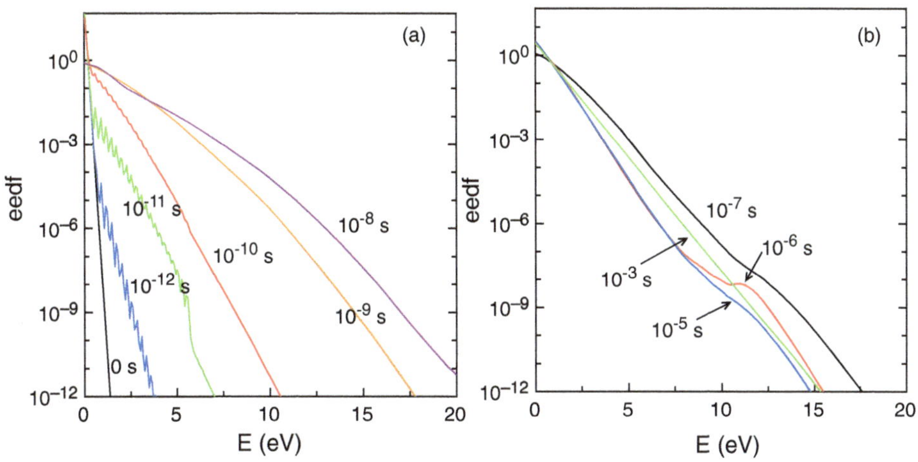

Figure 18.10. EEDF time evolution in case study 2.

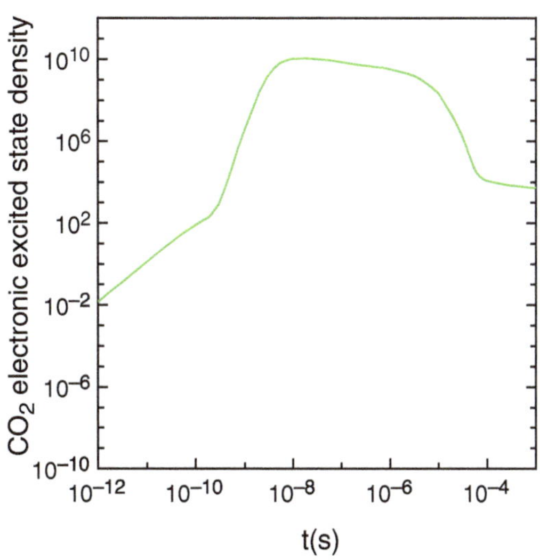

Figure 18.11. Density of the CO_2 electronic excited states at 10.5 eV.

EEDF cooling after 10^{-8}s is due to the decrease of the vibrational concentrations (not of the vibrational temperature) of CO_2 and CO species and the corresponding inelastic energy loss of electrons colliding with C and O atoms. The electronic excited states have a minor influence in affecting the EEDF through superelastic electronic collisions due to their small concentration. Only the electronic state of CO_2 at 10.5 eV (whose density is reported in figure 18.11) is able to produce small structures in the EEDF ($t = 10^{-6}$, $t = 10^{-5}$) (see figure 18.10(b)) as a consequence of

$$CO_2(10.5\,\text{eV}) + e(\epsilon) \rightarrow CO_2(0) + e(\epsilon + 10.5\,\text{eV}). \tag{18.44}$$

Figure 18.12 shows the vibrational distributions of CO_2 and CO species at different times. The form of these distributions generates the reported vibrational temperatures. From 10^{-8}s on, the absolute concentrations of the vibrational states for both molecules are reduced, decreasing their role in affecting the EEDF despite the corresponding high values of vibrational temperatures. Note also that the vibrational distributions result from the interplay of forward and backward V–T processes leading to Boltzmann distributions. The corresponding EEDFs, in contrast, present non-equilibrium forms that are much less pronounced than those presented in the previous case study. Figure 18.13 reports the different rate coefficients of the reactive channels as defined by equations (18.17) ($PVM_1'(CO)$), (18.18) ($PVM_2'(CO)$), (18.20) (DEM(CO)) and (18.21) (RES(CO)) as well as a direct electron-impact dissociation mechanism involving CO_2 (DEM(CO_2)). The new PVM rate coefficients (PVM′) do not contain the factor n_{CO}/n_e (see equations (18.35) and (18.36)). Inspection of the figure shows the prevalence of PVM rate coefficients due to the high translational temperature considered in the case study. The DEM rates follow the time evolution of T_e as reported in figure 18.9, as a consequence of the corresponding time evolution of the EEDF.

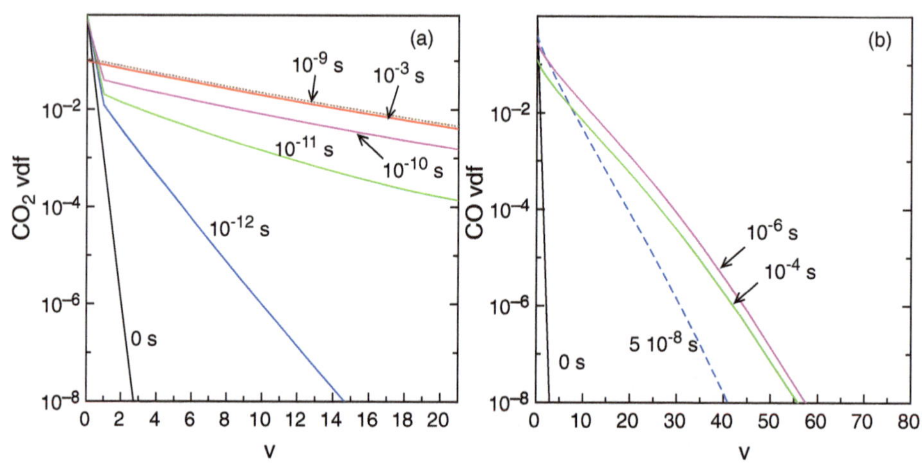

Figure 18.12. (a) CO_2 and (b) CO VDF time evolution in case study 2.

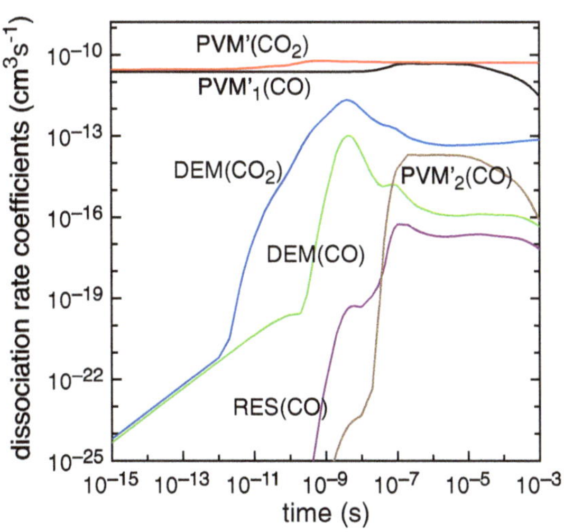

Figure 18.13. Dissociation rate coefficients for CO_2 and CO in case study 2.

18.3 Conclusions

The results presented in this chapter emphasize the role of vibrational excitation in affecting the rates of bimolecular endothermic reactions. First, we have discussed the role of vibrational excitation in affecting the Boudouard reaction [2] involving vibrationally excited CO molecules leading to CO_2 and C products. We have presented a recent quantum approach [2] to calculate the relevant E_a in the case of cold CO and the way to correct it in the presence of vibrationally excited molecules. A similar approach is also discussed to phenomenologically correct the E_a of the endothermic reactions occurring in a reacting CO_2 mixture. After the presentation of the basic approach, we have applied the described rates to two case studies. The first one refers to the non-equilibrium plasma kinetics occurring in cold CO plasma, emphasizing the role of vibrational excitation in enhancing the reactive channels. As a second example, we consider a CO_2 system suddenly heated to 20 000 K which relaxes, finally ending with C and O atoms through dissociation reactions controlled by vibrational excitation. In both cases, we discuss the non-equilibrium forms of the EEDF and VDF, a subject which should be considered in future advanced models for meteoroid entry physics (see chapters 6 and 7). The current results, even though they occur in CO and CO_2 reacting systems, are not so far from the corresponding results one can obtain for air under impact with a high velocity meteoroid. In this case, the chemical physics processes should be taken into account for describing the vibrational excitation of a N_2 and O_2 system by V–T collisions followed by the dissociation and ionization processes as well as by the formation of NO through collisions of vibrational excited N_2 and O atoms [27, 28]. In this last case, we could follow a similar approach to that presented in this chapter. The coupling between EEDF and VDF should also approximately follow the results shown in the reported case studies.

References

[1] Surzhikov S 2014 *Non-Equilibrium Radiative Gas Dynamics of Small Meteor* (Reston, VA: American Institute of Aeronautics and Astronautics)

[2] Barreto P R P, de O Euclides H, Albernaz A F, Aquilanti V, Capitelli M, Grossi G, Lombardi A, Macheret S and Palazzetti F 2017 Gas phase Boudouard reactions involving singlet–singlet and singlet–triplet CO vibrationally excited states: implications for the non-equilibrium vibrational kinetics of CO/CO_2 plasmas *Eur. Phys. J.* D **71** 259

[3] Essenhigh K A, Utkin Y G, Bernard C, Adamovich I V and Rich J W 2006 Gas-phase Boudouard disproportionation reaction between highly vibrationally excited CO molecules *Chem. Phys.* **330** 506–14

[4] Rusanov V D, Fridman A A and Sholin G V 1981 The physics of a chemically active plasma with nonequilibrium vibrational excitation of molecules *Sov. Phys. Usp.* **24** 447

[5] Rusanov V D and Fridman A A 1976 Reactions of vibrationally excited nitrogen molecules in nonequilibrium plasmochemical systems *Sov. Phys. Dokl.* **21** 739

[6] Igor B, Jens V, Morteza A, Sabine P and Annemie B 2017 Synthesis of micro- and nanomaterials in CO_2 and CO dielectric barrier discharges *Plasma Process. Polym.* **14** 1600065

[7] Capitelli M, Ferreira C M, Gordiets B F and Osipov A I 2000 *Plasma Kinetics in Atmospheric Gases* Springer Series on Atomic, Optical, and Plasma Physics vol 31 (Berlin: Springer)

[8] Macheret S O, Losev S A, Chernyi G G and Potapkin B V 2002 *Physical and Chemical Processes in Gas Dynamics: Cross Sections and Rate Constants* vol 1 (Reston, VA: American Institute of Aeronautics and Astronautics)

[9] Benedictis S D, Gorse C, Cacciatore M, Capitelli M, Cramarossa F, D'Agostino R and Molinari E 1983 Vibrational kinetics in He–CO reacting discharges *Chem. Phys. Lett.* **96** 674–77

[10] Gorse C and Capitelli M 1984 Kinetic processes in non-equilibrium carbon monoxide discharges. II. Self-consistent electron energy distribution functions *Chem. Phys.* **85** 177–87

[11] Benedictis S D, Capitelli M, Cramarossa F and Gorse C 1987 Non-equilibrium vibrational kinetics of CO pumped by vibrationally excited nitrogen molecules: a comparison between theory and experiment *Chem. Phys.* **111** 361–70

[12] Maksimov A I S A F, Polak L S and S D I 1979 Mechanism of the formation and decomposition of CO_2 molecules in a glow-discharge in carbon-monoxide *Khim. Vysokikh Energy (High Energy Chem.)* **13** 358

[13] Dunn O, Harteck P and Dondes S 1973 Isotopic enrichment of carbon-13 and oxygen-18 in the ultraviolet photolysis of carbon monoxide *J. Phys. Chem.* **77** 878–83

[14] Liuti G, Dondes S and Harteck P 1966 Photochemical production of C_3O_2 from CO *J. Chem. Phys.* **44** 4051–2

[15] Liuti G, Dondes S and Harteck P 1969 *The Photochemical Separation of the Carbon Isotopes* (Washington, DC: American Chemical Society) ch 5, pp 65–72

[16] Pietanza L D, Colonna G and Capitelli M 2017 Non-equilibrium plasma kinetics of reacting CO: an improved state to state approach *Plasma Sources Sci. Technol.* **26** 125007

[17] Pietanza L D, Colonna G and Capitelli M 2017 Electron energy and vibrational distribution functions of carbon monoxide in nanosecond atmospheric discharges and ms afterglows *J. Plasma Phys.* **83** 725830603

[18] Fridman A 2008 *Plasma Chemistry* (Cambridge: Cambridge University Press)

[19] Macdonald R L, Munafò A, Johnston C O and Panesi M 2016 Nonequilibrium radiation and dissociation of CO molecules in shock-heated flows *Phys. Rev. Fluids* **1** 043401

[20] Laporta V, Cassidy C M, Tennyson J and Celiberto R 2012 Electron-impact resonant vibration excitation cross sections and rate coefficients for carbon monoxide *Plasma Sources Sci. Technol.* **21** 045005

[21] Laporta V, Tennyson J and Celiberto R 2016 Carbon monoxide dissociative attachment and resonant dissociation by electron-impact *Plasma Sources Sci. Technol.* **25** 01LT04

[22] Pietanza L D, Colonna G, D'Ammando G and Capitelli M 2017 Time-dependent coupling of electron energy distribution function, vibrational kinetics of the asymmetric mode of CO_2 and dissociation, ionization and electronic excitation kinetics under discharge and post-discharge conditions *Plasma Phys. Controlled Fus.* **59** 014035

[23] Capitelli M, Colonna G, D'Ammando G, Hassouni K, Laricchiuta A and Pietanza L D 2016 Coupling of plasma chemistry, vibrational kinetics, collisional-radiative models and electron energy distribution function under non-equilibrium conditions *Plasma Process. Polym.* **14** 1600109

[24] Capitelli M, Colonna G, D'Ammando G and Pietanza L D 2017 Self-consistent time dependent vibrational and free electron kinetics for CO_2 dissociation and ionization in cold plasmas *Plasma Sources Sci. Technol.* **26** 055009

[25] Colonna G and Capitelli M 2001 The influence of atomic and molecular metastable states in high-enthalpy nozzle expansion nitrogen flows *J. Phys. D: Appl. Phys.* **34** 1812

[26] D'Ammando G, Colonna G, Capitelli M and Laricchiuta A 2015 Superelastic collisions under low temperature plasma and afterglow conditions: a golden rule to estimate their quantitative effects *Phys. Plasmas* **22** 034501

[27] Capitelli M, Celiberto R, Colonna G, Esposito F, Gorse C, Hassouni K, Laricchiuta A and Longo S 2016 *Fundamental Aspects of Chemical Physics of Plasmas: Kinetics* Springer Series on Atomic, Optical, and Plasma Physics vol 85 (Berlin: Springer)

[28] Colonna G, Tuttafesta M, Capitelli M and Giordano D 1999 Non-Arrhenius NO formation rate in one-dimensional nozzle airflow *J. Thermophys. Heat Transfer* **13** 372–75

www.ingramcontent.com/pod-product-compliance
Ingram Content Group UK Ltd.
Pitfield, Milton Keynes, MK11 3LW, UK
UKHW051908100926
8556IPUK00005B/809

9 780750 316668